SA09-16-76

BEAM - FOIL SPECTROSCOPY

VOLUME 2

Collisional and Radiative Processes

BEAM-FOIL SPECTROSCOPY

Volume 1:   Atomic Structure and Lifetimes
Volume 2:   Collisional and Radiative Processes

BEAM - FOIL SPECTROSCOPY

VOLUME 2

Collisional and Radiative Processes

Edited by

Ivan A. Sellin and David J. Pegg

University of Tennessee, Knoxville
and
Oak Ridge National Laboratory, Oak Ridge

PLENUM PRESS • NEW YORK AND LONDON

Library of Congress Cataloging in Publication Data

International Conference on Beam-Foil Spectroscopy, 4th, Gatlinburg, Tenn.,
  1975.
  Beam-foil spectroscopy.

  Includes bibliographical references and index.
  CONTENTS: v. 1. Atomic structure and lifetimes.—v. 2. Collisional and ra-
diative processes.
  1. Beam-foil spectroscopy—Congresses. I. Sellin, Ivan. II. Pegg, David, 1940-
    III. Title.
QC454.B39I57 1975   535'.8   76-10654
ISBN 0-306-37127-8 (v. 2)

Proceedings of the second half of the Fourth International Conference on Beam-
Foil Spectroscopy held in Gatlinburg, Tennessee, September 15-19, 1975

Published with the financial assistance of UNESCO,
UNESCO subvention – 1975 – DG/2.1/414/40.

©1976 Plenum Press, New York
A Division of Plenum Publishing Corporation
227 West 17th Street, New York, N.Y. 10011

Printed in the United States of America

Organizing Committee

T. Andersen
University of Aarhus
Aarhus, Denmark

S. Bashkin
University of Arizona
Tucson, Arizona

I. Dmitriev
Moscow State University
Moscow, U.S.S.R.

M. Dufay
Université de Lyon
Villeurbanne, France

R. Garstang
J.I.L.A.
Boulder, Colorado

A. Gabriel
Culham Laboratory
Abingdon Berks, England

H. Hay
Australian National University
Canberra, Australia

L. Heroux
Air Force Cambridge Laboratory
Bedford, Massachusetts

I. Martinson
Research Institute of Physics
Stockholm, Sweden

W. Meckbach
Com. Nac. de Energia Atomica
San Carlos de Bariloche
Rio Negro, Argentina

C. Moak
Oak Ridge National Laboratory
Oak Ridge, Tennessee

D. Pegg, Conference Secretary
University of Tennessee
Knoxville, Tennessee

F. Pipkin
Harvard University
Cambridge, Massachusetts

I. Sellin, Conference Chairman
University of Tennessee
Knoxville, Tennessee

W. Whaling
California Institute of Technology
Pasadena, California

Local Committee

Virgil Adams
Deborah Adams
Stuart Elston
John Forester
Wylene Guinn
Wendell Martin
Charles Normand, Conference Coordinator

Barbara Pack
Loni Pegg
Peggy Peterson
Randolph Peterson
Roy Pruett, Conference Treasurer
Helen Sellin
Betty Thoe
Robert Thoe

Preface

The appreciable evolution of the nearly teenaged branch of
atomic and molecular physics called beam foil spectroscopy is
clearly depicted in the present volumes, which are devoted to
publication of presentations at the Fourth International
Conference on Beam Foil Spectroscopy and Heavy Ion Atomic Physics
Symposium. The transition from childhood to adolescence parallels
human experience in that diffusion of interests and interactions
beyond the confines of the original family has most certainly
occurred. The pre-occupation with techniques and their development has been largely replaced by interest in the physics of the
widest possible array of atomic and molecular physics experiments,
in which spectroscopic study (visible, UV, XUV, X-ray, electron)
of collisional interactions of fast beams is the unifying theme.
The description "accelerator-based atomic physics" is perhaps
more representative of the subject today than is the original,
beam-foil spectroscopy," since so many experiments have nothing
to do with foils, and furthermore, employ spectroscopy mainly as
an incidental tool. What, then distinguishes beam-foil spectroscopy from overlapping fields of atomic collisions physics? In
an era where the boundaries are becoming ever more diffuse, there
can be no clear definition. A good functional definition was
recently conceived by Peter Erman, under the salubrious stimulus
of a large Tennessee bourbon: it is the tribal experience of the
community of scientists who have banded together to develop the
discipline over the past dozen years, as shared at the triennial
conferences devoted to it.

The Fourth Conference was the largest of those held to date,
attracting approximately one hundred fifty scientists. An
unusually large international participation was evident. Nearly
forty percent of the participants were European Nationals, a very
large proportion for a conference held in the continental United
States. The Conference was again an important arena for interaction
of astrophysicists, plasma spectroscopists, atomic structure
theorists, and those experimentalists using accelerators to measure

such important atomic parameters as lifetimes and transition probabilities.

Concerning the future, both the health and the growth of accelerator-based atomic physics will surely become even more dependent on governmental agency policies the world over than at present, as the emphasis in nuclear physics funding shifts to smaller numbers of large, central accelerator facilities. In this regard we call attention here to a survey of the field, commissioned by the U. S. National Academy of Sciences Advisory Committee on Atomic and Molecular Physics, to be issued in 1976. This report analyzes and speaks very well of the field. The editors express the hope that science policy advisory bodies in other countries will undertake similar efforts.

Ivan A. Sellin

University of Tennessee Knoxville

Acknowledgments

The editors wish to recognize the most generous support of the Conference by the National Science Foundation, Atomic, Molecular, and Plasma Physics Program, administered by Rolf Sinclair. Support by the Physics Program of the Office of Naval Research through the good efforts of John Dardis is gratefully acknowledged, as is a grant to the Conference from the International Union of Pure and Applied Physics. The institutional sponsorship provided by the University of Tennessee, Department of Physics (William M. Bugg, Head), and the Oak Ridge National Laboratory, Physics Division (Paul Stelson, Director) through the Energy Research and Development Administration was indispensable to the Conference. Support of cultural and social events by the Tennessee Arts Commission, Ortec, Inc. and Tennelec, Inc. provided congenial conditions at the Conference. The contributions of many individuals deserve appropriate recognition, especially the efforts of institutional secretarial staffs, Conference office personnel, and hard-working wives. A number of these individuals are cited in the accompanying listings of the memberships of the organizing and local committees.

Contents of Volume 2

Non-Characteristic X-Ray Production in Heavy Ion-Atom . . .   461
    H.-D. Betz, F. Bell, E. Spindler, M. Kleber, H. Panke,
    and W. Stehling

Angular Distribution Studies of Non-Characteristic
X-Radiation   477
    R. S. Thoe, I. A. Sellin, K. A. Liao, R. S. Peterson,
    D. J. Pegg, J. Forester, and P. M. Griffin

Problems of Quantum Electrodynamics in Heavy-Ion
Collisions.   483
    Berndt Müller

Differences in the Production of Noncharacteristic
Radiation in Solid and Gas Targets.   497
    R. S. Peterson, R. S. Thoe, H. Hayden, S. B. Elston,
    J. P. Forester, K.-H. Liao, P. M. Griffin, D. J. Pegg,
    I. A. Sellin, and R. Laubert

Energy Loss and Yield of Swift Molecular Clusters in
Solid Targets   505
    R. Laubert

K-Shell X-Rays From Ti, Ni, Ge, and Rb for Incident Ions
From ^1H to ^{35}Cl in the Energy Range from 1-3 MeV/Amu. . . .   519
    F. D. McDaniel and J. L. Duggan

L X-Ray Spectra of Chlorine and Sulfur.   539
    K. S. Roberts, W. L. Hodge, B. M. Johnson,
    D. Schneider, W. J. Braithwaite, L. E. Smith,
    and C. F. Moore

Collisional Quenching of Metastable X-Ray Emitting States
in a Fast Beam of He-Like Fluorine.   545
    D. L. Matthews and R. J. Fortner

Delayed X-Ray Emission Following Beam Foil Excitation . . .   553
     Forrest Hopkins, Jonathan Sokolov and
     Peter von Brentano

Measurement of X-Rays Emitted From Projectiles Moving
in Solid Targets.   559
     R. J. Fortner, D. L. Matthews, L. C. Feldman,
     J. D. Garcia, and H. Oona

L-Shell Vacancy Lifetime Effects on Kα X-Ray Satellites
Produced in Heavy-Ion-Atom Collisions   567
     R. L. Watson, T. Chiao, F. E. Jenson, and
     B. I. Sonobe

Secondary Electron Emission From Foils Traversed by
Ion Beams   577
     W. Meckbach

Spectroscopy of Electrons Accompanying the Passage of Heavy
Ions Through Solid Targets.   593
     Karl-Ontjes Groeneveld

KLM Radiative Auger Transitions   609
     W. L. Hodge, D. Schneider and C. F. Moore

High Resolution Ar K Auger Spectra Produced in 4 and 2
MeV H^+ on Argon Collisions.   615
     D. Schneider, K. Roberts, B. M. Johnson, J. Whitenton,
     C. F. Moore

Secondary Electron Emission From H^+ and H_2^+ Passage
Through Thin Carbon Foils   623
     M. G. Menedez and M. M. Duncan

Relative Multiple Ionization Cross Sections of Neon by
Different Projectiles   629
     C. P. Bhalla

Spectral and Electron Collision Properties of Atomic
Ions.   637
     D. K. Chao, J. L. Dehmer, U. Fano, M. Inokuti,
     S. T. Manson, A. Msezane, R. F. Reilman,
     C. E. Theodosiou

Distinctive Features of Capture and Loss of Electrons by
Excited Ions.   643
     I. S. Dmitriev

Multiple Electron Loss Cross Sections for 60 MeV I^{+10}
in Single Collisions with Xenon   657
      L. B. Bridwell, J. A. Biggerstaff, G. D. Alton,
      C. M. Jones, P. D. Miller, Q. Kessel, B. W. Wehring

Charge States of Backscattered He Ions.   665
      Allen Lurio and J. F. Ziegler

Charge-State Distributions in Single Atomic Collisions
of 2.5 MeV N^{+i} with N_2 at Small Impact Parameters   671
      F. W. Martin and R. K. Cacak

Stopping Power for Ions of Intermediate Atomic Numbers. . .   679
      B. W. Wehring and R. G. Bucher

Angular Behaviour of Stopping Powers of Carbon Foil for
Argon Ions Below 250 keV.   687
      Gilles Beauchemin and Robert Drouin

Vacuum Ultraviolet Emission Spectra From keV Energy Rare
Gas Ion-Atom Collisions   695
      W. W. Smith, D. A. Gilbert, and C. W. Peterson

Calibration of Spectrometer Detection Efficiency in the
Ultraviolet   705
      Ward Whaling

A High-Intensity Method for Beam-Foil Spectroscopy,
With Retained Spatial Resolution Along the Beam   719
      Karl-Erik Bergkvist

Nonadiabatic Spin Transitions: A Possible Source of
Polarized Electrons   727
      R. D. Hight and R. T. Robiscoe

Alignment and Orientation Production Measurement and
Conversion. . . . .   731 –
      M. Lombardi

Quantum Beats in the Electric Field Quenching of
Metastable Hydrogen . . . .   749 –
      G.W.F. Drake and A. van Wijngaarden

The Surface Interaction in Beam Foil Spectroscopy . . . . .   755 –
      H. G. Berry

The Effect of Tilted Foil Excitation on the Spacial
Decay of the 3^3P States of ^{4}He I in an Applied Magnetic
Field   773
        J. D. Silver and L. C. McIntyre, Jr.

Theoretical Aspects of Beam-Foil Collisions   781
        Joseph Macek

Hyperfine-Structure Measurements in Carbon-13   791
        J. L. Subtil, P. Ceyzeriat, J. Desesquelles,
        and M. Druetta

Quantum Beats in H_α and H_β After Beam-Foil Excitation . . .   799
        A. Denis, J. Desquelles, M. Druetta, and M. Dufay

Orientation and Alignment Changes Induced by Tilted
Foils   809
        R. M. Herman

Laser Resonance Spectroscopy on Excited States of High
Z Hydrogenic Atoms.   815
        D. E. Murnick

Photoexcitation of a Fast H(2s) Beam to Highly Excited
and Continuum States Using Doppler-Tuned CW Argon Ion
Laser UV Radiation.   829
        P. M. Koch, L. D. Gardner, and J. E. Bayfield

Laser Excitation in Fast Beam Spectroscopy.   835
        H. J. Andrä

On the Feasibility of Pulsed Laser Excitation of
Fast Atomic Beams   853
        M. Gaillard, H. J. Plohn, H. J. Andra, D. Kaiser

Cascade Free Lifetime Measurements by Laser Excitation
of Foil- or Gas-Excited Beams   859
        H. Harde

Lifetime of the 3d4p $z^1P_1^0$ Level in Sc II by Laser
Excitation of a Fast Ionic Beam   873
        John O. Stoner, Jr., L. Klynning, I. Martinson,
        B. Engman, and L. Liljeby

On The Possibility of a Precise Measurement of the
F VIII 1s2p $^3P_2$-$^3P_1$ Finestructure Splitting   877
        H. J. Andrä and J. Macek

EUV Solar Spectroscopy From Skylab and Some Implications
for Atomic Physics.   885
      E. M. Reeves and A. K. Dupree

Recent Advances in Ultraviolet Astronomy.   907
      Donald C. Morton

High Magnetic Field Spectroscopy.   919
      R. H. Garstang

Spectroscopy of Highly-Stripped Ions in Laser-Induced
Plasmas   925
      N. J. Peacock

Atomic Oscillator Strengths in Fusion Plasma Research . . . 951
      W. L. Weise and S. M. Younger

Plasma and Projectile Stripping: A Comparison 961
      D. J. Nagel

Spectroscopy of Plasmas for Short Wavelength Lasers 973
      R. C. Elton and R. H. Dixon

Subject Index   983

Contents of Volume 1

The Term Analysis of Atomic Spectra:
Present Status and Remaining Problems.   1
    Bengt Edlén

Recent Configuration Interaction Studies
in Atomic Lifetimes.   29
    A. Hibbert

Oscillator Strengths for Ac I, Sc II and Ti III.   43
    G. A. Victor, R. F. Stewart, and C. Laughlin

Transition Probabilities for Ionized Atoms   51
    A. W. Weiss

Correlation Effects and f-Values in the Sodium
Sequence   69
    Charlotte Froese Fischer

On the Possibility of Observing Nonexponential Decays
in Autoionizing States   77
    Cleanthes A. Nicolaides and Donald R. Beck

Coulomb Methods in Atomic Transition Probability
Calculations   83
    Richard Crossley and Susan Richards

Lamb Shift in Hydrogenlike Ions.   89
    Peter J. Mohr

Hyperfine Quenching of the 2^3P_0 State in Heliumlike
Ions   97
    Peter J. Mohr

Anomalies in the Fine and Hyperfine Structure. 105
    Donald R. Beck and Cleanthes A. Nicolaides

Relativistic Contributions to Transition Energies in
NiI and CuI Isoelectronic Sequences.   111
    C. P. Bhalla, C. L. Cocke and S. L. Varghese

Oscillator Strengths in N, N^+, O and O^+ Obtained From
the First Order Theory of Oscillator Strengths (Fotos) . . .   115
    Donald R. Beck and Cleanthes A. Nicolaides

Lifetimes and Fluorescence Yields of Three-Electron Ions . .   121
    C. P. Bhalla and A. H. Gabriel

Future Directions for Beam-Foil Spectroscopy   129
    S. Bashkin

Review of Experimental Lifetimes: Third Period
Elements   147
    M. E. M. Head, C. E. Head, and T. N. Lawrence

Review of Experimental Lifetimes: Fourth Period
Elements   155
    C. E. Head, T. N. Lawrence, and M. E. M. Head

Heavy-Element Beam-Foil Lifetime Measurements and Related
Experimental Problems.   165
    G. Sørensen

Measurements of He I Lifetimes and Fine Structure by a
Two-Spectrometer Method.   183
    G. Astner, L. J. Curtis, L. Liljeby, I. Martinson
    and J. O. Stoner, Jr.

Mean-Lives of RbII in the Visible and Vacuum Ultraviolet . .   191
    M. Czempiel and H. J. Andrä

Applications of High Resolution Measurements of Optical
Lifetimes.   199
    Peter Erman

Beam-Gas Studies of Cu II, Cl II, and As II.   217
    L. Maleki, D. B. King, C. E. Head and T. N. Lawrence

Beam-Foil Study of S III - S VI.   223
    B. I. Dynefors and I. Martinson

The Independent-Electron Model Applied to 100-600 keV
Sulfur Beam-Foil Population Functions.   231
    B. Dynefors, I. Martinson, and E. Veje

Beam-Foil Spectroscopy at the University of Alberta   235
    Eric H. Pinnington

Decay of the 2p3p^{3}S$_1$ Level of O III   251
    B. L. Cardon, J. A. Leavitt, M. W. Chang, and
    S. Bashkin

Profiles of the Spectral Lines Near 2363Å and 2577Å From
Foil-Excited He   259
    John O. Stoner, Jr. and I. Martinson

Relative Initial Populations of Foil-Excited He I States. .   263
    H. H. Bukow, N. v. Buttlar, G. Heine and
    M. Reinke

Radio Frequency Spectroscopy with a Fast Atomic Beam. . . .   271
    Francis M. Pipkin

X-Rays From Foil-Excited Beams at Tandem Energies   283
    C. L. Cocke

Lifetime Measurement of the ^{3}P$_1$ State of Heliumlike
Sulphur   299
    S. L. Varghese, C. L. Cocke, B. Curnutte, and
    R. R. Randall

Radiative Decay and Fine Structure of the 2^{3}P$_0$ and the 2^{3}S$_1$
States of Helium-like Krypton (Kr XXXV)   305
    Harvey Gould and Richard Marrus

Radiative Decay of the 2^{3}P States of Heliumlike Argon . . .   317
    William A. Davis and Richard Marrus

Extreme Ultraviolet Spectra of Highly Stripped Silicon
Ions.   321
    P. M. Griffin, D. J. Pegg, I. A. Sellin, K. W. Jones,
    D. J. Pisano, T. H. Kruse and S. Bashkin

Beam Foil Spectroscopy of Highly Ionized Fluorine, Silicon
and Copper Beams.   331
    L. C. McIntyre, J. D. Silver and N. A. Jelley

Beam-Foil Studies of Nitrogen, Sulfur and Silicon   339
    A. E. Livingston, P. D. Dumont, Y. Baudinet-Robinet,
    H. P. Garnir, E. Biemont and N. Grevesse

Recent Beam-Foil Mean-Life Measurements in Fluorine
V-VIII.   347
    D. J. G. Irwin and R. Drouin

Beam-Foil Spectroscopy of Highly-Ionized C, N, O and Ne
Atoms at 1 MeV/Nucleon.   355
    J. P. Buchet, A. Denis, J. Desesquelles, M. Druetta
    and J. L. Subtil

Spectroscopy of Heavy Ions Using the Beam-Foil Technique. .   367
    H. G. Berry and C. H. Batson

Satellite Lines in Highly-Stripped Ions of B,C,N,O, and F .   377
    E. J. Knystautas and R. Drouin

Doubly-Excited States in B III.   385
      K. X. To, E. J. Knystautas, R. Drouin and H. G. Berry

Doubly-Excited States in N V and N VI   393
    E. J. Knystautus and R. Drouin

The Atomic Physics Potential of New Accelerators.   401
    Paul H. Stelson

Autoionizing States in the Alkalis.   419
    David J. Pegg

Metastable Autoionizing Quartet- Quintet- and Sextet
States in B   437
      R. Bruch, J. Andrä, and G. Paul

Auger Electron Emission Spectra From Foil and Gas Excited
Carbon Beams.   445
      C. Fred Moore, D. Schneider, B. M. Johnson,
      L. E. Smith and W. Hodge

Autoionizing States in Highly Ionized Oxygen, Fluorine,
and Silicon   451
      J. P. Forester, R. S. Peterson, P. M. Griffin,
      D. J. Pegg, H. H. Haselton, K. H. Liao, I. A. Sellin,
      J. R. Mowat, and R. S. Thoe

Subject Index   xxi

NON-CHARACTERISTIC X-RAY PRODUCTION IN HEAVY ION-ATOM COLLISIONS

H.-D. Betz, F. Bell, E. Spindler, M. Kleber,
H. Panke, and W. Stehling

Universität München
D-8046 Garching, Germany

SUMMARY

During collisions between heavy ions continuous radiation is emitted. One may distinguish 3 types of radiation continua: radiative electron capture (REC), transitions between transiently formed molecular or quasimolecular orbitals (MO), and bremsstrahlung due to secondary (knock-on and delta-) electrons, nucleus-nucleus repulsion, various bound-free transitions of electrons, and quasimolecular dynamic effects. Not all of these processes are well understood and we comment on some of the open questions.

REC into projectile K shells presents the most distinguished radiation profile: there is a pronounced intensity maximum at a certain x-ray energy well beyond any characteristic projectile K x-ray lines, a distinct peak width, and a steep high-energy tail. The entire line profile reflects the momentum distribution of bound electrons prior to capture. When the collision velocity, v, is large compared to the orbital velocity, v_e, of target electrons to be captured, REC x-ray profiles become particularly clean, especially on the high energy side of the distribution and can be understood reasonably well. REC data for S and Cu ions with energies between 55- and 450 MeV is shown and compared with theoretical REC calculations.

K shell MO radiation continua occur for $v \lesssim v_e$ and

are examined near and beyond the so-called united atom
x-ray transition energy. It is demonstrated that the
width, H, of the structureless exponentially decreasing
line tail is well reproduced for all collision systems
where experimental data on tails is available, ranging
from 220-keV C to 450-MeV Cu, using a simple analytical
formula based on theory of dynamic collision broadening.
We discuss the observation that H seems to be practically independent of details in the MO level diagram. We
comment on vacancy production mechanisms and discuss
consequences of dynamic effects for MO x-rays.

It is suggested that sophisticated and detailed
calculations of MO levels as a function of internuclear
distance, obtained for stationary states in the adiabatic or diabatic case v→0, may be relatively insignificant
in fast collision systems which can be studied experimentally. We argue on simple grounds that dynamic
effects can not be ignored and, when included in theoretical treatments, may bring about significant changes
in calculated continua and in the interpretation of
experimentally observed quasimolecular radiation.

INTRODUCTION

In heavy ion-atom collisions continuous radiation
is produced which, at present, can not always be easily
accounted for. The reason for this difficulty in our
understanding is quite clear: heavy ion encounters are
very complex because large numbers of electrons with
many different velocities collide and many interactions,
electron-electron, electron-nucleus, nucleus-nucleus,
may occur at the same time. In particular, distortions
of electronic states can be tremendous so that first-
order perturbation techniques are not applicable. On
the experimental side, only limited interest can be
aroused by radiation continua which turn out to be
mostly exponential line tails without characteristic
structures. Nevertheless, some encouraging progress
has been achieved in the understanding of heavy ion
collisions. Initial hopes, though, to measure spectacular effects such as, for example, details of molecular
levels or quantum-electro-dynamical effects, are fading.
The most immediate questions, at this time, seem to
concentrate on three points: (1) the kind of processes
which contribute to continua and how one can separate
them experimentally, (2) proper theoretical treatment
of these processes, and (3) implications on our
fundamental understanding of heavy-ion collisions and

implications for questions of theoretical and practical interest.

We start out by distinguishing three types of continuum radiation: (1) bremsstrahlung due to several interaction mechanisms, (2) radiative electron capture (REC), and (3) line tails associated with molecular or quasimolecular effects (MO); REC and MO transitions are illustrated in Fig.1. Some of these phenomena are interconnected, giving rise to definition problems, and may occur simultaneously. In the following, we discuss these processes in some selected details. A more complete review will be given elsewhere[1].

BREMSSTRAHLUNG

Very satisfactory understanding has long been achieved for dipole bremsstrahlung from nucleus-nucleus collisions[2,3]. In the Born approximation, when the radiated energy, E_x, is very small compared to the projectile energy, E, the cross section is

$$\frac{d\sigma}{dE_x} = \frac{16e^6}{3\hbar c^3}(\frac{Z_p Z_T}{m_p v})^2 (\frac{Z_p}{A_p} - \frac{Z_T}{A_T})^2 \frac{\ln\left[(\sqrt{E_c}+\sqrt{E_c-E_x})^2/E_x\right]}{E_x} \quad , \quad (1)$$

where Z_p, A_p, Z_T, A_T are charge and mass number of projectile and target, respectively, m_p is the proton mass and E_c the center-of-mass energy.

Nuclear quadrupole bremsstrahlung can also be calculated in the Born approximation. The resulting cross section as derived[4] very recently is

$$\frac{d\sigma}{dE_x} = \frac{32e^6}{15\hbar c^5}(\frac{Z_p Z_T \mu}{m_p^2})^2 (\frac{Z_p}{A_p^2} + \frac{Z_T}{A_T^2})^2 \frac{\left[(E_c-E_x)/E_x\right]^{1/2}}{E_x} \quad , \quad (2)$$

where μ denotes the reduced mass of the nuclei. In case of identical nuclei an additional factor of 2 must be applied.

It can be seen from Eqs.(1) and (2) that nuclear bremsstrahlung gives rise to a spectral intensity which decreases weakly with E_x, roughly $d\sigma/dE_x \propto 1/E_x$, and represents sort of a background. It is important that both the radiation profile and absolute intensities can be calculated on safe grounds.

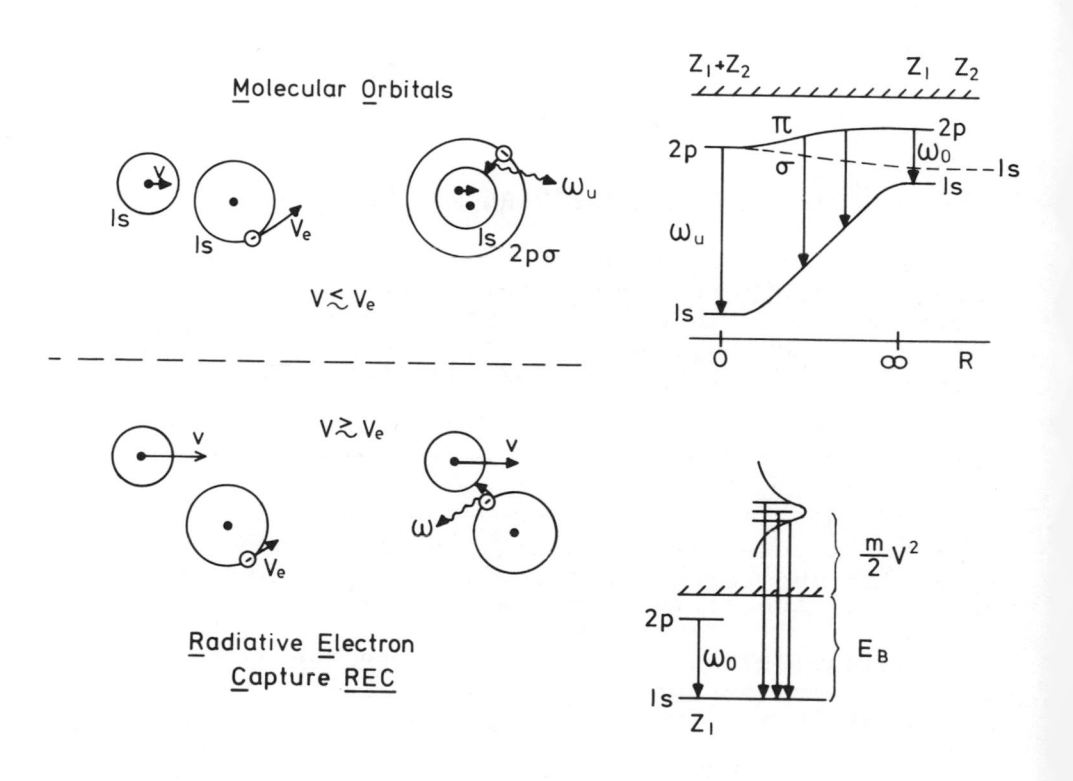

FIG.1. Schematic illustration of MO- and REC transitions.

Electron bremsstrahlung presents much more intricate problems, mainly because position and acceleration during collisions are not tracked as easily for electrons as for protons. We will comment on secondary electron bremsstrahlung (SEB) and inner bremsstrahlung (IB) which is also called radiative ionization.

SEB is a two step process; in a first collision an electron is ejected from either target or projectile, the liberated electron is accelerated and radiates in a second collision with target nuclei. Thus, SEB is density dependent and will be particularly important in solid targets. The first step, ionization of electrons, can be calculated in the classical binary-encounter approximation (see, for example, ref.5) and, especially for high-velocity collisions, with the Born approximation[6]. These procedures are expected to yield reasonable ionization cross sections as long as molecular collision effects are not important, i.e. for $v/v_e \lesssim 0.1$. Insufficient information on the electron energy spectrum is known for the case $v/v_e \lesssim 0.1$, but one may hope that the contributions to the radiation spectrum can be neglected for $E_x \lesssim 4E_K$, where $E_K = mv^2/2$ with the elctron mass m.

Intensities of SEB decrease weakly up to $E_x \simeq 4E_K$ and strongly thereafter; this predicted trend could be verified experimentally when very light particles are incident[5]. In radiation spectra from heavy ion collisions, however, a change of slope near $4E_K$ has not yet been observed; in these cases, either SEB is dominated by other radiation effects, or the overlapping contributions from inner electrons lead to a smear-out. In most experiments, the former explanation seems to hold (see below).

Inner bremsstrahlung[7] is a one step process: in a collision, an inner electron is ionized and radiates during its escape from the collision system in the electric field of this system. It is not clear, however, whether this effect can be strong enough to be observable in experiments. Theoretical treatment of IB and interpretation of spectral features in terms of IB as proposed in ref.7 can not be backed up. Recent thorough IB calculations[4] have shown that the effect is extremely small and does not noticeably contribute to observed spectra. The slowly decreasing spectral feature in ref.7 has also been attributed to SEB[4]; this raises a new problem: the feature in question was observed, for example, in 288-MeV Ar→Ne collisions. Recent measurements

of similar system, 250-350-MeV Ar→Al[8], do not show the
same effect but reveal exponentially decreasing radiation
tails (see Fig.2 in ref.9 and present Figs. 2a-4a).
Thus, some aspects in measurement and calculation of SEB
have not yet been handled satisfactorily, and the reason
for the discrepancies outlined above is not yet clear.

It is of some importance to point out another source
for background effects. Especially in high-energy collisions nuclear Coulomb excitation can be produced in
amounts significant enough to allow γ-rays with energies
in the MeV range, which result from subsequent deexcitation, to be seen with appropriate detectors. Such
γ-rays are difficult to absorb, and when they penetrate
through an x-ray detector Compton processes occur and
show up in the radiation spectrum as a very slowly
decreasing background.

RADIATIVE ELECTRON CAPTURE

The phenomenon of REC is schematically illustrated
in Fig.1 and has been frequently described[9,10]. Therefore, we limit the discussion to some fundamental aspects
and present some new experimental results.

Identification of REC is uncritical since the continuous radiation profile exhibits a distinct maximum at
a position which depends on v. Experimental maxima are
shown in Figs.2a-4a and are well reproduced theoretically. The REC distribution reflects the momentum distribution of target electrons, appropriately summed over all
electrons which can be captured. Clean and unambiguous
tails are obtained only for $v/v_e \gg 1$; in 346-MeV Ar→C
collisions[8,9], for example, an REC tail can be observed
over ∿3 orders of magnitude in intensity and agrees
exceedingly well with the profile calculated from C-1s
electrons; in this case, one could in turn determine
the C-1s Compton profile from REC measurements. It
remains to be seen to what extent such procedures can
become an important technique.

There are a number of difficulties which must be
mentioned. At low velocities, when target or projectile
electrons with $v/v_e \lesssim 1$ are present, distortions of
electronic states occur (molecular effects) and radiative transitions from or to such states overlap with REC
transitions involving target electrons with $v/v_e > 1$. It
has also been observed that the REC peak will be less
pronounced when Z_T becomes large. This trend is

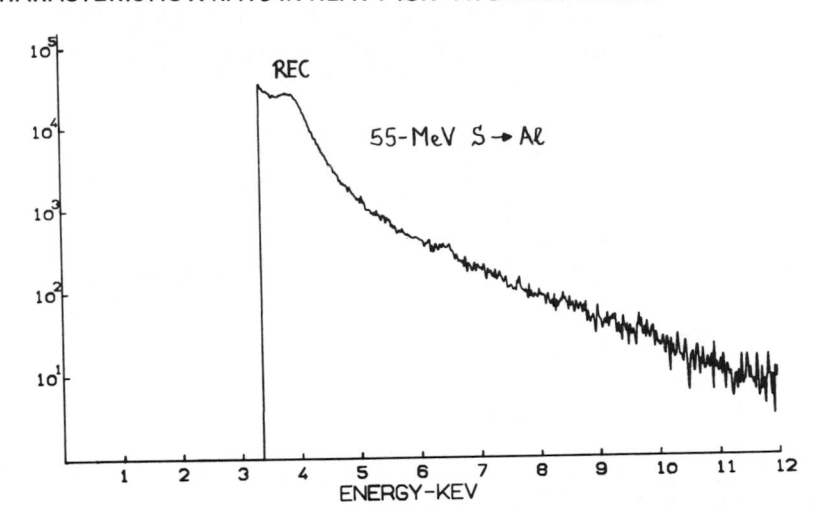

FIG.2a. Experimental radiation profile from collisions of 55-MeV sulphur ions in ∿100 μg/cm^2 aluminum.

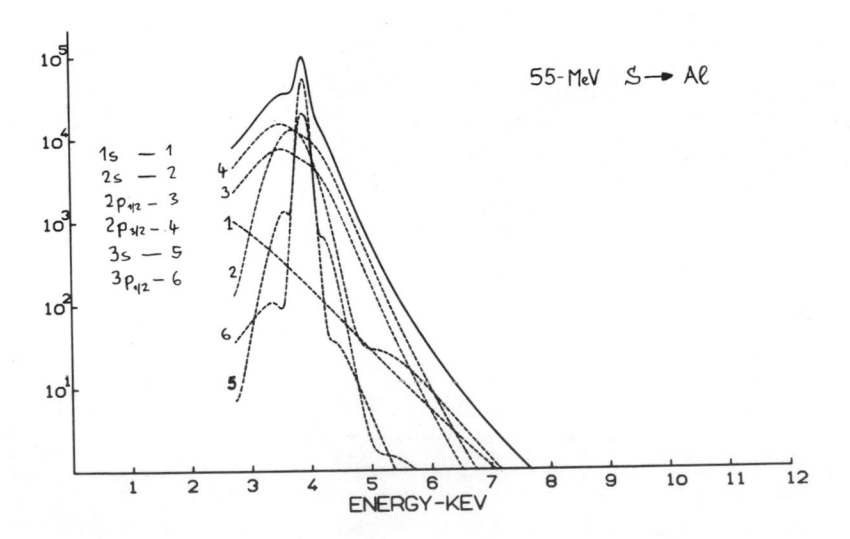

FIG.2b. Theoretical REC profile for 55-MeV S→Al collisions.

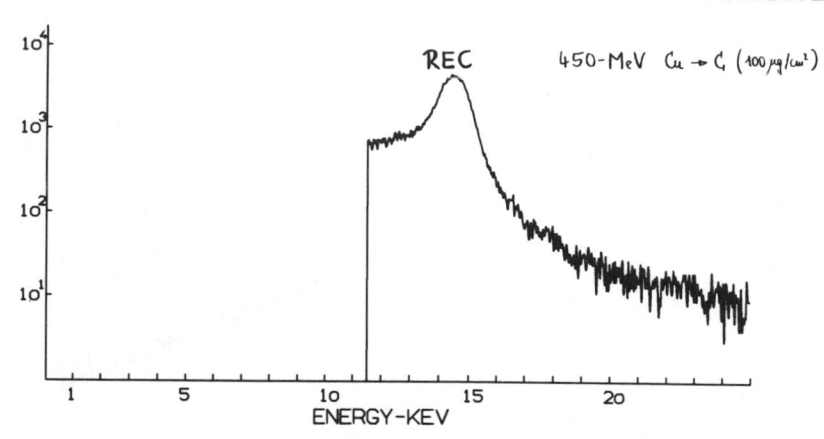

FIG.3a. Experimental radiation profile from collisions of 450-MeV copper ions in ∿100 μg/cm² carbon.

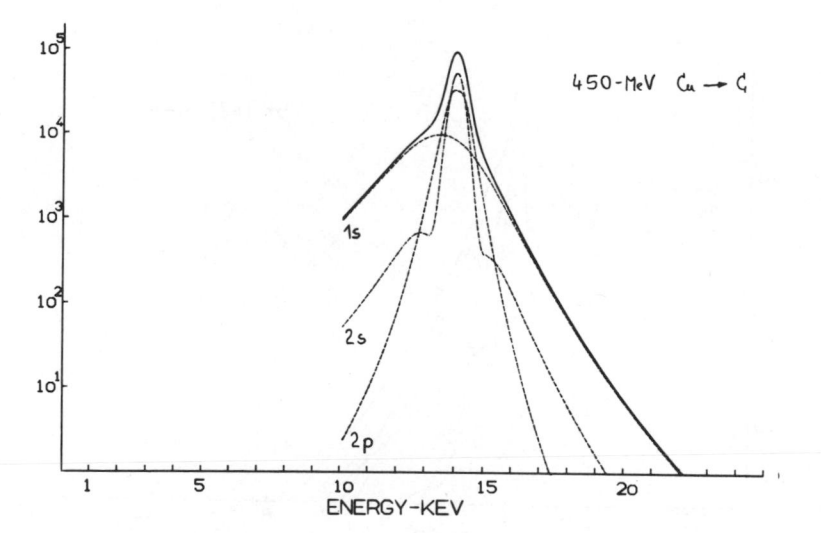

FIG.3b. Theoretical REC profile for 450-MeV Cu-C collisions.

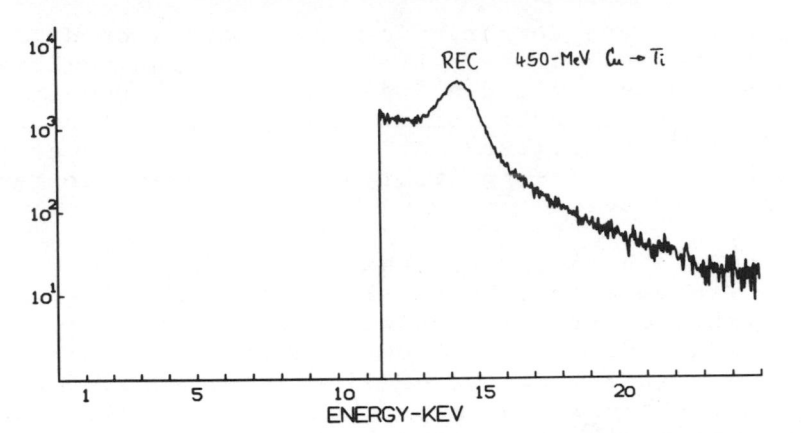

FIG.4a. Experimental radiation profile from collisions
        of 450-MeV copper ions in $\sim$100 µg/cm^2 titanium.

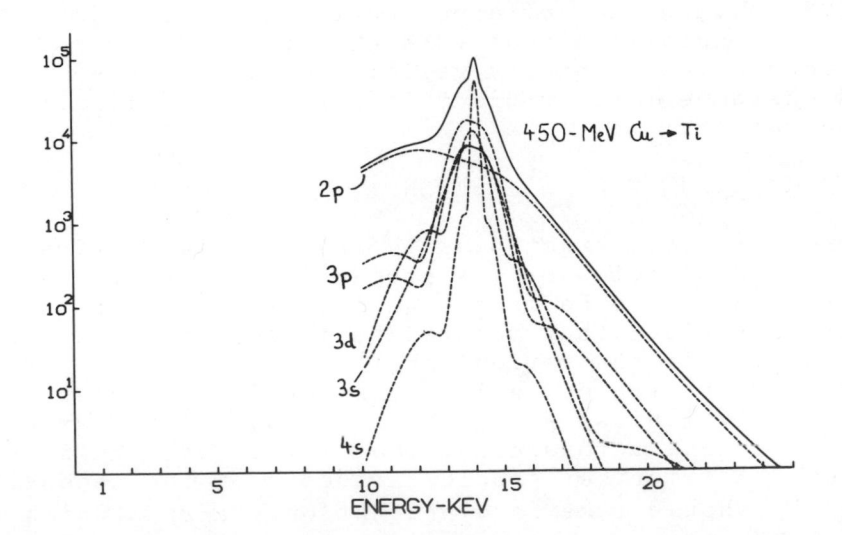

FIG.4b. Theoretical REC profile for 450- MeV Cu→Ti
        collisions.

obvious when one compares Figs.2a and 3a, or data from
Fig.2 in ref.9. In very heavy targets the maximum disappears completely and a tail is all what remains. We
note that our calculations of REC profiles[9] do not
explain this behaviour. Among possible explanations we
mention the increasing importance of molecular- and
screening effects.

The width of the REC maximum is found to be subst-
ancially broader than our calculations predict. We used
very accurate Hartree-Fock wave functions to derive
theoretical REC profiles, but the relatively sharp features due to C-2s, Al-2p, or Ti-4s electrons, for example,
are not visible in the experimental spectra (Figs.2-4).
It is realized that HF calculations for free atoms do
not describe electronic states of outer electrons in a
solid and one may tend to ascribe the discrepancy to
solid effects. Measurements[11] in gaseous targets, however,
reveal no better agreement except for He. Moreover, in
the case of carbon, the more tightly bound 2s electrons
($E_B \simeq 20$ eV) give a narrower peak than the less tightly
bound 2p electrons ($E_B \simeq 10$ eV for a free atom). Thus,
one must search for other than solid effects.

Absolute REC cross sections can be estimated from
the theory for capture of free electrons[12], and follow
with somewhat better accuracy from the formulation in
ref.9. Evaluation of experimental cross sections from
solid targets is almost straightforward. One possibility
is to use the intensity ratio, I_{REC}/I_K, between REC-
and characteristic projectile x-rays which fill the
same vacancy:

$$d\sigma_{REC}/d\Omega = 3I_{REC}/(8\pi I_K N v \tau) \quad , \qquad\qquad (3)$$

where N is the target density and τ the effective
radiative lifetime of the projectile vacancy. Determination of τ can be made experimentally[13], or theoretically provided that the relevant degrees of multiple
ionization in the various shells are known. At very
low v, where stripping is not extensive, tabulated values
for singly ionized atoms may be used, but at large v
one must take into account that τ will be significantly
reduced. A further problem arises from the necessity
that I_K should take into account only x-rays which are
produced inside the solid and not those which are
emitted after the ions have left the target; at large v
this effect can be significant[13]. The procedure resulting in Eq.(3) has the advantage that no absolute

normalization of the data is necessary; only relative
spectral intensities are needed, and one avoids more
complicated measurements in gaseous targets. For S→C,
we find reasonable agreement between experimental and
theoretical values; for example, ∿50 and 66 barn,
respectively, at 110 MeV and for an observation angle
of 90°.

REC may proceed into any empty final states. K
shell REC yields most distinguished ratiation profiles,
but REC into higher shells can also be observed. As an
example, REC into the Cu-L shell could be clearly identified in experiments with 300-450-MeV Cu ions[8].

Especially for light ions, REC into continuum
states of the projectile ions may occur with probabilities which are comparable with REC into bound states.
In this case, one obtains a radiation tail which can be
regarded as kind of bremsstrahlung[14].

QUASIMOLECULAR RADIATION CONTINUA

In sufficiently slow collisions, $v \ll v_e$, large dist-
urbances may occur to electrons within a time long
enough to allow the electrons to adjust adiabatically
and to follow molecular orbitals. For each internuclear separation, R, of the colliding systems stationary molecular states can be calculated and state diagrams,
energy levels of states as a function of R, can be constructed. Much effort has been spent in this direction
and diagrams are being produced with increasing accuracy
for almost any collision system, including very heavy
systems in which $Z_u = Z_p + Z_T$ exceeds the charge of known
nuclei[15].

Initially, there has been the hope that observation
of radiative transitions between molecular states allows
to map out these levels more or less directly (see
scheme Fig.1). Expectations of that sort have been un-
realistic, so far, and all experimental continua presently available and attributable to K- or L shell MO processes look very similar in the following sense: there is
simply a tail, relatively flat close to the relevant
characteristic x-ray line, and then decreasing exponentially for higher x-ray energies, with no characteristic
features such as peaks or a cut-off. An example for
the high-energy tail is presented in Fig.2a where, due
to presence of electrons with $v < v_e$ and $v > v_e$, both MO
and REC transitions become visible in the same spectrum;

above ∿6 keV, however, MO effects dominate the spectrum
because the REC profile falls off rapidly as is evident
from Fig.2b. Later on, we will discuss the MO tails in
detail.

        Understanding and identification of MO x-ray prof-
iles requires successful calculation of absolute production cross sections. In some cases, theoretical estimates
can indeed be obtained[16,17] which compare reasonably
well with experiments, though, in view of the complications involved, agreement of better than within a
factor of ∿2-4 is probably meaningless. Calculations
work best in cases of the so-called two-collision
mechanism[18], i.e. a vacancy from a first collision is
transferred along molecular orbitals into the actual
(second) collision where the radiative transition occurs.
The existence of such a vacancy in the lower lying
molecular orbital is a necessity for the kind of transitions considered. The question arises, however, whether
there are other ways to create this vacancy. As a
second possibility, the one-collision mechanism[16,19]
must be discussed, i.e. a vacancy is created and filled
during a single collision. So far, it has been assumed
that various excitation processes, Coulomb ionization
and/or molecular coupling, are effective. At present,
formulation of these processes is controversial and more
clarification on a fundamental basis has yet to be
worked out.

        MO x-ray spectra display one feature which, in our
opinion, is most intriguing: this is the width, H, of
the experimental tail near and beyond the x-ray transition energy of the statically united system, E_u. We
have measured H with good statistical accuracy in many
collision systems with incident S, Ar, and Cu ions, and
in a wide range of energies, 8- to 450 MeV. To our
surprise, we found remarkable regularities which, in
case of incident sulfur ions, have been reported recently[17]. Based on the work of Macek and Briggs[20] we
demonstrated how one obtains H numerically for any
collision system, symmetric or asymmetric:

$$H = 0.30(\hbar v \Delta E_x / \Delta R)^{1/2} , \qquad (4)$$

where ΔE_x is the difference of transition energies for
separated and united ion and $\Delta R = R_1(1+Z_2/Z_1)$ with index
1 representing the heavier one of the two colliding atoms
for ·convenience, we take for R the average between $<R>$
and $<1/R>^{-1}$ from Hartree-Fock calculations. This simple
procedure allows to reproduce H with an average error

of only ∿10 % for an impressive variety of systems,
yielding values of H between 60- und ∿5000 eV, as is
shown in Tab.1.

It must be emphasized that Eq.(4) does not contain
any details of the molecular state diagram. We want to
focus on two important consequences of the successful
reproduction of H.

Eq.(4) still works when v/v_e approaches unity, i.e.,
when one deals with collisions which are certainly not
adiabatic, not even quasi-adiabatic. In these relatively fast collisions one can not expect that electrons
are always able to move along rapidly changing molecular
orbitals, i.e. the projectile-target wave functions
are no longer able to adjust to the imposed perturbation[21]. As a consequence one might expect that Eq.(4)
works less and less well as v/v_e increases, contrary to
our experimental observations. To resolve this problem,
we speculate as follows. Independent of v, molecular
levels exist for all values of R as calculated from, say,
the two-center Schrödinger equation; these levels are
determined essentially from Z_p, Z_T, and R and, except
for screening effects, it is not really necessary to
occupy these levels with electrons. In a heavy ion
collision, therefore, molecular levels may exist and be
defined to the extent allowed by dynamic collision
broadening, but may not be occupied when the electrons
assigned to this level can not adjust fast enough.
Consequently, sort of dynamic vacancies are created
which may be filled by radiative transitions and give
rise to the observed MO x-ray spectrum.

There are certain probabilities that electrons do
follow molecular orbitals or are kicked out when v or
the energy gap between isolated and united levels becomes too large. One could then have a situation where
all 1s orbits are occupied before and after the collision, yet a radiative transition to a transiently
vacant "1sσ" state has occurred. Due to the dynamics
of the collision, the Pauli principle would not rule
out such processes.

Our hypothesis is backed up by experimental evidence. In O→Zr collisions, for example, the correlation
diagram does not allow transfer of a projectile K vacancy into the united K shell. Nevertheless, the width of
the observed x-ray tail can be accounted for when one
assumes that a K vacancy was present (cases marked by *

System	Shell	$\sqrt{\dfrac{\Delta E}{\Delta R}}$	Energy [MeV]	v	v/v_e	Width[eV] exp	Width[eV] th	Ref.
C → C	K	10.2	0.220	.86	.19	64	77	23)
S → C	K	27.1	16	4.5	.30	460	467	17)
			32	6.3	.43	510	556	17)
S → Aℓ	K	38.1	8	3.2	.21	540	551	17)
			16	4.5	.30	690	656	17)
			32	6.3	.43	740	780	17)
			55	8.3	.56	920	893	17)
			95	11.	.73	1040	1024	17)
Ni → Ni	K	101.	39	5.2	.21	2300	1874	24)
			70	7.0	.28	2600	2166	24)
Cu → Aℓ	K	74.6	450	17.	.66	2550	2506	8)
Kr → Ti	K	121.	200	9.8	.27	3280	3085	22)
Kr → Zr	K	178.	200	9.8	.27	4930	4547 *	22)
O → Zr	K	86.1	33	9.1	.25	2120	2116 *	22)
Ag → Ag	L	44.4	30	3.4	.21	900	662 ⊞	25)
J → J	L	54.5	25	2.8	.15	790	748 ⊞	25)

Tab. 1: Experimental and theoretical half-widths of MO tails near E_u for various collision systems. (*) see text for cases with $Z_T > Z_P$; (⊞) experimental widths uncertain due to poor statistics or background effects near E_u. ΔE and ΔR are explained in the text.

in Tab.1). Calculations of the MO x-ray yield based on
Zr K vacancies found after a collision give far too low
intensities[22]. Provided that experimental and theoretical results in ref.22 are correct, our model of transient vacancies in non-adiabatic collisions is strongly
supported.

Finally, we note that there are further possibilities
to produce radiation in heavy ion collisions. For
example, the acceleration of electrons must give rise
to emission of continuous x-ray. This process and others
have not yet been fully explored and await satisfactory
clarification.

REFERENCES

1) H.-D. Betz, M. Kleber, and F. Bell, Radiation Effects
   (to be published in 1976).
2) W. Heitler, The Theory of Radiation, London, Oxford
   University Press 1954.
3) K. Alder, A. Bohr, T. Huus, B. Mottelson, and A.
   Winther, Rev. Mod. Phys. $\underline{28}$ 432 (1956).
4) D.H. Jacubassa and M. Kleber, Z. Phys. A$\underline{273}$, 29 (1975).
5) F. Folkmann, C. Gaarde, T. Huus, and K. Kemp. Nucl.
   Instr. Meth. $\underline{116}$, 487 (1974).
6) J.D. Garcia, Phys. Rev. $\underline{159}$, 39 (1967).
7) P. Kienle, M. Kleber, B. Povh, R.M. Diamond, F.S.
   Stephens, E. Grosse, M.R. Maier, and D. Proetel,
   Phys. Rev. Lett. $\underline{31}$, 1099 (1973).
8) H.-D. Betz, C. Stéphan, B. Delaunay, E. Baron, E.
   Spindler, H. Panke, and F. Bell (to be published).
9) H.-D. Betz, M. Kleber, E. Spindler, F. Bell, H. Panke, and W. Stehling, Proc. IX International Conference on the Physics of Electronic and Atomic Collisions,
   Seattle 1975, to be published.
10) H.W. Schnopper, H.-D. Betz, J.P. Delvaille, K. Kala-
    ta, A.R. Sohval, K.W. Jones, and H.E. Wegner, Phys.
    Rev. Lett. $\underline{29}$, 898 (1972).
11) H.W. Schnopper and J.P. Delvaille, Proc. of the In-
    ternational Conference on X-Ray Processes in Matters,
    Helsinki 1974, p.19.
12) H.A. Bethe and E.E. Salpeter, Quantum Mechanics of
    One- and Two-Electron Atoms (Academic, New York,
    1957), p.320ff.
13) H.-D. Betz, F. Bell, H. Panke, G. Kalkoffen, M. Welz,
    and D. Evers, Phys. Rev. Lett. $\underline{33}$, 807 (1974).
14) H.W. Schnopper, J.P. Delvaille, K. Kalata, A.R. Soh-
    val, M. Abdulwahab, K.W. Jones, and H.E. Wegner,
    Phys. Lett. $\underline{47}$A, 61 (1974).

15) There is a tremendous amount of material on adiabatic
    and diabatic state calculations; see, for example,
    K. Helfrich and H. Hartmann, Theoret. Chim. Acta
    (Berlin) $\underline{16}$, 263 (1970); F. P. Larkins, J. Phys. $\underline{B5}$,
    571 (1972); B. Müller, J. Rafelski, and W. Greiner,
    Phys. Lett. $\underline{47B}$, 5 (1973).
16) W.E. Meyerhof, T.K. Saylor, S.M. Lazarus, A. Little,
    B.B. Triplett, L.F. Chase, and R. Anholt, Phys. Rev.
    Lett. $\underline{32}$, 1279 (1974).
17) H.-D. Betz, F. Bell, H. Panke, W. Stehling, E. Spind-
    ler, and M. Kleber, Phys. Rev. Lett. $\underline{34}$, 1256 (1975).
18) F.W. Saris, W.F. van der Weg, H. Tawara, and R.
    Laubert, Phys. Rev. Lett. $\underline{28}$, 717 (1972).
19) F. Bell, H.-D. Betz, H. Panke, E. Spindler, W. Steh-
    ling, and M. Kleber, accepted for publication in
    Phys. Rev. Lett.
20) J.H. Macek and J.S. Briggs, J. Phys. $\underline{B7}$, 1312 (1974).
21) R. Laubert, H. Haselton, J.R. Mowat, R.S. Peterson,
    and I.A. Sellin, Phys. Rev. $\underline{A11}$, 135 (1975).
22) R. Anholt and T.K. Saylor, to be published.
23) J.R. Macdonald, M.D. Brown, and T. Chiao, Phys.
    Rev. Lett. $\underline{30}$, 471 (1973).
24) J.S. Greenberg, C.K. Davis, and P. Vincent, Phys.
    Rev. Lett. $\underline{33}$, 473 (1974).
25) W. Wölfli, Ch. Stoller, G. Bonani, M. Suter, and
    M. Stöckli, to be published.

ANGULAR DISTRIBUTION STUDIES OF NON-CHARACTERISTIC X-RADIATION

R. S. Thoe, I. A. Sellin, K. A. Liao, R. S. Peterson,
D. J. Pegg, J. Forester, and P. M. Griffin

The University of Tennessee, Knoxville, Tenn. 37916
and
Oak Ridge National Laboratory, Oak Ridge, Tenn. 37830

We have recently measured the polarization of the non-characteristic x-radiation emitted from collisions between energetic (10-90 MeV) Al ions in thin foils. The motivation to persue this line of research originated from the calculations of Müller and Greiner[1] which predict that in addition to the normal spontaneous emission of molecular orbital x-rays, there also exists a mechanism which gives rise to induced emission of these radiations. This mechanism is a direct result of the coriolis forces which exist in the rest frame of the quasi-molecule, whose axis of quantization is rotating with angular velocity ω_{rot}. The cross sections derived by Müller and Greiner from the two center Dirac equations are

$$\frac{d\sigma_s}{d\omega d\Omega_K d\Omega_i} = \frac{\omega^3}{2\pi\hbar c^3} |d_{Fi}|^2 \sin^2\theta_K \left(\frac{d\sigma}{d\Omega_i}\right)_{Ruth} \tag{1}$$

$$\frac{d\sigma_i}{d\omega d\Omega_K d\Omega_i} = \frac{\omega}{2\pi\hbar C^3} |\vec{\omega}_{rot} \times \vec{d}_{Fi}|^2 \sin^2\theta_K \left(\frac{d\sigma}{d\Omega_i}\right)_{Ruth} \tag{2}$$

d_{Fi} is the ordinary dipole transition matrix element and $d\sigma_s$ is the differential cross section for detecting a photon of frequency ω at angle θ_K per solid angle $d\Omega_K$ for ions scattered into solid angle $d\Omega_i$. $\left(\frac{d\sigma}{d\Omega_i}\right)_{Ruth}$ is the differential Rutherford cross section.

477

These two equations are equal in magnitude when $\omega_{rot}^2 = \omega^2$. For photon frequencies near the united atom limit ω_{rot} is a maximum at the distance of closest approach and is $\sim \dfrac{T}{2m\,R_{\mu a}^{\,2}}$, in the straight line approximation for an impact parameter $b = 0.5R_{\mu a}$; $R_{\mu a}$ is the radius of the united atom K shell, T is the incident kinetic energy and m is the mass of the ion - m $\sim$ 2Z. Using the Bohr formula for Lyman alpha radiation gives:

$$\omega^2 = (1/4)\,\frac{(2Z)^2\,e^4}{\hbar^2\,R_{\mu a}^{\,2}}\,(3/4)^2 \qquad\qquad (3)$$

$\omega^2 = \omega_{rot}^2$ implies:

$$\frac{T_o}{2m\,R_{\mu a}^{\,2}} = (1/4)\,\frac{(2Z^2)\,e^4}{\hbar^2\,R_{\mu a}^{\,2}}\,(3/4)^2 \qquad\qquad (4)$$

$$T_o = \frac{Z^3}{144}\ \text{MeV}.$$

Therefore, at energies above T_o, the induced cross section should be larger than the spontaneous cross section. This expression compares very favorably with the value obtained by Betz, Bell $et\ al.$[2] - $T_o' = \dfrac{Z^3}{121}$ MeV, which was obtained from cross section measurements on the molecular-orbital radiation emitted from S-Al and S-Ne collisions.

The difference in the transition matrix elements in equations (1) and (2) also leads to differing polarizations between the induced and spontaneous radiation. The amount of polarization is also a strong function of subshell population, collision velocity, impact parameter, and photon frequency. In order to characterize this polarization it is necessary to consider the transformation between the laboratory rest frame and the rest frame of the quasi-molecule. In Fig. 1 the beam is moving in the -z direction, the internuclear axis is Z, the rest frame of the quasi-molecule is z', x, y', and the angle between z and z' is ξ. Therefore, $0 \leq \xi \leq \pi$, and the direction ω_{rot} is coincident with the x axis. The radiation pattern of the rotating quasi-molecule for united atom radiation can be calculated most simply since for this radiation the photon frequency is $\sim$ independent of the internuclear distance.

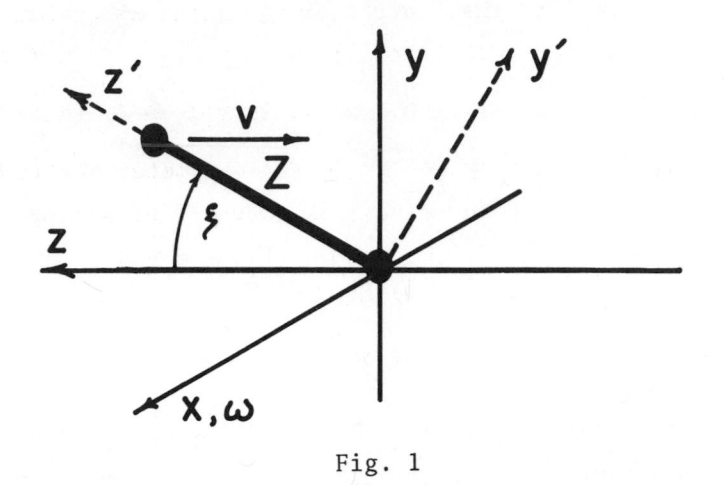

Fig. 1

Now this rotation can be represented by:

$$(\hat{x}_i) = R (\hat{x}')$$

$$R = \begin{pmatrix} 1 & 0 & 0 \\ 0 & \cos \xi & + \sin \xi \\ 0 & - \sin \xi & \cos \xi \end{pmatrix} \tag{5}$$

The two radiation patterns that have been considered are $I_1(\theta') = \sin^2 \theta'$ and $I_2(\theta') = \cos^2 \theta'$, where the primes indicate that these are the patterns as observed in the rotating quasi-molecular rest frame. In the lab frame these appear as

$$I_1(\theta) = x'^2 + y'^2 = x^2 + y^2 \cos^2 \xi + z^2 \sin^2 \xi \tag{6}$$

$$I_2(\theta) = z^2 = y^2 \sin^2 \xi + z^2 \cos^2 \xi. \tag{7}$$

Since this axis is rotating; (i.e., ξ goes from 0 to π), it is necessary to average these radiation patterns over all angles ξ and over all azimuthal angles θ. These averages yield:

$$\bar{I}_1(\theta) = 1/2 + 1/4 \sin^2 \theta \tag{8}$$

$$\bar{I}_2(\theta) = 1/4 + 1/4 \cos^2 \theta. \tag{9}$$

The radiation from the induced transition is somewhat different since it is proportional to ω_{rot}^2. For the collisions under
consideration the scattering angles are less than one degree for
impact parameters as small as 0.1 $R_{\mu a}$. Therefore ω_{rot} can be
approximated by: $\omega_{rot} \simeq \dfrac{v \sin \xi}{Z} = \dfrac{v \sin^2 \xi}{b}$. Then the average of
the induced radiation patterns must include the weighting factor
$\sin^4 \xi$. These averages yield:

$$\bar{I}_1{}'(\theta) = 2/32 + 9/32 \sin^2 \theta \qquad (10)$$

$$\bar{I}_2{}'(\theta) = 7/32 + 3/32 \cos^2 \theta, \qquad (11)$$

with the result that : $\bar{I}_1'(\theta) + \bar{I}_2'(\theta) = 12/32 + 6/32 \sin^2 \theta$. Defining β as the polarization fraction (i.e., $I(\theta) = \alpha + \beta \sin^2 \theta$,
$\alpha + \beta = 1$), we see that $\beta = 1/3$ even if no differential subshell
alignment exists. With complete alignment β can be as large as
9/11.

If we use the maximum polarization for induced transitions and
assume that the induced transition rate is proportional to the
incident energy T and is equal to the spontaneous transition rate
at $Z^3/121$ MeV = 18 MeV, we can describe rather accurately the
behavior of the polarization β as a function of incident ion
energy. The radiation pattern $I(\theta)$ is just the renormalized sum
of the induced and spontaneous radiation patterns. Therefore
assuming approximate isotropy of the spontaneous radiation yields

$$I(\theta) = \frac{1 + (2/11 + 9/11 \sin^2 \theta)\, T_o/18}{1 + T_o/18}$$

$$\beta(T) = \frac{9/11\, T_o}{18 + T_o} \qquad (12)$$

Figure 2 is a plot of equation 12 along with the measured
polarization fraction $\beta(T)$ for photons in the 5-6 keV energy interval (united atom limit) emitted from Al-Al collisions. The fit
with the experimental data is remarkably good considering that
equation 12 is derived from the theoretical results of Ref. 1 and
the cross section data of Ref. 2 and contain no adjustable parameters.

Although the data of Fig. 2 fit equation 12 extremely well,
there are still many unresolved problems. The effect of the beam
energy on incident charge state and, hence, alignment has been
completely ignored. Another major concern has to do with the rapid
rotation of the quasi-molecular axis. The energies and impact

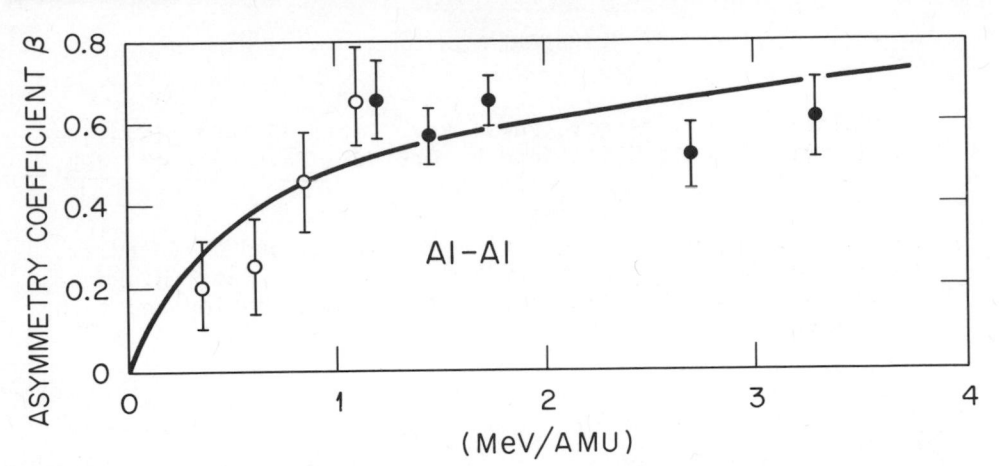

Fig. 2.   Plots of the polarization β for the photon energy bins
          between 5 and 6 keV (inclusive) as a function of incident
          beam energy. Error bars represent 1 sigma. The solid line
          corresponds to the expression β = (9/11)T/(T + 18 MeV),
          where the coefficient (9/11) is derived in the text, the
          T dependence is derived from ω_{rot}^{2} , and the cross over
          between induced and spontaneous radiation at $\sim$ 18 MeV
          corresponds to the results of Ref. 2.

parameter where the rotational mechanism is important are those
that give $\omega_{rot}^{2} \gtrsim \omega^{2}$. This means that the angular velocity of the
internuclear axis is larger than the angular velocity of even the
K-shell electrons, therefore it is difficult to understand how the
electrons could be in eigenstates that are approximated by molecular
orbitals. A final difficulty lies in that fact that while the
molecular orbital radiation is highly polarized, which implies considerable alignment, the characteristic radiation is thought to be
more nearly isotropic which implies a very small alignment.

ACKNOWLEDGMENTS

       We thank Professor Myron McKay for his helpful participation
in the data acquisition and the ORIC operating staff for their
valuable contributions.

REFERENCES

1. B. Müller and W. Greiner, Phys. Rev. Lett. 33, 464 (1974);
    B. Müller, R. Kent Smith, and W. Greiner, Physics Lett. 49B,
    219 (1974).

2. H. D. Betz, F. Bell, H. Panke, W. Stehling, and E. Spendler,
    Phys. Rev. Lett. 34, 1256 (1975); F. Bell, H. Betz, H. Panke,
    E. Spindler, and W. Stehling, private communication and to be
    published.

PROBLEMS OF QUANTUM ELECTRODYNAMICS IN

HEAVY-ION COLLISIONS[*]

Berndt Müller

Physics Division, Argonne National Laboratory

Argonne, Illinois 60439

Atomic physics has long been an important source for gaining fundamental insights. In the beginning of this century the measurements of the structure and fine structure of atomic spectra led to the invention of quantum mechanics, and around 1950 the Lamb-Shift and the (g-2) experiments started the development of quantum electrodynamics (QED) and field theory in general. Today the perturbative formulation of QED has been tested to several GeV/c momentum transfer in e^+e^- collisions, and also in binding energies of electrons and muons in heavy atoms. The agreement between experiment and theory is excellent. However, all these are cases where the coupling constant e^2 or Ze^2 is well below unity and it must be kept in mind that there is still no satisfactory theory of the strong interactions. We simply do not know what is the right way to handle coupling constants $g^2 \sim 10$. Moreover, there is the problem that the quark hypothesis explains many features of particle dynamics, but the quarks themselves seem to evade every effort of investigation. This has led many physicists to the conjecture of "quark confinement." It immediately confronts one with questions like: Can particles be bound so strongly they cannot escape? Can a system of ultrastrongly interacting particles shield itself so that it interacts only relatively weakly with its surroundings?

Most of these questions boil down to what happens when a particle is bound so strongly that the binding energy equals its rest mass (in case of a boson) or twice its rest mass (for fermions

which can only be produced in pairs). It would be very helpful if
this problem could be examined in a case where all the interactions
are known, i. e., in atomic physics. Therefore, let us have a look
at the binding energy of K-shell electrons as a function of nuclear
charge Z which is shown in fig. 1. Hartree-Fock calculations
predict that the binding curve reaches the negative Dirac continuum ($E_B = 2m_ec^2$) at Z = 172. Unfortunately, such an element does
not exist in nature and most likely, it will never exist in the laboratory. The alternative would be to take two very heavy nuclei,
e. g., U + U, and make a close collision between them, so that a
system of Z > 172 is created for a short instant of time. One can
calculate the energy states in such a superheavy quasi-molecule
as a function of internuclear distance R. Figure 2 shows that the
situation of critical binding is reached at R $\sim$ 35 fm in a U-U
collision.

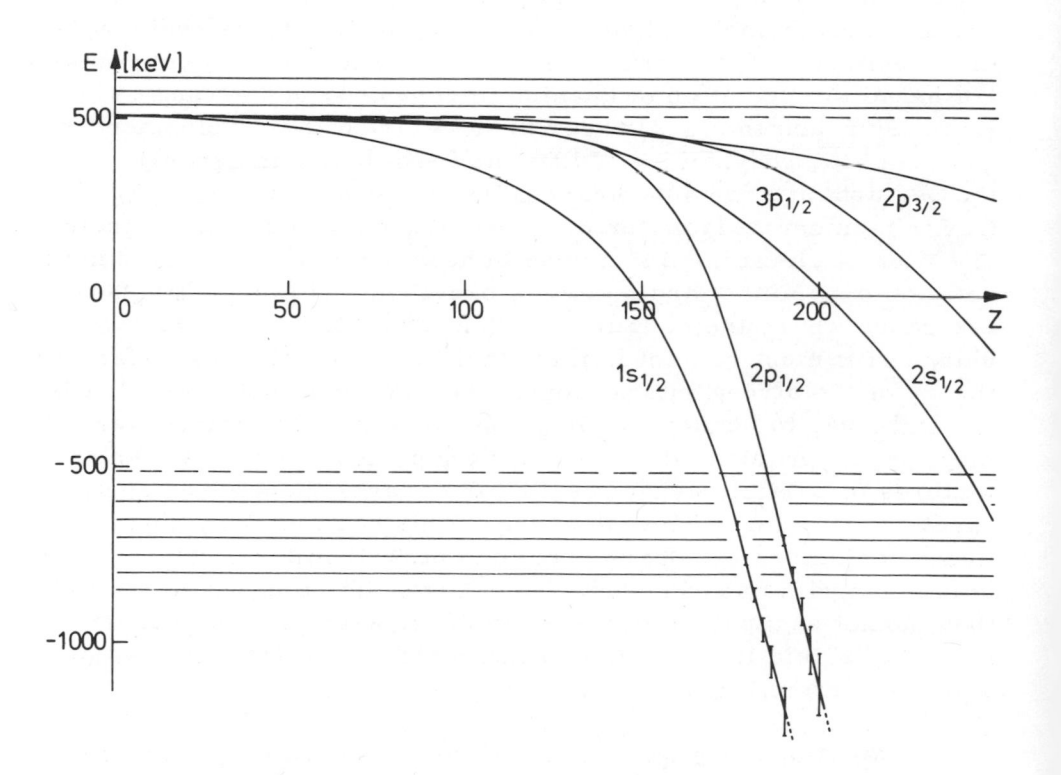

Fig. 1. Atomic binding energies as a function of Z.

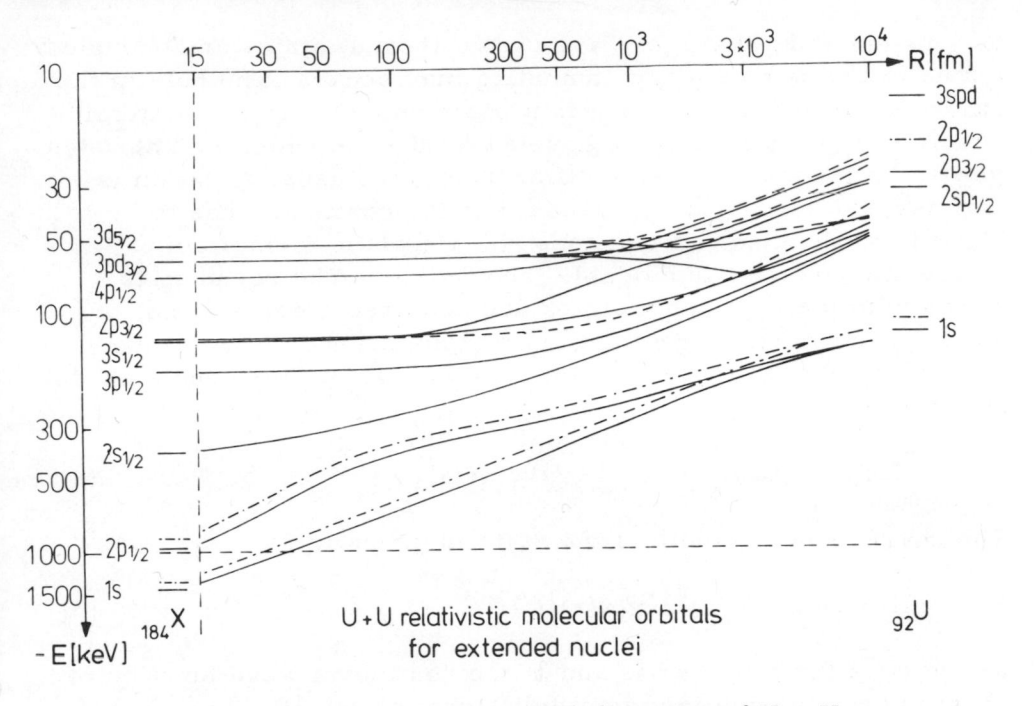

Fig. 2. Molecular orbital diagram of U + U.

Before I come to the experiment, let me first outline our ideas on what will happen in the supercritical region.[1—3] If any bound state is vacant, it can in principle be occupied by an electron, if at the same time a positron is created into a continuum state. The energy difference between these two situations is:

$$\Delta E = 2m_e c^2 - E_B. \tag{1}$$

Obviously, $\Delta E > 0$ if the binding energy is smaller than $2m_e c^2$, i.e., the vacant bound state is energetically more favorable. On the other hand, if $E_B > 2m_e c^2$, the second situation is favored and we speak of a supercritical bound state. The state, in which a positron is spontaneously produced, then is the true ground state of the system. In the language of field theory it is the vacuum state, which now is charged since it contains an electron (the positron eventually will escape).

This process may be viewed intuitively in a variety of ways. In terms of solid state physics there is a gap of $\Delta E = 2m_e c^2$ between unoccupied and occupied states, and when it is deformed

by a potential so that there are vacant (bound) states and occupied
states of the same energy, tunneling may occur. The hole in the
otherwise filled electron sea is a positron. Or, from an atomic
physicist's point of view we have a bound state which is embedded
into a continuum. The only difference to the usual situation is
that now the bound state is vacant and the continuum has to be
viewed as occupied. If one does so, the whole formalism of autoionization becomes appplicable (see fig. 3). The bound state
mixes with the continuum states and acquires a certain energy
spreading. Starting from the critical situation $E(Z_{cr}) = -m_e c^2$
and adding charge to the nucleus, the energy of the bound state is
lowered to

$$E(Z) = -m_e c^2 + (Z - Z_{cr}) \langle \varphi | v(r) | \varphi \rangle. \qquad (2)$$

The corresponding width of the state is given by

$$\Gamma = 2\pi (Z - Z_{cr})^2 | \langle \varphi | v(r) | \psi \rangle |^2, \qquad (3)$$

where φ is the bound state and ψ the continuum wave function of
equal energy, normalized to a delta function $\langle \psi_E | \psi_{E'} \rangle = \delta(E - E')$.

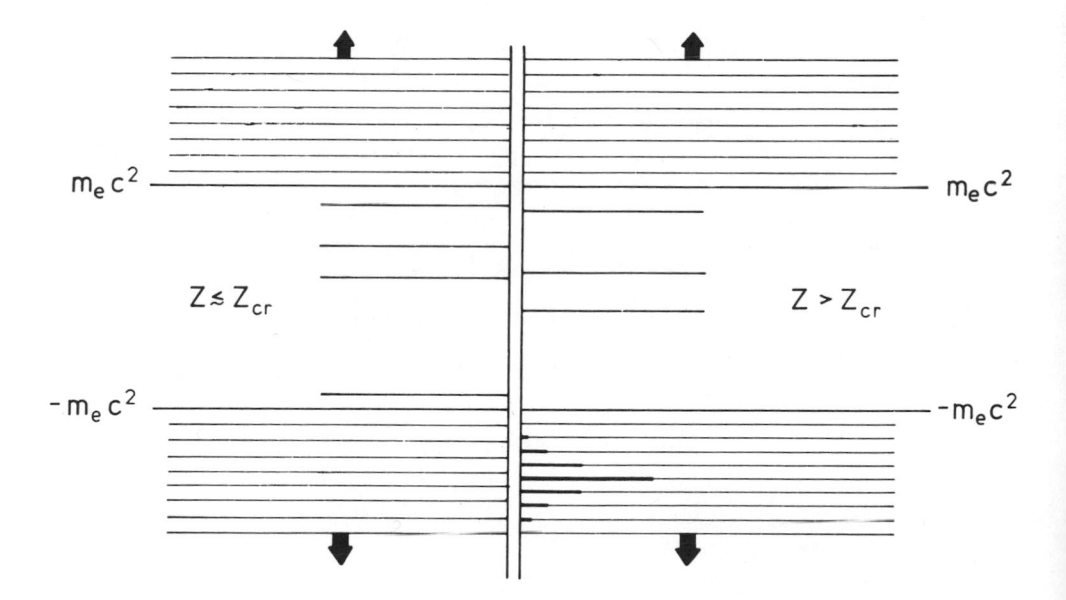

Fig. 3. Schematic view of the change from 1s
bound state to continuum resonance.

The numerical way to obtain accurate results is to make a phase shift analysis of the continuum wave functions for $Z > Z_{cr}$. Figure 4 shows the energy of the 1s and $2p_{1/2}$ states as a function of Z, compared with the linear approximation eq. (2) (dashed lines). Similarly, fig. 5 shows the Breit-Wigner distribution of the 1s state in the hypothetical element $_{184}X$ (U + U) which exhibits a width of about 5 keV. This number should be kept in mind for the discussion of the experiment.

Let me come to a subtle point in the argument. When the bound state has joined the continuum, the vacuum becomes charged twice (because of the two spin states), but how can one see this afterwards? There is always one continuum state for every energy, how can there be one more? To see this, one has to enclose the super-heavy atom in a large box of radius R. Then the continuum solutions become discrete and they are determined by the boundary condition

$$(pR + \delta + \delta_R) = 0 \qquad (4)$$

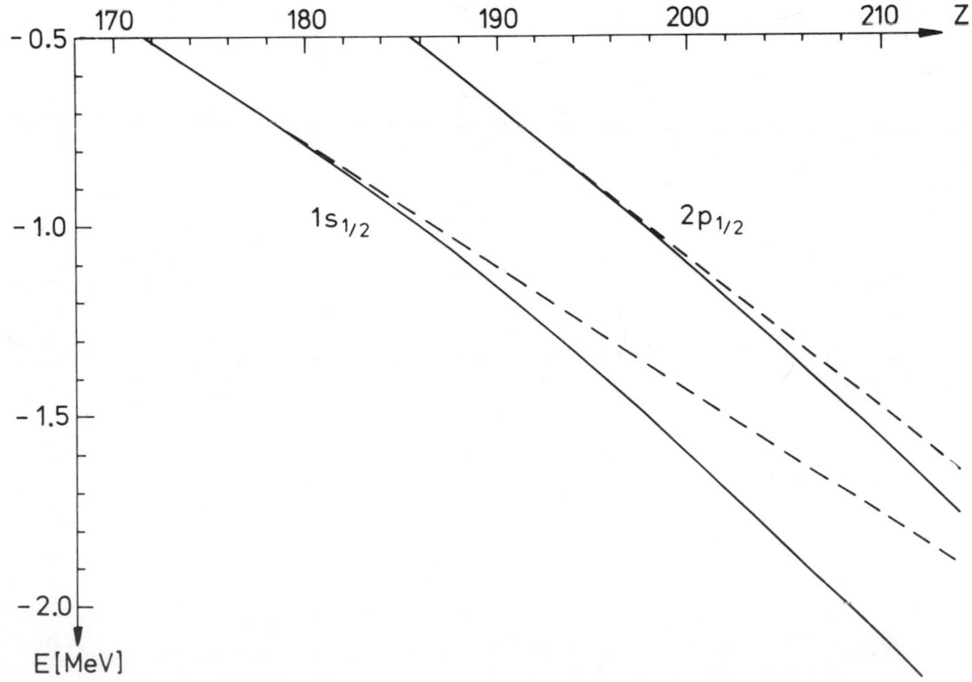

Fig. 4.   1s- and $2p_{1/2}$-resonance energies as a function of Z. Exact calculation (full lines), linear approximation (dashed lines).

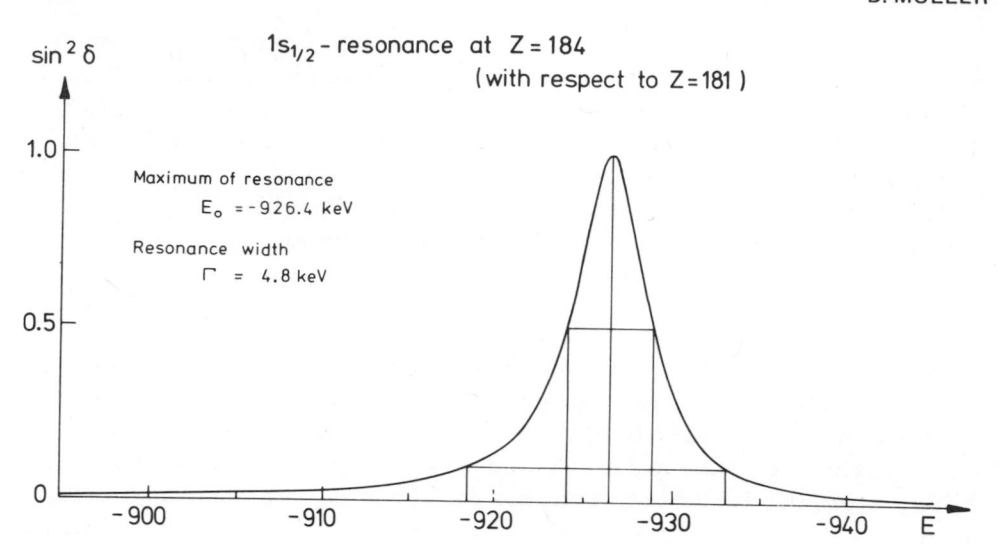

Fig. 5.   Resonance shape of the 1s state in $_{184}$X.

where I have split the phase shift into a smooth part and the resonance from the bound state.   Obviously, we have

$$n\pi = pR + \delta + \delta_R \tag{5}$$

and

$$\frac{dn}{dE} \approx \frac{1}{\pi}[R\frac{dp}{dE} + \frac{d\delta_R}{dE}]. \tag{6}$$

In eq. (6) the derivative of the smooth part of the phase shift has
been neglected.   (This is only a very crude argument, but it can
be shown exactly to give no contribution to the number of states.)
If we integrate over the second part of the density of states, we
find indeed:

$$n_R = \frac{1}{\pi} \int dE \frac{d\delta_R}{dE} = \frac{1}{\pi}\Delta(\delta_R) = 1. \tag{7}$$

Similarly, the charge distribution of the bound resonance can be
extracted from the continuum.   Figure 6 shows that it is (for
Z = 184) the natural continuation of the charge distribution of the
critical ($Z_{cr} \sim 172$) bound 1s state.

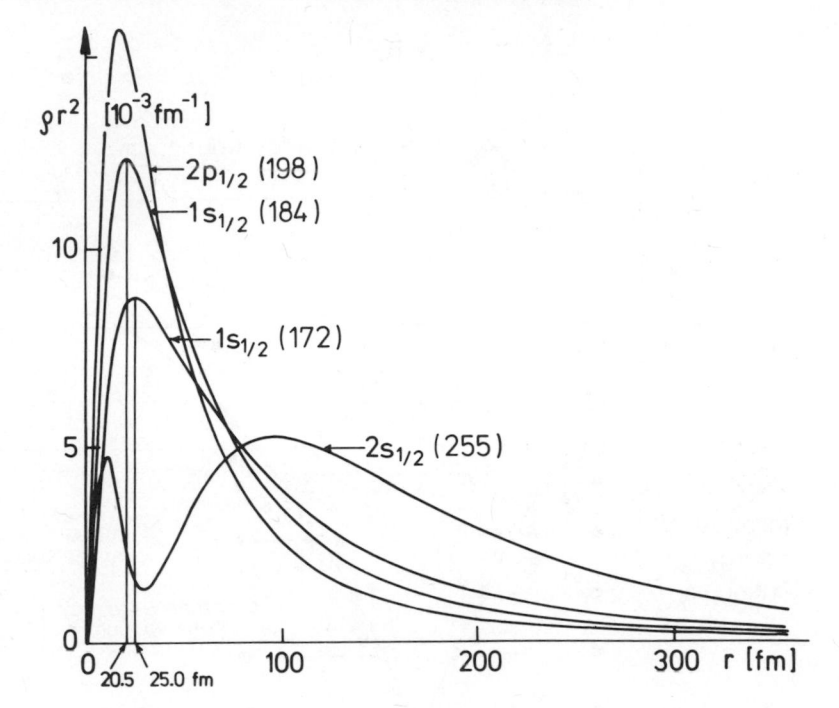

Fig. 6.   Charge distribution of various supercritical states.

        After this theoretical introduction let us have a look at the
possibilities for experiments. Figure 7 shows a schematic view of
the collision between two uranium atoms.   As I already have mentioned, it is necessary to have a vacancy in the molecular 1s state
at the point of closest approach for the positron creation to occur.
These vacancies may be created at large distances where the K-
shell ionization energy is relatively small.   The predictions for
the probability of this ionization vary at present over a wide range,
from 5×10^{-6} to 10^{-1}.   The reason for this ignorance is that the
impact parameter dependence of direct K-shell ionization is not
yet well understood (the above range only corresponds to an uncertainty of a factor 5 in the average ionization impact parameter).
Evidently, it is a crucial number for every experimental effort,
and it should be measured in collisions between medium heavy

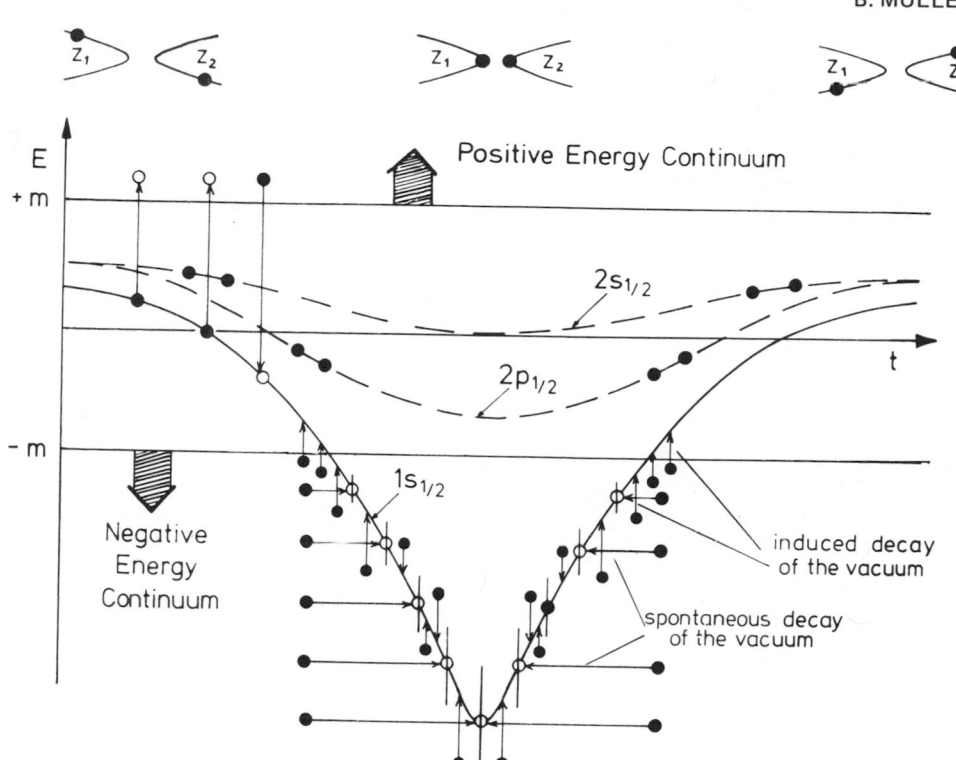

Fig. 7. Schematic view of atomic processes in a U + U collision at ca. 5 MeV/nucl.

atoms. In our calculations we have assumed (!) a probability of 10^{-2}. (It should be mentioned at this point that there is a process which could best be described as virtual vacancy excitation during the collision, which does not finally give rise to real ionization. In adiabatic collisions it can be many times larger than the ionization probability. How much it can contribute for our purposes remains to be seen.)

Then there follows the region—at $R \leqslant 50$ fm—where the bound $1s\sigma$ state becomes a member of the continuum and spontaneous positron production will occur. A rough estimate of the efficiency of this process is obtained from the collision time $\tau_c \sim 4 \times 10^{-21}$ sec in comparison with the vacancy decay time $\tau_{pos} \sim (5 \text{ keV})^{-1} \sim 2 \times 10^{-19}$ sec. Thus one can expect to see one out of fifty vacancies escape as a positron. Figures 8 and 9 show the results of more rigorous calculations, taking into account

the full dynamics of the collision process.[4] The e^+ distribution
will be broadly peaked around 400 keV which should make the posi-
trons easy to detect.   The cross section falls off steeply when the
impact parameter grows beyond 50 fm or the collision energy falls
below 300 MeV (in the CM system).   This is reflected in the energy dependence of the total cross sections.   However, because of
the collision dynamics, no sharp cutoff can be expected.

The background for the experiment can originate from
mainly two sources: (1) the usual e^+e^-pair production in collision

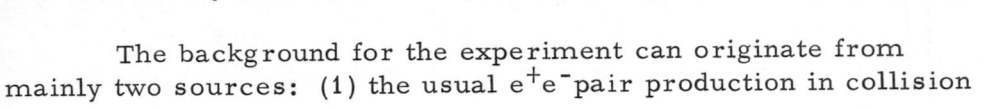

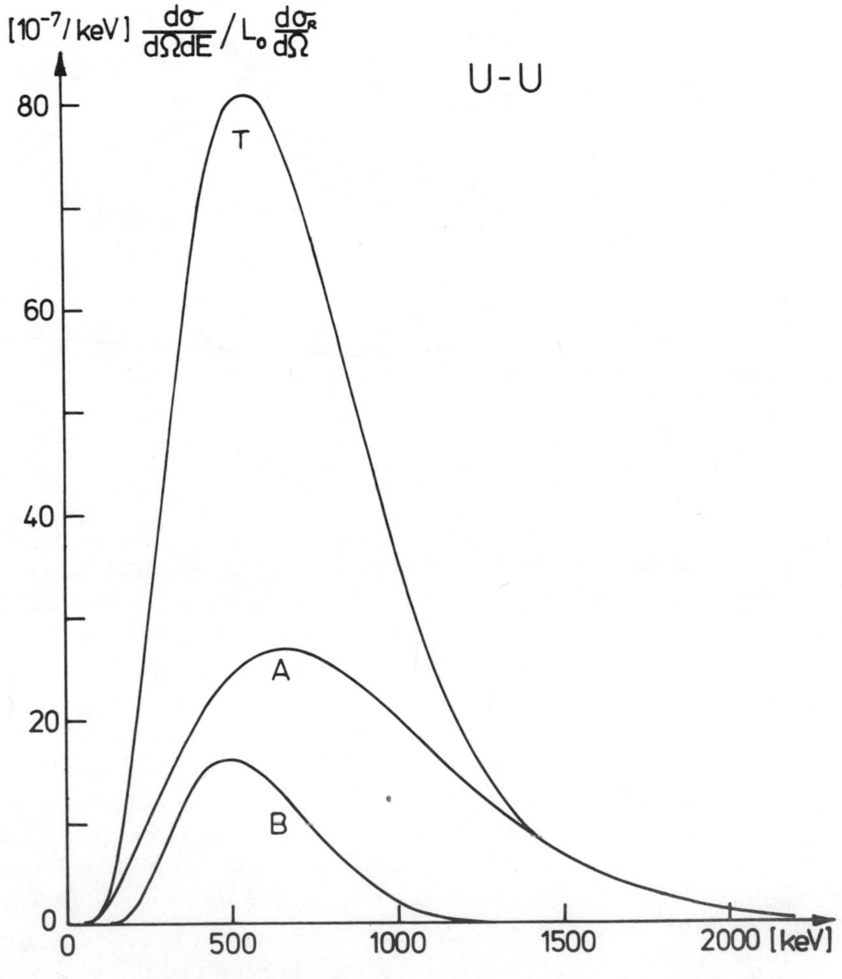

Fig. 8.   Differential positron cross section in 1600 MeV
        U + U backward scattering (T = total).

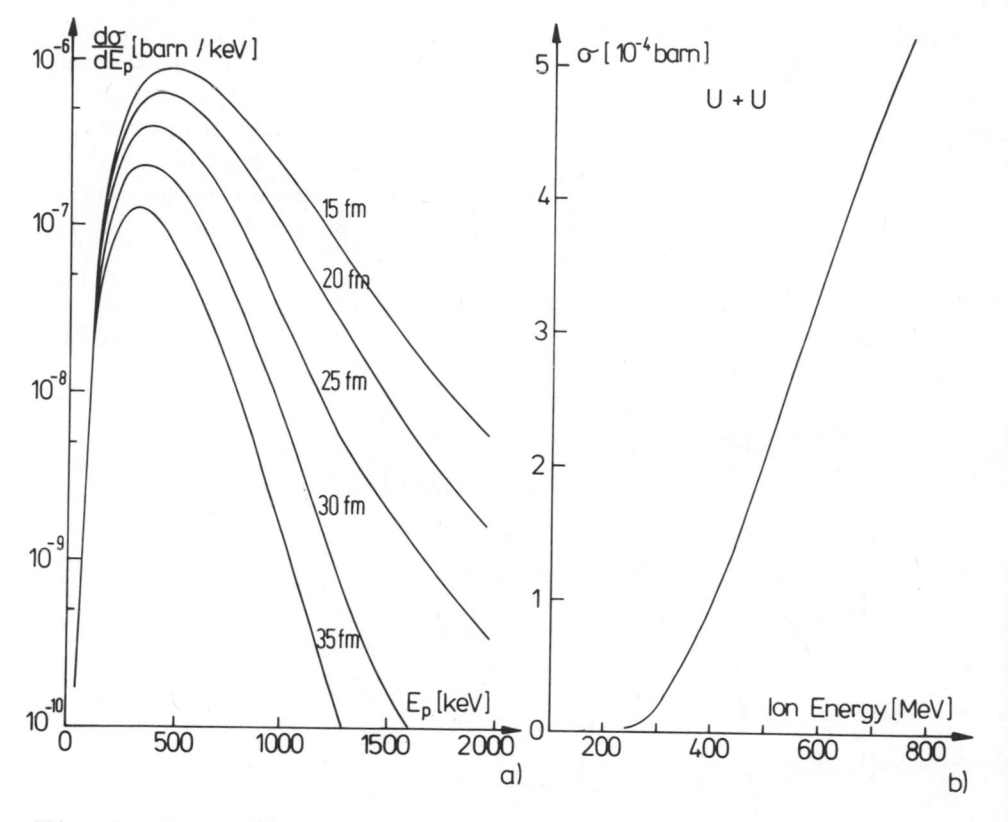

Fig. 9.   Positron cross sections in U + U collisions as a function
of projectile energy (CM system).

between charged particles. It must be of much smaller cross section since the matrix element are rather small and the energy
necessary is at least 1 MeV, whereas the spontaneous process is
almost energyless. Computations show that it should be smaller
by about 3 orders of magnitude. (2) Coulomb excited nuclear
states may decay via pair production if the transition energy is
above 1 MeV. Fortunately, uranium is a rather soft nucleus and
has very low excitation energies for the collective states and only
a very small portion of Coulomb excited states will decay in that
way. G. Soff and V. Oberacker[5] have calculated that this background effect would become serious if the K-shell ionization probability is smaller than 10^{-3}.

If the experiment is successful and supports our ideas
about positron emission in U + U collisions, it will also be a piece

of evidence for a physical situation of supercritical coupling. In
the meantime it may be permitted for a theorist to apply the same
ideas to more unfamiliar situations. If one could increase the
charge of a nucleus more and more, other electronic states
($2p_{1/2}$, $2s$, etc.) will join the continuum and be spontaneously
filled (fig. 10). Thus the vacuum will become highly charged,
until it really begins to neutralize the bare nuclear charge distribution. [6] One may view this purely hypothetically, but it could
be a real occurrence if such speculative things like collapsed
nuclear states do exist.

Figure 11 shows how the apparent screeened charge γ of
such an object would be connected to the number Z of protons it
contains. It is seen that the asymptotic coupling constant γa
reaches a value of ca. 15 even for $Z = 10^6$.   That this is of
the order of magnitude of the strong interactions coupling constant,
may be just an accidental fact. But it provides a model for a situation where an extremely strongly interacting object screens itself,
so that its apparent coupling to other particles becomes of moderate

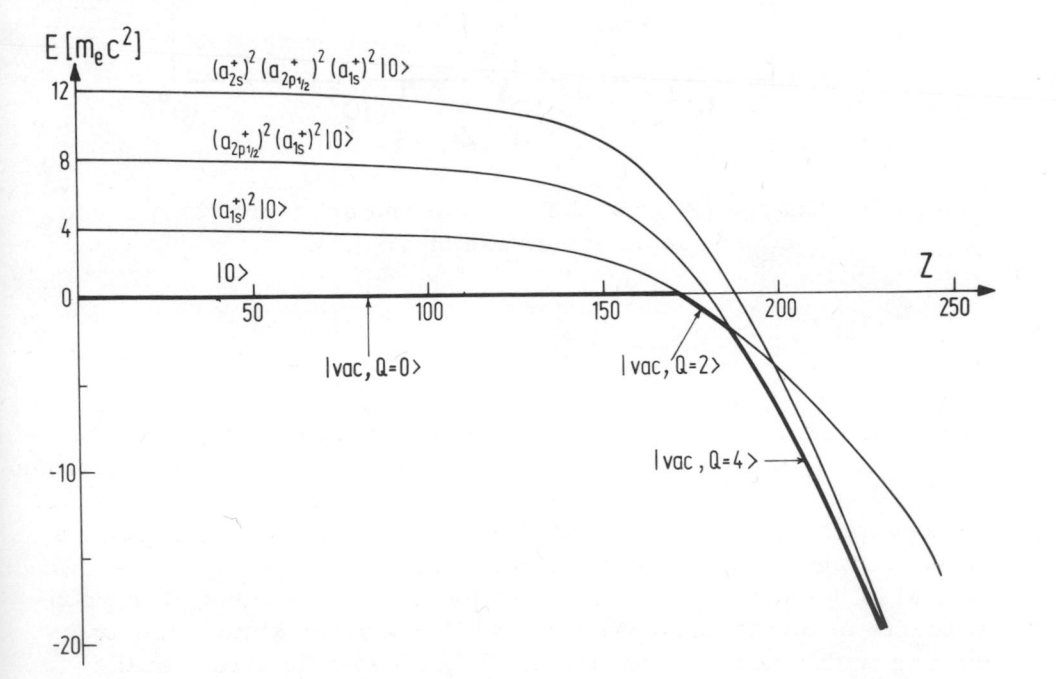

Fig. 10. Higher charged ground states (vacua) are developing
as Z increases.

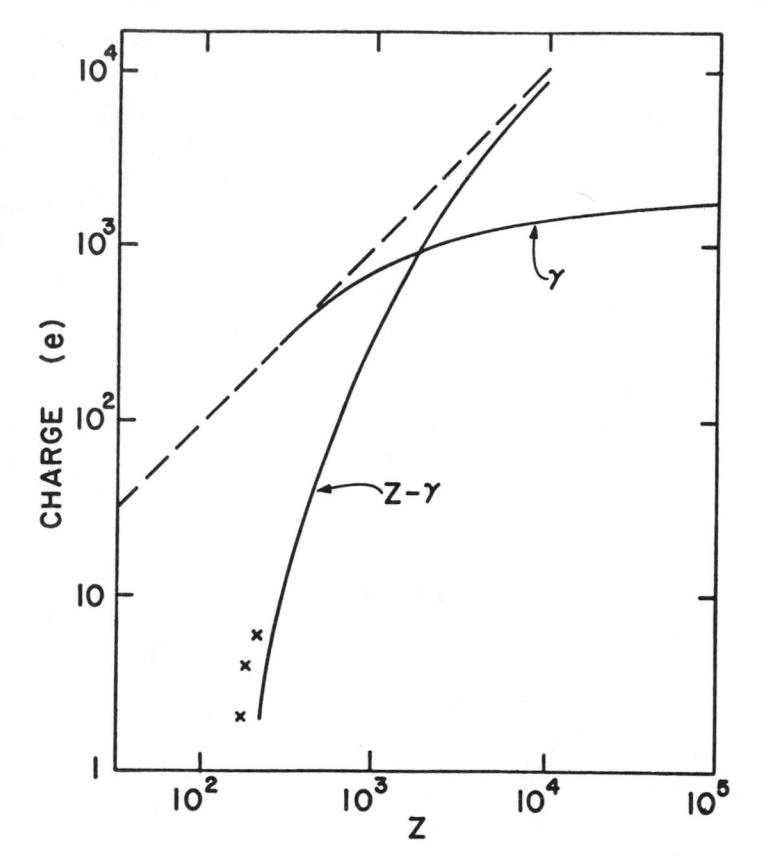

Fig. 11. Charge (Z-γ) of the vacuum and apparent charge γ of
the nucleus as a function of Z.

strength. Moreover, if we look at the potential of this hypernu-
cleus (fig. 12), we encounter a square well, in which the electrons
can move almost freely yet they are strongly bound. If one would
ionize one of the screening electrons its place would immediately
be reoccupied spontaneously under positron emission. Thus, we
may also be on the way to a model for the confinement of the constituents of elementary particles while they are almost freely
moving within their boundaries. This closes the circle to the
questions I have raised at the beginning of my talk. And, maybe,
atomic physics can provide some of the necessary evidence in
this way.

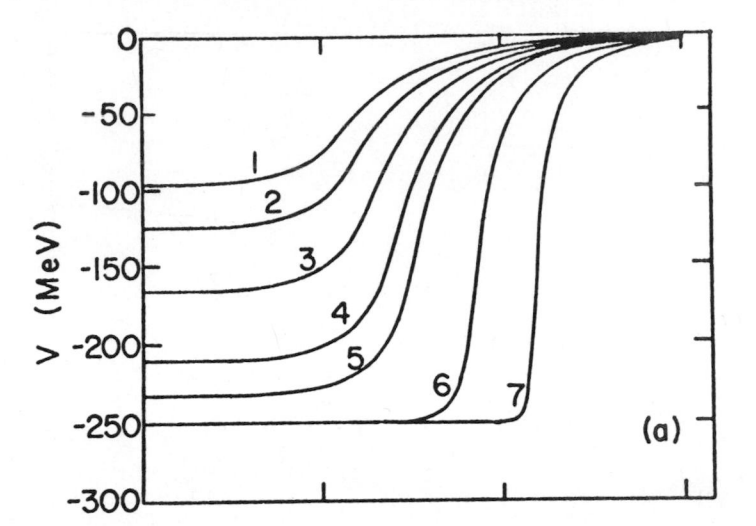

Fig. 12.   The potential inside an abnormal nucleus (numbers
correspond to increasing Z).

REFERENCES

*Work performed under the auspices of the U.S. Energy Research
and Development Administration.
1. B. Müller, H. Peitz, J. Rafelski, W. Greiner, Phys. Rev.
Lett. 28, 1235 (1972); J. Rafelski, B. Müller, W. Greiner,
Nucl. Phys. 88, 585 (1974).
2. L. Fulcher, A. Klein, Phys. Rev. D8, 2455 (1974).
3. Ya. B. Zeldovich, V. S. Popov, Usp. Fiz. Nauk 105, 403
(1971).
4. K. Smith, H. Peitz, B. Müller, W. Greiner, Phys. Rev.
Lett. 32, 554 (1974).
5. V. Oberacker, G. Soff, W. Greiner, Internal pair creation
induced by nuclear Coulomb excitation in heavy-ion collisions,
preprint, Frankfurt 1975.
6. B. Müller, J. Rafelski, Phys. Rev. Lett. 34, 349 (1975).

DIFFERENCES IN THE PRODUCTION OF NONCHARACTERISTIC RADIATION IN SOLID AND GAS TARGETS[*]

R. S. Peterson, R. S. Thoe, H. Hayden, S. B. Elston,
J. P. Forester, K.-H. Liao, P. M. Griffin, D. J. Pegg,
and I. A. Sellin

University of Tennessee
Knoxville, Tennessee 37919
and Oak Ridge National Laboratory
Oak Ridge, Tennessee 37830

and

R. Laubert

New York University
New York, New York

Recent experimental results of Bell et al. [1] for 55 MeV
S → Al and 48 MeV S → Ne collisions indicated that the production
of noncharacteristic radiation (NCR) was similar for gas or solid
targets when normalized to the characteristic line of the projectile ion. Such a result would indicate that a one-collision mechanism for NCR production, in which a vacancy in the K shell is
produced and filled during the collision, is as important as the
two-collision model proposed by Saris et al. [2]. Experiments have
been completed using more nearly symmetric collision partners, and
it is found that the yield of x rays near the combined-atom K x-ray
limit (E_u) for 40 MeV Si^{6+} → SiH_4 is significantly smaller than the
yield of NCR for 40 MeV Si^{6+} → Al.

Beams of silicon ions from the Oak Ridge National Laboratory
6.5 MV tandem Van de Graaff were passed through a gas target cell,
shown in Figure 1, and the x rays produced in the resulting collisions

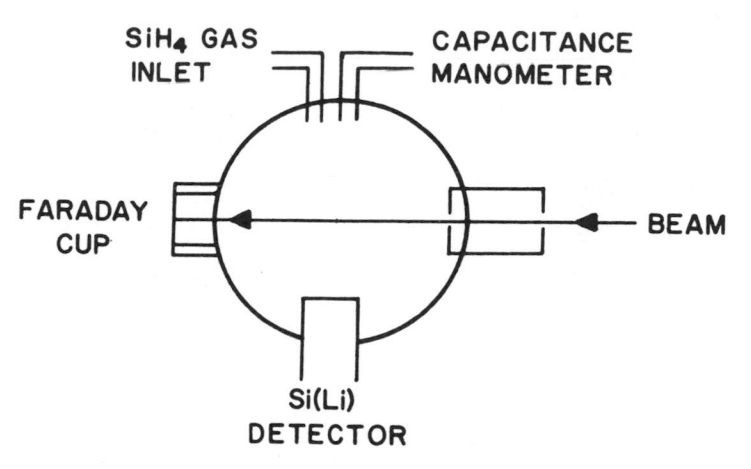

Figure 1.   Gas Target Schematic.   Ion beams of various energies
entered the gas cell through collimators and were collected in the
Faraday cup.   A capacitance manometer was used to measure the gas
pressure and to control the gas flow, maintaining the target gas
pressure at 300 mTorr.   A Si(Li) detector was used to observe the
x-ray spectrum at 90 degrees to the ion beam.

were viewed at 90° by a Si(Li) detector with FWHM of 160 eV at 5
keV. A two mil Mylar window was used with the Si(Li) detector to
preferentially absorb the Si and Al characteristic K x rays with
respect to the higher energy x rays.   The beam current, which was
collected in a Faraday cup, was kept low enough that pulse pile-up
was avoided.

        The gas pressure was monitored by a capacitance manometer which
also controlled the gas flow valve, keeping the target pressure constant at 300 mTorr.   When thin foils were used, no gas was admitted
to the chamber and a foil was inserted in the beam path.

        X-ray spectra were recorded for 10 - 40 MeV Si^{q+} on Al (50
$\mu g/cm^2$) foils and 40 MeV Si^{6+} on SiH_4 (300 mTorr).   The spectra for
40 MeV Si^{6+} on SiH_4 and Al are shown in Figure 2, where the spectra
were normalized to the gas target K x-ray characteristic lines.
Even though no corrections for absorption were made in this figure,
it is obvious that the x-ray yield near the combined-atom limit (E_u)
is greater for the solid target (Al) than for the gas target (SiH_4).

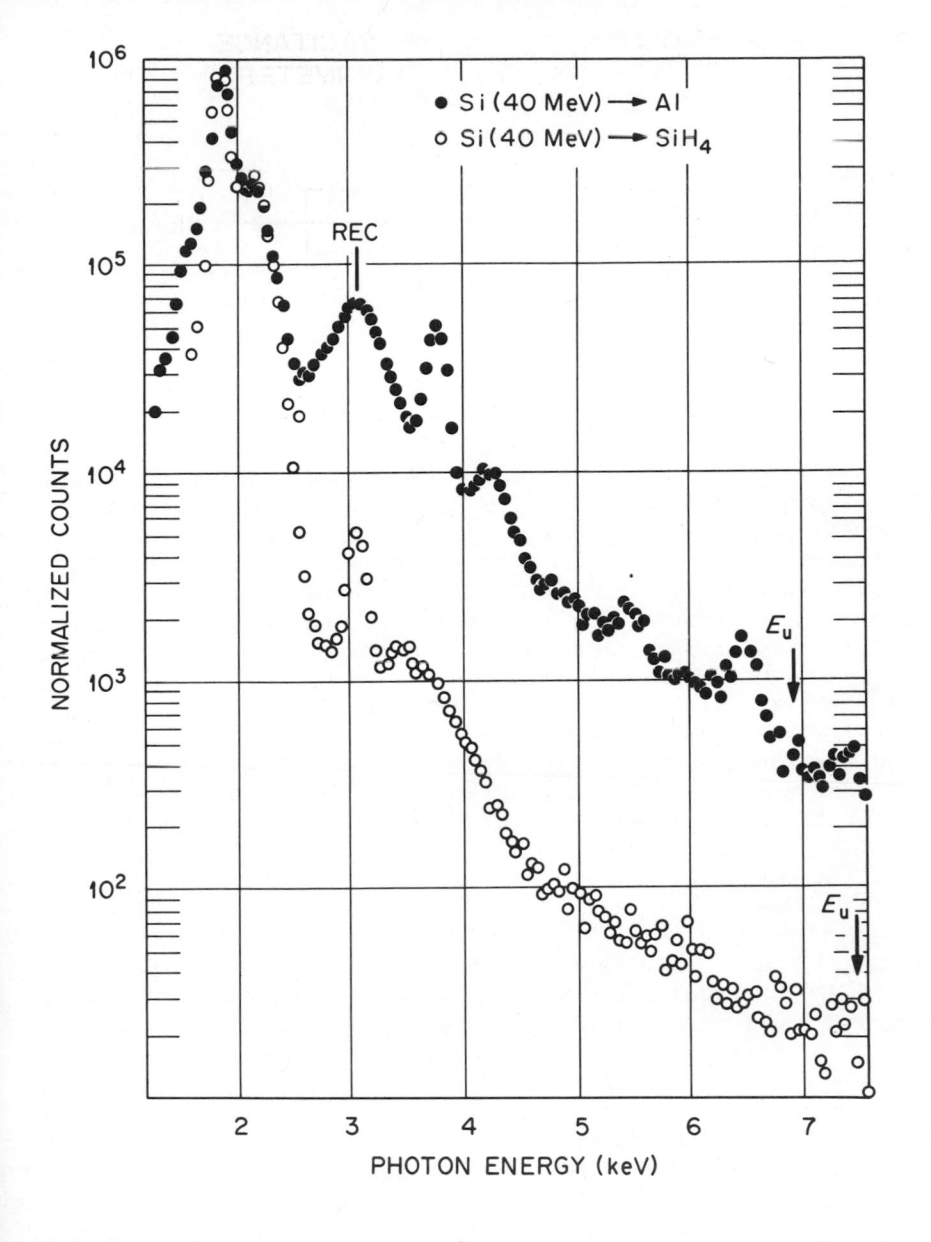

Figure 2. X-ray spectra for Si^{6+} (40 MeV) on Al and SiH_4. The
x-ray spectra have been normalized to the Si K x-ray lines. Impurity peaks at 3.7 keV, 5.4 keV, and 6.4 keV in the solid Al
target are probably due to K x-rays of calcium, chromium, and iron.
The peak at 3.1 keV is due to REC in the solid target and argon K
x rays in the gas target.

Although several impurities were present in the Al foil, their contribution to the NCR yield was found to be negligible. The large
peaks in the Si-SiH$_4$ data are due to a residual argon impurity in
the gas.

Corrections for absorption of x rays in the Be window and silicon dead layer of the Si(Li) detector, for absorption in the Mylar
attenuator, and for absorption in the target were made from available x-ray absorption data [3]. The corrected spectra for Si (40
MeV) on SiH$_4$ and Al are shown in Figure 3. Again the spectra have
been normalized so that the areas of the Si K x-ray characteristic
lines are equal. The yield of x-rays near the combined atom limit
(E$_u$) is greater for the solid target than the gas target.

In order to study the yield of NCR in solid targets for different projectile velocities, silicon ion beams at 10, 20, and 30
MeV were used on thin Al (50 µg/cm^2) foils. The x-ray spectra for
the photon energy region 4.5 keV - 8 keV, normalized to the Si
characteristic lines, are shown in Figure 4. The x-ray yields for
20 and 30 MeV Si ions on Al fall between the 10 and 40 MeV Si^{q+} on
Al data. In all cases the yield of x-rays near the combined atom
limit for solid targets is greater than that for the gas target.

The possibility that these observed differences in gas vs.
solid target x-ray yields may be due to bremsstrahlung can be dis-
counted for the low beam energy data. An approximate energy cutoff
for secondary electron bremsstrahlung would be the maximum kinetic
energy that can be transferred to the target's bound electrons [4]:

$$E_{max} \approx \frac{4mE}{MA} + 2mv_1v_2$$

where MA, E, and v$_1$ are the projectile mass, kinetic energy, and
velocity, respectively, and m, v$_2$ are the electron mass and velocity
of target atom K-shell electron. These energies are listed in
Table I for the collision systems used in this experiment. Cutoff
energies for the 10 and 20 MeV Si on Al collisions are much lower
than the combined-atom energy (E$_u$) of 6.9 keV, thus rendering negligible bremsstrahlung contributions to the NCR yields. Yet, the
NCR yields are an order of magnitude greater than the gas target
yields. Although the bremsstrahlung cutoff energies for 30 and 40
MeV Si on Al collisions are higher, there does not appear to be
any significant increase in the measured NCR yield at these higher
beam energies. The small x-ray yield near the combined-atom limit
(E$_u$) for the gas target collisions and the high bremsstrahlung
energy cutoff does not preclude the possibility that these x rays
were due to secondary electron bremsstrahlung.

The fact that the yield of x rays near the combined-atom limit
(E$_u$) for Si on SiH$_4$ was small can be emphasized by comparing the

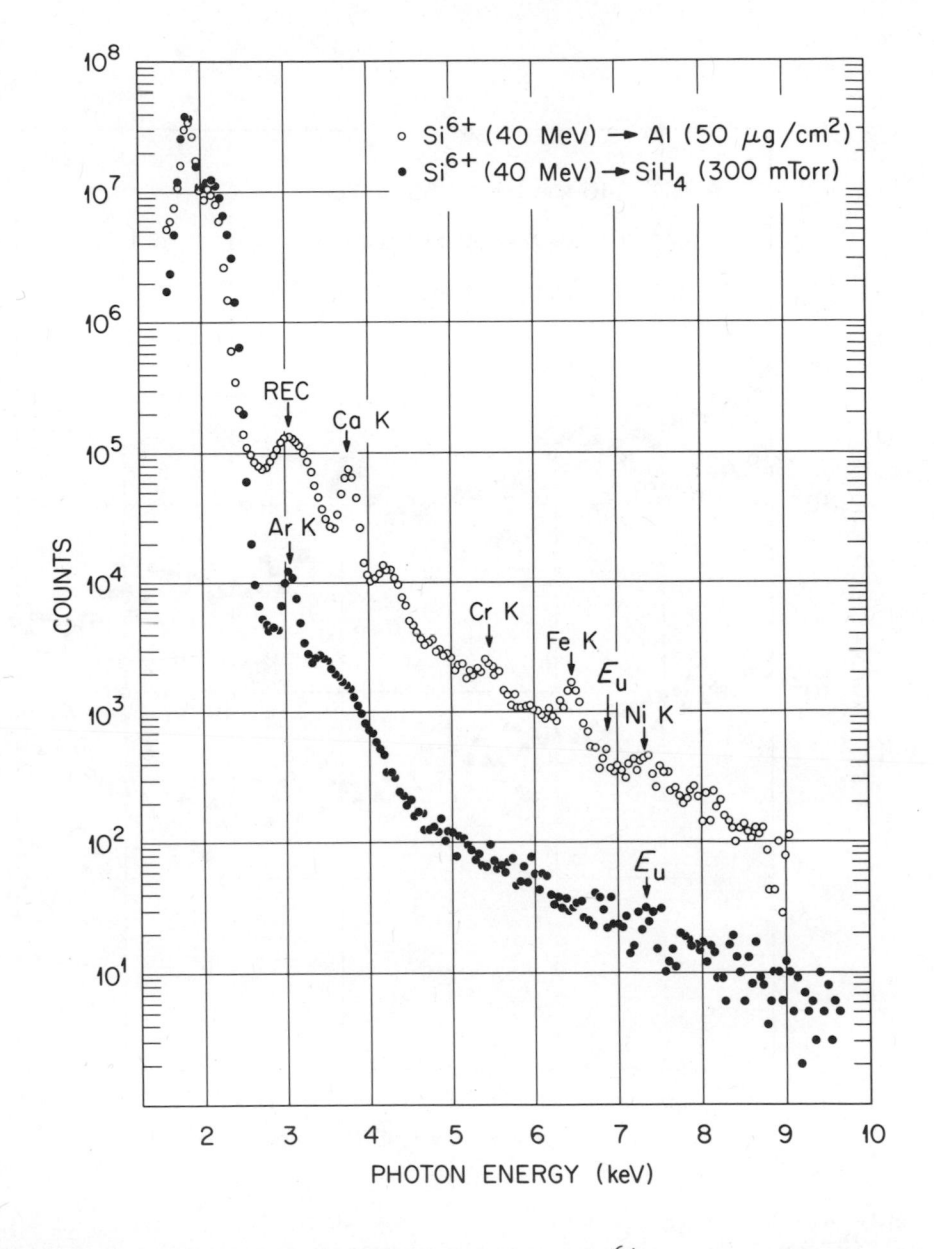

Figure 3.   The x-ray spectra for 40 MeV Si^{6+} on Al and SiH$_4$ have
been corrected for window absorption and normalized to the Si K
x-ray peak.   The x-ray yields near the combined-atom limits (E$_u$)
are larger in the solid target than in the gas target.

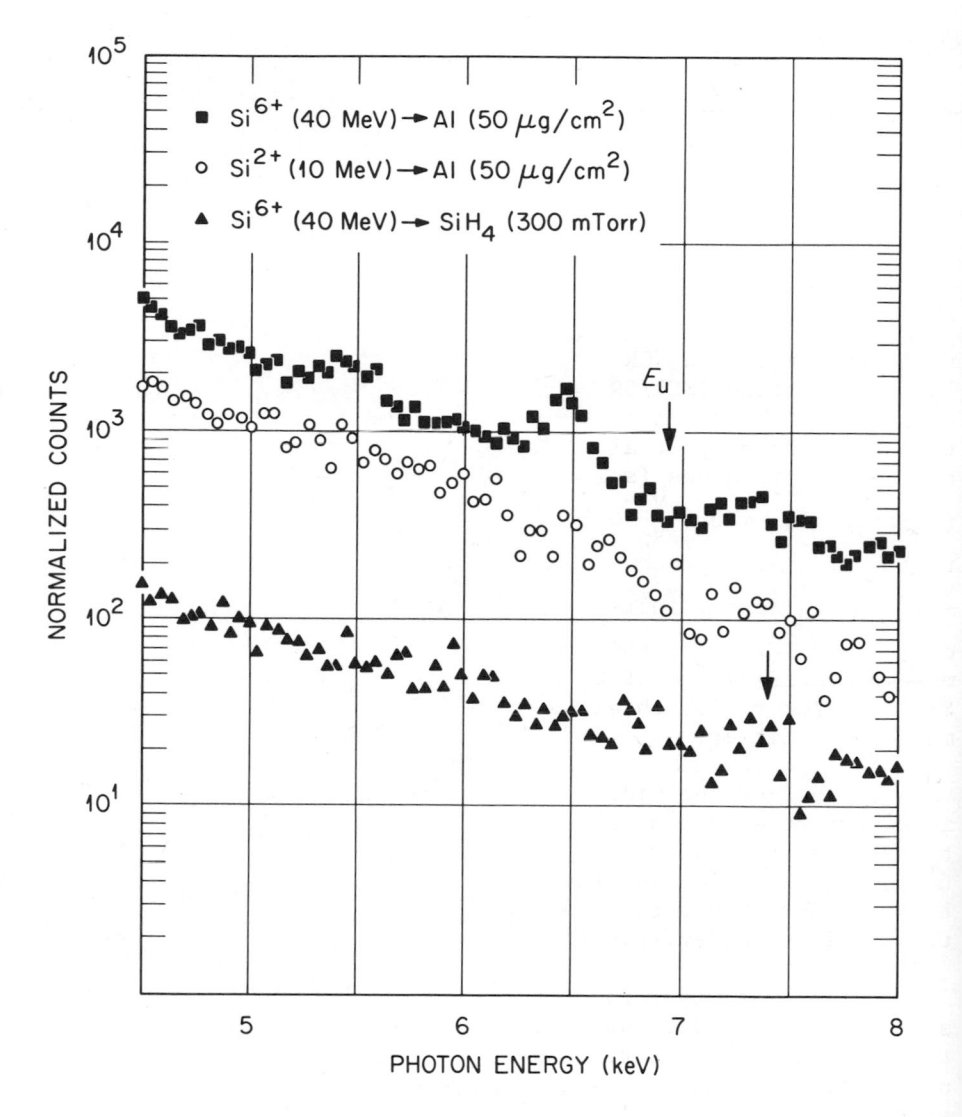

Figure 4. NCR x-ray spectra for 10 and 40 MeV Si^{x+} on Al and 40 MeV Si^{6+} on SiH$_4$. The NCR spectra obtained for 20, 30 MeV Si^{x+} on Al (not shown) lie between the 10 and 40 MeV solid target data. All the spectra above were normalized to the Si K x-ray lines.

TABLE I

Collision	Beam Energy (MeV)	E_{max} Maximum Kinetic Energy Transferred to Bound Electron (keV)
Si–Al	10	2.9
	20	4.6
	30	6.1
	40	7.5
Si–Si	40	7.8

ratio of integrated NCR intensities to the characteristic K x-ray line intensities (NCR/CR) for different symmetric collisions where the ratio of beam velocity to K-shell electron velocity (V/V_k) is the same. Laubert et al. [5] measured the ratio of NCR/CR for C–C and Al–Al collisions (solid targets) to be $\sim 2 \times 10^{-4}$ and $\sim 7 \times 10^{-4}$, respectively, where $\frac{V}{V_k} \sim 0.5$. In the present experiment the NCR/CR ratio for Si–Si (gaseous target) is found to be $\sim 2 \times 10^{-5}$ for $\frac{V}{V_k} \sim 0.5$. These results show that the NCR/CR ratios for symmetric collisions involving thin solid targets are of the same order of magnitude for different Z, whereas the present data shows that the Si–Si NCR/CR ratio, obtained with a gas target, is an order of magnitude smaller.

Because of the wide range of velocities of the projectile Si ions used on Al foils, there were correspondingly large changes in the charge state distributions of the projectile. Any changes in fluorescence yields due to these differing charge state distributions do not appear to have influenced the NCR yield with respect to the Si characteristic lines.

It appears, therefore, that the large yield of NCR in the solid targets is not due to secondary-electron bremsstrahlung or to any projectile fluorescence yield effect. The yield of NCR is not similar for gas and solid targets for the present collision systems, where the solid target collisions give NCR yields at least an order of magnitude greater than the gas target collisions.

Atomic densities in gas targets (for pressures less than 1 Torr) are several orders of magnitude smaller than atomic densities in thin, solid foils. Consequently, the inverse lifetime of a K-shell vacancy produced in a projectile traversing a gas target is much less than the frequency of collisions, whereas this inverse

lifetime for low Z projectiles is comparable to the collision frequency in solid targets.

If a single-collision model is important for NCR production, the yield of x rays near the combined-atom limit (E_u) for ion-atom collisions in gases should be comparable to the NCR yields in solid target collisions. This not being true for the collision systems studied implies that the proposed two-collision mechanism for NCR production in solid targets dominates over any single-collision processes.

REFERENCES

*Research supported in part by NSF, ONR, NASA, and by Union Carbide Corporation under contract with ERDA.

[1]. F. Bell, H.-D. Betz, H. Panke, E. Spindler, W. Stehling and M. Kleber, Phys. Rev. Lett. 35, 841 (1975).
[2]. F. W. Saris, W. F. van der Weg, H. Tawara, and R. Laubert, Phys. Rev. Lett. 28, 717 (1972).
[3]. B. L. Henke, Norelco Reporter, Vol. XIV, N 3-4 (1967), p. 127.
[4]. F. Folkmann, C. Gaarde, T. Huus, and K. Kemp, Nucl. Instr. Meth. 116, 487 (1974).
[5]. R. Laubert, H. H. Haselton, J. R. Mowat, R. S. Peterson, and I. A. Sellin, Phys. Rev. A 11, 1468 (1975).

ENERGY LOSS AND YIELD OF SWIFT MOLECULAR CLUSTERS IN

SOLID TARGETS*

R. Laubert[+]

New York University

New York, NY   10003

Projectiles moving as clusters through solid targets lose energy more rapidly than isolated particles of the same velocity. This result can be explained if one considers the proximity of the atoms of the cluster and the concomitant changes in the electronic stopping power. The possibility of forming "wake-riding states" is investigated by measuring the yield of molecules after their penetration through carbon foils.

When a charged particle, of velocity $v > v_o = e^2/\not{h} = 2.8 \times 10^8$ cm/sec, moves through a dense medium, a coherent electron displacement occurs in its wake.[1] The resulting potential distribution behind the ion is shown in Fig. 1.[2] The periodicity in space is determined by v/ω_o, where ω_o is the plasmon frequency of the medium, and the amplitude is proportional to $Z\omega_o/v$, where Z is the effective charge of the ion in the medium. When a spatially correlated cluster of atoms, i.e. a molecule, is injected into the material the resulting wakes will interfere and hence cause changes in the stopping power.[3] There also exists the possibility[2] that the trailing ion of the cluster can be trapped in the potential wake of the leading ion resulting in a "wake-riding state". These two possibilities are investigated in this progress report.[4]

STOPPING POWER OF CLUSTERS

The electronic stopping power due to the valence electrons, S^v, can be separated into single particle and resonant collisions, suggesting[5] that one can write the valence stopping number

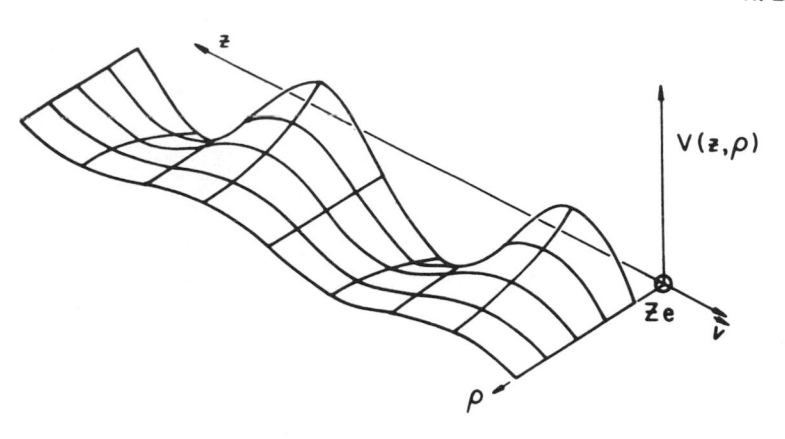

Figure 1   The potential distribution behind a moving positively
              charged particle Ze [2].

L^V ($S^V = e^2 \omega_o^2 z^2 L^V / v^2$; $L^V = \ln(2mv^2/\hbar\omega_o)$ where e and m are the electronic charge and mass respectively) in the form

$$L^V = L_s^V + L_r^V \tag{1}$$

where

$$L_s^V = C + L_r^V \tag{2}$$

and C is a constant depending on the target material and is typically the order of unity.[5] Hence

$$L_s^V = \frac{1}{2} (1 + \frac{C}{L^V}) L^V$$
$$\tag{3}$$
$$L_r^V = \frac{1}{2} (1 - \frac{C}{L^V}) L^V$$

In the high velocity limit, when $C/L^V \ll 1$, Eq. (3) is usually referred to as the equipartition rule.[1] The boundary between single
and resonant collisions is $r_s = \hbar/2mv$, while resonant collisions extend to $r_r = v/\omega_o$.

    When spatially corellated clusters, in the form of molecules
composed of atoms $Z_1 M_1$, $Z_2 M_2$, etc., having a velocity $v > v_o$ are incident on a dense medium the bonding valence electrons are stripped
in the first few atomic layers of the target.[6] The ions in the
cluster repell each other, due to the coulombic force, and their
interatomic distance will increase from $R(o) = R_o \approx 1\text{Å}$ (the internuclear

distance of typical molecules) to $R(t) > R_o$ at time t given by[3]

$$t = t_o \{ \xi^{1/2} (\xi-1)^{1/2} + \ell n [\xi^{1/2} + (\xi-1)^{1/2}] \} \qquad (4)$$

where $t_o = (\mu R_o^3 / 2 Z_1 Z_2 e^2)^{1/2}$ and $\xi = R/R_o$ with $\mu = (M_1^{-1} + M_2^{-1})^{-1}$ being the reduced mass.

From the above discussion, we note that if the internuclear distance $R(t)$ is less than $r_s = \hbar/2mv$ the ions of the cluster will have an effective atomic number for single and resonant collisions of $Z^2 = (\Sigma Z_i)^2$ and hence will act as a united atom. This condition is not attainable for real molecules where $R_o \simeq 1 \overset{o}{A}$, but might be realized for muonic molecules. For the condition $r_s < R(t) < R_r = v/\omega_o$, the ions of the molecule act as independent particles for single collisions, $Z^2 = \Sigma Z_i^2$, and as a united atom for the resonant collisions with an effective atomic number $Z^2 = (\Sigma Z_i)^2$. It is this domain of internuclear separations that is amenable to experiments. For $R(t) > r_r = v/\omega_o$, the ions act as independent particles for all collisions and the total stopping power is the sum of the individual stopping powers.

The stopping power of clusters of interatomic separation $r_s < R(t) < r_r$ can be written as[4]

$$Z_c^2 L^v = \Sigma Z_i^2 L_s^v + (\Sigma Z_i)^2 L_r^v$$

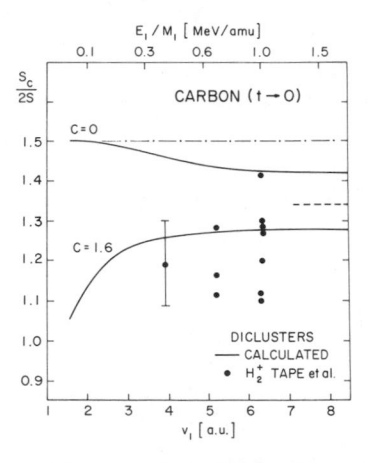

Figure 2   The stopping power ratio for molecular and atomic hydrogen as a function of the projectile velocity. The upper dashed line is calculated for $C=0$ and $S^v/S=1$. The upper solid curve shows the variation of S^v/S with v in Eq. (8). The lower solid curve shows Eq. (8), with Eq. (7), for $C=1.6$, a value appropriate for carbon. The experimental results are from [7].

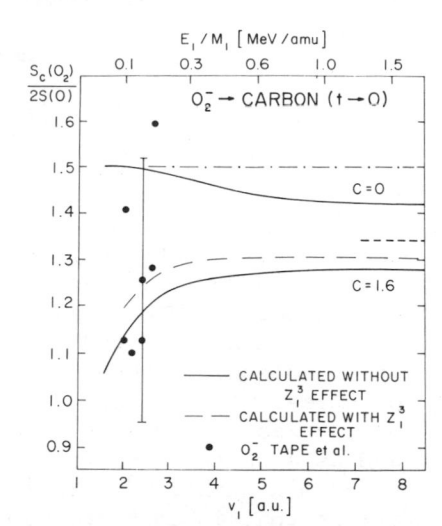

Figure 3   The stopping power ratio for oxygen ions in carbon
              as a function of the projectile velocity. The lower
              dashed curve is calculated according to Eq. (8) with
              Eq. (12) and incorporates z_1^3 term. Other curves are
              the same as in Figure 2. The experimental results are
              from [7].

or using Eq. (3)

$$z_c^2 L^v = \frac{1}{2} [\Sigma z_i^2 (1 + \frac{C}{L^v}) + (\Sigma z_i)^2 (1 - \frac{C}{L^v})]L^v \qquad . \tag{5}$$

It is convenient to define a vicinage function[3] $\zeta^2 = z_c^2/\Sigma z_i^2$, in
terms of which Eq. (5) becomes for homonuclear clusters $(z_i = z)$

$$\zeta^2 = \frac{z_c^2}{nz^2} = \frac{n+1}{2} [1 - \frac{n-1}{n+1} \frac{C}{L^v}]. \tag{6}$$

Specifically for diclusters n=2

$$\zeta^2 = \frac{3}{2} (1 - \frac{1}{3} \frac{C}{L^v}) \qquad . \tag{7}$$

The ratio of the cluster stopping power, S_c, to the sum of the
stopping powers of the component parts, S_i, becomes[3]

$$\frac{S_c}{\Sigma S_i} = 1 + \frac{S_p^v}{S_p} (\zeta^2 - 1) \tag{8}$$

where S_p is the stopping power for protons. Figure (2) compares
Eq. (8), with Eq. (7), to the experimental results of Tape et.al.[7]

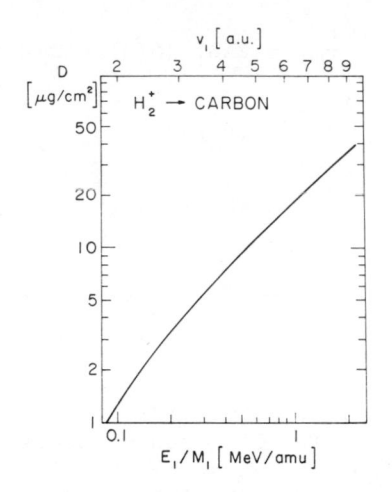

Figure 4   The maximum carbon foil thickness at given cluster velocity v, to keep the internuclear separation, R(t), less than the extent of resonance collisions.

for H_2^+ ions incident upon carbon films. The dashed line at 3/2 corresponds to C=0, or $C/L^v \ll 1$, and would be the result if equipartition applies and the target is considered as an electron gas. The solid line merging to 3/2 for $v \to 1$ a.u. results from the term S_p^v/S_p in Eq. (8) and is due to the decreased contribution of core electrons to the electronic stopping power. It merges with the dashed line on the right hand side of the figure for large velocities. For low velocities $S_p^v/S_p \approx 1$ and the stopping power ratio measures the vicinage function or the effective atomic number of the molecule. The data follows the trend of the curve marked C=1.6, which is the constant appropriate for carbon.[5]

In clusters of ions with large atomic numbers, the effective charge can exceed unity and the z_1^3 term in the stopping power[8] must be considered in the discussion of the vicinage effect. This can be accomplished by writing

$$S^v = z_1^2(1+\alpha z_1)S_p^v \tag{9}$$

where

$$\alpha(a) = \frac{e^2 \omega_o}{m v^3 L^v} I\left[\frac{\omega_o a}{v}\right] \tag{10}$$

in terms of a tabulated function I[8] where a is the minimum impact parameter. We add in Eq. (5) the z_1^3 contribution for impact parameters, b, in the range $a<b \leq R_o$ to single collisions, L_s, and the z_1^3 contribution in the range $b \geq R_o$ to resonant collisiosn, L_r, so that

$$z_c^2 L^V = \frac{L^V}{2} \{ \Sigma z_i^2 (1+\frac{C}{L^V}) + 2\Sigma z_i^3 (\alpha(a) - \alpha(R_o))$$

$$+ (\Sigma z_i)^2 (1-\frac{C}{L^V}) + 2(\Sigma z_i)^3 \alpha(R_o) \} \tag{11}$$

For homonuclear diclusters $z_i = z$ and $n=2$, the vicinage function becomes

$$\zeta^2 = \frac{z_c^2}{2z^2(1+\alpha(a)z)} = \frac{3}{2} \; \frac{1 - \frac{1}{3}\frac{C}{L^V} + \frac{2}{3}\alpha(a)z(1+3\alpha(R_o)/\alpha(a))}{1 + \alpha(a)z} \tag{12}$$

Fig. (3) exhibits Eq. (8), with Eq. (12), for O_2^- and O^- ions
incident on carbon films. The difference between the broken and
solid curves marked C=1.6 indicates the contribution of z_1^3 term
for oxygen (for protons it is negligible, <1%). The data of Tape
et.al.[7] has too large a scatter to indicate more than $\zeta^2 < 3/2$ as
would be the case for C=0, or equipartition.

In this sense the stopping power measurements of molecules
and atoms has provided the first quantitative experimental confirmation of the partition rule of stopping powers[3] and offers the hope
of measuring the partition constant of real materials. The restrictions on the experiment are rather stringent. For example,
Fig. (4) displays the maximum carbon target thicknesses for internuclear distances $R(t) \leq R_r$ for H_2^+ ions, where Eq. (8) and (12)
apply.

WAKE RIDING STATES

The possibility of capturing the trailing ion of the molecule
in the potential wake of the leading ion, and hence form wake riding
states[2], was investigated by measuring the yield of molecules after
their penetration through carbon foils.[9] The experimental arrangement is shown in Fig. (5). A momentum analyzed molecular beam
(H_2^+, D_2^+, DH^+, $^3HeD^+$, $^4HeH^+$,...) was incident on thin self supporting carbon targets (6-30 $\mu g/cm^2$). A fraction of the emerging beam,
typically 10^{-3} to 10^{-4}, was selected by a collimator located approximately 15 cm. from the target and centered on the beam axis.
The charged and neutral components of the beam were separated by
an analyzing magnet and detected with a silicon surface barrier
particle detector centered on the beam axis for neutral projectiles
and rotated to 30°, in the deflection plane of the magnet, for
charged projectiles. Normalization to the total beam flux was
accomplished by a detector positioned at 45° with respect to the
beam and located $\sim$ 6 cm from the target. Using this procedure we
were able to determine the neutral and charged components of the
atomic and molecular beam.

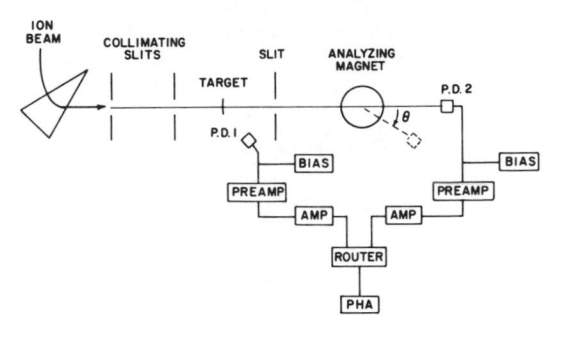

Figure 5   Experimental arrangement employed in measuring the
                fraction of the beam that emerges as molecules.

        Table I shows the neutral and charged fraction of the emerging
atoms from a 6 $\mu g/cm^2$ carbon target for incident projectiles of
300 keV molecular hydrogen and deterium.   Table II shows similar
data for incident molecules where at least one atom of the molecule
is helium.   The results are in reasonable agreement with neutral
and charged fractions measured with atomic beams of the same
velocity.[10] Meggit et.al.[11] have reported a slightly lower (∿5%)
charged fraction for molecules as compared to protons.   Their
results are in agreement with the data in Table I if the experimental uncertainties are included.

        The fraction, ϕ, of the emerging beam that is in a molecular
state is shown in Fig. 6 as a function of the target thickness for
300 keV incident projectiles of H_2^+, D_2^+, and DH^+.   The curves are
drawn to aid the eye.   The data points with arrows pointing down
indicate an upper limit measurement. Included in Fig. (6) are the
results for 2 MeV H_2^+ of Poizat and Remillieux.[12]   Their coincidence with the 300keV data is fortuitous, but indicates that the
molecular fraction is independent of the incident energy.   The
curves separate, as shown in Fig. 7, when the H_2^+ data are plotted
as a function of the dwell time, $\tau = D/v$ in atomic units, of the
cluster in the target of thickness D.   In the lowest order the
curves reflect the electron capture probabilities[10] of the spatially corellated clusters (at equal dwell times the clusters will
separate the same amount due to the mutual coulombic force).

        The emerging molecular fraction for helium molecules are
shown in Fig. 8 for various He molecules of 300 keV energy incident
on carbon targets.   For the thinnest targets employed (∿ 6$\mu g/cm^2$)
the molecular yield is, compared to hydrogenic molecules, about an
order of magnitude lower and decreases faster with increasing target
thickness.   This reflects the requirement that two, or even three,
electrons have to be captured by the emerging cluster in order to
form a molecule.

Table I. The neutral and charged fraction of the emerging beam for 300 keV projectiles incident on 6 µg/cm^2 carbon targets. The experimental uncertainty is ±10%.

Incident Projectile

φ	H_2^+	D_2^+	DH^+	H_3^+	D_3^+
H^o	0.17		0.3	0.27	
H^+	0.82		0.7	0.73	
H^-	6×10^{-4}		1.5×10^{-3}		
D^o		0.45	0.3		0.55
D^+		0.55	0.7		0.45
D^-		5×10^{-3}	1.5×10^{-3}		6×10^{-3}

Table II. The neutral and charged fraction of the emerging beam for 300 keV projectiles in which at least one atom has Z=2. The carbon target thickness is 6 µg/cm^2. The experimental uncertainty is ±10%.

Incident Projectile

φ	$^4HeH^+$	$^4HeD^+$	$^4He_2^+$	$^3HeD^+$
D^o		0.5		0.5
D^+		0.5		0.5
D^-		8×10^{-3}		
$^{3,4}He^o$	0.45	0.5	0.5	0.45
$^{3,4}He^+$	0.55	0.5	0.5	0.55
$^{3,4}He^{+2}$	4×10^{-2}	4×10^{-2}	1.5×10^{-2}	
$^{3,4}He^-$	1.5×10^{-4}	1×10^{-4}	1.3×10^{-4}	

    These experiments were performed to ascertain if wake riding increases the fracion of molecules emerging. Towards this end we must investigate alternate explanations, in particular, explosion, differing retarding forces, and small angle scattering.

    The bonding valence electron of the molecule will be lost in the first few atomic layers of the target. The internuclear distance will increase with time due to the mutual coulombic repulsion

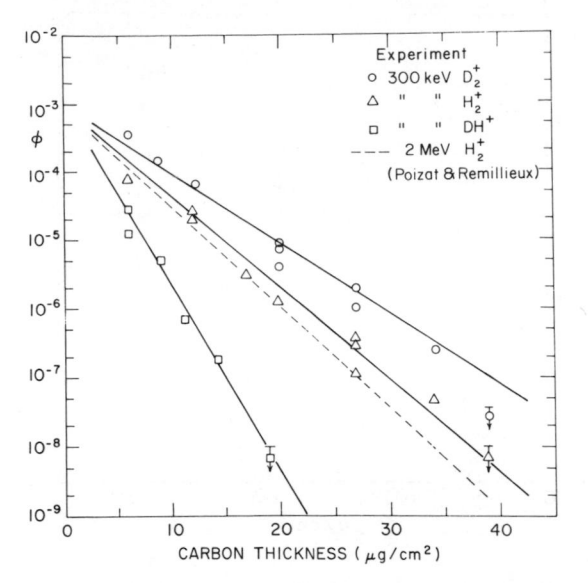

Figure 6   The molecular ion yields, from the projectiles
            listed, as a function of the carbon foil thickness
            [9]. The solid lines are to guide the eye. The
            dashed curve summarizes the data of [12].

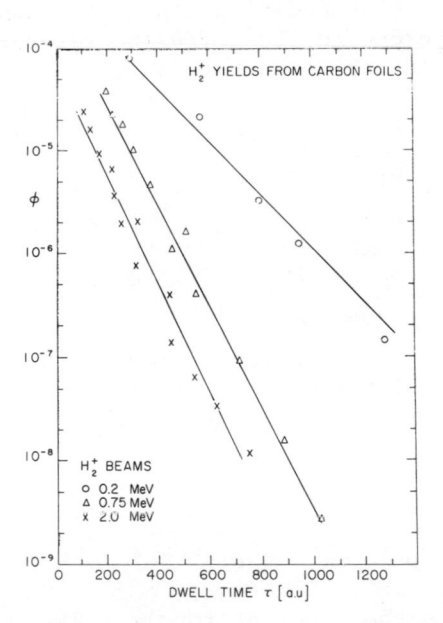
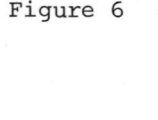

Figure 7   The molecular ion yield as a function of the cluster
            dwell time in carbon foils. The curves are drawn
            to aide the eye. The data is from [9 and 12].

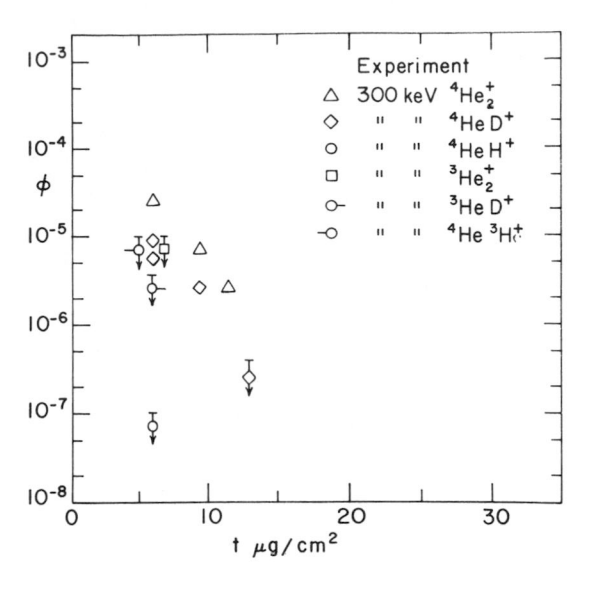

Figure 8   The molecular ion yields from incident projectiles
                where at least one atom has Z=2.

as given by Eq. (4). In addition, for heteronuclear clusters, the
atoms will separate due to the differing retarding forces.   The
particles of the cluster decelerate as

$$\ddot{z}_1 = -\frac{S_1}{M_1} = -\frac{z_1^2}{M_1}\,S_p$$

$$\ddot{z}_2 = -\frac{S_2}{M_2} = -\frac{z_2^2}{M_2}\,S_p \tag{13}$$

where S_1 and S_2 are the respective stopping powers. If one particle is in the wake of the other particle it will experience an additional force to the stopping power.   The resulting deceleration
is given as[2]

$$\ddot{z}_2 = -\frac{z_2^2}{M_2}\,S_p + \frac{z_1 z_2}{M_2}\,S_p^v\,\cos\left(\frac{\omega_o z}{v}\right)\,. \tag{14}$$

Wake riding implies that the internuclear distance, z, remains
constant.   Hence

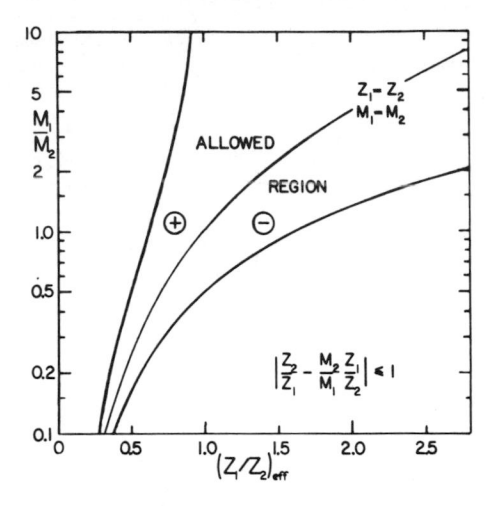

Figure 9   The compatibility plot for wake riding diclusters as given by Eq. (16).

$$z = z_1 - z_2 = [-\frac{z_1^2}{M_1} + \frac{z_2^2}{M_2} - \frac{z_1 z_2}{M_2} \frac{S_p^v}{S_p} \cos(\frac{\omega_0 z}{v})]s_p = 0 \quad (15)$$

Wake riding is possible[2] in the first mode of the wake when

$\frac{\pi}{2} \le \frac{\omega_0 z}{v} \le \pi$ and when $2m\pi \le \omega_0 z/v \le (2m+1)\pi$, m=1, 2, ... Eq. (15) requires as a necessary condition for wake riding that the ions are matched to comply with the restriction[4]

$$\left| \frac{z_2}{z_1} - \frac{z_1}{z_2} \frac{M_2}{M_1} \right| \le \frac{S_p^v}{S_p} \quad (16)$$

which is shown in Fig. 9 for $S_p^v = S_p$. If wake riding is to be identified through molecule formation by the clusters after electron capture on leaving the solid, only wake riding in the domain of closest possible distances should contribute. This further restricts the compatibility condition, Eq. (16), and is represented by the area marked (-) in Fig. 9.

The compliance of various diclusters with the wake riding compatibility condition are summarized in Table III. Wake riding with compatibility (-) is most conducive to molecule formation after cluster transmission. Comparing this with the experimental results shown in Fig. 7 we note that $\varphi(DH^+) << \varphi(H_2^+)$ or $\varphi(D_2^+)$ compatible with the results in Table III. However the explosion model, Eq. 6 with Eq. 13, predicts a similar ordering. The data in Fig. 8 were all obtained with E=300 keV where $Z_{eff}(He) \simeq 1.5$.

Table III. Wake riding compatibilities of various clusters. A
cluster is compatible with the negative branch of compatibility
relation, Eq. (16), when marked −, with the positive branch when
marked +, and does not match Eq. (16) when marked 0. Symbols in
( ) refer to low-velocity conditions, where $Z_{eff}(He)\simeq1$; the others
refer to $Z_{eff}(He)=2$.

Trailing	Leading ion			
ion	^1H	^2D	^3He	^4He
^1H	−	+	− (+)	− (+)
^2D	−	−	− (+)	− (+)
^3He	+ (0)	0 (−)	−	+
^4He	− (0)	+ (−)	−	−

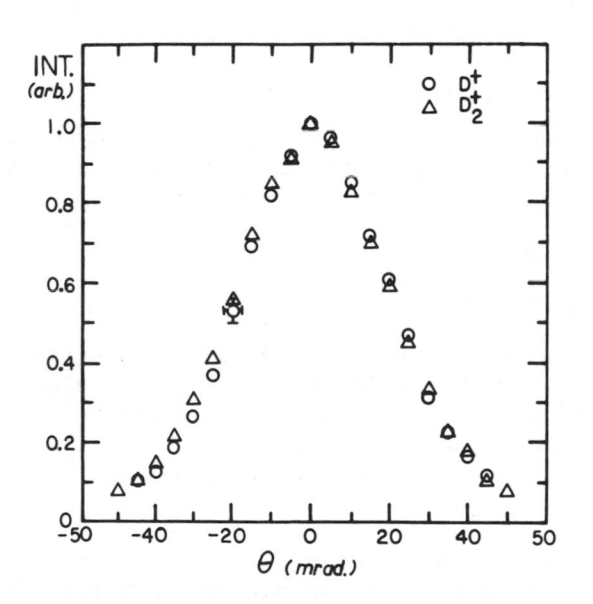

Figure 10   The angular distribution, in millradians, of the
              emerging beam from an $18\mu g/cm^2$ carbon foil for mole-
              cular, D_2^+, and atomic D^+, projectiles. The incident
              energy was 100keV/atom and the acceptance angle of
              the detector was 3mrad.

Under these conditions $(^4He^1H)^+$ and $(^1H^4He)^+$ (the leading ion is written first) both have the compatibility 0 or + and indeed no $(^4He^1H)^+$ yield was observed with a sensitivity limit of 10^{-7}. However, again, the explosion model would predict a similar ordering.

To ascertain whether explosion or small-angle scattering determines the mean separation between the atoms at the exit surface of the foil, we measured the angular distribution of the emerging beam when a $18\mu g/cm^2$ carbon foil was bombarded by clusters, D_2^+, and atoms, D^+, at the same velocity. The results are shown in Fig. 10 and there is no apparent difference. The atomic results agree with previous measurements.[13] Eq. (4) predicts that the explosion distance is equivalent to ~6mrad. Hence, at low incident cluster energies, small-angle scattering is dominant in determining the mean separation between the atoms at the exit surface of the foil.

From the above discussion, it is evident that from these molecular yield data one is unable to conclude whether wake riding is instrumental in determining the final yield of molecules observed. Further work, which is now in progress, is required to describe the behavior of the molecules in a solid. Wake riding should be evident in any experiment that is sensitive to the internuclear distance between the atoms of the molecule, and we hope that the data presented here, can serve as a guide and stimulation for such experiments.

REFERENCES

*   Work supported in part by the United State Energy Research and Development Administration.
+   This is a progress report of a collaborative effort with W. Brandt and A. Ratkowski of New York University and R. H. Ritchie of Oak Ridge National Laboratory.

1. N. Bohr, Kgl. Danske Videnskab. Selskab, Mat. Fys. Medd. 18, No. 8 (1948).
2. V. N. Neelavathi, R. H. Ritchie, and W. Brandt, Phys. Rev. Lett. 33, 302 (1974); ibid. 33, 670 (1974) and 34, 560 E (1975).
3. W. Brandt, A. Ratkowski, and R. H. Ritchie, Phys. Rev. Lett. 33, 1325 (1974) and 35, 130 E (1975).
4. See also W. Brandt and R. Ritchie, Intl. Conf. Atomic Collisions in Solids, Amsterdam, Holland, September 1975, Paper E1. To be published in Nucl. Inst. Methods.
5. W. Brandt and J. Reinheimer, Phys. Rev. B2, 3104 (1970) and B4, 1395 E (1971).
6. W. Brandt, in Atomic Collisions in Solids, Eds. S. Datz, B.R. Appleton, and C. D. Moak (Plenum, New York, 1975) p. 261.
7. J. W. Tape, W. M. Gibson, J. Remillieux, R. Laubert, and H. E. Wegner, Intl. Conf. Atomic Collisions in Solids, Amsterdam.

Holland, September, 1975, Paper E5. To be published in Nucl.
Inst. Meth.

8. J. C. Ashley, R. H. Ritchie and W. Brandt, Phys. Rev. $\underline{B5}$,
2393 (1972); $\underline{A8}$, 2402 (1973); $\underline{A10}$, 731 (1974). The function
$I(\xi)$ is tabulated in J. C. Ashley, V. E. Anderson, R. H.
Ritchie and W. Brandt, Z_1^3 Effects in the Stopping Power of
Matter for Charged Particles: Tables of Functions. National
Auxiliary Publication Service Document No. 02195, to be ordered
from Microfiche Publications, 305 East 46th Street, New York,
NY 10017, USA.

9. W. Brandt, R. Laubert, and A. Ratkowski, Intl. Conf. Atomic
Collisions in Solids, Amsterdam, Holland, September 1975,
Paper E2. To be published in Nucl. Inst. Meth.

10. A. Chateau-Thierry and A. Gladieux, in Atomic Collisions in
Solids, Eds. S. Datz, B. R. Appleton and C. D. Moak (Plenum,
New York, 1975) p. 307; A. Chateau-Thierry, A. Gladieux, and
B. Delaunay, Intl. Conf. Atomic Collisions in Solids, Amster-
dam, Holland, September 1975, Ppaer H4. To be published in
Nucl. Inst. Meth.

11. B. T. Meggit, K. G. Harrision, and M. W. Lucas, J. Phys. B:
Atom. Molec. Phys. $\underline{6}$ (1973) L362.

12. J. C. Poizat and J. Remillieux, Phys. Lett. $\underline{34A}$, 53 (1971)
and private communication (1975).

13. G. Hogberg, H. Norden, and H. G. Berry, Nucl. Inst. Meth. $\underline{90}$,
283 (1970).

K-SHELL X-RAYS FROM TI, NI, GE, AND RB FOR INCIDENT IONS FROM

^{1}H TO ^{35}CL IN THE ENERGY RANGE FROM 1-3 MEV/AMU *

F. D. McDaniel and J. L. Duggan

North Texas State University

Denton, Texas 76203

ABSTRACT

A systematic study of ion-induced x-ray production for the K-shell is described. X-ray production cross sections, K_α/K_β ratios, and x-ray energy shifts are presented for representative targets of Ti, Ni, Ge, and Rb for incident ion species of ^{1}H, ^{4}He, ^{7}Li, ^{12}C, ^{14}N, ^{19}F, ^{28}Si, and ^{35}Cl. Measurements over an incident ion energy range from 1-50 MeV were made using a 2.5 MV and a 6 MV single-ended Van de Graaff accelerators and a 7 MV tandem accelerator. X-rays were detected with a lithium drifted silicon x-ray detector with a fwhm resolution of 175 eV at 6.0 keV. Trends observed in the experimental results as a function of incident ion species, incident ion energy, and target atom are presented. The experimental cross sections are compared to theoretical models of Coulomb ionization which are used to describe ion-atom interactions. These models include the Binary Encounter Approximation (BEA) and the Plane Wave Born Approximation (PWBA) with corrections for Coulomb deflection of the incident projectile and increased electron binding. Calculations of polarization and relativistic effects are also included for some of the results. The departures of the predictions of Coulomb ionization theories from the experimental cross sections with increasing projectile mass and the effects of multiple ionization are discussed.

* Work at NTSU is supported in part by the Robert A. Welch Foundation, the Faculty Research Fund (NTSU), and Research Corporation.

INTRODUCTION AND DISCUSSION OF EXPERIMENTAL PROCEDURE

Over the past few years considerable progress has been made
in the study of inner shell ionization by light-ion bombardment.
The earlier results have been summarized in review articles by Garcia, Fortner, and Kavanagh[1] and Rutledge and Watson.[2] These results
have been compared quite successfully to a number of theories of
direct Coulomb ionization of inner shell electrons. Two primary
theoretical approaches which have been employed are the quantum
mechanical Plane Wave Born Approximation (PWBA) and the semiclassical Binary Encounter Approximation (BEA).

While these theories provide a description of direct Coulomb
ionization for light incident ions, it is not clear that they should
also be applicable for heavy ion-induced inner shell ionization.
As the mass of the incident ion is increased (Z1 of the projectile
$\leqslant$ Z2 of the target) the assumption that the incident projectile is
a bare nuclear charge is no longer valid. For some incident energies
the projectile carries its electronic cloud into the interaction
and Pauli excitation is also possible.

It is the purpose of this work to study systematic trends in
experimental x-ray cross section measurements and to provide a comparison between experiment and theories of Coulomb ionization for
a range of projectiles from ^{1}H to ^{35}Cl.

The experiments outlined in this paper were done on four Van
de Graaff accelerators. Some of the low energy alpha particle
measurements were made on the 2.5 MV machine at NTSU while the
remainder of the data was taken on the 6.0 MV vertical machine at
ORNL and the Tandem accelerators at ORNL and at The Florida State
University. In all cases the beam was tightly collimated so that
the resultant spot on the thin foil target was ~1.5mm in diameter.
The atomic foil wheel scattering chamber that was used for most of
the experiments is shown in Figure 1. The chamber is designed in
such a manner that twenty four targets can be placed in it and
manually rotated into the beam position. All of the data discussed
in this paper were taken with a Si(Li) X-ray detector (~170ev at
5.9Kev) which was mounted at 90° with respect to the incident beam
direction. The foil wheel is designed so that the targets are
positioned at 45° with respect to the beam axis as shown in Figure
1.

The Bremsstrahlung background associated with most ion-atom
experiments of this nature is a continuum which extends out to an
energy of approximately 10 KeV. The intensity of the continuum
relative to that of the characteristic x rays is dependent on the
incident energy, the incident ion species, and the material and
thickness of both the target and substrait. Most of the targets
used for these experiments were evaporated onto carbon foils that

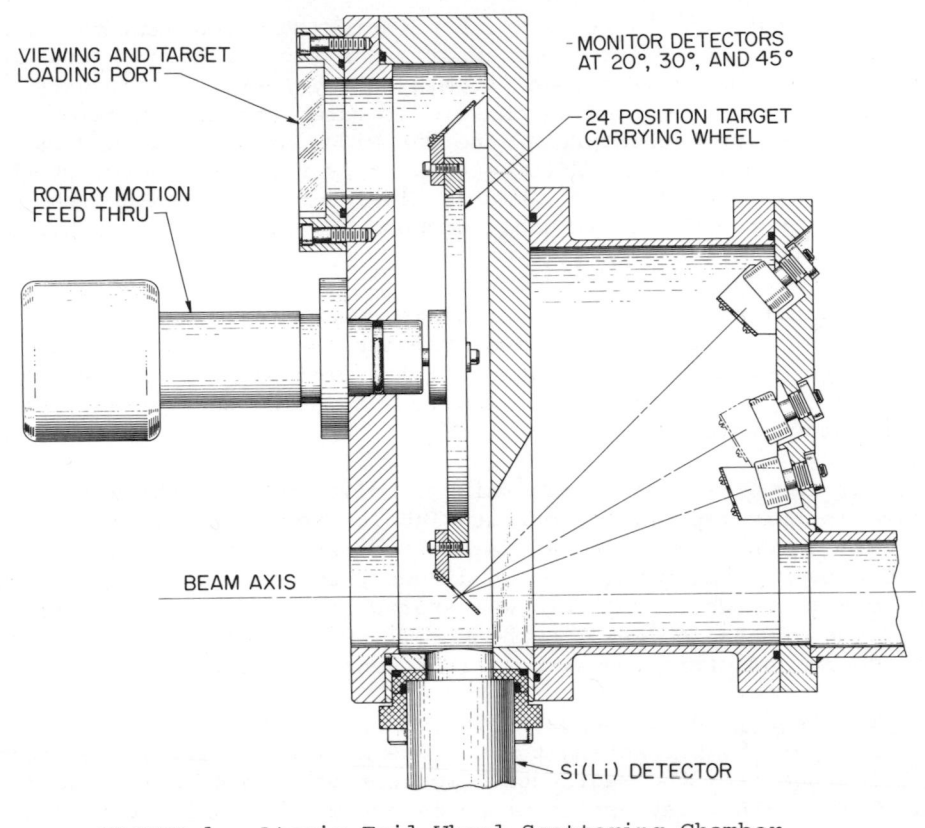

VIEWING AND TARGET
LOADING PORT

MONITOR DETECTORS
AT 20°, 30°, AND 45°

24 POSITION TARGET
CARRYING WHEEL

ROTARY MOTION
FEED THRU

BEAM AXIS

Si(Li) DETECTOR

FIGURE 1:   Atomic Foil Wheel Scattering Chamber

were from 7.0 to 30 μgm/cm^2. The target thicknesses varied from
5.0 to 100 μgm/cm^2 depending on the above mentioned conditions.
For some of the measurements, for heavier ions at lower energies,
the energy loss of the projectile in the target was large enough
(For example, ~150 Kev out of 7 Mev) to warrant a correction in the
cross sections for energy loss. In order to make this correction
the formalism of Laubert et al.[3] was used where applicable.

The Bremsstrahlung background was filtered by the use mylar
foils (from 3-20 X 10^{-3}in) that were placed between the target spot
and the Si(Li) detector. The resultant efficiency of the detector
was carefully measured by placing calibrated radioactive sources
at the target spot. The sources were designed so that the radioactive spot size was about 1.5mm which, as was mentioned above,was
the diameter of the beam spot.

The sources used were ^{51}Cr, ^{54}Mn, ^{57}Co, ^{65}Zn, ^{88}Y, and ^{241}Am.
The techniques associated with this efficiency measurement, the de-
cay schemes and virtues of the above sources has been described in

the literature.[4,5,6] It is worth pointing out, however, that most
of the sources used for these experiments were primary standards
from the NBS or Interantional Atomic Energy Agency. Secondary
standards were also used which were prepared at the Radiation
Standardization Laboratory of the Oak Ridge Associated Universities.
The uncertainty in the efficiency curve is the main source of absolute experimental error. For the low energy end (<5KeV) the efficiency of the detector should be good to $\pm$ 15% while for higher
energies it is known to $\pm$ 5%.

     In order to overcome the problems associated with current inte-
gration and target thickness measurements the Rutherford scattering
of the projectile was simultaneously measured with the x-ray spectra.
For some of the measurements, elastic scattering data was collected
simultaneously for 30^0 and 45^0.

     For the ORNL experiments which involves most of the data dis-
cussed in this report, the output signals from all amplifiers were
fed into a Tennecomp Pace Data Aquisition and Processing System.
The Tennecomp System was coupled directly into the ORNL Control Data
3200 computer, where the data was stored on disk. Data from the disk
could then be called back to the PDP 11 on the Tennecomp system for
processing and stripping with a light pen assembly.

     The energy shifts reported in this paper were established for
the most part by bracketing the runs periodically with energy calibration runs from the radioactive sources mentioned above. We found
that ORTEC 739A X-ray amplifier to be extremely stable over long
periods of time. For some of the runs, and in order to double
check our absolute energy shifts, the photons from a (25mCi) Cd-109
radioactive source was used to source excite the target material
being studied. This X-ray Fluorescence technique compared very well
with the standard source method mentioned above. From these measurements, energy shifts for $K_\alpha \pm$ 7ev and $K_\beta \pm$ 15ev were established.

                         RESULTS AND DISCUSSION

     We have measured x-ray intensities for a large number of tar-
get species and incident ions for the purposes of determining systematic trends in the experimental x-ray cross sections and for a
detailed comparison to existing theories of Coulomb ionization. We
have measured x-ray production cross sections, shifts in the characteristic energies of emitted x rays, and ratios of K_α/K_β intensities.

     Figure 2 shows the x-ray production cross sections for Ti for
a number of incident ions as a function of E/M (MeV/amu) of the projectile. The different incident ions employed are given in the
legend of the figure. For the Ti target the x-ray production cross

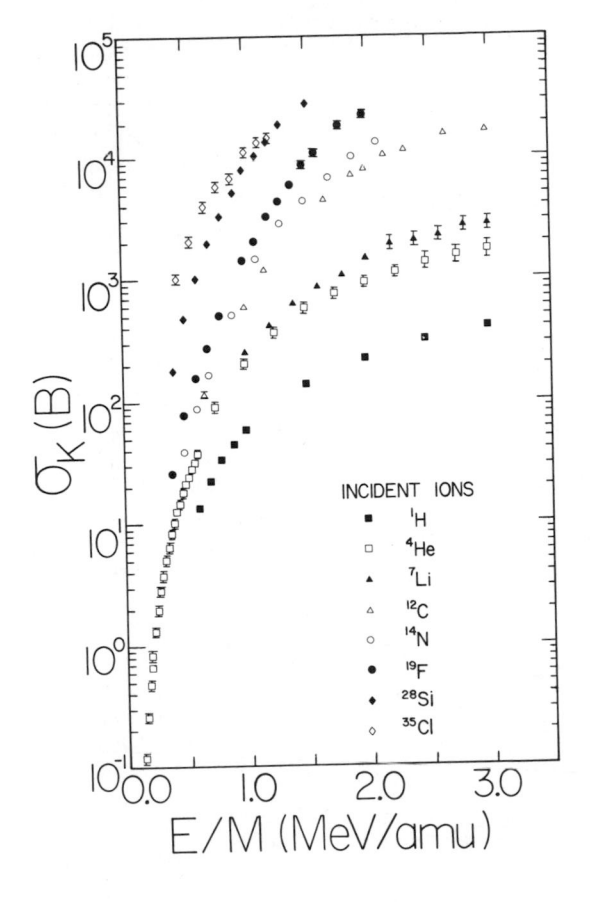

FIGURE 2:   Experimental X-ray Production Cross Sections for Ti for
            a Number of Incident Ions.

sections are found to increase with increasing Z1 and energy of the
the incident ion.   Ti is the lightest target for which data is presented.

     The uncertainties shown with the data are relative uncertainties
only and are derived from quantities which vary with energy.   Relative uncertainties vary from 5-13%.   Normalization uncertainties
are those which do not vary with energy.   These uncertainties are
primarily due to efficiency and solid angle measurements and range
from 5-15%.   Total uncertainties are determined from the square
root of the sum of the squares of the relative and normalization
uncertainties and range from 9-20%.

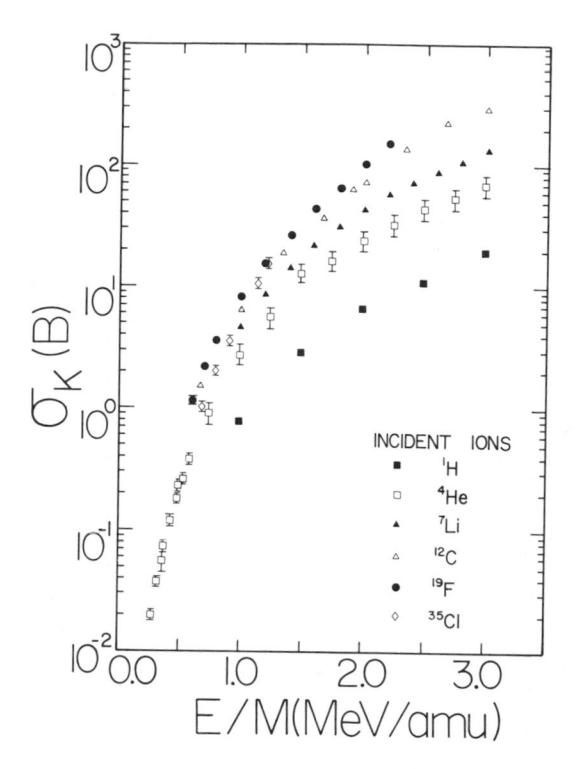

FIGURE 3:   Experimental X-ray Production Cross Sections for Rb
            for a Number of Incident Ions.

Figure 3 presents x-ray production cross sections for the
heaviest element investigated, Rb, for a number of projectiles.
The x-ray production cross sections still exhibit an increase
with Z1 and energy of the incident ion although this increase is
not quite as dramatic as for the lighter Ti target.

Figure 4 presents x-ray production cross sections as a function of E/M (MeV/amu) for Ti, Ni, Ge, and Rb for incident ^{19}F ions.
The x-ray production cross sections are found to decrease with increasing Z2 of the target.   The x-ray production cross sections
should approach a maximum where the velocities of the incident ion
and the K-shell electron are equal.

Having presented some of the systematic trends in the experimental x-ray production cross sections as a function of projectile Z1, energy, and target Z2, one would like to compare these

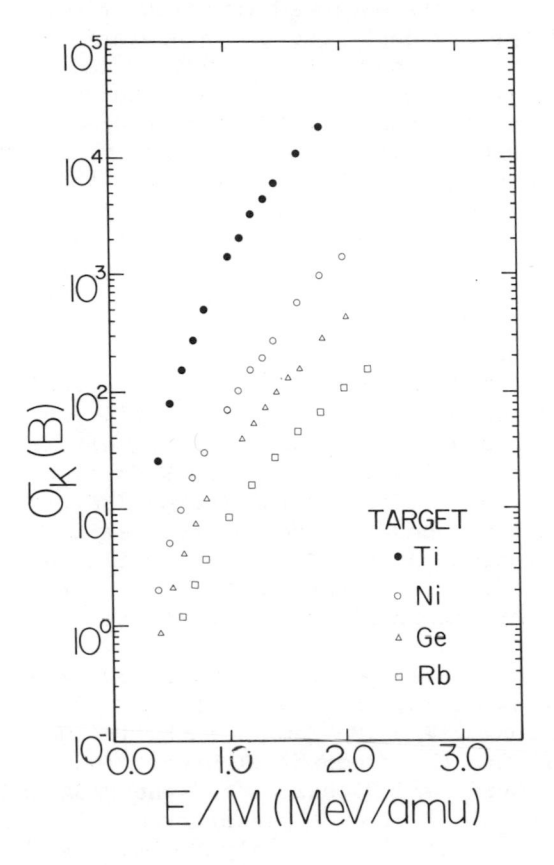

FIGURE 4:   Experimental X-ray Production Cross Sections for Ti,
            Ni, Ge, and Rb for Incident [19]F Ions.

results to existing theories of Coulomb ionization.   The two pri-
mary approaches are the semiclassical Binary Encounter Approximation
(BEA) of Garcia et al.[7] and the quantum-mechanical Plane Wave Born
Approximation (PWBA) of Merzbacher et al.[8,9,10]

     The BEA assumes the interaction between the incident projectile
and the target electron is through a direct energy exchange.   The
calculation of BEA ionization cross sections after procedures given
by Hansen[11] and McGuire and Richard[12] have been found to compare
quite favorably with light-ion experimental data over limited incident energy ranges.[13]

     The PWBA treatment of the incident ion as a plane wave has been
described in a review article by Merzbacher and Lewis.[8]   A number of

simplifying assumptions are employed in calculations of the PWBA.
The electronic states are described by non-relativistic hydrogenic
wave functions which are assumed to be undisturbed by the Coulomb
interaction between the incident ion and the target electrons. The
screening of the nuclear charge of the target atom is approximated
by replacing the nuclear charge Z by an effective charge Z_e = Z-0.3.
Khandelwal, Choi, and Merzbacher have tabulated the results of these
calculations.[9]   The PWBA predictions have been found to provide good
agreement with the cross sections at energies greater than a few MeV
for ^{1}H and ^{4}He projectiles[13,14,15] and ^{7}Li projectiles.[16]   The PWBA
has been found to over predict the experimental cross sections at the
lower energies by as much as an order of magnitude.[16]

  Basbas, Brandt, and Laubert have made corrections to the PWBA
for Coulomb deflection of the incident projectile by the nuclear
charge of the target and for increased binding of the target electrons due to the penetration of the target K shell by the incident
ion.[17]   The PWBA corrected for binding energy (BE) and Coulomb deflection (CD) has been found to give the best estimate of the experimental cross sections at the lower projectile energies for the
lighter incident ions.[14,16,18] For the heavier Rb target both BE
and CD corrections are expected to be important.

  Figure 5 presents cross section results for a heavier target,
Rb, for incident ions of ^{1}H, ^{7}Li, ^{12}C, ^{19}F, and ^{35}Cl.   These are
plotted as a function of E/M (MeV/amu) of the incident projectile.
Shown in the figure are the theoretical predictions of the BEA as
calculated after McGuire and Richard[12] and the PWBA and the PWBA
with BE and CD corrections after a prescription given by Basbas,
Brandt, and Laubert[17] employing an analytic form of the universal
function, F, as given by Brandt and Lapicki.[19] In making a comparison between theory and experiment, theoretical x-ray production
cross sections were obtained using fluorescence yields obtained
from Bambynek et al.[20]   The use of single hole parameters in view
of the multiple ionization present for the heavier projectiles
is a possible source of uncertainty.

  For higher Z2 targets relativistic effects are also expected
to be important at the lower incident ion velocities.[21]   Relativistic effects are expected to raise the experimental x-ray cross
sections at lower incident ion energies.[22] In lieu of a rigorous
theoretical treatment in which relativistic atomic electron wave
functions are employed, the theoretical cross sections were corrected by an approximate method suggested by Hansen.[11]   The correction consists of a multiplicative factor applied to the non-relativistic theoretical cross section.   The curves labeled PWBABCR have
been corrected for relativistic effects by this approximate technique.   One can see the magnitude of the correction in the panel
of Figure 5 for ^{19}F incident ions on Rb.   The results presented
for Rb are found to be in better agreement with the PWBA with

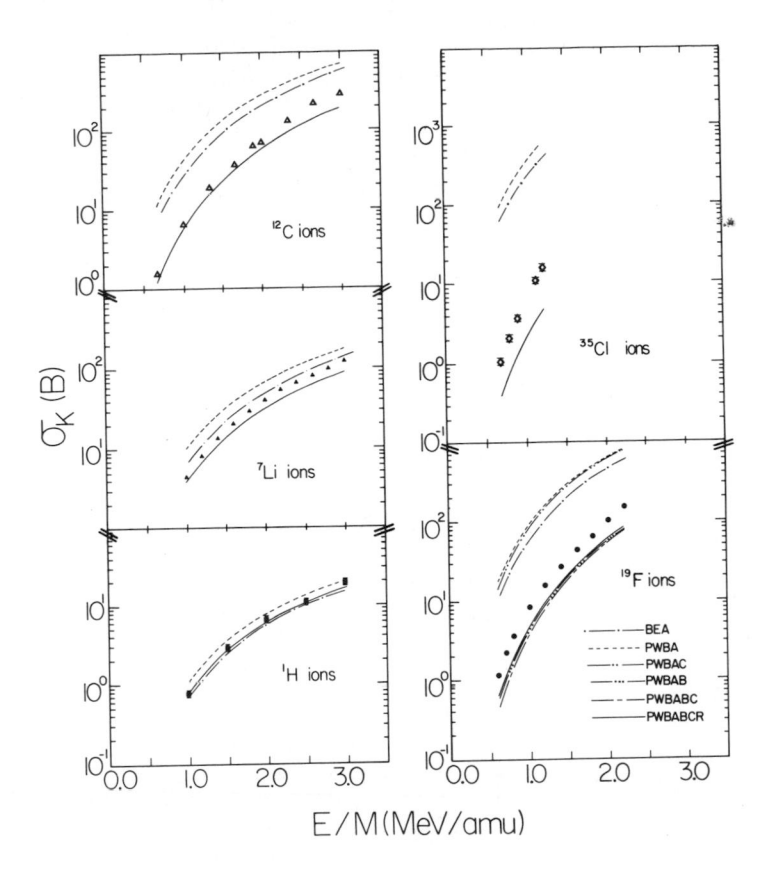

FIGURE 5:   X-ray Production Cross Sections of Rb for a Number of
            Incident Ions are compared to Theories of Coulomb Ion-
            ization.   Theoretical Curves are presented for the Binary
            Encounter Approximation (BEA), the Plane Wave Born Approxi-
            mation (PWBA) with Corrections for Coulomb Deflection
            (PWBAC) and Increased Electron Binding (PWBAB) and Rela-
            tivistic Effects (PWBABCR).

corrections for BE, CD, and relativistic effects.

    Figure 6 presents the experimental cross sections of Ti for
a number of different incident ions. The theoretical descriptions
presented are again the BEA, the PWBA, and the PWBA with corrections
for BE, CD, and relativistic effects. For the lighter Ti target,
the correction for CD and relativistic effects are negligible except
for ^{1}H and ^{7}Li. The corrected PWBA consists of primarily the BE
correction. The general trends in the data exhibit a movement away
from the corrected PWBA toward the PWBA for the lighter incident ions

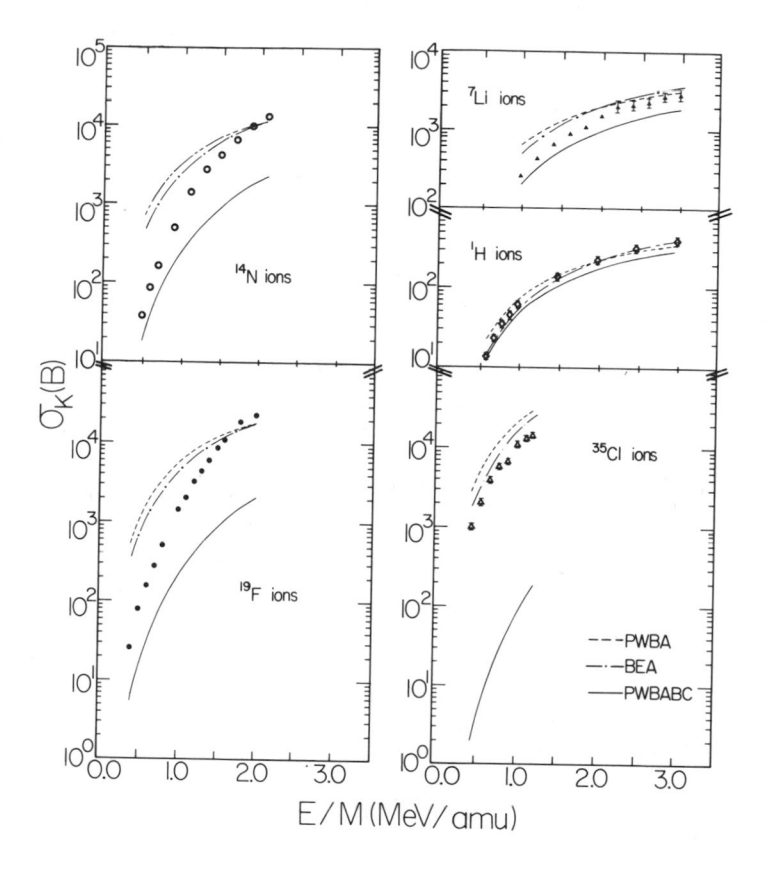

FIGURE 6:   X-ray Production Cross Sections of Ti for a Number of
            Incident Ions are Compared to Theories of Coulomb Ion-
            ization.

at the higher energies. For the heavier incident ions the deviations from the corrected PWBA become very large.

    The deviations at the higher energies have been attributed to
a high velocity polarization effect which has been discussed by
others.[14,23-25] The polarization involves the distortion of the
target electron wave functions due to the electro-magnetic field
of the passing projectile and results in an increase in the ioni-
zation probability for values of $0.05 < \eta_K/(\varepsilon_K\Theta_K)^2 < 1.0$. η_K is the reduced velocity parameter, Θ_K is the reduced binding energy, and ε_K
is the binding correction factor in the PWBA formalism.

    While the methods for making PWBA calculations with polari-
zation corrections have not been published, Dr. Basbas and Laubert

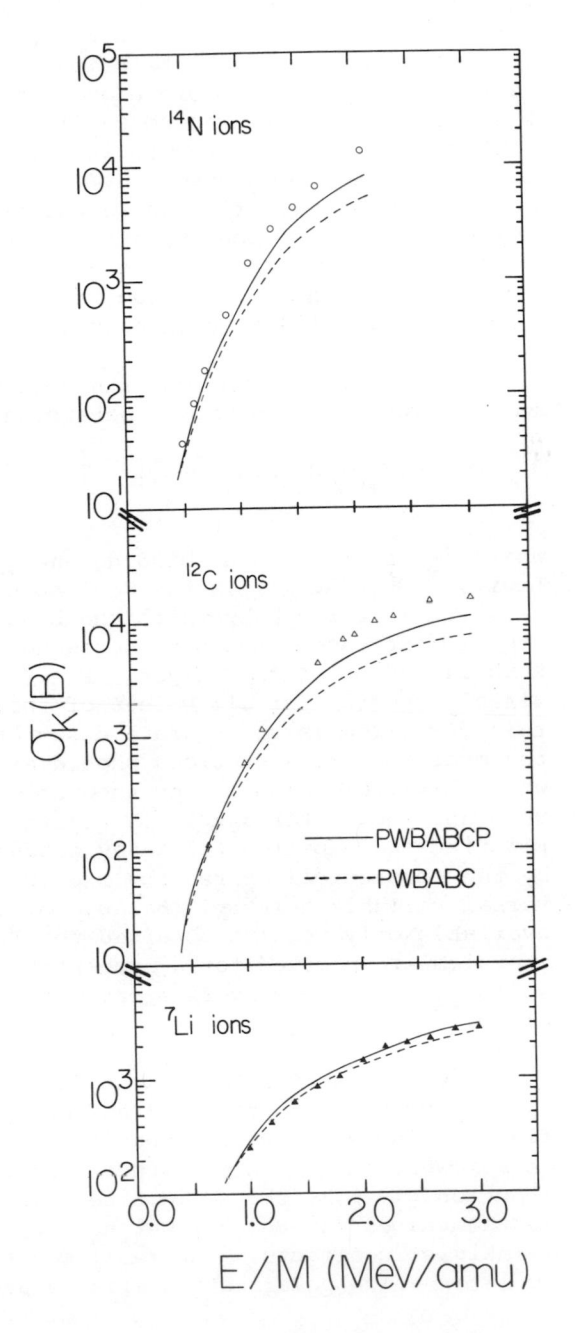

FIGURE 7: X-ray Production Cross Sections of Ti for ^{7}Li ^{12}C, and ^{14}N Incident Ions. Theoretical Curves are presented for the PWBABC with corrections for Polarization Effects (PWBABCP).

were kind enough to perform some calculations for us for ^{7}Li, ^{12}C, and ^{14}N projectiles.[26] Figure 7 presents x-ray production cross sections for Ti for ^{7}Li, ^{12}C, and ^{14}N incident ions. The dashed curve is the theoretical prediction of the PWBA with CD and BE corrections.[26] It might be pointed out that these calculations have been improved and are higher in magnitude at the larger energies than calculations presented earlier. The solid curve includes the correction for the high-velocity polarization effect and is seen to provide very good agreement with the experimental data except at the higher velocities where the data for ^{12}C and ^{14}N exceed the theoretical predictions by a few percent. The small deviations may be due to changes in the fluorescence yields because of multiple ionization of the target atom or to inner-shell ionization processes which are not Coulombic in nature.

    In the PWBA formalism with corrections for BE and CD the ioni-zation cross section may be determined by the expression

$$\sigma_I = 9E_{10}(\Pi dq_0 \varepsilon) \left| \frac{\sigma_{oK}}{\varepsilon \theta_K} \right| F \left(\frac{\eta_K}{\varepsilon^2 \theta_K^2} \right)$$

where θ_K is the reduced binding energy defined by Merzbacher and Lewis,[8] $9E_{10}(\Pi dq_0 \varepsilon)$ is the Coulomb deflection correction factor, ε is the factor ≥ 1 by which the binding energy of the target elec-tron is increased due to the passage of the incident ion within the K shell radius of the target. $F(\eta_K/\varepsilon^2 \theta_K^2)$ is an approximate uni-versal function for all values of the parameters η_K and θ_K. If this formalism is an accurate description of the ionization process the measured cross sections should exhibit a universal behavior when they are scaled by the theoretical Coulomb deflection factor $9E_{10}$ and the factor $\sigma_{oK}/\varepsilon \theta_K$. Figure 8 presents the results of plotting the experimental x-ray production cross sections divided by the theoretical expression σ_{oK} $9E_{10}(\Pi dq_0 \varepsilon) \omega_K/\varepsilon \theta_K$ versus the uni-versal variable $\eta_K/\varepsilon^2 \theta_K^2$. Because of the large quantities of data available only results obtained for Ti and Rb at three or four energies are plotted for a number of incident projectiles. The solid line is the universal parameter $F(\eta_K/\varepsilon^2 \theta_K^2)$ in the PWBA form-alism.[17]

    The results for Rb for the different incident ions are for $\eta_K/\varepsilon^2 \theta_K^2$ values less than 0.1 and occupy the lower portions of the figure. The Rb results appear to fall on a universal curve which is somewhat larger in magnitude than the theoretical universal function given by the solid line. The deviation of the experi-mental values for Rb at lower $\eta_K/\varepsilon^2 \theta_K^2$ values can be attributed to relativistic effects. The results for Rb are for $\eta_K/\varepsilon^2 \theta_K^2$ values <0.1 for which polarization effects are expected to be small.[14]

    The results for Ti for the various incident ions are for $\eta_K/\varepsilon^2 \theta_K^2$ values greater than 10^{-2} and occupy the upper portions of

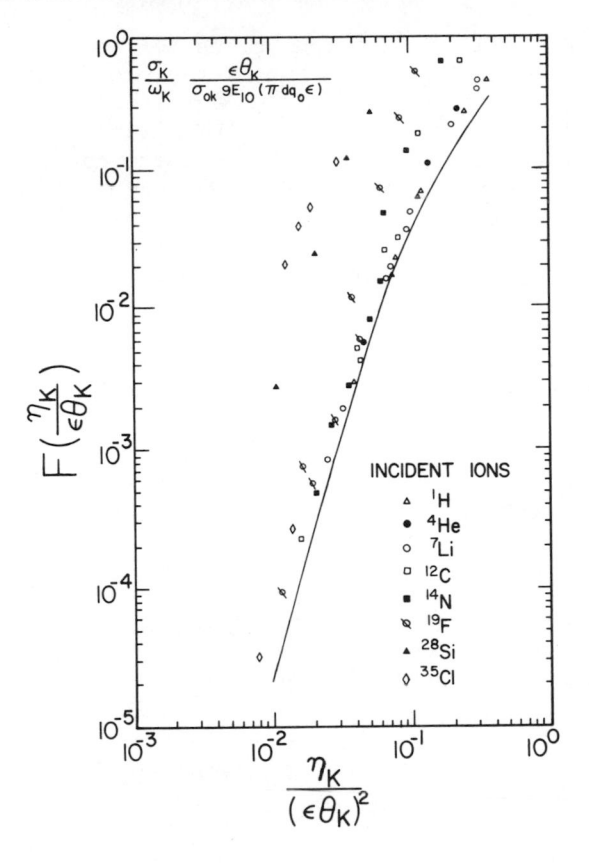

FIGURE 8:   X-ray Ionization Cross Sections of Ti and Rb for a
            Number of Incident Ions Scaled by the Theoretical
            Coulomb Deflection Factor $9E_{10}$ and the Factor $\sigma_{0K}/$
            $\epsilon\theta_K$. The Solid Curve is the Universal Parameter F
            $(\eta_K/\epsilon^2\theta_K^2)$ in the PWBA Formalism.

Figure 8.   The Ti results do not fall on a universal curve as found
for Rb except for ^{1}H, [4]He, and [7]Li incident ions.   The deviation of
the Ti results from a universal curve becomes greater with increasing
Z1 of the incident ions.   The deviation of the results for ^{19}F,
[28]Si, and [35]Cl incident ions cannot be explained by polarization
effects alone, since they occur for values of $\eta_K/\epsilon^2\theta_K^2 < 0.05$.   The
deviations may result from changes in the fluorescence yields due
to multiple ionization and/or a general inapplicability of a Coulomb ionization description of the ionization process.   For incident ions lighter than or equal to ^{14}N the deviation of the experimental results from the universal theoretical curve at the
higher energies can be explained by polarization effects as shown

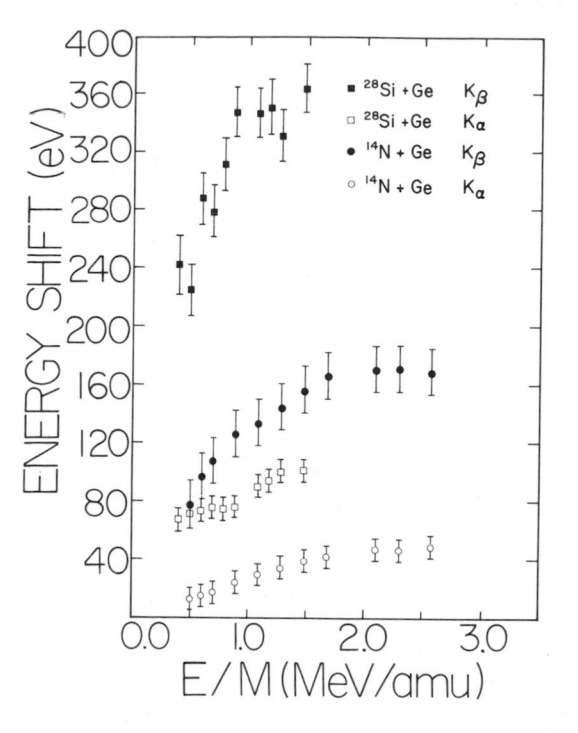

FIGURE 9:   Energy Shifts of Ge K_α and K_β X-rays for ^{14}N and ^{28}Si
            Incident Ions.

in Figure 7 since the deviation occurs for values of $\eta_K/\varepsilon^2\theta_K^2 > 0.05$.

Shifts in the characteristic energy of x-ray lines and the
deviations of K_α/K_β ratios from the zero defect configuration are
indications of the amount of L and M shell vacancy production simultaneous with K vacancy production. For ^{1}H, ^{4}He, and ^{7}Li projectiles
the energy shifts were less than 15 eV at all incident ion energies.
For the incident projectiles heavier than ^{7}Li, shifts in the characteristic energies of x-ray lines were determined. Figure 9 presents
the K_α and K_β energy shifts for ^{14}N and ^{28}Si ions on Ge as a function
of E/M (MeV/amu).

The maximum energy shift for a particular incident ion occurs
where the probability is greatest for producing vacancies in the
L or M shells. The probability is greatest where the velocity of
the incident ion and the velocity of the L or M shell electrons
are equal. The incident ion energy corresponding to a velocity
that is equal to the average L-shell electron velocity is given by

$$E_{14_N} = \lambda \bar{U}^L_{Ge} = 32.8 \text{ MeV}$$

where λ is the ratio of projectile mass to electron mass and $\bar{U}^L_{Ge}$ is
the average L-shell electron binding energy for Ge. This corre-
sponds to an incident ^{14}N ion energy/amu of ~2.3 MeV/amu. The
energy shift for Ge K_α and K_β lines for ^{14}N ions does appear to
reach a maximum between 2.0-2.5 MeV/amu in Figure 9. The maximum
energy shift for ^{28}Si should also occur in the region of ~2.3 MeV/
amu. Unfortunately the present data does not extend this far. However, the fact that the maximum energy shifts for ^{14}N ions does
occur at ~2.3 MeV/amu indicates that the energy shifts are due
primarily to K vacancy production in the presence of L vacancy
production.

A comparison of the energy shifts produced by ^{14}N and ^{28}Si
ions reveals that both the K_α and K_β energy shifts are greater for
^{28}Si than ^{14}N as one might expect. Figure 10 shows the energy
shifts of Ti and Ni K_α and K_β x rays at an energy of 1 MeV/amu as
a function of Z1 of the incident ion. The energy shifts are indeed
found to increase with increasing Z1 of the projectile.

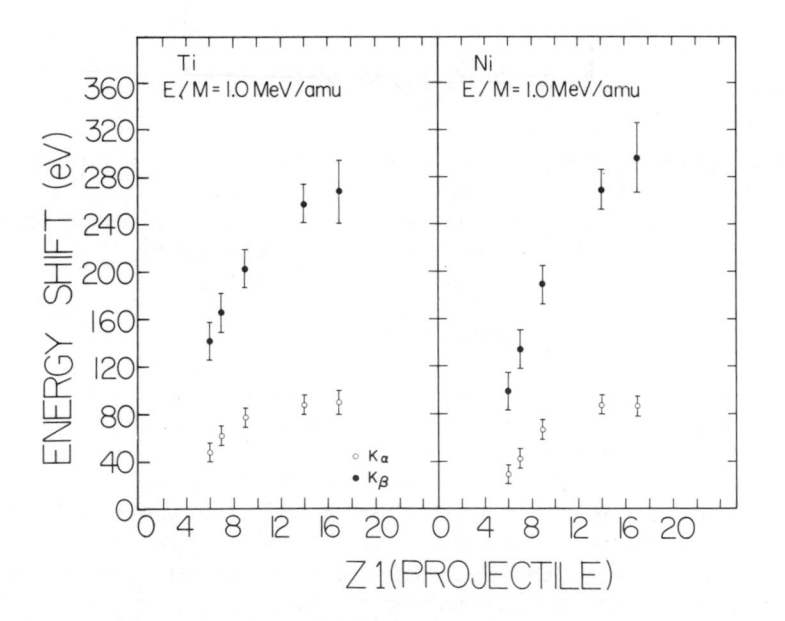

FIGURE 10: Energy Shifts of Ti and Ni K_α and K_β X-rays as a
            Function of Projectile Z1.

From the magnitudes of the measured energy shifts and the calculated shifts per L or M vacancy it is apparent that the amount of multiple ionization decreases as the target elements become heavier. Also, as the targets become heavier the deviations of the K_α/K_β intensity ratios from that determined for no multiple ionization are smaller. Figure 11 presents the ratios of K_α/K_β results for ^{14}N ion bombardment to that for no multiple ionization verses V_L, the velocity of the incident ^{14}N ion to that of the average L shell electron. The values for no multiple ionization are those of Salem et al.[27] The ratios exhibit approximately the same V_L dependence. It is clear from Figure 11 that the K_α/K_β ratios for Ge deviate further from the no defect ratios than those for Rb.

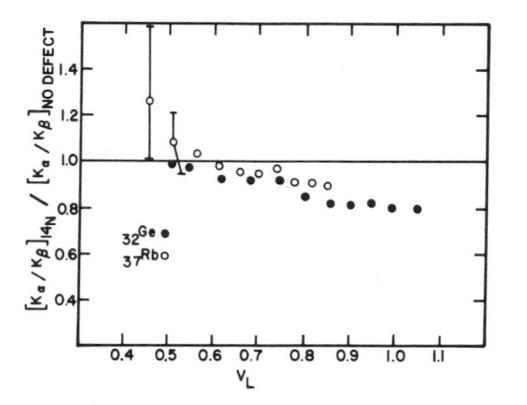

FIGURE 11:   Ratios of K_α/K_β Intensities for [14]N Ion Bombardment
to K_α/K_β Intensities for no Multiple Ionization. V_L
is the Ratio of the Velocity of the Incident [14]N Ion
to that of the Average L-Shell Electron.

CONCLUSIONS

A systematic study of K-shell ionization is presently in progress. At this stage in the study we may conclude that for incident ions lighter than or equal to ^{14}N incident upon targets heavier than Ti the PWBA with corrections provides good estimates of the experimental cross sections. Even at higher velocities where polarization effects are important and at the lower velocities where relativistic effects are important, the experimental results are predicted very well by the theories of Coulomb ionization.

However, as one attempts to describe ion-atom interactions by theories of Coulomb ionization for heavier projectiles, the theorical predictions begin to deviate from the experimental values. This is expected since the incident ion no longer appears as a bare

nuclear charge, which is a primary assumption in Coulomb Ionization theories.

Multiple ionization effects are very important in heavy-ion atom interactions. Multiple ionization produces shifts in the characteristic energies of emitted x-rays, changes in K_α/K_β ratios, and possible changes in the fluorescence yields. These effects are found to increase with increasing Z1 of the incident ion and decreasing Z2 of the target material. For light ions of ^1H, ^4He, and ^7Li at these incident energies, shifts in the characteristic x-ray energy were <15eV and were not determined. For the heavier incident ions, the K_α and K_β energy shifts were a maximum where $V_L = v_i/\bar{v}_L$ which indicates that these shifts are primarily due to multiple ionization of the L-shell.

ACKNOWLEDGEMENTS

The authors would like to acknowledge the many people who have contributed to the results presented either by participating in the data acquistion or analysis.

These people are P. D. Miller, ORNL; R. P. Chaturvedi and R. M. Wheeler, SUNY Cortland; T. J. Gray, NTSU; J. Lin, Tenn. Tech.; L. A. Rayburn, U. Texas Arlington; K. A. Kuenhold, Univ. of Tulsa; S. J. Cipolla, Creighton Univ.; and a large number of graduate and undergraduate students.

We would like to thank J. H. McGuire for helpful discussions concerning relativistic effects, and express our appreciation to G. I. Gleason of Oak Ridge Associated Universities for calibration of our radioactive sources.

Finally, we wish to graciously acknowledge the assistance of the entire staff of the High Voltage Laboratory at Holifield National Laboratory and Oak Ridge Associated Universities for travel support.

REFERENCES

1. J. D. Garcia, R. J. Fortner, and T. M. Kavanagh, Rev. Mod. Phys. 45, 111 (1973).

2. C. H. Rutledge and R. L. Watson, Atomic Data and Nuclear Data Tables 12, 195 (1973).

3. R. Laubert, H. Haselton, J. R. Mowat, R. S. Peterson, and I. A. Sellin, Phys. Rev. A 11, 135 (1975).

4. L. B. Magnusson, Phys. Rev. 107, 161 (1957).

5.   J. S. Hansen, J. C. McGeorge, D. Nix, W. D. Schmidt-Ott, I. Unus, and R. W. Fink, Nucl. Instr. and Meth. 106, 365 (1973).

6.   R. J. Gehrke and R. A. Lokken, Nucl. Instr. and Meth. 97, 219 (1971).

7.   J. D. Garcia, E. Gerjuoy, and J. E. Welker, Phys. Rev. 165, 66 (1968); J. D. Garcia, Phys. Rev. A 1, 1402 (1970); Phys. Rev. A 1, 280 (1970).

8.   E. Merzbacher and H. Lewis, Encyclopedia of Physics, edited by S. Flugge (Springer-Verlag, Berlin: 1958), Vol. 34, p. 166.

9.   G. S. Khandelwal, B. H. Choi, and E. Merzbacher, Atomic Data 1, 103 (1969).

10. B. H. Choi, E. Merzbacher, and G. S. Khandelwal, Atomic Data 5, 291 (1973).

11. J. S. Hansen, Phys. Rev. A 8, 822 (1973).

12. J. H. McGuire and P. Richard, Phys. Rev. A 8, 1374 (1973).

13. J. Lin, J. L. Duggan, and R. F. Carlton, Proceedings of the International Conference on Inner-Shell Ionization Phenomena and Future Applications, April 1972, Atlanta, Georgia, edited by R. W. Fink, S. T. Manson, J. M. Palms, and P. V. Rao (USAEC Report No. CONF-720404, Oak Ridge, Tennessee, 1973), p. 998.

14. W. Brandt, Proceedings of the International Conference on Inner-Shell Ionization Phenomena and Future Applications, April 1972, Atlanta, Georgia, edited by R. W. Fink, S. T. Manson, J. M. Palms, and P. V. Rao (USAEC Report No. CONF-720404, Oak Ridge, Tennessee, 1973), p. 948.

15. R. Akselsson and T. B. Johansson, Z. Phys. 266, 245 (1974).

16. F. D. McDaniel, T. J. Gray, R. K. Gardner, G. M. Light, J. L. Duggan, H. A. Van Rinsvelt, R. D. Lear, G. H. Pepper, J. W. Nelson, and A. Zander. To be published in Phys. Rev. A 12, (Oct. 1975).

17. G. Basbas, W. Brandt, and R. Laubert, Phys. Rev. A 7, 983 (1973

18. F. D. McDaniel, T. J. Gray, R. K. Gardner, Phys. Rev. A 11, 160 (1975).

19. W. Brandt and G. Lapicki, Phys. Rev. A 10, 474 (1974).

20.   W. Bambynek, B. Crasemann, R. W. Fink, H.-U. Freund, H. Mark,
      C. D. Swift, R. E. Price, and P. V. Rao, Rev. Mod. Phys. $\underline{44}$,
      716 (1972).

21.   D. Jamnik and C. Zupancic, Mat. Fys. Medd. Dan. Vid. Selsk. $\underline{31}$,
      No. 2 (1957).

22.   B. -H. Choi, Phys. Rev. A $\underline{4}$, 1002 (1971).

23.   G. Basbas, W. Brandt, R. Laubert, A. Ratkowski, and A.
      Schwarzschild, Phys. Rev. Lett. $\underline{27}$, 171 (1971).

24.   G. ,Basbas, W. Brandt, and R. Laubert, Phys. Lett. $\underline{34A}$, 277
      (1971).

25.   K. W. Hill and E. Merzbacher, Phys. Rev. A $\underline{9}$, 156 (1974).

26.   G. Basbas and R. Laubert, private communications.

27.   S. I. Salem, S. L. Panossian, and R. A. Krause, Atomic Data
      and Nuclear Data Tables $\underline{14}$, 91 (1974).

L X-RAY SPECTRA OF CHLORINE AND SULFUR[*]

K S Roberts, W L Hodge, B. M Johnson, D Schneider[+]
W J Braithwaite, L E Smith and C F Moore
Dept of Physics, Univ of Texas , Austin Texas 78712 USA
[+]Hahn-Meitner-Institut, West Berlin, Germany
[*]Supported in part by the R A Welch Foundation, A F O S R
and E R D A

In the absence of external electric and magnetic fields an atom
in a specific state of total J, L and S : $^{2S+1}(L)_J$ is degenerate
with respect to the magnetic quantum number M_J. The number of different M_J sublevels is 2J+1 and is called the statistical weight
of the level. The statistical weight of a term $^{2S+1}(L)$ with all
possible J is (2S+1)(2L+1). The calculations of transition probabilities for the decay of atoms excited by energetic projectiles
in the absence of external electric and magnetic fields assume that
the different terms are populated statistically (Bhalla 1975, Chen
and Craseman 1975, Matthews et al 1975). Recently, measurements
of atoms excited by energetic particle bombardment have indicated
that this assumption is valid in some cases (Fortner 1974), and
false in others (Bhalla et al 1974).

In a recent paper (Fortner et al 1975) the L X-rays from 40
MeV $Cl^{n+} \rightarrow Cl_2$ collisions were reported. The lines observed were attributed to $1s^22s^22p^53s^1$ $(^{1,3}P) \rightarrow 1s^22s^22p^6$ (^1S) and $1s^22s^22p^43s^1$
$(^2S^2D^2,^4P) \rightarrow 1s^22s^22p^5(^2P)$ L x-ray transitions. Their initial states
are metastable with respect to Auger decay, therefore they have a
fluorescence yield of unity. The $^2P \rightarrow ^4P$ transitions are also metastable with respect to x-ray transitions in pure LS coupling,
however their lifetimes are still shorter tnan the competing metastable Auger transition (Pegg et al 1973). If the initial states
are statistically populated the relative intensities within the
configuration $2p^43s^1$ $(^4P_{5/2 \, 3/2 \, 1/2}$ $^2P_{3/2 \, 1/2};$ $^2D_{5/2 \, 3/2};^2S_{1/2})$
would be 18/10/2. Previously reported data (Fortner et al 1975)
indicates the multiplets of these configurations are statistically
populated in chlorine plus chlorine collisions.

This letter reports further L x-ray measurements for chlorine and a measurement for sulphur. The chlorine data do not substantiate the usual assumption of statistical population within a given multiplet; however, the sulphur data are in essential agreement.

A 29.4 MeV $^{35}Cl^{4+}$ beam from a model EN Tandem Van de Graaff accelerator bombarded a Ne gas target maintained at a pressure of 30 mTorr (lower and higher pressure did not alter the relative intensities). In a second measurement, a 500 keV Ar^+ beam from a model JN Van de Graaff accelerator impinged on a H_2S gas target. The x-rays were analyzed with a curved lead steriate crystal (2d=100.4Å) spectrometer using a flow proportional counter for the detector. The spectrometer was controlled by a PDP-7 computer and the data were normalized by the accumulated charge of the collected beam. Further experimental details can be found in reference (Matthews et al 1973).

Table 1 shows the energies of the different transitions, the relative intensities to the $^1S \rightarrow ^{1,3}P$ transition and the relative population within a configuration for the reactions 29 MeV $Cl^{4+} \rightarrow Ne$; 40 MeV $Cl^{n+} \rightarrow Cl_2$; and the reaction 500 keV $Ar^+ \rightarrow H_2S$. Figure 1 displays the data, The multiplets within the $2p^{-2}3s$ configuration are the same transition as observed in the $Cl \rightarrow Cl_2$ data (Fortner et al 1975), however the transition labelled $2p^{-3}3s$ may possibly be due to other L x-ray transitions. A comparison of peak intensities indicates the $Cl \rightarrow Cl_2$ collision populates the multiplets statistically while the collision $Cl \rightarrow Ne$ does not (with reservations for the $2p^{-3}3s$ configuration). Within the $2p^{-2}3s$ configuration, the intensity of the P initial state is reduced in the $Cl \rightarrow Ne$ reaction. The reason for this difference is not clear. Obviously, there is substantial difficulty explaining the population of the $2p^{-3}3s$ configuration at the higher energy. This is compounded in that the multiplets within this configuration overlap. In contrast to chlorine, the x-ray data for the decay of the $2p^{-2}3s$ configuration in sulphur essentially agree with the statistical population. It is likely that the collision breaks up the H_2S and that chemical effects do not play an important role for these observed decays in multiply ionized sulphur.

In conclusion, these experiments indicate that in some cases processes are involved which selectively populate members of multiplets within a given configuration in atom-ion reactions. These processes are seen to be large and dominate the interaction and thus they are fundamentally important to the understanding of ion-atom reactions.

Table 1

Energy (eV)	Configuration	Relative Intensities				
		A	B	C	B†	C†
Chlorine						
200		(0.35)			(0.13)	
210	$\{p^6-2p^53s\}$ $^1S-^{1,3}P$ $\{2p^5-2p^43s\}$	---	1.00	---	1.00	---
225		---	0.05	---	0.03	---
235	$^2P-^4,^2P$	18	0.72	11.7	0.57	18.0
242	$^2P-^2D$	10	0.70	11.4	0.32	10.1
250	$^2P-^2S$	2	0.42	6.9	0.06	1.9
Sulphur						
155			0.01			
163			0.08			
171	$\{2p^6-2p^53s\}$ $^1S-^{1,3}P$ $\{2p^5-2p^43s\}$		1.00			
185			---			
195	$^2P-^4,^2P$	18	0.41	15.4		
202	$^2P-^2D$	10	0.31	11.6		
208	$^2P-^2S$	2	0.08	3.0		

A, relative intensities within a single configuration assuming
statistical population.
B, relative intensities of peaks in spectra normalized to that of
the $^1S-^{1,3}P$ transition.
C, relative intensities within a single configuration normalized
to statical weight.
†, results from Fortner,Matthews and Scofield.

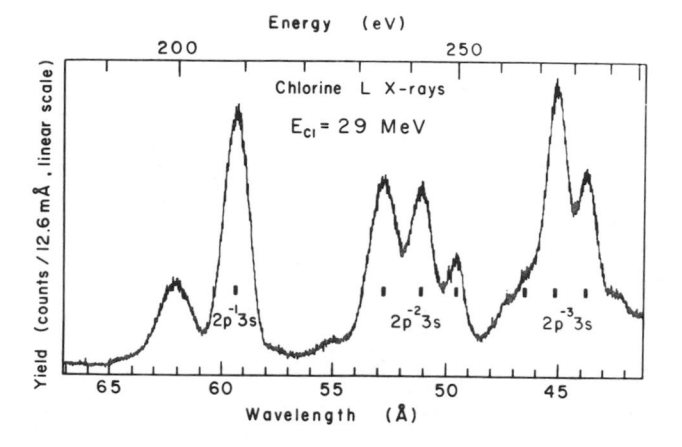

Figure 1a. Chlorine L x-ray spectra induced by a 29.4 MeV chlorine
(+4) bombardment of a neon gas target under single collision
conditions. Markers indicate the positions of multiplets for
charge state 7, 8, and 9, as determined in reference (Fortner et.
al. 1975).

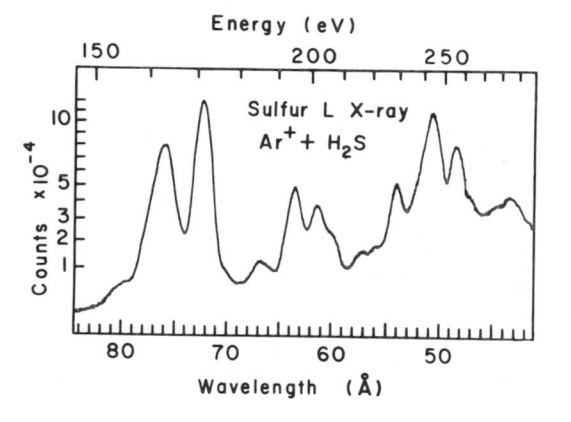

Figure 1b.   Sulphur L x-ray spectra induced by a 400 keV Ar+
        beam bombardment of a hydrogen sulphide gas target under
single collision conditions. X-ray yield above 220eV is due
in part to Ar beam decay.

R E F E R E N C E S

1   Bhalla C P,1975,Proc. 9th Conf.on Physics of Electronic and
     Atomic Collisions,Seattle Abstract 943
2   Bhalla C P,Matthews D L,and Moore,1974,Phys. Lett.46A,336
3   Chen M H and Craseman B,1975,Proc. 9th Int. Conf. on Physics
     of Electronic and Atomic Collisions,Seattle Abstract 937
4   Fortner R J,1974,Phys. Rev.A 10
5   Fortner R J,Matthews D L,and Scofield J H,1975,Phys. Lett.53A,
     336
6   Matthews D L,Braithwaite W J,Wolter Hermann H and Moore C Fred,
     Phys.Rev.,A8,1397 (1973).
7   Matthews D L,Fortner R J,Chen M H and Craseman B,1975,Proc.
     9th Int. Conf. on Physics of Electronic and Atomic Collisions,
     Seattle Abstract 941
8   Pegg D,Sellin I,Peterson R,Mowat J R,Smith W W,Brown M D and
     MacDonald J R,1973,Phys. Rev. A8,1350

*Supported in part by the Robert A. Welch Foundation, AFOSR and ERDA.

COLLISIONAL QUENCHING OF METASTABLE X-RAY EMITTING STATES IN A FAST BEAM OF He-LIKE FLUORINE[*]

D. L. Matthews and R. J. Fortner

Lawrence Livermore Laboratory

Livermore CA 94550

ABSTRACT

High resolution x-ray spectral measurements are used to determine the relative intensity of He-like fluorine x-ray transitions, $2\ ^3P_1 \to 1\ ^1S_0$ and $2\ ^1P_1 \to\ ^1S_0$, produced fluorine collisions with Ne, Ar, and Kr gas targets at various pressures and in thin carbon foils and a thick carbon slab. The relative intensities are observed to be strong functions of both target density and incident charge Z. These effects are attributed to strong collisional quenching of both initial states by subsequent large impact parameter collisions. The data permit extraction for the first time of the total quenching cross sections (σ_Q) for fast fluorine ions in both states. A strong enhancement of the relative intensity of the $2\ ^3P_1$ is observed for F^{7+} projectiles. This strong enhancement is attributed to selective excitation of metastable states in the beam, i.e. $1s2s\ ^3S_1$, into the $1s2p\ ^3P_1$ state. Finally, the data for foil and solid targets are used to obtain new information on the excitation states of ions moving in solids. High resolution measurements of the radiative electron capture (REC) peak are reported and analyzed for the first time.

[*]Work performed under the auspices of the U. S. Energy Research and Development Administration.

The study of collisional quenching of atomic states in ion-
atom collisions has received considerable attention. However,
the extension of this type study to metastable x-ray emitting
states in high velocity helium-like ions (i.e. Z > 2) has received
little or no attention. In the following we will show that the
cross section for collisional quenching of metastable x-ray emitting states for high velocity ion-atom collisions is appreciable.
In fact, we demonstrate that the effects of collisional quenching
can dominate the spectral intensities of x-ray lines formed under
single inner shell collision conditions.   First, measurements of
gas target spectra will be reported for different gas pressures
and the collisional quenching cross section, σ_Q, will be determined
for the 1s2p $^3P_1$ state in fluorine. Then the data for thin carbon
foils will be used to determine σ_Q for the 1s2p $^1P_1$ state in fluorine. New information on the "dynamic fluorescence yield" for individual x-ray transitions are obtained for ions moving in solids.
Finally, high resolution measurements of the radiative electron
capture (REC) x-rays of conduction electrons are also obtained and
analyzed.

Fig. 1 demonstrates the principal effects observed in this
measurement.   The transitions 1s2p $(2\ ^3P_1) \rightarrow 1s^2\ (1\ ^1S_0)$ and

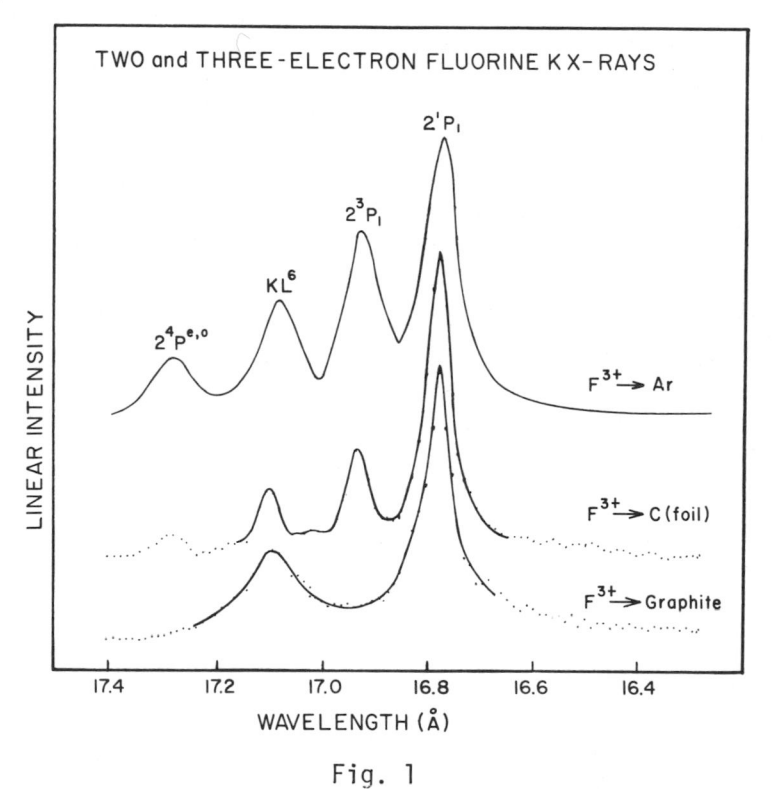

Fig. 1

$1s2p$ $(2\ ^{1}P_{1}) \rightarrow 1s^{2}$ $(1\ ^{1}S_{0})$ from the decay of two electron fluorine ions excited in gas, foil and solid targets are shown. The $^{3}P_{1}/^{1}P_{1}$ variation with target gas density in F^{5+} + Ne, Ar, and Kr colli- sions is shown in Fig. 2. In going from 10 to 200 microns, a mono- tonic decrease in the ratio is observed. To explain this variation, a theoretical formulation for the intensity of a given atomic state in a fast moving ion beam is required.

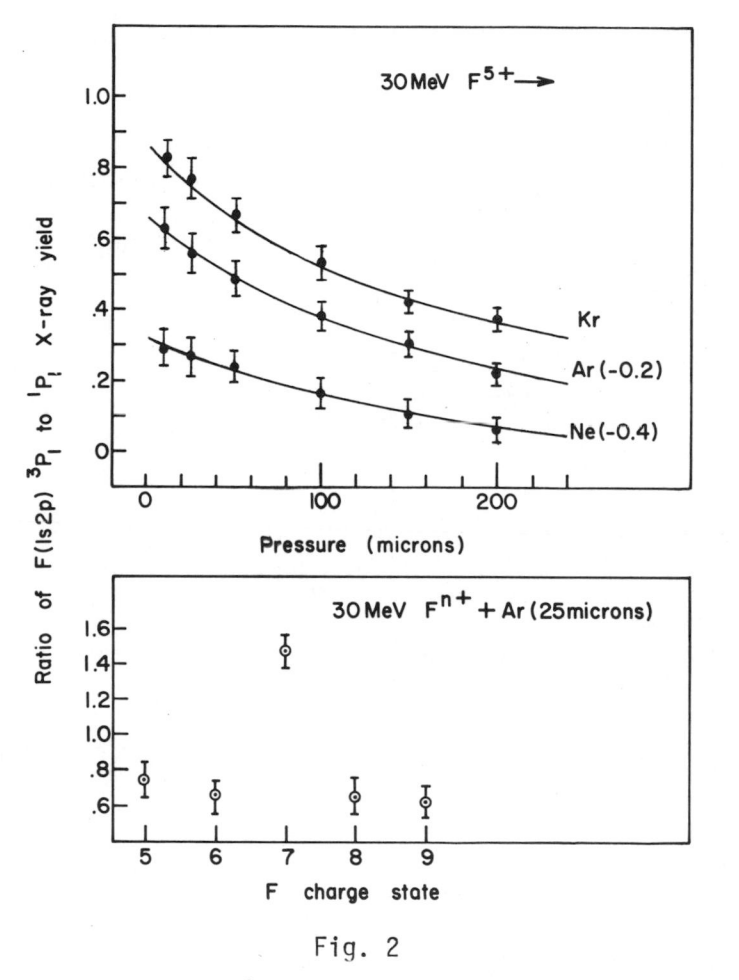

Fig. 2

The number density $n(\mathrm{cm}^{-3})$ of fast excited ions in state j at a given distance x from the entrance to the target satisfies the dif- ferential equation

$$\frac{dn_{j}}{dx} = \frac{-n_{j}}{v\tau} - n_{j}N\sigma_{Q} + \frac{\sigma_{0}N(I-n_{j}v)}{v} \tag{1}$$

where v is the ion velocity, I is the beam flux (cm^{-2}-sec^{-1}), σ_0 is the cross section for producing j, N is the target gas density, τ is the atomic lifetime of state j and σ_Q is the total effective cross section for collisional quenching or extinction of excited state j. This σ_Q may be the result of several loss terms due to collisional excitation/de-excitation out of state j. If at x = 0, n_j = 0, then Eq. 1 has the solution

$$n_j = \frac{\sigma_0 IN}{v\beta} [1 - e^{-\beta x}] \tag{2}$$

where $\beta = [N(\sigma_Q + \sigma_0) + 1/v\tau]$. Consider for an example a 30 MeV fluorine beam and the 1s2s $^3P_1$ state (τ = 0.54 x 10^{-9} s). In a solid 1/$\beta \sim$ 10 Å and in a gas 1/β < 1.1 cm. In both cases equilibrium is easily obtainable, thus

$$n_j = \frac{\sigma_0 IN}{v\beta} \tag{3}$$

In a gas the number of photons emitted is

$$R_j = \int_0^L \frac{\sigma_0 IN}{1 + N\sigma' v\tau} \, dx = \frac{\sigma_0 INL}{1 + N\sigma' v\tau} \tag{4}$$

For prompt transitions such as from the 1s2p 2 $^1P_1$ state (τ = 1.8 x 10^{-13} s) Nσ'vτ << 1 for meaningful σ' and gas densities of interest. Eq. 4 then becomes simply $R_j(^1P_1) = \sigma_0 INL$. For the 2 $^3P_1$ state Nσ'vτ is not necessarily small compared to 1. Under these circumstances we can get an expression for the relative 2 $^3P_1$ to 2 $^1P_1$ x-ray yields which is given by

$$R' = \frac{R_0}{1 + N\sigma' v\tau} \tag{5}$$

where R_0 is the ratio of x-ray production cross sections for the $^3P_1$ to $^1P_1$ states. The expression for R' represents a simple formula for predicting the reduction in $^3P_1/^1P_1$ relative intensities with increasing gas density, and can be fit to experiment thus determining R_0 and σ'. In the analysis R_0 was determined by extrapolating the data to near zero gas pressure while σ' was determined by the best fits to the data shown in Fig. 2. The fitted R' values are shown in Fig. 2 by the solid line. Over a pressure change of a factor of 20 we get good agreement with experiment. The values we obtain for $\sigma' = \sigma_0 + \sigma_Q$ are much too large in comparison with measured inner shell vacancy production cross sections thus $\sigma' \simeq \sigma_Q$. The values obtained for σ_Q and R_0 are shown in Table 1.

TABLE 1

Target	$R_0 (\pm 0.10)$	$\sigma_0 (10^{-16} \text{ cm}^2)$	$\sigma(x \ 10^{-16} \text{ cm}^2)$
Ne (Z = 10)	0.71	0.7 ± 0.2	0.9
Ar (Z = 18)	0.86	1.5 ± 0.4	2.4
Kr (Z = 36)	0.87	2.0 ± 0.6	3.5

Another interesting aspect of these data is the dependence of the relative intensity of the $^3P_1$ and $^1P_1$ x-ray lines on the incident charge state of the fluorine projectile as can be seen in Fig. 2b. For all charge states except 7^+ the relative yields are constant. For charge state 7^+, the $^3P_1$ line is enhanced by almost a factor of two. For 7^+ projectiles, a predominant mode for production of the $^3P_1$ and $^1P_1$ lines is electron excitation since the number of electrons in initial and final states is the same. In this type of collision the long range coulomb interaction is expected to populate selectively into excited states not requiring a spin change (i.e. spin flip). At first this seems to be contradicted by our data which indicates that the $^3P_1$ state is selectively populated from a ground state $1s^2 \ ^1S_0$. However, we attribute the selective enhancement of the $^3P_1$ state to be due to excitation of the metastable $1s2s \ ^3S_1$ states and not the $1s^2 \ ^1S_0$ state. The presence of metastable $1s2s \ ^3P_1$ states in the incident F^{7+} beam is a result of their long mean free path $v\tau = 2.3 \times 10^5$ cm as compared to 1.4×10^3 cm path length from post stripper foil to target gas cell.

The quenching cross section for the $1s2p \ ^1P_1$ state can be obtained from the thin foil data. Before starting this discussion, however, some general conclusions obtained from the total x-ray yield from these targets should be discussed. The x-ray spectra were measured for several foil thicknesses ranging from 5 to 200 $\mu g/cm^2$. Except for small differences in going from 5 to 10 $\mu g/cm^2$ in the hydrogen-like lines the x-ray yields and relative intensities were constant--independent of foil thickness. From these results we can conclude that nearly all the x-ray emission observed when viewing beam-foil interactions comes from outside the foil. This general conclusion appears true for all the normal x-ray lines. The REC peak (see below) was observed for only the thickest foils indicating that it is a strong function of foil thickness. It appears that these x-rays come from transitions taking place predominantly in the foil consistent with the REC mechanism.

Comparisons of the relative intensity of individual x-ray
lines in foil and solid targets is instructive. Consider for example the normal 2p → 1s transitions for the helium- and lithium-
like lines. These are compared for gas, foil, and thick targets
in Fig. 1. In this figure we can have 4 different x-ray transitions in order of increasing energies, the ^4P transitions which
includes transitions from the 1s2s2p and 1s2p^2 configurations and
J components of 1/2, 3/2, and 5/2 (these states have lifetimes
ranging from 1 to 12 ns), the next line comes predominantly from
the ^2P states again from the 1s2s2p and 1s2p^2 configurations
(these states have lifetimes of approximately 10^{-13} s), the third
line is the $^3P_1$ state from the 1s2p configuration (0.54 ns lifetime), and the last line is the $^1P_1$ state from the 1s2p initial
configuration (1.8 x 10^{-13} s lifetime). In the gas target spectra
we see that the two lines in the same series are about equally
populated. In the foil produced spectra the relative intensities
are quite different. The longer lived components in the spectra
appear considerably reduced. For solid spectra the longer lived
components are reduced even further in fact within the limits of
our detection system they are completely gone.

Inside a solid the probability that a state will decay giving
an x-ray is reduced. Consider, for example, the helium-like states,
$^3P_1$ and $^1P_1$ which have essentially unit atomic fluorescence yield.
Inside the foil the probability that a state will emit an x-ray
is given by

$$\frac{1/\tau_R}{1/\tau_R + N\sigma_Q v} \tag{6}$$

where τ_R is the intrinsic radiative lifetime. For the $^3P_1$ state
in helium-like fluorine, $N\sigma_Q v \approx 10^{15}$ s^{-1}, thus the probability
that this state emits an x-ray in the solid, i.e. dynamic fluorescence yield, is $\approx 10^{-6}$. σ_Q for the $^1P_1$ helium-like state can be
estimated from the relative intensities of the $^3P_1$ and $^1P_1$ lines
in the foil produced spectra. σ_0 has been measured to be approximately the same for both the $^3P_1$ and $^1P_1$ lines in gas target measurements. Since essentially all the x-rays are emitted outside
the foil and both lines have unit fluorescence yields the relative
intensities are obtained from Eq. 2 as

$$\frac{Y(^3P_1)}{Y(^1P_1)} = \frac{N\sigma_0(^1P_1) + N\sigma_0(^1P_1) + 1/v\tau(^1P_1)}{N\sigma_Q(^3P_1)} \tag{7}$$

Using estimates of σ_0 of 10^{-18} cm^2 and $\sigma_Q(^3P_1)$ of 10^{-16} cm^2
and a measured $Y(^3P_1)/Y(^1P_1)$ of $\sim$0.1 we obtain an estimated
$\sigma_Q(^1P_1)$ of 8 x 10^{-18} cm^2. Thus, the probability that this state

will x-ray decay in the solid is ∿7%. These observations are consistent with the solid target data in Fig. 1. In the solid target
spectra the $^3P_1$ state is not observed and the small yield of x-
rays obtained from the solid graphite target (the yield of characteristic x-rays from ions impinging on a thick target slab was
only 20% of that observed for a 5 µg/cm^2 foil) is then due to the
small probability for x-ray decay in the solid.

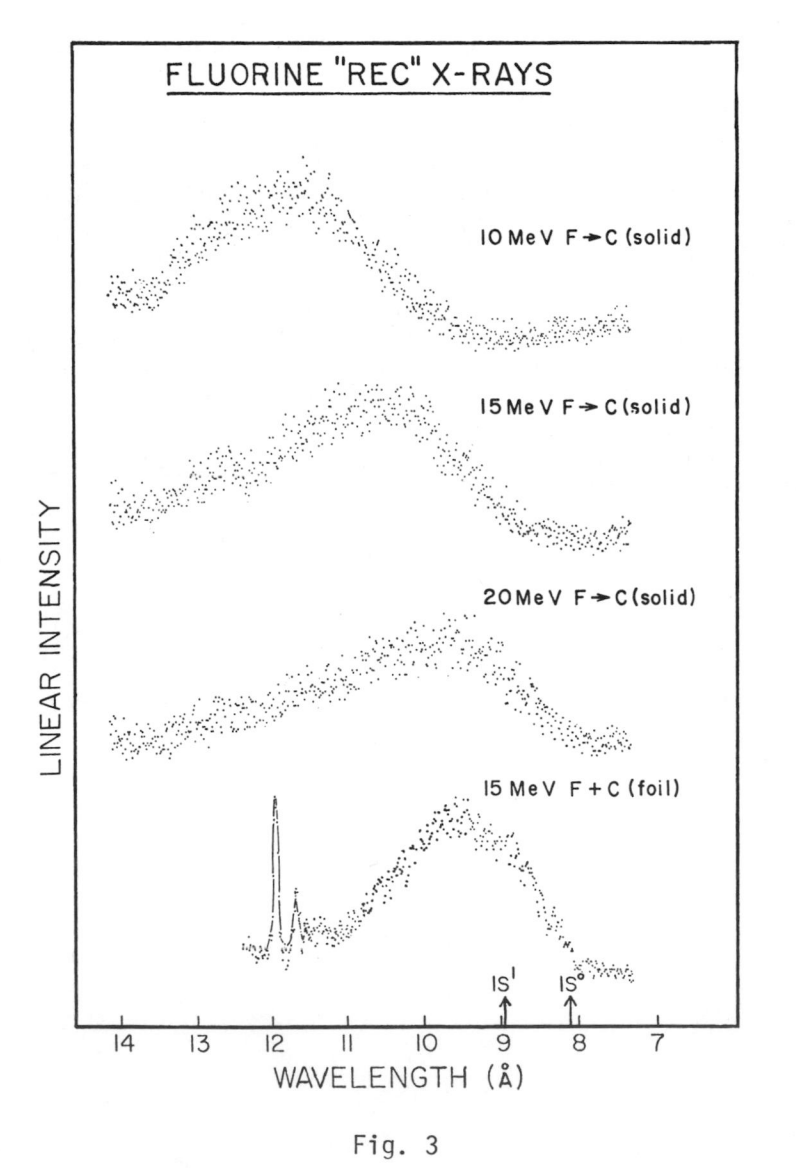

Fig. 3

In Fig. 3 the REC peaks for various energy fluorine ions incident on a solid and 15 MeV ions incident on a 100 μg/cm^2 foil are shown. The existence of the large REC peak indicates that there are substantial amounts of zero electron and one electron ground state (1s) atoms moving in the solid. For these types of atoms REC is the predominant mechanism for filling of the K vacancies since there are no outer shell electrons present. The centroid energy of the REC peak moves as the incident energy changes. The value of the centroid energy is still slightly lower than one would expect, consistent with other observations and probably due to the electronic screening of the ion due to the presence of electrons in the solid. For the solid the REC peak is skewed to lower energies. This is due to REC in ions which have lost considerable energy in the solid. The measured half width for the REC peak in the 100 μg/cm^2 foil is 260 eV. The contribution to the width of this line due to energy loss in the foil is negligible (<20 eV). The width of the REC peak assuming a Fermi distribution is $4 \sqrt{T_R T_f}$ where T_R is the kinetic energy of the captured electron relative to incident ion and T_f is the width of the conduction band. Our measured width was inconsistent with the assumption of a single REC peak and a reasonable Fermi energy, T_f. The REC line was then analyzed by assuming it was composed of two components of equal width involving capture into both the 1s^0 and 1s^1 cores. The energy separation was assumed to be the theoretical value of 120 eV. Assuming Gaussian line shapes this resulted in a REC width of 180 ± 20 eV implying a value of T_f of 4.7 ± 1.1 eV.

DELAYED X-RAY EMISSION FOLLOWING

BEAM FOIL EXCITATION*

Forrest Hopkins, Jonathan Sokolov and Peter von Brentano[†]

Department of Physics
State University of New York
Stony Brook, New York 11794

The recent observation of delayed 1s-2p hydrogen-like oxygen[1] and $1s^2(^1S_0)-1s2p(^1p)$ helium-like fluorine[2] K x rays in a time interval from ∿0.5 nanoseconds to several nanoseconds after passage of ∿1 MeV/amu ions through thin carbon foils implies cascading into the short-lived states from states of high principal quantum number n with much longer lifetimes. We consider here various possibilities of high n state population in a hydrogenic ion and show that it is possible to explain the experimentally observed $t^{-3/2}$ power law dependence on time of the delayed radiation in terms of equally simple power law dependences on n for the primary population.

We first limit ourselves to two specific populations, which yield simple cascade schemes, and obtain computational solutions: s states only or yrast (n,n-1) states only. We assume an equal $(m_\ell=0)$ primary population of the form $P_n=P_o n^{-a}$, with a constant P_o, both for the s-state and yrast case. In addition we will cite approximate results for statistical $(2\ell+1)$ weighting of all subshells.

For pure s states, cascading to the 2p shell is taken to be the direct branches,

$$I^s_{2p}(t) = \sum_{n=3}^{n_f} \lambda_{no} P_{no} e^{-\lambda_{no} t} \tag{1}$$

where λ_{no} is the total decay rate for the nth s state and n_f is an appropriate cutoff value. A calculation of the El branching ratios with a program by Lennard[3] indicates that indirect cascading is negligible. Population of the yrast states leads to sequential cascading, e.g., 5g→4f→3d→2p. The decay from the kth level due to

an initial population of level 1 is given by[4]

$$I_k(t) = A_{k1} e^{-\lambda_1 t} + A_{k2} e^{-\lambda_2 t} + \ldots + A_{kk} e^{-\lambda_k t} \tag{2}$$

where $A_{k,i} = A_{k-1,i} \dfrac{\lambda_{k-1}}{\lambda_k - \lambda_i}$ and $I_k(0) = \delta_{1k} P_1.$

The total yield is then calculated by summing the contributions from various states using the primary population P_n given above.

A useful sidelight in this work has been the determination of a simple, yet reasonably accurate expression for the decay constant $\lambda_{n\ell}$, i.e., the lifetime $\tau_{n\ell} = (\lambda_{n\ell})^{-1}$, which holds over the region of interest. Such approximate expressions have long been of interest. Bethe and Salpeter pointed out that the average lifetime τ_{av} of all ℓ states of a given n increases roughly as $n^{4.5}$. More recently, Curtis noted[6] that the lifetimes of high n yrast states increase as n^5. We find that the calculated lifetimes of the states over a wide range of n and ℓ, excluding s states, can be fit surprisingly well by the expression $\tau_{n\ell} = \tau_0 \, n^b \ell^c$, where τ_0, b and c are suitably chosen constants.

From a calculation of the lifetimes of the levels (n,ℓ) in hydrogenic oxygen using a program due to Lennard[3] we find the following approximate formulas,

$$\tau(n,\ell) = (Z/8)^{-4} \cdot 10^{-4} \cdot (n-1/2)^{2.74} \cdot \ell^{1.75} \text{sec}; \quad \ell \geq 1 \ \ 2 \leq n \leq 20; \ (<27\%) \tag{3}$$

$$\tau(n,0) = (Z/8)^{-4} \cdot 7.8 \cdot 10^{-4} \cdot n^{2.8} \text{sec}; \quad 6 \leq n \leq 20; \ (<10\%) \tag{4}$$

where (3) and (4) have an accuracy of better than 27% and 10% respectively in the specified ranges. We note that the formula for $n \leq 20$ does not yet have the asymptotic form n^5 found by Curtis for the lifetime of yrast, i.e., (n,n-1) levels.

Summing eqs. (1) and (2) over the intervals of n specified above using the above approximations for $\tau_{n\ell}$, we obtain various curves for the time dependence of the intensity of the delayed K x rays for different values of the population exponent "a". These are displayed in Fig. 1 along with the data of oxygen[1] and fluorine.[2] The curves and the oxygen points have been normalized to 1.0 at 0.5 nsec and by a $t^{-3/2}$ dependence elsewhere, with the fluorine points normalized at $\sim$0.8 nsec. The n dependences of 4.0 for the yrast states and 2.0 for s states yield reasonably good agreement with the data. The curves for statistical population of the yrast band are very similar to those in Fig. 1, with the value of a

increased by 1.0 in each case. It should be mentioned that similar conclusions have been arrived at independently by Schectman[7].

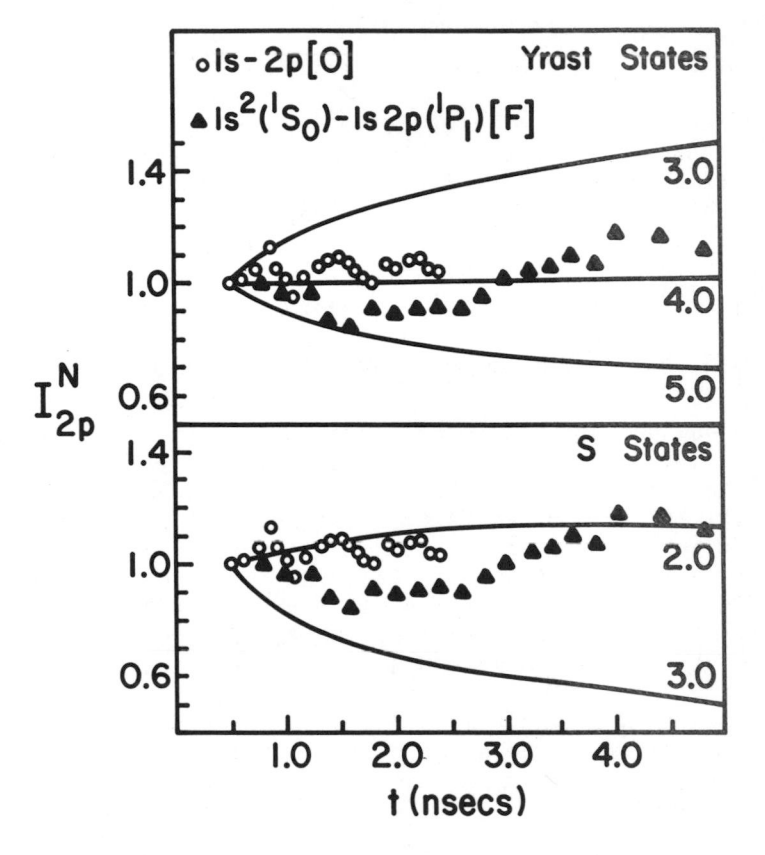

Fig. 1 Curves for delayed K x ray emission for various values of the population exponent a, all normalized by a $t^{-3/2}$ dependence, compared to data. The primary population used is $P_n = P_o n^{-a}$.

It has been reported elsewhere[8] that the exact solutions in Fig. 1 are approximated by the expressions,

$$I_{2p}(t) = \text{const. } t^{-\{1 + \frac{a-1}{d}\}} \tag{5}$$

where d=2.8 for s states and d=4.5 for yrast states. Further,

including yrast like cascades from other subshells in an approximate
way one obtains[8]

$$I_{2p}(t) = \text{const. } t^{-\{1+\frac{a-1}{4.5}\}} \tag{6}$$

which equations hold for arbitrary foil and ion.

The determination of a specific population mechanism for beam
foil excitation is difficult. One possible procedure is to use
the competing cross section approach from Betz et al.[9] Assuming
that rearrangement at the surface is negligible, the relative
probability at equilibrium of finding a hydrogenic ion in the state
(n,ℓ) is given by

$$P_{n\ell} = \frac{\sigma_c^{n\ell}}{\sigma_c^{n\ell} + \sigma_v^{n\ell}} \tag{7}$$

where $\sigma_c^{n\ell}$ and $\sigma_v^{n\ell}$ are the total cross sections for population and
depopulation, respectively, of the level in the solid. For our
purposes we take $\sigma_c^{n\ell}$ to be the direct charge exchange into the subshell (n,ℓ) and $\sigma_v^{n\ell}$ to be the ionization of an electron in that
level. We neglect rearrangement due to excitation to bound states
and also due to transitions of electrons into and out of the subshell, since the collision times are much shorter than the lifetimes
involved. In the limit of high n, the Binary Encounter Approxima-
tion theory[10] yields a simple n dependence for ionization, $\sigma_v^{n\ell} = \sigma_o n^2$
where σ_o is a constant. According to available data[9], a more
correct form is a slightly less rapidly increasing cross section as
$n^2/\ln n$, analogous to the asymptotic $\ln E/E$ dependence predicted by
the Plane Wave Born Approximation. A theory for capture into
individual subshells of high n levels is not presently available,
apart from a prediction of $\sigma_c^{no} \alpha n^{-3}$ for s states[11]. Thus for s states
the n dependence can be given as

$$P_{no} \approx P_o n^{-5}/\ln n \approx P_o n^{-4.5}; 4 \leq n \leq 33 \tag{8}$$

This finding together with the results of Fig. 1, which imply
a n^{-2} law for a pure s- state population, seem to rule out a pure
s- state population. Thus the above experiments seem to imply that
a high ℓ state population is important insofar as cascading into
the 2p shell is concerned. However, the theory is far too crude
and incomplete to make final conclusions, other than that power law
dependences for the population appear quite reasonable. It is
interesting to note in this connection that previous workers[12] have
observed very strong yrast transitions in the optical emission
spectra from few-electron oxygen ions for similar velocities and
time intervals.

References

*Supported in part by the National Science Foundation.
†On leave from the Institute fur Kernphysik, Universitat zu Koln, Koln, Germany.

1. W. J. Braithwaite, D. L. Matthews and C. F. Moore, Phys. Rev. A11, 1267 (1975).

2. P. Richard, Phys. Letters A45, 13 (1973).

3. W. N. Lennard, private communication.

4. E. Segre, Experimental Nucl. Phys., (John Wiley and Sons, New York, 1959), p.3.

5. H. A. Bethe and E. E. Salpeter, Quantum Mechanics of One and Two Electron Atoms, (Academic, New York, 1957), p. 269.

6. L. J. Curtis, J. Opt. Soc. Am. 63, 105 (1973).

7. R. M. Schectman , Phys. Rev. A (to be published).

8. F. Hopkins and P. von Brentano, submitted for publication.

9. H. D. Betz, F. Bell, G. Panke, G. Kalkoffen, M. Welz and D. Evers, Phys. Rev. Letters 33, 807 (1974).

10. J. D. Garcia, E. Gerjuoy and J. E. Welker, Phys. Rev. 165, 66 (1968).

11. J. van den Bos and F. J. de Heer, Physica (Utr.) 34, 333 (1967).

12. R. Hallin, J. Lindskog, A. Marelius, J. Pihl and R. Sjodin, Physica Scripta 8, 209 (1973).

MEASUREMENT OF X-RAYS EMITTED FROM PROJECTILES MOVING IN SOLID TARGETS[*]

R. J. Fortner and D. L. Matthews
Lawrence Livermore Laboratory
Livermore, California   94550

L. C. Feldman
Bell Telephone Laboratories
Murray Hill, New Jersey   07974

J. D. Garcia and H. Oona
University of Arizona
Tucson, Arizona   85721

ABSTRACT

In this paper the results of three separate experiments all
dealing with the production of x-rays in projectiles moving in
solids will be discussed. The first experiment[1] deals with the
measurement of line widths of x-rays emitted from projectiles
moving in solid targets. The effects of collisional   broadening
of x-rays is found to dominate the line widths giving greater
than an order of magnitude increase in the measured line widths.
The second experiment[2] studies "solid target effects" in producing
non-binomial distributions of characteristic K x-ray spectra in
heavy ion-atom collisions. The third experiment[3] studies aluminum
K x-ray production in $Ar^+ \rightarrow Al$ collisions in very thin aluminum foils
as a function of foil thickness. Parameterization of the observed
non-linear dependence enables us to measure the lifetime of the
argon 2p vacancy and total ionization cross sections for the argon
L-shell in $Ar \rightarrow Al$ collisions.

[*]Work performed under the auspices of the United States Energy
Research and Development Administration.

1. COLLISIONAL BROADENING OF X-RAYS

In Fig. 1b and 1c the neon K x-ray spectra for 3 types of collisions, 1.3 keV electrons incident on a neon gas target, and 90 keV
neon ions incident on a neon gas target and a solid copper target.
The gas target (Fig. 1c) spectra exhibit clean well resolved peaks
and the origins of these peaks have been discussed extensively
in the literature. Basically each line corresponds to an initial
state involving one K-shell vacancy and fixed ionization in the L-
shell. The peaks are designated as such in the figure where KL^n
corresponds to the x-ray transition from an initial state having
one K-shell vacancy and n L-shell vacancies. However, these well
resolved discrete peaks are not observed for the neon K x-ray
spectra emanating from ions moving in solid copper targets (Fig. 1b).
The peak of this spectra occurs in the region of the normal KL^0 peak;
however, the peak extends in energy over a very broad energy range
with essentially no structure detectable in the spectra. This
spectra is skewed toward the high energy side, probably indicating
the existence of states of excitation higher than KL^0. On the low
energy side of KL^0 the spectra extends to lower energies indicating
that the individual lines are considerably broadened.

In Fig. 2 the boron K x-ray spectra for 2 types of boron-carbon
collisions are presented. The first spectra is for a methane gas
target. Basically the x-ray lines are from boron coming primarily
from electron transitions in one, two, and three electron atoms.
The second spectra is for a solid diamond target. The most obvious
difference is the large carbon K line which dominates the solid target spectra but is not observed in a gas target. This peak occurs
in the solid targets because of recoiling carbon atoms and subsequent C-C collisions. A more important difference with respect to
this paper is the x-ray transitions emanating from the boron projectile. For the solid target only two broad continuous x-ray bands
in the region where boron K x-ray transitions are expected are
observed. The identification of the two bands is straightforward.
The low energy band, centered near 190 eV, is consistent with calculations of the 2p → 1s x-ray transitions in atoms having initial
states involving one K-shell vacancy and one, two or three electrons
in the L-shell. The second higher energy band, centered near 245 eV
on the low energy tail of the carbon K x-ray is consistent with
2p → 1s transitions in atoms having two K-shell vacancies in the
initial state and one, two or three L-shell electrons. The significant point to be extracted from both Fig. 1b and Fig. 2 concerns the
widths of the x-ray lines obtained from the projectiles moving in
solid targets. In both figures the data obtained from gas target
spectra exhibit clearly resolved peaks. However, when the x-ray
spectra are measured using the same instrumental resolution for
ions moving in a solid target all the sharply resolved x-ray structure disappears and only broad x-ray bands are resolved. This

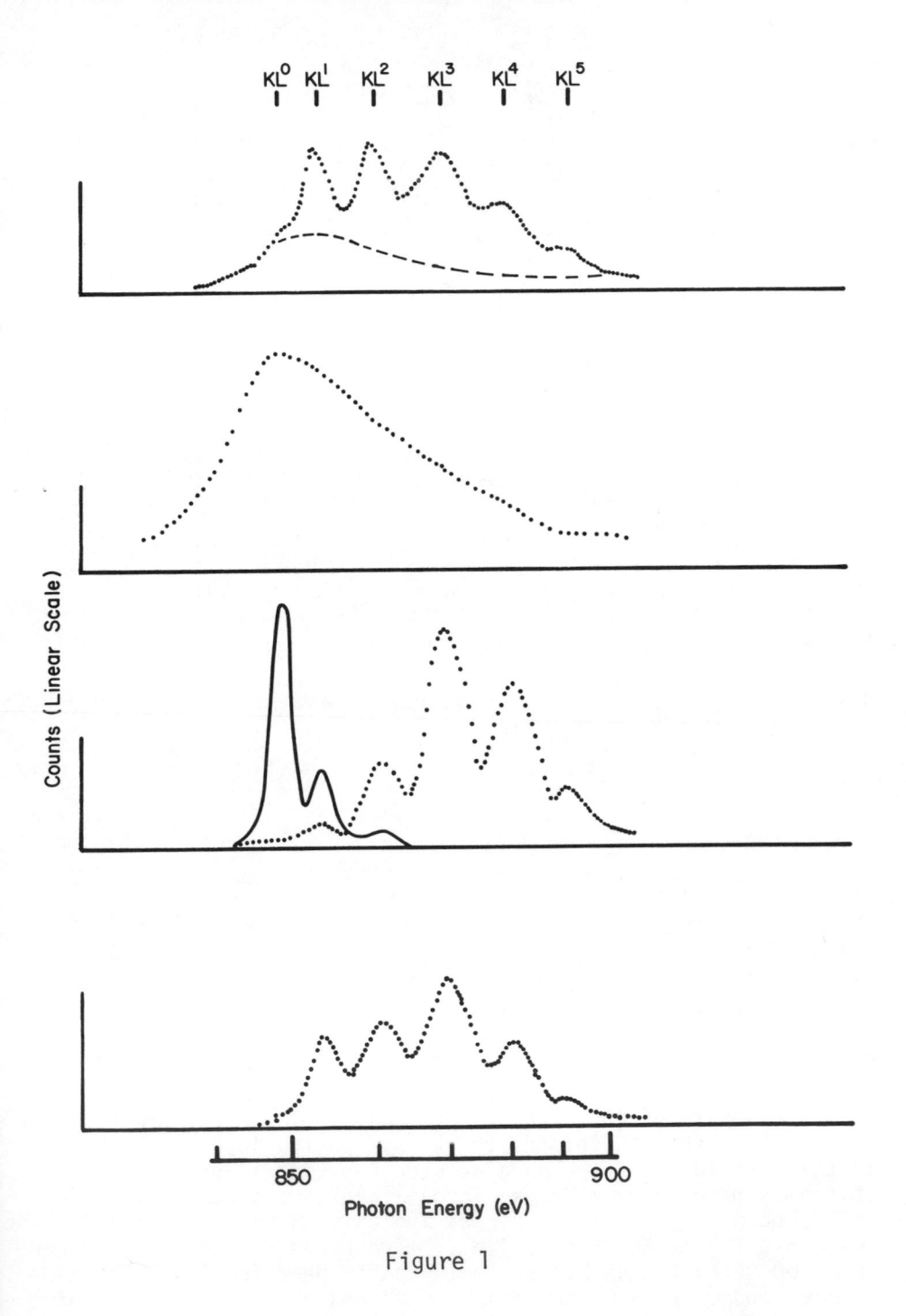

Figure 1

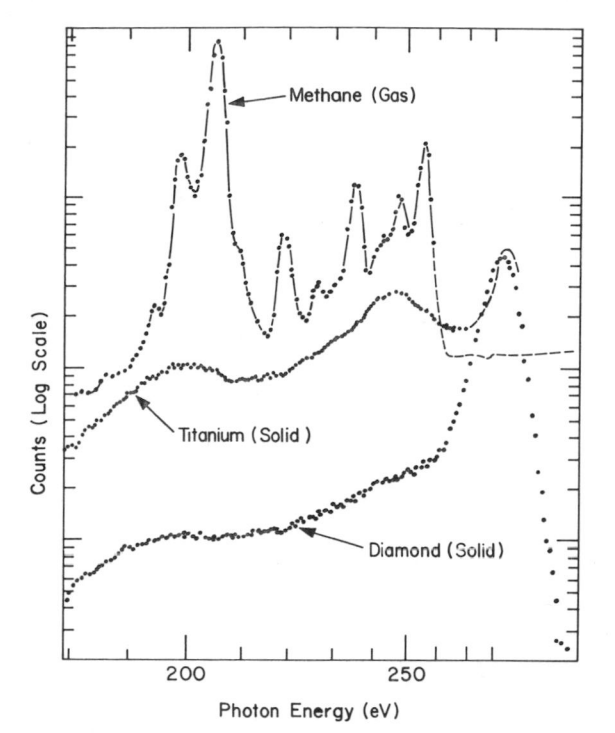

Figure 2

broadening of the x-rays in the solid is the x-ray analog to the collisional broadening of spectral lines observed in dense plasmas. It is the result of a disruption in the radiation field of the inner shell vacancy induced by multiple collisions which take place prior to the filling of the x-ray transition.

An estimate of the size of the collisional x-ray broadening can be obtained by unfolding the x-ray spectra in Fig. 1b and 2. In this unfolding of the data we used Gaussian line shapes and centroid energies as observed in gas target measurements. This unfolding of the $Ne^+ \rightarrow Cu$ spectra indicated a width of 12 ± 3 eV for the peaks KL^0 to KL^4 respectively. The boron K x-ray spectra were unfolded using the double K-shell vacancy peak from Fig. 2b. The spectra indicated a width of 12 ± 5 eV and equal intensities of the K^2L^0 and K^2L^1 x-ray peaks.

2. NON BINOMIAL DISTRIBUTIONS OF HEAVY-ION INDUCED
NEON K X-RAY SPECTRA

In Fig. la the neon K x-ray spectra produced in collisions of neons ions incident on a graphite (carbon) thick target. The spectra was observed only after long bombardment of neon ions and thus is attributed to Ne - Ne collisions taking place in the carbon solid. This spectra is expected to have two components: one from a moving neon projectile, and the second from implanted neon atoms at rest in the graphite. The projectile spectra is broadened as in Fig. lb and the target spectra shows peaks. The dashed line in Fig. la indicates our estimate of the projectile contribution obtained by normalizing the spectra shown in Fig. lb to the low energy part of the spectra in Fig. la. The difference spectra which would be associated with the implanted neon target is shown in Fig. ld. Notice the difference in the two neon-neon spectra for a gas target (lc) and a solid target (ld). The x-ray line intensities from the gas target spectra are binomial in character and for the solid target spectra they are not. We can attribute this difference to the capture of target electrons into excited states when the neon is in the carbon solid. These electrons can auto-ionize filling L-shell vacancies prior to the K x-ray decay which then results in fewer number of L-shell vacancies as observed in the x-ray spectra and a non-binomial distribution.

3. LIFETIME STUDIES OF Ar-2p VACANCIES TRAVELING THROUGH SOLIDS

We report studies on the production of aluminum K x-rays in Ar-Al collisions using varying thicknesses of thin aluminum foils. It has been suggested that the mechanism for aluminum K vacancy production requires two collisions; the first collision produces an argon 2p vacancy in the argon projectile and the second collision transfers this 2p vacancy to the 1s level of the aluminum. The reasons for interest in these studies is that in very thin foils, i.e., thickness less than $v\tau$ where v is the velocity of the ion and τ is the lifetime of the argon L vacancy, the aluminum K yield is not a linear function of thickness and the shape of the curve will depend on the lifetime of the argon 2p vacancy. Thus, parameterization of this data enables us to infer an average lifetime for the argon 2p vacancy traveling through a solid.

The experiments used thin foils of carbon of thickness $\sim$10 μg/ cm^2 with thin evaporated layers of aluminum ranging from 60 Å to 333 Å. In the first experiment the yield of aluminum K x-rays was measured for the argon beam incident, first on the carbon side and then on the aluminum side of the foil. Those yields are displayed in Fig. 3 as a function of energy. The aluminum yield for argon first traveling through carbon is considerably higher since there

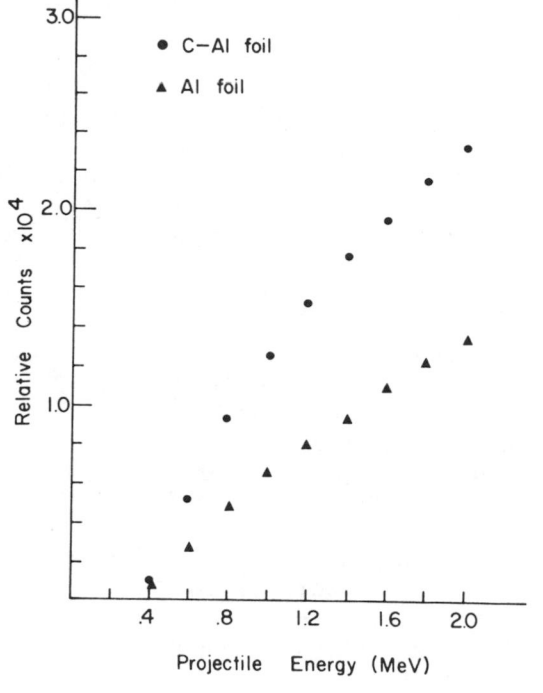

Figure 3

is a large steady state distribution of argon L vacancies established in this beam prior to penetration into aluminum. This
observation confirms the basic explanation of Al-1s vacancy production mechanism discussed above.

In a second set of measurements the yield of aluminum K x-rays
was measured as a function of aluminum thickness (T) by using different foils and various angles to the beam. In this case the
fraction of beam with L vacancies, $f_1(t)$ at depth t, now due to
Ar-Al collisions may be written as

$$f_1(t) = \frac{N\sigma v\tau}{1+N\sigma v\tau} \left(1-e^{-(N\sigma+\frac{1}{v\tau})t}\right)$$

where N is the atom density of Al, σ the cross section for argon L
vacancy production. Then the total yield of Al, Y_{Al}, is proportional to

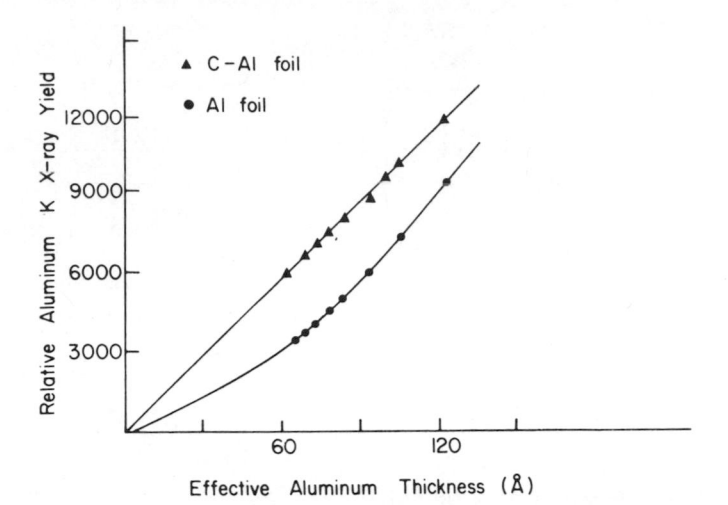

Figure 4

$$Y_{Al} \propto \int_0^T f_1(t)dt \propto \left[T + \frac{e^{-(N\sigma+\frac{1}{V\tau})T} -1}{N\sigma + \frac{1}{V\tau}} \right] .$$

In Fig. 4 the yield of aluminum K x-rays as a function of foil thickness is shown. The data for the aluminum foils clearly shows the non-linear dependence of the aluminum K x-ray yield on foil thickness. Parameterization of these data enabled us to determine independently both σ and τ. A summary of the values of τ and σ obtained for a variety of projectile energies is shown in Table 1.

TABLE 1

Energy (Mev)	σ (x10^{-17}cm^2)	τ (x10^{-15}sec)
.3	.35	1.50 (±.30)
.4	.59	2.15 (±.44)
.5	.90	4.8 (±.5)
1.0	1.58	11.0 (±2.5)
1.5	3.10	20. (±20)
2.0	4.10	-----

REFERENCES

1. R. J. Fortner, D. L. Matthews, J. D. Garcia and H. Oona (to be published).

2. R. J. Fortner and D. L. Matthews (to be published).

3. L. C. Feldman, R. A. Levesque, P. J. Silverman and R. J. Fortner (to be published).

L-SHELL VACANCY LIFETIME EFFECTS ON Kα X-RAY SATELLITES PRODUCED IN HEAVY-ION-ATOM COLLISIONS*

R. L. Watson, T. Chiao, F. E. Jenson, and B. I. Sonobe

Cyclotron Institute and Department of Chemistry, Texas A&M University, College Station, Texas 77843

ABSTRACT

The spectra of Kα x-ray satellites produced by 32 MeV oxygen-ion bombardment have been measured for a number of sulfur, silicon and aluminum compounds. The satellite intensity distributions are found to vary with chemical environment. This effect is attributed to L-vacancy filling transitions which alter the initial vacancy distribution prior to K x-ray emission. The temperature dependence of this effect has also been investigated.

INTRODUCTION

High-resolution measurements of the spectra of Kα x-rays emitted following heavy-ion-atom collisions provide information concerning the states of excitation produced. For example, it is now well known, from such measurements, that multiple L-shell vacancies are commonly produced in K-shell ionizing collisions. From an analysis of the intensity distribution of Kα x-ray satellites, one would like to determine the distribution of L-shell vacancy configurations produced in the collision process. The conversion of one to the other, however, requires a detailed knowledge of transition lifetimes for multiply-ionized atoms. We have been conducting experiments aimed at delineating the extent to which L-shell vacancy filling processes change the original vacancy distribution prior to K x-ray emission. Recently, we have observed an appreciable effect on the intensity distribution of Kα x-ray satellites caused by varying the chemical environment of the atom under study.[1] Our measurements indicate that a considerable alteration of the original vacancy distribution can occur

and that interatomic transitions from the valence levels of neighboring atoms provide an important mechanism for filling the L-shell
vacancies.

Kα SATELLITE INTENSITY DISTRIBUTIONS

Examination of the spectra of Kα x-rays produced by heavy-ion
bombardment using a Bragg crystal spectrometer, in general, reveals
a series of satellite peaks, each of which is composed of the
overlapping lines arising from configurations having the same
number of initial-state L-shell vacancies. For the purpose of
making quantitative comparisons of the overall intensity distributions from one spectrum to the next, it is useful to define a
parameter which we shall refer to as the average L-vacancy fraction. Let f_n represent the fraction of the total Kα x-ray intensity contained in the nth satellite peak where n is the number
of L-shell vacancies. Then the average L-vacancy fraction, p_L, is
defined as

$$p_L = \frac{\bar{n}}{N_L} = \frac{1}{N_L}\Sigma f_n n \qquad (1)$$

where N_L is the number of L-shell electrons in the ground state
atom. It has previously been observed[2,3] that, in general, the
satellite intensity distributions are almost binomial, in which
case

$$f_n \approx \binom{N_L}{n} p_L^n (1 - p_L)^{N_L - n} . \qquad (2)$$

Figure 1 shows the dependence of the average L-vacancy fraction (p_L) on projectile atomic number. It is seen that p_L displays essentially the same trend over a range of elements - namely
a rapid rise with a leveling off around $p_L = 0.4$. This leveling
off feature is rather surprising. Consideration of the mechanism for the production of the initial L-shell vacancies, which
is thought to be multiple Coulomb ionization, leads one to expect
p_L to increase approximately as the square of the projectile atomic
number. If the He ion result is scaled in this manner, p_L is predicted to go to unity at Z = 9 (see dashed curve in Fig. 1).
Instead, what is observed experimentally suggests the existence
of a saturation effect - as though a point is reached at which the
x-ray intensity distribution no longer accurately reflects the
initial vacancy distribution. This effect could be an indication
that L-vacancy filling transitions cause appreciable alteration of
the initial vacancy distribution prior to x-ray emission. The
reason for this can be understood as follows. Consider an atom
which has one K-shell vacancy and n L-shell vacancies as a result
of a heavy ion collision. Now if the K vacancy decays first with
the emission of an x-ray, the initial state is correctly represented
but if one or more of the L-shell vacancies decays first, then an

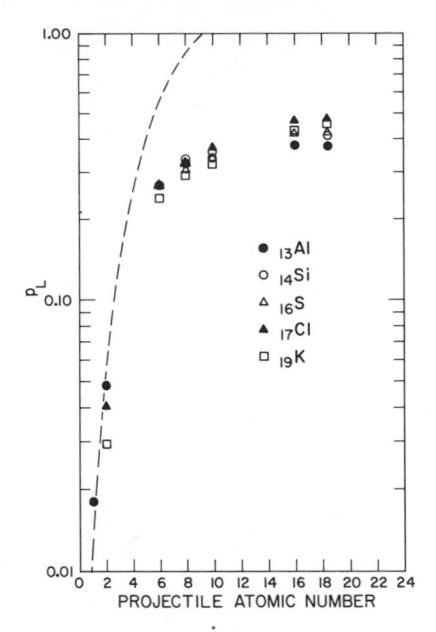

Fig. 1   Dependence of the average L-vacancy fraction on projectile atomic number.

x-ray which is emitted subsequent to this will no longer directly
reflect the initial number of L-shell vacancies. The net result
is that the number of x-rays falling in a given satellite peak is
related (in a rather complicated way) not only to the number of
those particular states formed initially but also to the number
of higher order states formed as well.

Since this relationship involves the lifetimes of transitions
from the valence levels in the case of third row elements, it is
reasonable to expect a dependence on chemical environment if these
transition rates are sufficiently fast so as to effectively compete
with the K transitions which have mean lifetimes of the order of
10^{-15} sec in singly-ionized atoms.

DEPENDENCE ON CHEMICAL ENVIRONMENT

We have recently performed measurements on a number of S, Si,
and Al compounds using 32 MeV O-ions for the purpose of delineating
the extent to which the Kα satellite intensity distributions are
affected by chemical environment. In Figure 2, spectra for Na_2SO_4,
S_8, and Na_2S and spectra for SiO_2, SiO, and Si are compared. As
can be seen, the intensity distributions change markedly in going
from Na_2SO_4 to Na_2S and from SiO_2 to Si. In Na_2SO_4 the satellite

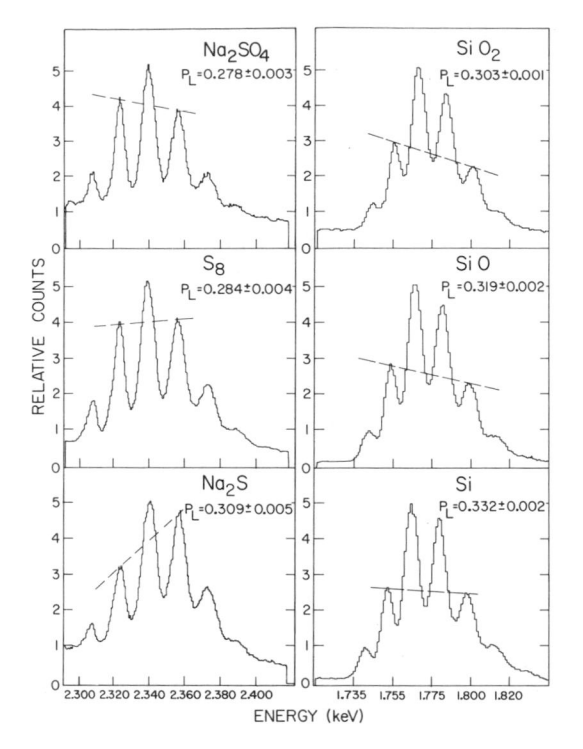

Fig. 2   Kα x-ray spectra for 32 MeV oxygen-ions incident on
           several different sulfur and silicon targets.

intensities are skewed toward the <u>lower</u> order satellites while
in Na_2S they are skewed toward the <u>higher</u> order satellites. This
seems to indicate that the initial vacancy distribution is altered
to a greater extent in Na_2SO_4 than it is in Na_2S, which is just
the opposite of what we expected based upon an analysis of the
problem in terms of <u>intraatomic</u> transitions. That is, since
the oxidation state of S in Na_2SO_4 is +6 one might expect that
these S atoms would have fewer valence electrons available to
participate in transitions to the L-shell than would S atoms in
Na_2S where the oxidation state is -2, and yet the p_L value for
Na_2SO_4 is <u>lower</u> than that for Na_2S. A similar trend is observed
in the Si series.

     Before proceeding further, it should be mentioned that there
are two other effects which could conceivably cause intensity
variations similar to those observed. The first is x-ray absorption in the target. However, this effect is quite small because
mass absorption coefficients do not change much over the small

energy range spanned by the satellite lines. We have checked
the accuracy of our absorption corrections experimentally by
placing absorbers between the target and the spectrometer.   The
second effect is associated with the fact that p_L varies with projectile energy. As a charged particle enters the target, it
loses energy and as a result the K-shell ionization cross section
decreases with increasing penetration depth. At the same time,
the transmission probability of those x-rays produced also decreases. These two factors determine the depth profile for x-ray
production plus transmission which in turn determines the effective average energy of the ion beam in the target. In the case of
32 MeV oxygen-ions incident on a thick SiO target, the average
depth for Kα x-ray production plus transmission is calculated to
be about 1 mg/cm^2 which translates into an average energy loss
of 6 MeV.

     In order to correct for the second effect mentioned above, we
have measured the dependence of p_L on projectile energy. The results of these measurements for oxygen-ions incident on Si and SiO
targets are shown in Fig. 3. The two open circles show the results
of measurements using 32 MeV oxygen-ions incident on SiO targets
of two different thicknesses. The excellent agreement exhibited
by these points with the rest of the curve gives us confidence
that we are able to quite accurately correct for variations of
p_L due to projectile energy loss in the target.

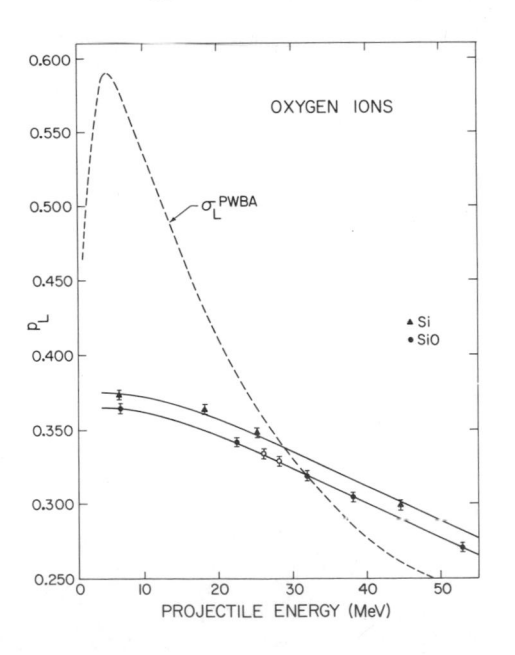

Fig. 3   Variation of p_L with oxygen-ion energy for Si and SiO
           targets.

Thus we must conclude that the relative intensity variations observed in the sulfur spectra are really caused by a chemical effect. Moreover we are led to consider the possibility that interatomic rather than intraatomic transitions play a dominant role. By interatomic transitions we mean transitions from the valence levels of neighboring atoms to the L-shell of the atom of interest. Such transitions have been postulated by various workers over the years, but evidence for their existence has been rather indirect.[4,5] In order to test this idea more quantitatively, one would like to seek a correlation between the average L-vacancy fraction and some parameter relating to interatomic transition probabilities. Following a suggestion by Citrin[6] who reasoned that such transitions must depend upon the overlap of the valence orbitals with the core-hole state, which in turn is related to the valence electron density, we have compared the p_L values with average valence electron densities, D_V, calculated directly from available mass densities or lattice volumes. In Fig. 4 is shown the dependence of p_L on D_V for compounds of sulfur and silicon.

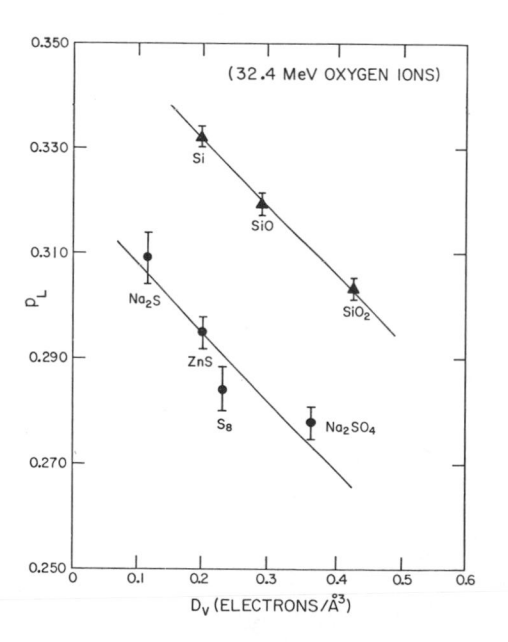

Fig. 4   The dependence of the average L-vacancy fraction on average valence electron density for sulfur and silicon compounds.

It is seen that both sets of data correlate approximately
linearly with average valence electron density and that the trend
is in the direction expected for interatomic transitions. As the
average valence electron density increases, the lifetimes of
interatomic transitions decrease thereby decreasing the value of
p_L.

In contrast to the trend observed for sulfur and silicon compounds is the data shown in Fig. 5 for aluminum compounds. Here
it is seen that the p_L values for Al_4C_3, AlN, and Al_2O_3 increase
with increasing average valence electron density. It is well
known that the bonding in these aluminum compounds is much more
ionic than in the sulfur and silicon compounds and hence one expects that the valence electrons are much more localized on the
ligand atoms. The data appear to indicate that interatomic transitions do not dominate in aluminum compounds but rather intraatomic transitions are more important as is suggested by the fact
that p_L increases with increasing ligand electronegativity. The
anomalously high p_L value for aluminum metal may reflect the fact
that conduction electrons are extremely efficient in shielding
localized charges.

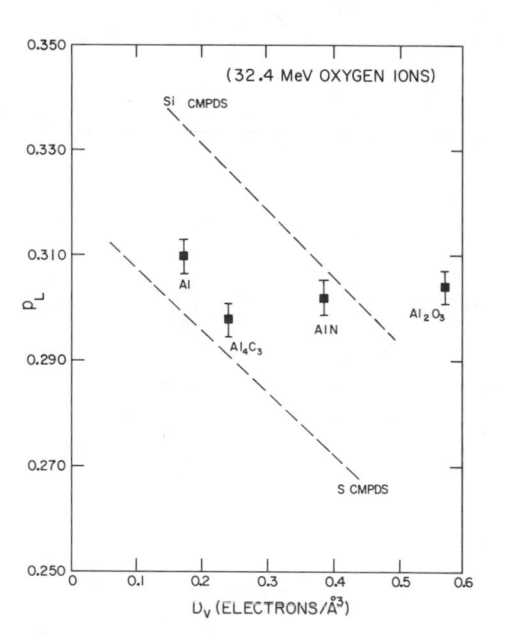

Fig. 5   The dependence of the average L-vacancy fraction on
          average valence electron density for aluminum compounds. The
          dashed lines show the trends displayed by the data for sulfur
          and silicon.

DEPENDENCE ON TEMPERATURE

Measurements have also been performed on aluminum metal and crystalline silicon as a function of temperature. The results of these measurements are shown in Fig. 6. The expected behavior for Al is shown by the solid line. This line (which is normalized to p_L at 298°K) simply shows the predicted relative change in p_L due to variation in average valence electron density as a consequence of thermal expansion. Both the Al and the Si data show large and unexpected increases in p_L at liquid nitrogen temperature. The explanation of this interesting effect is not understood at the present time.

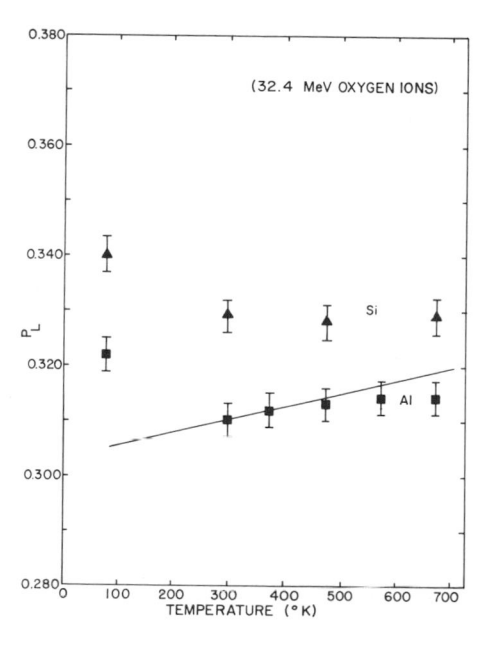

Fig. 6   Temperature dependence of the average L-vacancy fraction for aluminum and silicon. The solid line shows the behavior expected for aluminum due to thermal expansion.

CONCLUSIONS

These experiments have demonstrated that L-vacancy filling transitions cause appreciable alteration of the initial vacancy distribution prior to K x-ray emission. As a result of this, it is apparent that these processes will have to be taken into account in the interpretation of Kα satellite spectra. Moreover, the present results indicate that free atom transition rates are not likely to be adequate for this purpose, at least not in the case of molecular or solid targets. It would appear that Kα

satellite production by heavy-ion–atom collisions offers a particularly suitable means for studying the characteristics of interatomic
processes.

REFERENCES

*This work was supported in part by the U. S. Energy Research and
Development Administration and the Robert A. Welch Foundation.

1.   R. L. Watson, T. Chiao, and F. E. Jenson, Phys. Rev. Lett.
      35, 254 (1975).

2.   R. L. Kauffman, J. H. McGuire, P. Richard, and C. F. Moore,
      Phys. Rev. A 8, 1233 (1973).

3.   R. L. Watson, F. E. Jenson, and T. Chiao, Phys. Rev. A 10,
      1230 (1974).

4.   Richard D. Deslattes, Phys. Rev. 133, A390 (1964).

5.   P. H. Citrin, J. E.ectron. Spect. Relat. Phen. 5, 273 (1974).

6.   P. H. Citrin, Phys. Rev. Lett. 31, 1164 (1973).

SECONDARY ELECTRON EMISSION FROM FOILS TRAVERSED BY ION BEAMS

W. Meckbach

Centro Atomico Bariloche
Comision Nacional de Energia Atomica
S.C. de Bariloche, Argentina
and
Centre for Interdisciplinary Studies in Chemical Physics
The University of Western Ontario
London, Canada

ABSTRACT

Experimental and theoretical work on ion-induced kinetic secondary electron emission from solid targets, as well as on collisional production of energetic electrons in gaseous targets is reviewed.

It is deduced that the possibilities in obtaining information about secondary electron emission from solids are potentially increased by including the study of energy transfer to electrons from ions traversing a solid from inside the target. This implies the use of beam-foil techniques. In the beam direction charge exchange to the continuum gives rise to a characteristic peak, also observed in gaseous targets. In Bariloche experimental studies of secondary electron emission from both sides of foils traversed by 25-250 keV protons suggest a substantial contribution of direct ion-electron collisions to the secondary electron emission mechanism.

1. INTRODUCTION

Ion induced kinetic secondary electron emission (SEE) from solid targets is a three-step process. First energetic electrons are produced inside the target by collisional energy transfer from the ion. Second these "internal" secondary electrons migrate to the surface. Third, they have to traverse the surface barrier to be finally emitted.

The ion-electron collisional energy transfer gives rise to an internal excitation function or distribution $S(X, E_e, \theta)$ in distance from the surface, initial energy and angle with respect to the beam direction[1].

In the course of their migration to the surface, the internal SE suffer large-angle scattering from elastic collisions with nuclei, they lose energy by inelastic collisions with electrons or exciting plasmons. These energy losses may render it impossible for internal SE to surmount the surface potential barrier. We then speak of absorption. The inelastic cross sections and the associated energy losses being large, few internal SE will escape; the effective escape depth is small, typically of the order of 5, up to perhaps 50 atomic layers.

It follows that the energy and angular distribution close to the surface is likely to retain properties of the initial excitation function. This is expected to be valid particularly for SE resulting from forward energy transfer by ions traversing the target from inside outward. Indeed, to be finally emitted, these forward internal SE do not have to suffer large-angle deflections, which induce them to forget about their origin, particularly if the target material is chosen such, that electron elastic scattering is of minor importance, compared with absorption. From back-scattering experiments with incident electrons[1,2] it has been shown that this is the case for light materials, of low atomic number.

The energy and angular distribution j_s (E_e, ϕ) of the SE-flux measured outside is further affected by passage through the surface barrier, which for oblique incidence reduces only the energy of the electrons, associated with their internal momentum component normal to the surface. This results in a refraction of the electron path, away from the surface normal. Total reflection then leads to an internal emission cone, which, with decreasing electron energy shrinks toward the surface normal. If the internal SE-distribution is isotropic within this escape-cone the emitted SE flux j_s (E_e, ϕ) obeys a cosine distribution in ϕ, the angle with respect to the surface normal[3]. We understand that, if the surface barrier height is appreciable, few emitted SE of low energy are observed and, irrespective of the momentum distribution inside, their angular distribution will be a cosine distribution[3].

Thorough understanding of kinetic SEE is hampered by the fact that we are obliged to measure outside the solid material and are unable to study the processes of SE-production, migration to and passage through the surface separately. In particular we are unable to measure directly the internal excitation function $S(x, E_e, \theta)$.

Almost all the extensive experimental information about ion-induced kinetic SEE, providing the basis for theoretical discussions, has been obtained from backward-emitted electrons, the primary beam hitting the front surface of a thick target. The subject has been reviewed by Kaminski[4], Krebs[5], Medved and Strausser[6], Carter and Colligon[7], Arivov[8], Dettmann[10] and McCracken[11].

New and more information, and, consequently a better insight in the mechanisms leading to SEE can be obtained if measurements on SE emitted in the forward direction, from the rear surface of thin foils traversed by the ion beam are included. Measuring and comparing forward and backward SEE simultaneously, asymmetries in the distribution of internal SE are likely to manifest themselves in differences between the total forward and backward emissions as well as in differences in the energy and angular distributions of SE emitted from both foil surfaces.

II. EXPERIMENTAL EVIDENCE FROM BACKWARD EMISSION

The secondary electron emission coefficient γ, defined by the total number of SE emitted per ion, as a function of primary energy E_b for incident protons is seen in Fig. 1[5]. γ increases with increasing E_b, passes through a broad maximum and then decreases at higher energies. As, in the energy range considered, the penetration depth of the protons is large compared wtih the escape depth of SE, and the excitation function of internal SE can be considered uniform along the beam path, there is a meaningful resemblance of this dependence with that of a total ionization cross section (Fig. 2), as measured in gaseous targets[9], which is governed by the Massey adiabatic criterion. This led to the concept employed in all accepted theoretical discussions of SEE induced by ionic projectiles, that the internal SE are originated by ionizing collisions from the valence bands. Contrary to SEE induced by primary electrons, conduction electrons are not believed to contribute to SEE induced by ionic beams.

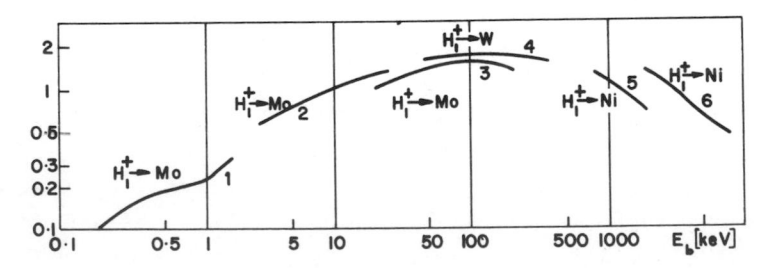

Fig. 1   Total secondary electron emission yield γ as a function of beam energy for incident protons (from: K.H. Krebs, ref. 5).

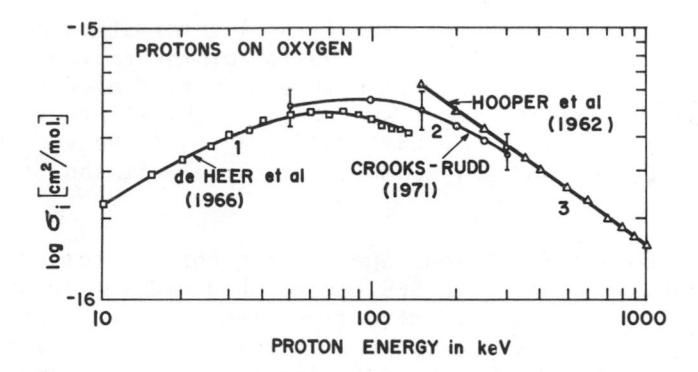

Fig. 2   Total cross section for electron ejection from oxygen gas
for incident protons (from: J.B. Crooks and M.E. Rudd, ref. 9).

Fig. 3 shows typical energy distributions of SE[12]. As may be
expected, the high energy tails extend to larger E_e as the projectile energy increases.   The low energy tail and position of the
maximum is seen to be independent of the projectile energy, they
being mainly determined by the escape mechanism through the surface barrier.

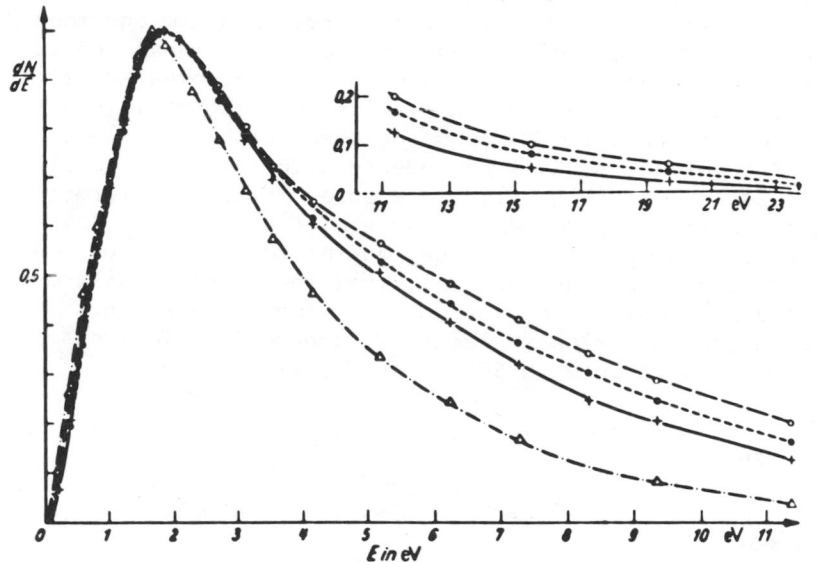

Fig. 3   Energy distributions of secondary electrons ejected by
He[+] on Mo; He -energies: Δ, 2 keV; +, 5 keV; ●, 10 keV; o, 15 keV
(from: G. Wehner, ref. 12).

Little has been measured on angular distributions of SEE. Klein[13] obtained cosine-angular distributions for 4 keV He$^+$, Ne$^+$ and Ar$^+$ incident of W. No information about well resolved doubly differential measurements of j_s (E_e, ϕ) is reported.

There is considerable scattering in the results on SEE, obtained by different authors which is mainly attributed to un-controlled surface conditions [5,6,8], a problem which will not be discussed here.

III.   THEORETICAL ASPECTS OF ION INDUCED SECONDARY
ELECTRON EMISSION

Accepted theories are those of Parilis and Kishinevskii[14], Sternglass[15], Gosh and Khare[16]. Without giving details about the excitation function $S(x, E_e, \theta)$, in the former two theories the excitation of energetic electrons is computed from the col-lisional energy loss of the projectiles, in the latter from total ionization cross sections.

In the theory of Parilis and Kishinevskii[14] the ion-electron energy transfer mechanism is that described by Firsov[17] and attributed to overlap and electron exchange between the outer electron shells of beam ions and lattice atoms during collisions which are considered adiabatic. This theory, therefore, is valid in the lower energy range.

No electron, excited by this mechanism, is considered to possess sufficient energy to be emitted through the surface barrier, a somewhat arbitrary assumption[6]. However, electron hole pairs, formed in the excitation process, recombine, giving rise to the production of Auger electrons which finally may be emitted. From this mechanism an isotropic angular distribution of the internal excitation function is obviously expected.

In the theory of Sternglass[15] the energy transfer to lattice electrons is computed from stopping power theories of N. Bohr[18] and Bethe[19]. This theory, therefore is valid at higher projectile energies.

According to N. Bohr[18] electronic stopping is due in equal parts to distant or weak "glancing" collisions with atoms as a whole and close or violent "knock on" collisions with individual electrons. Sternglass assumes that the former produce a large number of slow internal SE (Fig. 4a), the latter a small number of forward peaked fast internal SE, the so-called ∂ electrons. These, then create higher order internal SE (Fig. 4b) whose number, according to the above mentioned equipartition of energy deposi-tion, should equal that of the first order slow internal SE. We

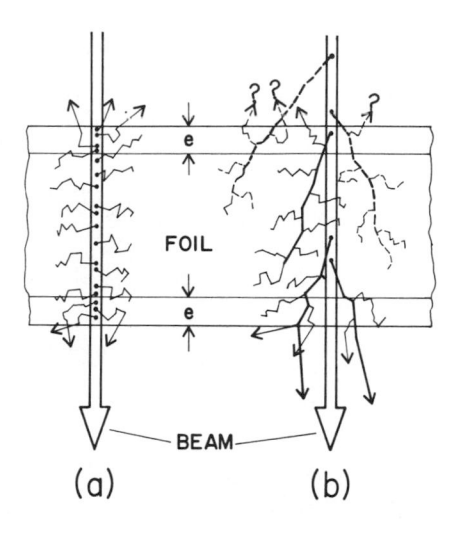

Fig. 4   Model of Sternglass (ref. 15) for Creation of: a) slow
secondary electrons; b) fast, forward peaked ∂-electrons and slow
tertiary electrons.

deduce, however, from Fig. 4b, that close to the (front) surface
of the target a considerable part of the forward peaked ∂ electrons,
whose range is large compared to the escape depth of slow SE, is
absent because their "would be" origin is seen ahead of the surface, in vacuo.   For forward emission from rear surface of a foil
there is no such "goemetrical" reduction of the production of slow
higher order SE.   Also preferred emission of ∂ electrons is to be
expected.

        Neglecting the higher order contribution to SEE from ∂-
electrons altogether, and using a lower-energy approximation for
stopping power of N. Bohr[18], which considers only the participation of outer electron shells of target atoms in stopping,
Sternglass arrives for the purpose of a rough test of his theory
at $\gamma \propto Z_i^2 \, E_b^{-\frac{1}{2}}$.   Here Z_i is the atomic number of the beam ions,
E_b the projectile energy.   In applying this relationship, ex-
tensively used as the final result of the theory of Sternglass,
its limitations must be considered carefully.

        In all the above mentioned theories the transport and
absorption of the internal energetic SE is described by a simple
exponential dependence of the form $e^{-x/L}$, the attenuation length
L being considered constant and independent of electron energy.

IV. EVIDENCE FROM ION INDUCED IONIZATION IN GASEOUS TARGETS

The initial collisional excitation of internal SE being considered to be ionization, let us look at cross sections, for
electron ejection differential in energy and angle σ (E_e, θ), as
obtained by ion induced ionization in gaseous targets, where they
can be measured under single collision conditions. Such studies
have been performed extensively since 1963, the corresponding work
being reviewed by Ogurtsov[20], Kim[21], Manson[22] et al. and in a
forthcoming report of Rudd[23].

Interaction of structureless projectiles (H^+, He^{++}...) with
free electrons leads to the so-called binary peaks, SE energy
distributions for a certain angle and angular distributions for a
certain energy being δ-functions. If the electrons are bound in
atomic shells, the collisional energy and momentum transfer is
affected by the initial momentum distribution of the atomic
electrons, their correlation, binding as well as the interaction
of the ejected electron with the target nucleus and the passing
ion. Consequently, an electron can, in principle, be ejected in
any direction. The mentioned effects have been partially accounted for by applying a classical binary encounter approximation[24] and the plane wave Born approximation (PWBA)[25].

In Fig. 5 measured angular distributions of 150 eV electrons[26]
ejected from H_2 and Ne by 300 keV H^+ are compared with scaled Born
approximation calculations, using hydogenic wave functions[25].
These calculations show the expected broadened binary peak. In
the case of H_2 good agreement is observed from the peak maximum up
to about 90° where the experimental cross section levels off.
Madison[27], using proper final state wave functions in the PWBA,
obtained a substantial improvement at large angles. For the
heavier target, neon, the binary peak is almost totally smeared
out.

The characteristic disagreement at small angles led to the
understanding of a new mechanism for collisional electron ejection:
charge exchange into the continuum. This effect, due to the fact
that the finally ejected electron is moving not only in the field
of the ionized target atom, but also in the field of the passing
ion[28], has been calculated by Salin[29] and Macek[30], using different formal approaches. This explains the observed and calculated steep rise of σ (E_e, θ) at small θ. Obviously the
effect is sharply dependent on how much the velocity of the
ejected electron differs in direction and magnitude from that of
the projectile. Correspondingly, in Fig. 6, a spectacular cusp-
shaped peak is seen for σ (E_e, $\theta = 0° \pm 1.4°$) as a function of E_e,
as measured by Crooks and Rudd[31], and compared with the theory of
Macec[30] for $\theta = 0°$ and $1.4°$.

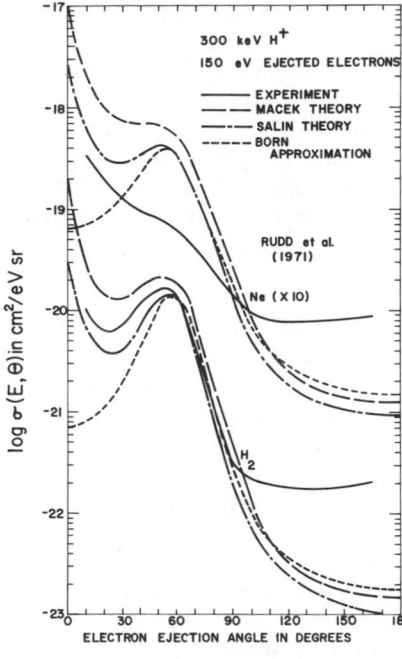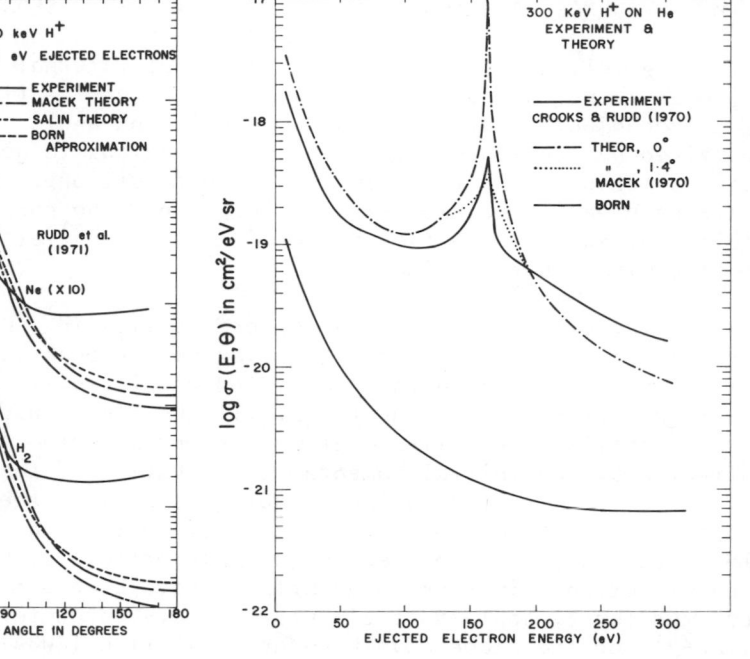

Fig. 5   Double differential cross sections σ (E_e, θ) of 150 eV electrons as a function of ejection angle θ, for 300 keV H[+] on H_2 and Ne (from: M.E. Rudd et al., ref. 26; Born approximation, ref. 25; Macek theory, ref. 30; Salin theory, ref. 29).

Fig. 6   Energy distribution of electrons, ejected at $\theta = 0°\pm$ 1.4°, in collisions of 300 keV H[+] on He (from: J.B. Crooks and M.E. Rudd, ref. 31). Macek theory (ref. 30) at $\theta = 0°$ and 1.4°.

V.   BEAM - FOIL FORWARD ELECTRON EMISSION

Almost simultaneously with the discovery of charge exchange into the continuum in gaseous targets, Harrison and Lucas observed the same cusp-shaped peak for beam-foil emitted electrons[32]. Fig. 7 shows results corresponding to a more recent work, which includes a theoretical discussion of the effect by Dettmann[33]. Typically the peak is shifted towards lower electron velocities with increasing mass of the ionic projectiles of equal incident energy.

Recent Neelavathi et al.[34] suggested a typical solid state effect, also causing forward electron emission with the velocity of the emerging ion, in terms of wake riding states produced by

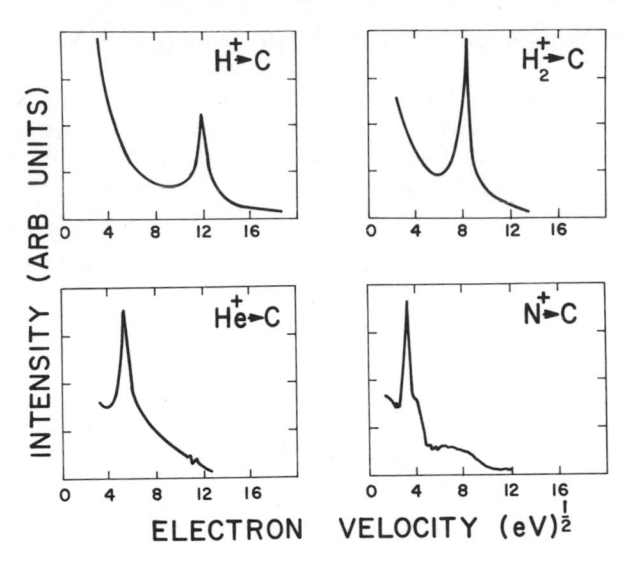

Fig. 7   Velocity spectra of electrons, ejected at $\theta = 0^{o} \pm 4.6^{o}$
from the rear suface of a carbon-foil traversed by 269 keV, H^{+},
H_2^{+}, He^{+} and Ne^{+}-ions (from: K. Dettmann et al., ref. 33).

Auger-type capture into potential throughs originating from
periodical valence electron desity fluctuations accompanying the
projectile. The binding energy of an electron trailing a $\simeq$ 300
keV proton in the first depression is $\simeq$ 10 eV in a typical metal.
These potential throughs vanish when the ion emerges into vacuum.

        Another mechanism for ion-accompanying electron emission
could be electron loss from bound states. According to a model of
Betz and Grodzins[35] this effect is expected to be strongly enhanced in the case of heavy ions emerging from a solid. The
electron shells of these ions, excited collectively inside the
solid, are allowed to shake off electrons by autoionization after
having emerged through the surface.

        The experiment of Harrison and Lucas[32] was the first, in
which ion-induced SE emitted from the rear surface of a foil
were observed, in the special case of emission at $\theta = 0^{o} \pm 4.6^{o}$.
A particular property of the internal excitation function was seen
to be strikingly conserved.

        In Frankfurt, Groeneveld et al.[35] measured spectra of ion
induced SE from foils, emitted at 42° with respect to the beam
direction. He will report about these and further results at
this Conference.

VI.   MEASUREMENTS PERFORMED IN BARILOCHE[39)]

In Bariloche we studied kinetic SEE from both surfaces, that is in the backward and forward half spaces, when protons of energies between 20 and 250 keV traversed carbon foils of 5 and 10 $\mu g/cm^2$ thickness. Total SEE coefficients γ_{backw} and γ_{forw} were measured simultaneously and compared. Energy distributions of SE, integrated over almost all angles of emission into each half space, were obtained by applying a retarding field.*

Fig. 8 shows the measured SEE coefficients γ_{forw} and γ_{backw}

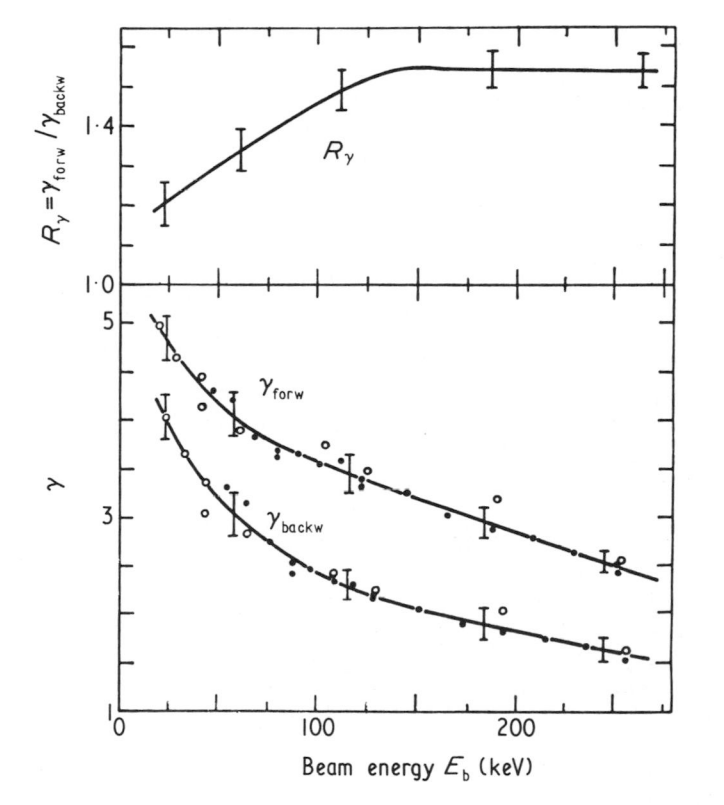

Fig. 8   Total secondary electron emission coefficients for emission into the forward and backward half spaces and their ratio as a function of the energy of protons traversing carbon foils (foil thickness: o, 5 μgcm^{-2}; •, 10 μgcm^{-2}).

as well as their ratio R_γ as a function of the proton beam energy E_b. We observe that forward emission predominates. R_γ increases with increasing proton energy and saturates at $R_\gamma = 1.55$ above 140 keV. Partial neutralization of the beam and a slightly

larger SEE for H^o as compared to H^+, observed by Stier et al.[36] and Chambers[37] was estimated to contribute only little to the observed enhancement of forward emission, specially at higher E_b where the neutral beam component is small. The observed pre-dominance of forward emission is to be attributed to a forward peaked asymmetry in the angular distribution of the internal excitation function S. Such an asymmetry is observed in gaseous targets. Crooks and Rudd[9] found strongly enhanced forward emission in the singly differential cross section $\sigma(\theta)$ for ioniz-ation of Ar-gas by protons.

Typical integral spectra of SE ejected into the forward and backward half spaces by 254 and 257 keV protons, respectively, are seen in Fig. 9 as a function of the retarding potential U. Abrupt downward brakes at U = +0.8V indicate the surface potential difference between foil and electron collectors. Predominance of forward emission is seen to increase with increasing electron energy.

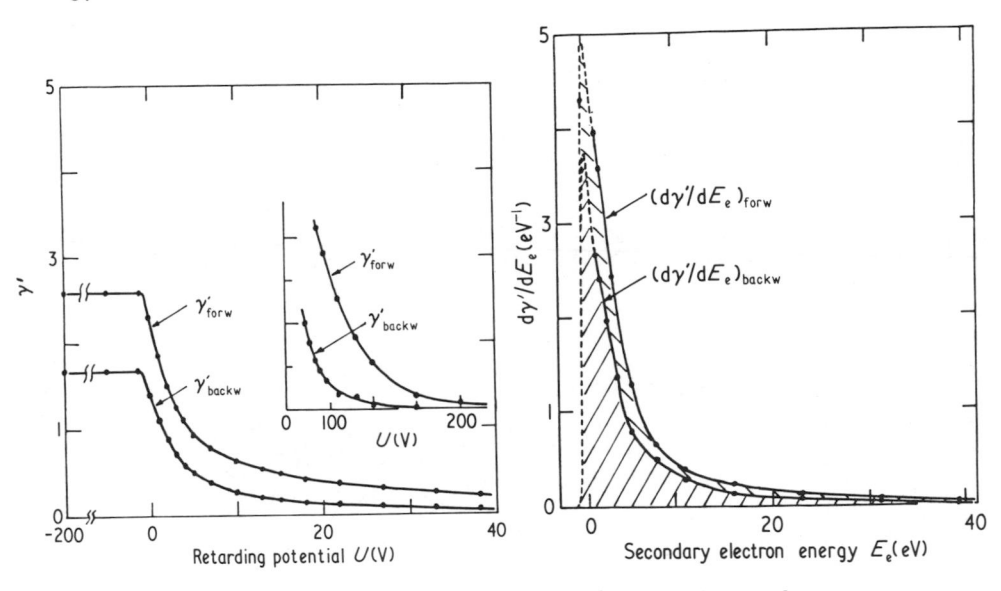

Fig. 9   Integral energy distributions of secondary electrons emitted in the forward and backward directions from a carbon foil. Beam energies: forward: E_b = 254 keV; backward: E_b = 257 keV. Fig. 10   Differential spectra as obtained by graphical differ-entiation of the spectra in Fig. 9.

Differential energy distributions, as obtained graphically from the measured spectra, are shown in Fig. 10. These spectra are peaked at about 0.6 eV only. The form and height of the sharp peaks could not be obtained with precision from graphical dif-

ferentiation, it is, however, clear that also for SE of these low
energies forward emission predominates, approximately in accordance with an estimation of Sternglass geometrical reduction of
the production of slow tertiary electrons by fast, forward
peaked binary electrons(∂-rays), which was seen to be valid only
for backward emission.

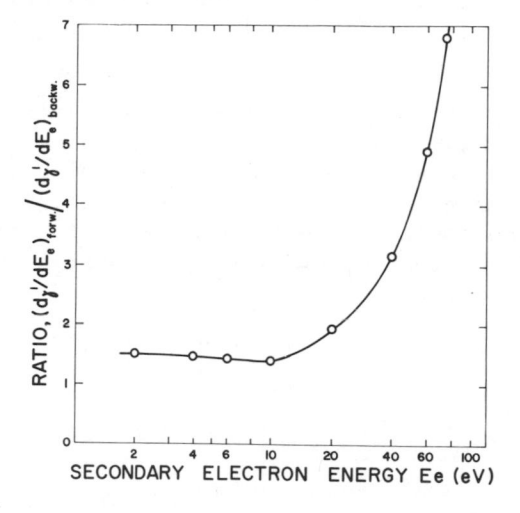

Fig. 11 Ratio of emission into forward half space, divided by
emission into backward half space, for electrons of a given E_e,
as extracted from Fig. 10.

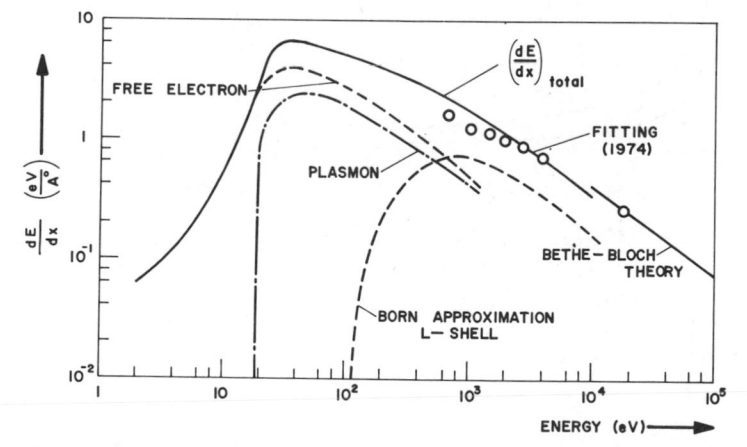

Fig. 12 Stopping power of electrons in Al as a function of their
energy, calculated by Ritchie et al. (ref. 38); o, experimental
data of Fitting (ref. 2).

The ratio of forward SEE to backward SEE, for electrons of a given energy E_e, is shown in Fig. 11. This ratio is approximately constant up to about 10 eV; above this energy it is seen to increase quickly. This behaviour is meaningful, if we compare with Fig. 12 where the stopping power of electrons as a function of their energy, as computed by Ritchie et al.[38] is seen to experience a sharp rise at 10 eV, the mean free path for absorption being quickly reduced compared with the mean free path of elastic scattering. Consequently, forward peaked internal SE have increasing difficulty not to be absorbed on their way back to the front surface of the foil for being emitted backwards.

Finally, in Fig. 13 the high-energy tails of the differential spectra of Fig. 12b are represented in doubly logarithmic scale. For forward emission straight lines with slopes close to - 3/2 result, independently of beam energy. This is shown for 41.5 and 254 keV. For backward emission no such simple dependence is observed.

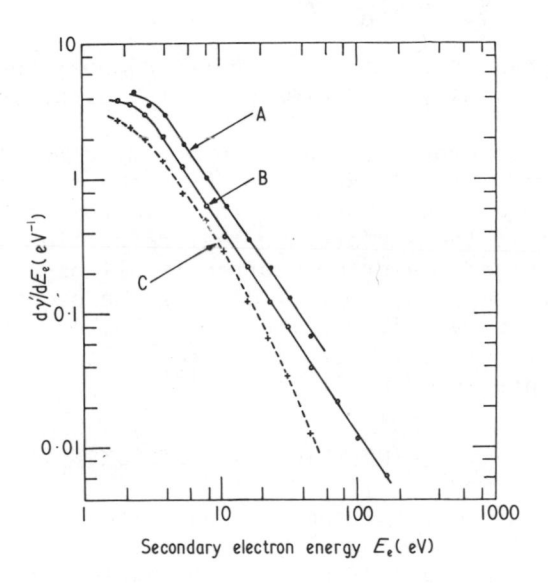

Secondary electron energy E_e(eV)

Fig. 13 Differential energy distributions represented in double logarithmic scale. Curve A, $(d\gamma'/dE_e)_{forw}$, E_b = 41.5 keV, slope = 1.51: curve B, $(d\gamma'/dE_e)_{forw}$, E_b = 254 keV, slope = 1.53: curve C, $(d\gamma'/dE_e)_{backw}$, E_b = 257 keV.

We conclude that, the collisional excitation function of internal SE being forward peaked, their transport to the rear surface of the foil obeys to a more direct and simple mechanism, since no backward scattering is needed for the SE to be emitted forward. We try a model based on the following assumptions:

We assume that the attenuation of internally excited SE obeys simply to a $\exp(-\frac{r}{L}) = \exp(-\frac{x}{L\cos\theta})$ dependence, x being the normal distance to the surface and L = const. If the internal collisional excitation function $S(x, E_e, \theta)$ is independent of x, integration along the beam path leads to a distribution at the surface given by $S(E_e, \theta) \cos\theta$.

We further assume that internally energetic electrons are produced by a simple proton-free electron binary collision mechanism. This leads to the Rutherford - distribution, proportional to $\cos^{-3}\theta$ and, by energy conservation to the distribution in electron energy E_e given by $S(E_e) \propto E_b^{-1} E_e^{-2}$; E_b being the beam energy.

From above we see that at the surface the angular distribution is proportional to $\cos^{-2}\theta$ and, correspondingly, applying energy conservation we obtain for the energy distribution at the surface $S'(E_e) \propto E_b^{-3/2} E_e^{-3/2}$.

This is the distribution in energy proportional to $E_e^{-3/2}$, observed experimentally for forward emitted electrons.

It cannot be expected that such a simple model, which, besides the assumed initial excitation by Rutherford collisions, implies that there is no change of the momentum of the electrons when travelling to the surface and that refraction at the surface barrier is neglected, describes correctly all aspects of experimental evidence. As a matter of fact the proportionality of S' to $E_b^{-3/2}$ is not verified. The total SEE, observed experimentally bends at higher beam energies E_b rather towards the $E_b^{-1/2}$ dependence, as predicted by Sternglass[15].

VII.   CONCLUSION AND OUTLOOKS

Resuming our results, there is experimental evidence that the internal excitation function of ion induced SE is forward peaked and that this forward distribution is enhanced for electrons of energies larger than about 10 eV. Even down to lower electron energies (≈ 6 eV) the energy distribution of forward emitted electrons obeys to a simple power law, not verified for backward emission.

This evidence suggests that an essential contribution of SE-emission induced by fast protons originates from binary collision processes with individual lattice electrons. The participation of free, conduction electrons in this mechanism may not be discarded.

In this context Crooks and Rudd[9] and Rudd et al.[26] have given experimental and theoretical evidence that in gaseous targets

ion induced ejection of SE is mainly from outer shells and that, within the frame of the Born approximation there is an enhanced contribution to forward emission from the least tightly bound outer shell electrons.

We conclude that beam-foil techniques proportionate a tool for investigating fundamental aspects of ion induced SEE from solids. Measurements will have to be extended by varying beam projectiles and energies, as well as targets.

In order to obtain more information about the internal excitation function it is necessary to study well resolved doubly differential distributions $j_s(E_e, \phi)$ in energy and angle for emission from both sides of foils. The preparation of such experiments is in progress.

* Experimental details of this work are described in a forth-coming letter in Journ. Physics B (Atomic Molec. Physics)[39].

VIII. REFERENCES

1) O. Hachenberg and W. Brauer, Adv. Electr. Electr. Phys. XI (1959) 413
2) H.J. Fitting, Phys. Stat. Sol. (a) 26 (1974) 525
3) H. Stolz, Ann. Physik. [7] 3 (1959) 196
4) M. Kaminsky, Atomic and Ionic Impact on Metal Surfaces (Berlin, Heidelberg, New York; Springer Verlag) (1965) 300
5) K.H. Krebs, Fortschritte Phys. 16 (1968) 419
6) D.B. Medved and Y.E. Strausser, Adv. Electr. Electr. Phys. 21 (1965) 101
7) G. Carter and J.S. Colligon, Ion Bombardment of Solids (London, Heinemann) (1969) 38
8) V.A. Arivov, Izdatel'stvo Akademii Nauk Vzbekskoi SSR Tashkent; AEC-tr-6089 (1961)
9) J.B. Crooks and M.E. Rudd, Phys. Rev. A1 (1971) 1628
10) K. Dettmann, Proc. Conference on Interaction of energetic charged particles with solids, Istanbul (1972) BNL 50336
11) G.M. McCracken, Rep. Progr. Phys. 38 (1975) 241
12) G. Wehner, Z. Phys. 193 (1966) 439
13) H.J. Klein, Z. Phys. 188 (1965) 78
14) E.S. Parilis and L.M. Kishinevskii, Sov. Phys. -Solid St. 3 (1960) 885
15) E.J. Sternglass, Phys. Rev. 108 (1957) 1
16) S.N. Gosh and S.P. Khare, Phys. Rev. 125 (1962) 1254
                                    Phys. Rev. 129 (1963) 1638
17) O.B. Firsov, Soviet Physics - Jetp 9 (1959) 1076
18) N. Bohr, K. Danske Vidensk. Selsk., Mat. - Fys. Meddr. 18 no. 8 (1948)
19) H.S. Bethe, Am. Physik 5 (1930) 325

20) G.N. Ogurtsov, Rev. Mod. Phys. 44 (1972) 1
21) Y.K. Kim, ANL - Report 7960 (1972)
22) S.T. Manson, L.H. Toburen, D.H. Madison, N. Stolterfoht, Phys. Rev. A12 (1975) 60
23) M.W. Rudd, To be published in Radiation Research, Oct. 1975
24) M. Gryzinski, Phys. Rev. 115 (1959) 374
                    Phys. Rev. 138 (1965) A322
    T.F.M. Bonsen and L. Vriens, Physica 47 (1970) 307
                    Physica 54 (1971) 318
25) D.R. Bates and G. Griffing, Proc. Phys. Soc. (London) A66 (1953) 961
    C.E. Kuyatt and T. Jorgensen, Jr., Phys. Rev. 130 (1963) 1444
    M.E. Rudd, C.A. Sautter and C.L. Bayley, Phys. Rev. 151 (1966) 20
26) M.E. Rudd, D. Greoire and J.B. Crooks, Phys. Rev. A3 (1971) 1635
27) D.H. Madison, Phys. Rev. A8 (1973) 2449
28) W.J.B. Oldham, Phys. Rev. 140 (1965) A1477
                    Phys. Rev. 101 (1967) 1
29) A. Salin, J. Phys. B 2 (1969) 631 and 1255
              J. Phys. B 5 (1972) 979
30) J. Macec, Phys. Rev. A1 (1970) 235
31) G.B. Crooks and M.E. Rudd, Phys. Rev. Lett. 25 (1970) 1599
32) K.G. Harrison and M.W. Lucas, Phys. Letters 33A (1970) 142
33) K. Dettmann, K.G. Harrison and M.W. Lucas, J. Phys. B 7 (1974) 269
34) V.N. Neelavathi, R.H. Ritchie and W. Brandt, Phys. Rev. Lett. 33 (1974) 302
35) K.O. Groeneveld, R. Mann, W. Meckbach and R. Spohr, Vaacuum, 25 (1974) 9
36) P.M. Stier, C.F. Barnett and G.E. Evans, Phys. Rev. 96 (1954) 973
37). E.S. Chambers, Phys. Rev. 133 (1964) A1202
38) R.H. Ritchie, F.W. Garber, M.Y. Nakui and R.D. Brikhoff, Adv. Rad. Biology 3 (1969)
39) W. Meckbach, G. Braunstein and N. Arista, Letter to be published in J. Phys. B.

SPECTROSCOPY OF ELECTRONS ACCOMPANYING THE PASSAGE

OF HEAVY IONS THROUGH SOLID TARGETS

Karl-Ontjes Groeneveld

Institut für Kernphysik der Universität

6 Frankfurt/M., Germany

The recent installation of heavy ion accelerators has revived the interest in the study and understanding of electron emission accompanying the passage of heavy ions through solid targets. The interest has been stimulated by a number of current problems as: What is the magnitude of electron induced bremsstrahlung production in heavy ion collisions in solids? The answer is important as well to the sensitivity of trace element analysis with heavy ion induced characteristic x-rays /1/ as to the interpretation of the non-characteristic part (e.g. the molecular orbital x-rays) of x-rays /2,3/.

The electron energy spectra and their angular distribution may give important information on the heavy-ion-solid reaction mechanism and e.g. interesting information on transiently formed collision molecules /4/. In addition, for a detailed study of Augertransitions from heavy ion - solid interactions (i.e. beam-foil spectroscopy) it is a necessity to know the ratio of the continuous electron energy distributions to the superimposed discrete Augerlines /5,6,7/.

The subject has been reviewed at "lower" incident energies e.g. by Carter and Colligon /8/, by Krebs /9/ and - extensively on this conference - by Meckbach /10/. In the present paper these phenomena are treated at "higher" energies; the term "high" /this paper/ and "low" /10/ has been chosen rather arbitrarily, the dividing energy being defined to be 0.5 MeV

– with an appropriate diffuseness.

The reaction mechanisms in both energy ranges are basically the same and have been reviewed in detail by Meckbach /10/ on this conference. The application of existing theories for interactions with gaseous targets to interactions with solid targets is of very limited value and fails to reproduce the details of the observed phenomena.

NUMBER OF ELECTRONS PER PROJECTILE

Solid foils of sufficient thickness bombarded with heavy ions have been studied by Werlein et al. /11/ in the incident energy range from 4 to 20 MeV with He-, N- and Ne-projectiles and by Clerc et al. /12/ with He-, O-, S-and I-projectiles up to 40 MeV energy. The average number $\bar{n}$ of electrons produced per projectile was found in both experiments to be proportional to the differential energy loss of the projectiles in the target foil (fig. 1). Since the energy loss in this energy range is dominated by electronic stopping power, i.e. the interaction of the projectile with the electrons of the target, this result may not be surprising. Projectiles passing through two successive thin carbon foils /12/ yielded more than two times the number of secondary electrons since, for the given geometry, high energy secondary electrons could produce tertiary electrons in the other foil. Since both the energy loss and the increase of $\bar{n}$ with more than one target foil is strongly Z-dependent (Z atomic number of projectile) it has been suggested /12/ to use such a device as an Z-detector in heavy-ion-reaction telescopes. The secondary electrons are frequently used in heavy-ion time-of-flight telescopes as a start signal.

Following Sternglass /13/ the probability P(x) for an electron produced at a point x inside a solid target to leave the surface of this target is separated in two main factors:

$$P(x) \propto T \quad \bullet \quad \exp(-x/\lambda) \tag{1}$$

Here λ is the average mean free path of the electrons in the solid /14/ and T is the transmission of the electrons through the surface, which, in turn, depends on the surface potential and the work function /15/. This relationship, eq.(1), is illustrated by the experimental result /11/ in fig. 2 : For a given material

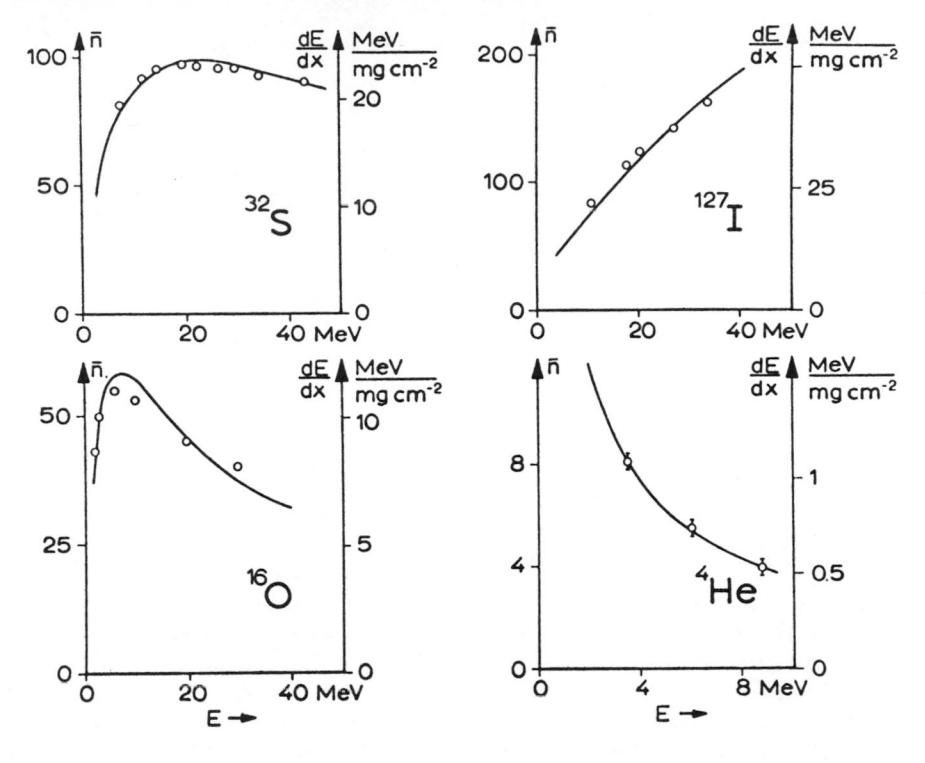

Fig. 1 Average number $\bar{n}$ of electrons (left scale) emitted from a carbon target foil (6 μg/cm^2) bombarded with He-, O-, S- and I-projectiles as function of projectile energy E. The solid lines represent the normalized differential energy loss dE/dx (right scale) /12/.

the secondary electron yield $\bar{n}$ follows an exponential function of the target thickness; moreover, the electron yield $\bar{n}$ increases by a factor of two to three if a material with a low work function is used.

Another interesting property is the probability P(n) of ejection of n electrons per projectile in a heavy-ion solid collision. At lower projectile energies this problem is of practical interest e.g. for multipliers used in mass spectrometers as particle detectors. The knowledge of the distribution function is needed for the quantitative correction of signals

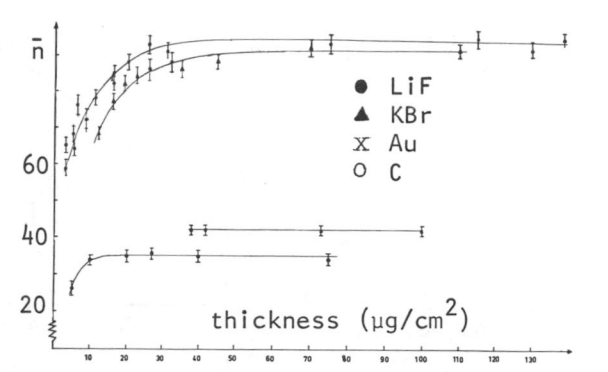

Fig. 2 Average number n̄ of electrons emitted per N-projectile (14 MeV incident energy) as function of target-foil thickness. Parameter is the target material as indicated /11/.

with P(n = o). In an coincidence experiment /12/ between He-projectiles and electrons ejected into a solid state detector the pulse height distribution of the detector reflects this probability P(n). Achtstein's results in a similar experiment /16/ with a surprisingly low background from incident Cs-projectiles with energies up to 30 keV impinging on metal surfaces (Ni, Mo, Ta) indicate that this distribution P(n) cannot be described analytically by a Poisson distribution as well as by a Pólya-distribution.

ENERGY AND ANGULAR DISTRIBUTIONS

In most experiments, which were performed to study the energy and angular distributions of the electrons only gas targets were used /17/. A comparison of these data with existing theories as the binary-encounter-approximation (BEA) /18/ and the Born-approximation yield satisfactory agreement /19/ under single collision conditions.

Only few data, however, from ion-solid interactions have been published or analysed. To approach a quantitative understanding here /5/ it is assumed that the dominant interaction process in atomic collisions is a direct energy exchange between the projectile and the target atoms. In this model the binary-encounter-theory /20/ can be applied to calculate the double differential cross section for electron emission from ion bombardment of an atom under single collision condition.

To calculate the electron spectra from the bombardment of solid targets the energy loss of the scattered electrons during their penetration through the target has to be considered. The Bethe stopping power relation was used to calculate the energy loss for each electron energy and for each target thickness t. Integrating over the target thickness one gets the yield of secondary electrons Y. In fig. 3 the calculated Y-values are shown for carbon foils with different

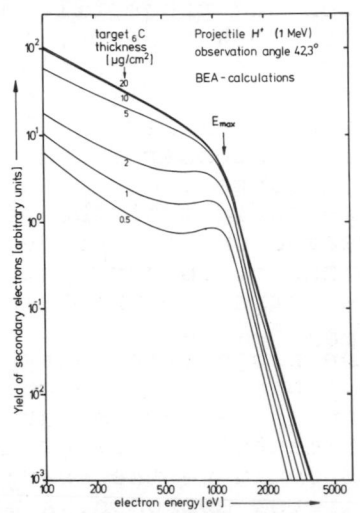 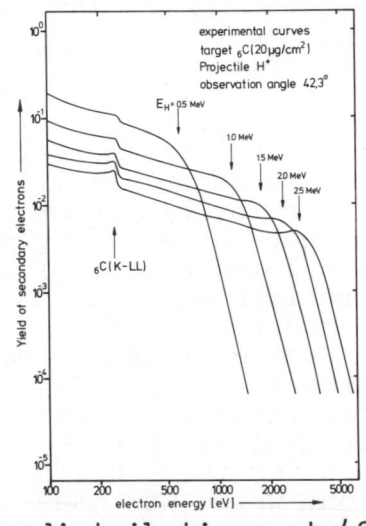

Fig. 3 (left) Electron energy distributions at 42.3° from H (1 MeV) + C(solid)-collisions calculated with the BEA-theory. Parameter is the solid carbon foil thickness.

Fig. 4 (right) Experimental electron energy distributions at 42.3° from H⁺ + C(solid)-collisions. The arrow indicates at each incident energy E_H the value E_{max} as given by eq. (2).

thicknesses and bombardment with H projectiles. The shape of the electron distribution changes significantly with the target thickness; the yield increases with the thickness and saturates for thicknesses larger than 10 µg/cm².

To start with the simplest case, measured electron spectra from a carbon foil bombarded with H are shown in fig. 4 /5/. On the continuous part of the spectra a weak peak at 255 eV energy which becomes more prominent with increasing incident energy E_P is produced from

K-LL Augertransitions of target carbon atoms. The
sharp drop in intensity at $E > E_{max}$ is caused by the
kinematic conditions of the binary collision process,
where the projectile transfers a maximum momentum under
an observation angle of $\vartheta = 42.3°$ to a target electron
at rest:

$$E_{max} = 4 \; (m_e/m_p) \; E_p \; \cos^2 \vartheta \qquad\qquad (2)$$

Here, m_e and m_p are the electron and projectile masses,
resp. At the energy E_{max} a broad peak appears, the "binary collision peak". With increasing E_p the binary
collision peak becomes more prominent. This is demonstrated in fig. 5, where the double differential cross
sections of electron scattering by bombarding H on carbon gas are calculated. It is a consequence of the energy loss of the electrons in the solid that the binary
collision peak is strongly smeared out in the measured
and calculated spectra fig. 4 and 6. In contrast to
the experimental results the electron distribution thus
calculated exhibit for $E_e < E_{max}$ and for $E_p < 2$ MeV
only a weak dependence from the projectile energy E_p.
Furthermore, at $E_e > E_{max}$ the slope of the calculated
electron intensity distribution is larger and at $E_e <$
E_{max} is smaller than the experimental distributions.
A reason for these differences could be the energy loss
relation, which was used as a first approach. A more
realistic energy loss relation for solids including e.g.
plasmon excitation at low electron energies should be
applied in future studies. It was found that the slope
of the distributions for $E_e > E_{max}$ depends on the atomic number Z of the target (compare fig. 10).

The continuous electron spectra observed from the
same target by bombardment with protons and heavy ions
of equal velocity are, in the framework of this theory,
expected to be similar except that the total intensity
is proportional to Z_p^2. To test this point a solid C
target was bombarded with 0.5 MeV/amu H and Ne ions. The
observed spectra are shown in fig. 7 together with the
calculated BEA result for the 20 μ/cm^2 C target. Within an experimental uncertainty of 2 the ratio of the
overall yield for neon bombardment to the yield for H
bombardment was found to be about 100, which is in
agreement with the expected ratio of Z_{Ne}^2/Z_H. The main
intensity observed is, at least for $E_e \gtrsim 1$ keV, accounted for by the BEA-model. In detail, however, especially at low energies, the measured spectra differ signifi
cantly from both experiment and from BEA calculations
(fig. 7). Augerelectrons ejected in flight from the

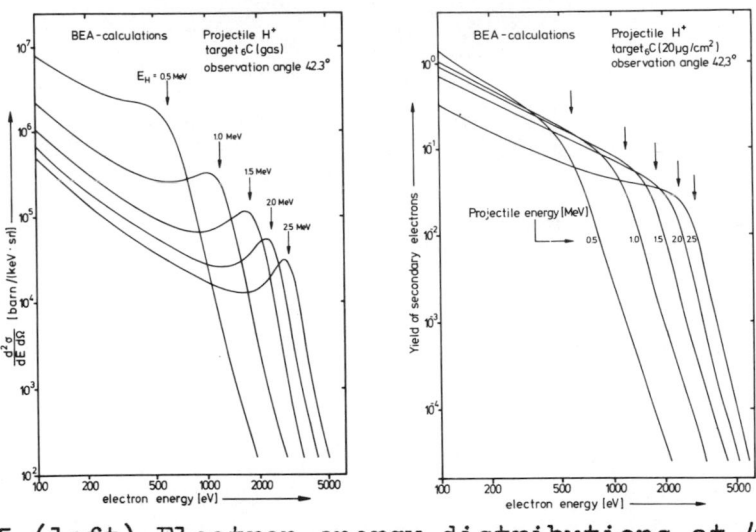

Fig. 5 (left) Electron energy distributions at 42.3°
from H+ + C(gas)-collisions calculated with the BEA-
theory. The arrow indicates at each incident energy E_H
the value E_{max} as given by eq. (2).

Fig. 6 (right) Electron energy distributions at 42.3°
from H+ + C(solid)-collisions calculated with the BEA-
theory. The arrow indicates at each incident energy E_H
the value E_{max} as given by eq. (2).

Ne-projectiles are clearly seen as characteristic lines
in the spectrum (see below) /6/. To elucidate these
phenomena further, more data of energy and angular
distributions of electrons ejected from heavy-ion-solid
collisions are needed.

The energy dependence of the electron energy
spectra observed under 42° of Ne + C collisions is
shown in fig. 8 /6/. The sharp drop in electron intensity at E_{max} (eq. (2)) and its incident projectile
energy dependence is clearly seen.

Augertransitions excited by the projectile in the
target atoms are weak compared to the intensity of
continuous electron contributions for 42.3°; at the
highest studied incident energy the carbon KLL Augertransitions become barely visible. The energies of
Augerelectrons $E_e^{c.m.}$ emitted in flight from the foil
excited Ne beam are shifted in the laboratory frame
(E_e^{lab}):

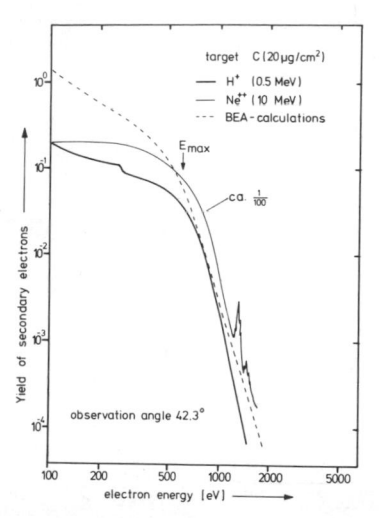

Fig. 7 Electron energy distributions at 42.3° from
H + C(solid)-collisions (thick line) and Ne + C (solid)-
collisions (thin line) at equal velocity (0,5 MeV/amu).
The intensity is approximately divided by Z_{Ne}^2 whereby
the shape is expected to be given by one theoretical
curve (dashed line) only.

$$E_e^{c.m.} = E_e^{lab.} + E_p(m_e/m_p) - 2 \cos (\vartheta)(E_e^{lab.} \cdot (m_e/m_p)E_p)^{1/2}($$

where m_e, m_p are the masses of the electron and projectile respectively. In fig. 8 these Augertransitions
emitted in flight from the projectiles are clearly seen
between $E_e^{lab.}$ = 1 and 2 keV. They have the projectile
energy dependence given by eq. 3.

    The conditions under which these transitions can
be studied with a reasonable peak to background ratio,
can be learned from fig. 8: The projectile energy may
not be too low since the ionization cross section then
becomes too small; on the other hand, the projectile
energy may not be too high since the continuous background fills in under the Augerlines. In addition, at
low projectile energies a broader velocity distribution
of the projectiles introduced by energy and angular
straggling /21/ in the exciting carbon foil contributes
considerably to the measured electron line width.

    Since the projectiles penetrating through the
solid foil undergo many successive collisions (typical

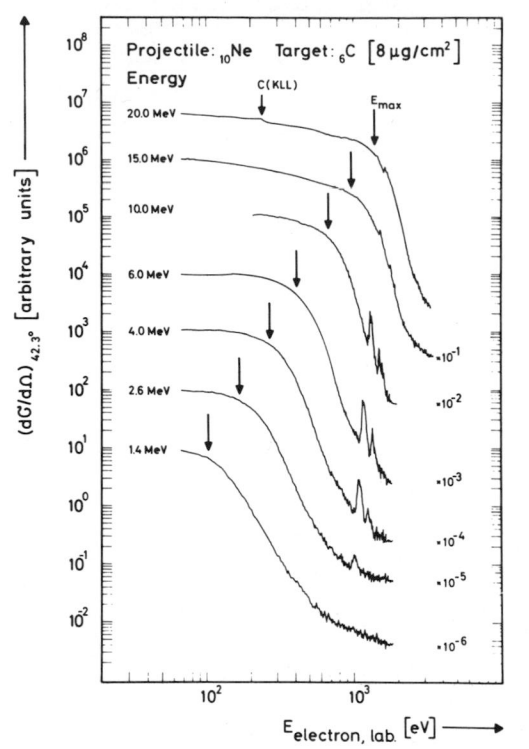

Fig. 8. Energy distributions of electrons emitted under 42° by bombarding solid carbon targets with neon-
projectiles at incident energies between 1.4 and
20 MeV /6/.

time between two successive collisions is 10^{-15} sec.),
the ions emerging from the foil are, in the first moment, in equilibrium excitation and equilibrium charge
states; the incident charge state of the projectiles
is irrelevant, i.e. the spectra are independent of the
incident charge state for the foil thicknesses used in
this experiment /22/.

Fig. 9 shows in detail the N Augerspectrum obtained from the beam foil experiment at 1.0 MeV projectile energy. The measured lines may be attributed
to transitions in N^{2+}, N^{3+} or N^{4+} which are the
preferably populated charge states for 1.0 MeV Ne in
having passed through a solid foil /23/. The Auger-
spectra obtained by the beam-foil-technique differ
significantly from those excited by heavy ion collisions under single collision conditions /24/.

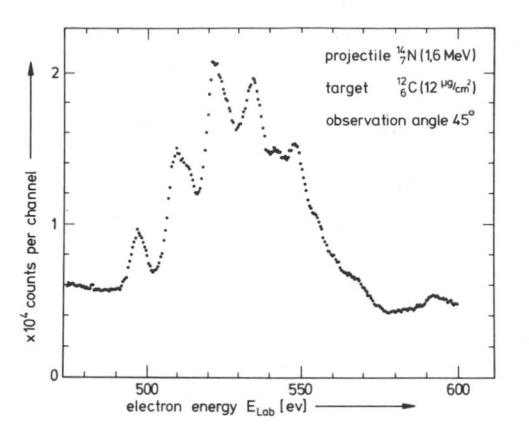

Fig. 9 Augerspectrum from foil-excited N emitted in flight from the projectile /23/.

     High energy-resolution Auger electron spectro-
metry has been applied successfully e.g. in ref. /25/.
This technique gives very precise values of the Auger
transition energies but does not characterize the type
of transition by a different parameter as e.g. the
initial charge state of the excited projectile prior
to the Auger transition. Several different approaches
have been applied to identify the electronic configurations involved in the Auger transitions by such
an additional parameter:

1. By proper selection of the projectile energy a
particular charge state can be preferentially populated since the dominant charge of the projectile emerging from the exciter foil is determined by its energy /26/. This method, however, has a limited applicability, as has been discussed in the context of
fig. 8. The use of a gas target as exciter of the projectiles can extent somewhat the range of applicability of this method.

2. By separating the locus of the beam excitation by
the target foil and the locus of electron emission for
detection in the spectrometer, transitions which are
metastable are selected. This technique has been
applied very successfully to the measurements of life
times of autoionizing states inisoelectronic series
/7,27/.

3. Very recently coincidence experiments between the
emitted Auger electrons and the projectiles of a par-

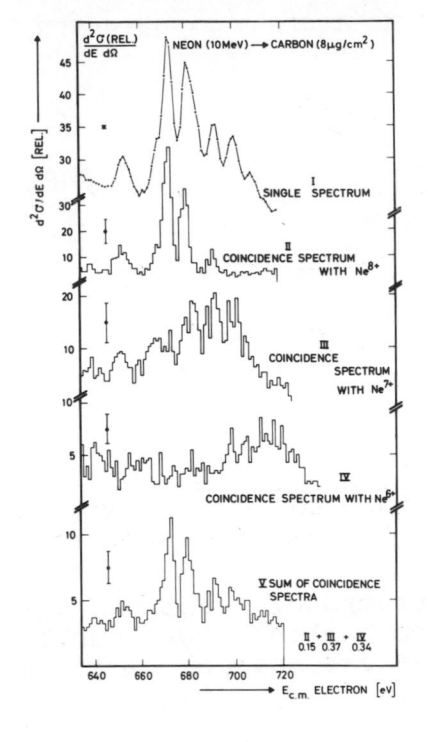

Fig. 10 Energy spectra of electrons emitted in coincidence with Ne-projectiles of charge 6+, 7+ and 8+
/33/

ticular charge state have been performed in Frankfurt
/33/: The charge analysis has been achieved by electrostatic deflection, the projectiles are detected by a
Si-surface barrier detector, its pulses served as time
signals (1). The time signal (2) was produced in a
channel electron multiplier where the Auger electrons
were detected after energy analysis in a coaxial cy-
lindersymmetric analyser /6/. Fig. 10 gives a preliminary result. For a particular electron peak, its
energy and, as additional parameter, its final charge
state are now known. This technique permits a very
detailed identification of the electronic configurations taking part in an Auger transition.

The measurement of angular distributions will
elucidate the different reaction mechanisms contribu-
ting to measured electron energy spectra. A first experiment with energy spectra at two different angles

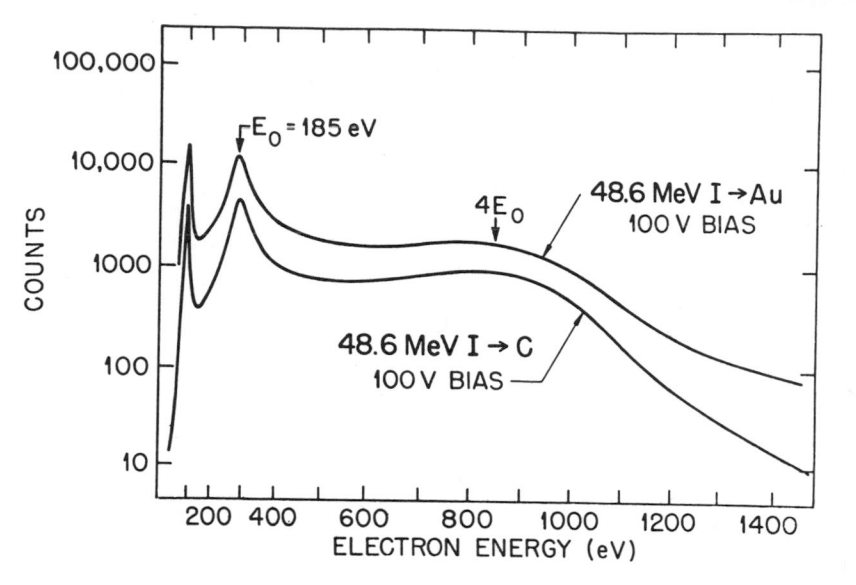

Fig. 11 Energy spectra of electrons emitted in forward direction from carbon (lower curve) and gold
(upper curve) targets bombarded by I (48.6 MeV energy)
/24/.

has been reported by Pferdekämper et al. /28/ for light
fission products from a ^{252}Cf-source. The data are
discussed in the framework of the BEA-theory, and total
yields and mean electron energies are derived.

Energy spectra of electrons ejected in the forward direction from I + (C, Au and Air)-collisions have
been measured at Oak Ridge National Laboratory /29/.
In this experiment (Fig. 11) a large contribution of
electrons with essentially zero energy has been observed. Besides the other features, already mentioned
above, these spectra exhibit a prominent intensity of
electrons released from the moving projectile with almost zero energy in the projectile rest frame (E =
$E_{max}/4$), and which is attributed to the charge exchange to a continuous state /30,31/. A very interesting
experiment, which will only be mentioned here, will be
reported by the Oak Ridge Group /32/; electrons produced by oxygen and copper projectiles emerging from
thin Au crystals in channeled and in random direction

have been studied.

We can resume that the characteristics of the energy distributions of electrons emitted from ion-solid collisions are:

a. intensity decrease with electron energy,
b. intensity dependence on the atomic number Z of the target,
c. a sharp drop in electron energy distributions starting at electron velocities equal twice the velocity of the projectile (see eq.(2)),
d. dependence of the slope of the electron energy distributions on the target electron binding energies,
e. superimposed Auger electrons characteristic for the target and/or projectile,
f. in forward direction a prominant peak at $v_e = v_p$.

Theoretical approaches within the framework of the binary encounter model have been attempted taking into account the energy loss of the electrons inside the target material. The results of these calculations are in reasonable agreement with the experimental data for light projectiles; a quantitative understanding, however, of the various physical processes has not been archieved.

The collaboration in the results reported from Frankfurt with F. Folkmann (GSI), R. Mann, G. Nolte, S. Schumann and R. Spohr is gratefully acknowledged.

References

/1/   F. Folkmann, Journal of Physics E, Sc. Instr. $\underline{8}$, 429 (1975)

/2/   P. Armbruster, G. Kraft, P. Mokler, B. Fricke, H.J. Stein, Physica Scripta, $\underline{10A}$, 1053 (1974)

/3/   B. Müller, W. Greiner, Phys. Rev. Lett. $\underline{33}$, 469 (1974)

/4/   P. Kienle, M. Kleber, Ch. Kozhuharov, Martin, B. Povh, D. Schwalm, to be published

/5/   F. Folkmann, K.O. Groeneveld. G. Nolte, R. Mann, R. Spohr, Z. Physik to be published

/6/   K.O. Groeneveld, R. Mann, G. Nolte, S. Schumann, R. Spohr, B. Fricke, Z. Physik (1975)

/7/   K.O. Groeneveld, R. Mann, G. Nolte, S. Schumann, R. Spohr, Phys. Lett. (1975)

/8/   G. Carter, J.S. Colligon, Ion Bombardment of Solids (Heinemann Educational Books Ltd., London 1969)

/9/   K.H. Krebs, Fortschritte der Physik $\underline{16}$, 419 (1968)

/10/ W. Meckbach, this Conference

/11/ U. Werlein, R. Bass, E. Dietz, K.O. Groeneveld, Verhandl. DPG (VI) $\underline{8}$, 79 (1973) and

      U. Werlein, Diploma Thesis, Frankfurt/M. 1973

/12/ H.-G. Clerc, H.J. Gehrhardt, L. Richter, K.H. Schmidt, Techn. Hochschule Darmstadt/ Germany, Report IKDA 73/4

/13/ E.J. Sternglass, Phys. Rev. $\underline{108}$, 1 (1957)

/14/ H. Seiler, Z. Angew. Physik $\underline{22}$, 249 (1967)

/15/ A. Eberhagen, Fortschritte der Physik $\underline{8}$, 245 (1960)

/16/ D. Achtstein, Univ. Frankfurt/M., Germany, Annual Report IKF-$\underline{32}$, 92 (1973), private communication and to be published

/17/ D. Burch, N. Stolterfoht, Ninth International Conference on the Physics of Electronic and Atomic Collisions, Seattle/Wa., USA, July/August 1975

/18/ J.D. Garcia, Phys. Rev. $\underline{A1}$, 280 (1970)

/19/ Yong Ki Kim, Radiation Research 61, 21 (1975) and
      S.T. Manson, L.H. Toburen, D.H. Madison,
      N. Stolterfoht, Phys. Rev. A12, 60 (1975)

/20/ T.F.M. Bonsen, L. Vriens, Physica 47, 307 (1970)

/21/ G. Spahn, K.O. Groeneveld, Nucl. Instr. Meth.
      124, 425 (1975)

/22/ H.D. Betz, Rev. Mod. Phys. 44, 465 (1972)

/23/ K.O. Groeneveld, R. Mann, R. Spohr, IKF-Frank-
      furt/M and GSI-Darmstadt, unpublished

/24/ K.O. Groeneveld, R. Mann, G. Nolte, S. Schumann,
      R. Spohr, Physica Fennica 9S1, 31 (1974)

/25/ C.F. Moore, J.J. Mackey, L.E. Smith, J. Bolger,
      B.M. Johnson, D.L. Matthews, J. Phys. B7, L 302
      (1974) and
      N. Stolterfoht, H. Gabler, U. Leithäuser,
      Phys. Lett. 45A, 351 (1973) and
      K. Körber, W. Mehlhorn, Z. Physik 191, 217 (1966)

/26/ R. Hallin, J. Lindskog, A. Marelius, J. Pihl,
      R. Sjödin, Physica Scripta 8, 209 (1973)

/27/ H.H. Haselton, R.S. Thoe, J.R. Mowat, P.M. Grif-
      fin, D.J. Pegg, I.A. Sellin, Phys. Rev. A11, 468
      (1975) and references quoted there

/28/ K.E. Pferdekämper, H.-G. Clerc, submitted for
      pub. to Z. Physik

/29/ S. Datz, B.R. Appleton, J.A. Biggerstaff,
      M.G. Menendez, C.D. Moak, Bullt. Am. Phys. Soc.
      18, 662 (1973) and
      B.R. Appleton, private communication

/30/ J. Macek, Phys. Rev. 1, 235 (1970)

/31/ K. Dettman, K.G. Harrison, M.W. Lucas,
      J. Phys. B7, 269 (1974)

/32/ S. Datz, B.R. Appleton, J.A. Biggerstaff,
      T.S. Noggle, H. Verbeek, Amsterdam, September
      1975, to be published in Surface Science

/33/ K.O. Groeneveld, G. Nolte, S. Schumann, to be publ.

KLM RADIATIVE AUGER TRANSITIONS *

W.L. Hodge, D. Schneider, and C.F. Moore

University of Texas at Austin
143 ENS University of Texas
Austin, Texas 78712

The radiative Auger (RA) process is a double electron transition in which one electron fills an inner vacancy and another electron is excited into a bound or continuum state with the emission of a photon. Åberg has suggested the mechanism responsible is both shake-off and configuration interaction (1). In the 1930's the RA process was discussed as a possible explanation of satellite lines observed in x-ray spectra (2), and recently the process has been analyzed in more detail (1) concerning mainly the KLL, KMM and LMM RA transitions. The KLM RA transitions have been postulated earlier (3-7), however nothing conclusive was reached.

In recent experiments (6,7,8) heavy ion excition of inner shell electrons, transitions were observed lower in energy than the $K\alpha_{12}$. These transitions have been classified by various physicists as volume plasmons (8), surface plasmons, local plasmons, Compton scattering, or absorption effects. This talk presents the first conclusive evidence that the transitions are KLM radiative Auger transitions.

In photon, electron and proton excitation the KLL (RA) transition dominates the lower energy part of the spectrum (Figure 1), however in heavy ion excitation the KLM (RA) transitions are more intense than the KLL (RA). The KLL (RA) are generally a collection of unevenly spaced broad peaks followed by continuous background in light particle excitation and washed out very broad peaks in heavy ion excitation. The KLM (RA) are typically four fairly evenly spaced peaks on top of edge of the $K\alpha_{12}$ transition. Figure 1 shows the x-ray spectra of Aluminum following 3 MeVproton, 20MeVOxygen

excitation. The proton data is identical with photon or electron
excitation (4),and the oxygen induced spectra indicate 3 peaks
in poor statistics data and 4 peaks in good statistics data.
Figures 2-5 show spectra of sodium, silicon, calcium, and scandium.
Theoretical energies of the RA transition were calculated using
Hartree-Fock (9) average energies with a relativistic correction
(10).

The notation $2p^{-n}3p^{-1}$ indicates a $2p \to 1s$ and $3p \to$ continuum
transition with various holes in the 2p sublevel and Table 1
is a sample calculation.

Table I
$2p^{-n}3p^{-1}$ for Al

Initial State	Final State	$E_{HF}+E_{REL}$ (keV)
$1s^1 2s^2 2p^6 3s^2 3p^1$	$1s^2 2s^2 2p^5 3s^2 3p^0$	1.472
$2p^5$	$2p^4$	1.468
$2p^4$	$2p^3$	1.464
$2p^3$	$2p^2$	1.459
$2p^2$	$2p^1$	1.454
$2p^1$	$2p^0$	1.448

Notice the RA from different L shell vacancies are nearly equal
and the energy spread is about 20 eV (the same as FWHM of the 4
peaks). For aluminum and silicon the calculations don't completely
predict the peak position, however for calcium and scandium the
comparison is closer and the discrepancy is essentially removed
in scandium. In the lower Z atoms, the 3s and 3p electrons (which
are involved in the transition) are valence electrons and the
binding energy is strongly dependent upon the chemical environment
of the atom. In calcium and scandium the 3s and 3p electrons are no
longer valence electrons and therefore there is better agreement
between Hartree-Fock calculations and the RA transitions.

Some of the 4 peaks that we observe are very weak in photon
and electron excitation data (4,5). Several possible explanations
for the increased KLM (RA) intensity in heavy ion excitation are
(A) the KLL (RA) transitions are spread over a wider energy range
than the KLM (RA) and therefore appear less intense (B) the multiply
ionized final states in oxygen excitation may mix easier with other
configurations, or (C) the RA transtiion probability due to shake-
off is increased with Oxygen excitation.

In conclusion, the 3 or 4 small peaks observed lower in energy
than the $K\alpha_{12}$ x-ray are KLM radiative Auger transitions. The peaks
are shifted in lower Z atoms because of the chemical environment.

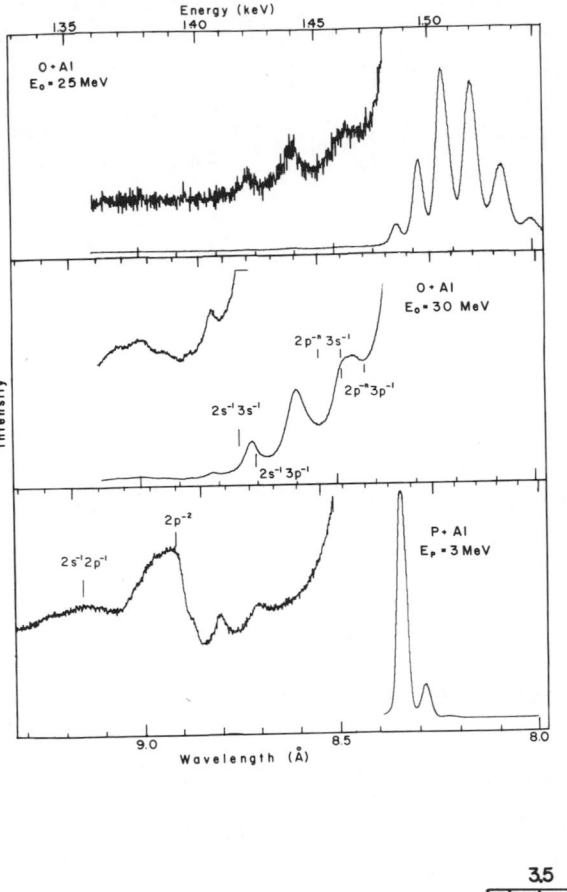

FIGURE 1

X-Ray Spectra of
Aluminum following ion-
impact excitation.

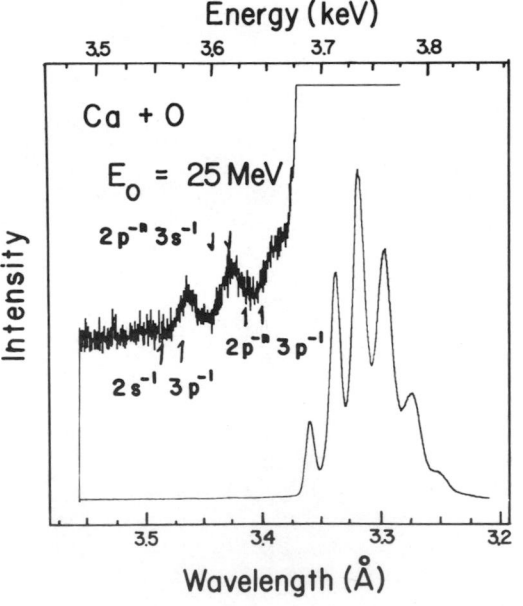

FIGURE 2

X-Ray Spectrum of Calcium
following oxygen
bombardment.

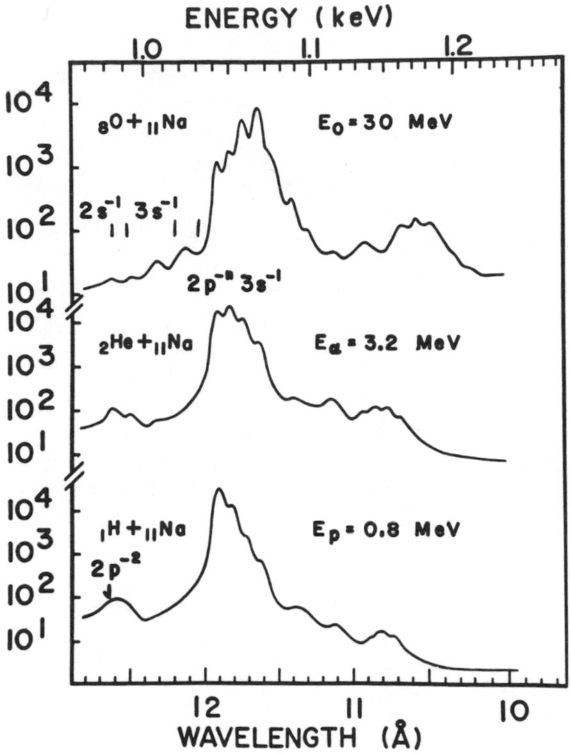

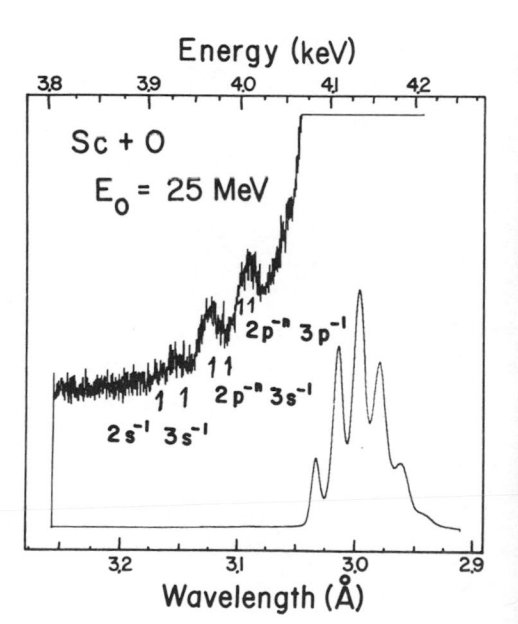

FIGURE 3

X-Ray xpectrum of Sodium
following ion
bombardment.

FIGURE 4

X-Ray spectrum of Scandium
following oxygen
bombardment.

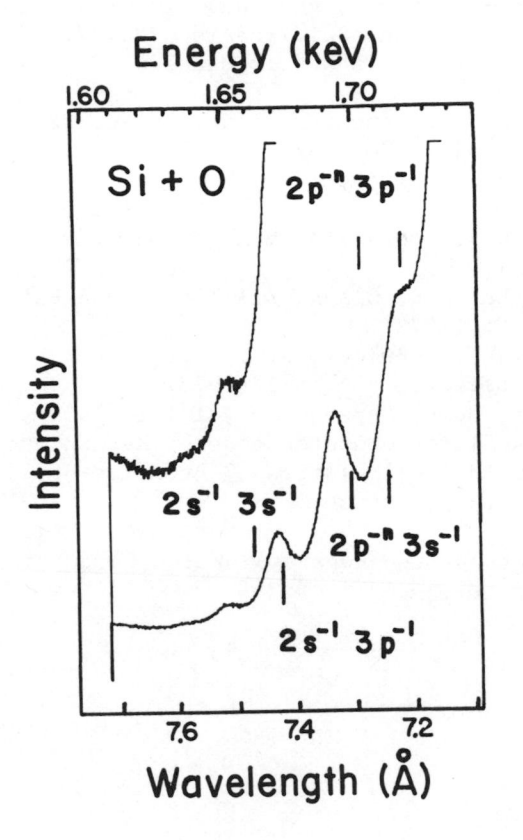

FIGURE 5

X- Ray spectrum of Silicon following oxygen bombardment.

References

1. T. Aberg, In "Atomic Inner-Shell Processes",(B. Crasemann, ed.)
    Vol. I, Academic Press,   N.Y.(1975) and references contained
    within.
2. E. G. Ramberg,Phy.Rev. 45, 389 (1934).
    F. Bloch, Phy. Rev. 48,187 (1935).
    F. Bloch and P.A. Ross , Phy. Rev. 47, 884 (1935).
3. H. Hulubei, C.R. Acad. Sci. (Paris) 224, 770 (1947).
    H. Hulubei, Y. Cauchois and I. Manescu, C.R. Acad. Sci. (Paris)
    226, 764 (1948).
4. J. Siivloa, J. Utriainen, M. Linkoaho, G. Graeffe, and T. Aberg
    Phy. Letters, 32A, 438 (1970).
5. T. Aberg and J. Utriainen, J. Phy, (Paris), 32, Suppl. C4, 295
    (1971).
6. P. Richard, C.F. Moore, and D.K. Olsen, Phy. Letters, 43A 519
    (1973).
7. C.F. Moore, David K. Olsen, Bill Hodge and Patrick Richard, Z.
    Physik 257 288 (1972).
8. J. McWherter, J.E. Bolger, H.F. Wolter, D.K. Olsen and C.F.
    Moore, Phy. Letters, 45A, 57, (1973).
9. C.F. Fischer, Comp. Phys. Comm., 1 151 (1969).
10. Hartree-Fock-Slater computer code by Herman and Skillman,
    "Atomic Structure Calculations," Prentice-Hall, Englewood Cliffs
    New Jersey.

*Supported in part by the Robert A. Welch Foundation, AFOSR and ERDA.

HIGH RESOLUTION AR K AUGER SPECTRA PRODUCED IN 4 AND 2 MEV H[+]

ON ARGON COLLISIONS

D. Schneider,[+] K. Roberts, B. M. Johnson, J. Whitenton,

C. F. Moore*

Department of Physics

The University of Texas at Austin

Austin, Texas   78712

Relative Intensities of satellite Auger transitions in neon have been studied for excitation by photons (Krause et al. {1}), electrons (Körber and Mehlhorn {2}; Krause et al {1}) protons (Edwards and Rudd {3}; Schneider et al. {4}) and heavy ions {10,11} at various energies. It has been shown that Ne K Auger spectra excited by photons, electrons, and protons of sufficiently high projectile energy are identical. It is shown in earlier measurements that with decreasing proton energy the satellite intensity increases to 80% of the total Auger intensity. The determination of relative line intensities in the Auger spectra provides a sensitive test of theoretical models as was shown by Krause {8} and Mackey et al. {9}. The change of the spectroscopic structure in an Auger spectrum caused by different projectiles and different energies, allows the identification of specific satellite line structures and the determination of the degree of the multiple ionization. In order to extend a detailed study as yet only carried out for neon, we measured proton induced highly resolved Ar K Auger-electron spectra at two different projectile energies.

* Research supported in part by the Robert A. Welch Foundation, and The Energy Research and Development Administration.

+ On leave from the Hahn-Meitner-Institut, West Berlin, Germany.

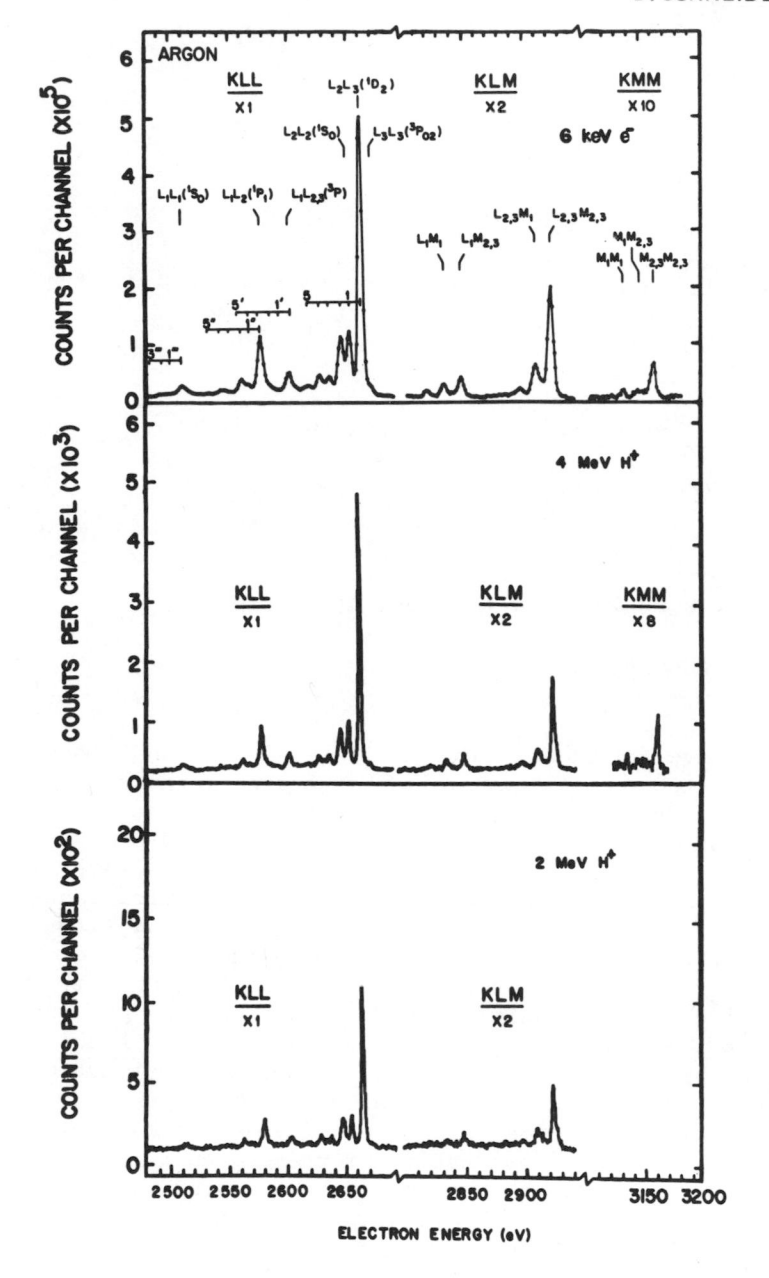

Figure 1

The KLL, KLM and KMM Auger spectra of argon, excited by 6 keV
electrons and 4 and 2 MeV protons. The electron excited spectrum
was measured by Krause et al. {8}. According to the notation of
Krause the satellite Auger lines are denoted by 1 to 5.

   Ar K Auger electrons produced in proton on Ar collisions were
measured by using the crossed beam method {5,6}. A tandem Van de
Graff accelerator was used to bombard Ar in an atomic beam by 4
and 2 MeV protons. Auger electrons ejected from the target atoms
at an angle of 90° with respect to the incident beam were analyzed
using a double focussing electrostatic analyzer (McPherson ESCA
36). A value of 2 eV(FWHM) was observed for the Ar KL_2L_3 ($^1D_2$) line.
Taking into account the natural width {7} of the line and the
kinematical (Doppler) broadening due to target recoil, an instrumental resolution of 1.4 eV is estimated. This value is larger
than expected; however, there may be broadening effects such as
small shifts of the electron energy during a scan of the spectrum.
The peak to background ratio in the Auger spectrum was improved in
this measurement by using a gas jet instead of a gas cell in
earlier measurements {9}. In order to define the scattering center
accurately an improved collimation system was used which collimated the protons to a beam of 2 mm in diameter. To insure single
collision conditions the target gas pressure was maintained at
roughly a few mTorr in the collision region, and at 1×10^{-4} Torr a
few cm from the gas get, and less than 5×10^{-6} Torr in the analyzer
region. The experimental apparatus has been reported earlier {6}.

   In fig. 1, Ar K Auger electron spectra obtained by 4 and 2
MeV proton impact are displayed and compared to spectrum resulting from 6 keV electron impact {8}. Since a modified beam collimation system and a gas jet instead of a gas cell as gas target
were used, the resolution could be improved over earlier measurements {9}. The high signal to background ratio obtained in these
measurements makes the proton and electron induced spectra look
identical. It was possible to correct some of the results obtained
from a previous proton induced spectrum. The line energies and
relative line intensities are given in Table I.

   The lines in the Auger electron spectra are superimposed on
an electron background resulting from electrons knocked out in
binary ion-electron collisions. This background which is not subtracted in the displayed spectra, is very low at the impact
energies and observation angle used here. In the present work the
background was determined by fitting a polynomial to the electron
continuum on each side of the Ar K Auger groups. As pointed out
in an earlier paper {9} and as it is shown from Krause {8} there
is a good agreement between the measured line intensities for all
identifiable KLL transitions and the recently calculated transition rates for the spectra resulting from 4 MeV proton and 6 keV
electron bombardment.

   In fig. 2, a detailed view of the KL_2L_3($^1D_2$) and KL_3L_3($^3P_{02}$)
line is shown. The relative intensities for KL_3L_3($^3P_{02}$) are found
to be somewhat different than those deduced from the electron induced spectrum {8}. However, this might be due to an increased

Group	Line	EXPERIMENTAL						THEORY	
		4 MeV H^+		2 MeV H^+		6 keV e^- [a]			
		Energy(eV)	Rel.Inten.	Energy(eV)	Rel. Inten.	Energy(eV)	Rel.Inten.	Energy(eV) [b]	Rel.Inten. [c]
KLL	KL_1L_1 ($^1S_0$)	2513.4	0.09	2513.6	0.084	2508.9	0.085	2514	0.09
	satellite	2563.8	0.10	2563.9	0.012	--	--	--	--
	KL_1L_2 ($^1P_1$)	2580.4	0.32	2580.7	0.28	2575.6	0.31	2577	0.34
	KL_1L_{23} (^3P)	2605.5	0.12	2605.6	0.10	2599.4	0.13	2601	0.09
	satellite	2648.7	0.29	2648.7	0.33	--	—	--	--
	KL_2L_2 ($^1S_0$)	2656.3	0.14	2656.4	0.11	2650.6	0.13	2651	0.12
	KL_2L_3 ($^1D_2$)	2662.2	1.00	2662.2	1.00	2660.6	1.00	2662	1.00
	KL_3L_3 ($^3P_0$)	2671.3	0.004	2671.2	0.004	2666.8	0.007	2668	0.002
	KL_3L_3 ($^3P_2$)	2674.1	0.020	2674.3	0.001	2669.1	0.017	2670	0.018
KLM	KL_1M_1	2837.8	0.29	2837.9	0.27	--	--	--	--
	KL_1M_{23}	2852.9	0.32	2852.8	0.30	--	--	--	--
	satellite	2902.1	}0.39	2902.3	}0.43	--	--	--	--
	satellite	2915.2		2915.4		--	--	--	--
	$KL_{23}M_1$	2917.2	0.53	2917.4	0.50	--	--	--	--
	$KL_{23}M_{23}$	2927.7	1.00	2927.9	1.00	2923.5	--	—	--
KMM	KM_1M_1	3140.4	0.51	--	--	--	--	--	--
	KM_1M_{23}	3150.6	0.34	--	--	--	--	--	--
	satellite	3158.1	0.39	--	--	--	--	--	--
	$KM_{23}M_{23}$	3166.6	1.00	--	--	3162.7	--	--	--

Table I

Line energies and relative line intensities deduced from the 4
and 2 MeV proton induced Ar K Auger lines and groups. Intensities
are normalized to the KL_2L_3 ($^1D_2$) line in the KLL group, to the
$KL_{23}M_{23}$ line in the KLM group and to the $KM_{23}M_{23}$ line in the KMM
group. The energy of the KL_2L_3 ($^1D_2$) line is normalized to Sieg-
bahn et al. (1969). The accuracy for the line energies is ±0.2
eV (a. Ref. 8; b. Ref. 13; c. Ref. 12 with configuration interaction).

probability for KLL transitions during which a 1s electron is
excited into the 3d or 4s level where it remains as a spectator
of the transition. The structure caused by the decay of those

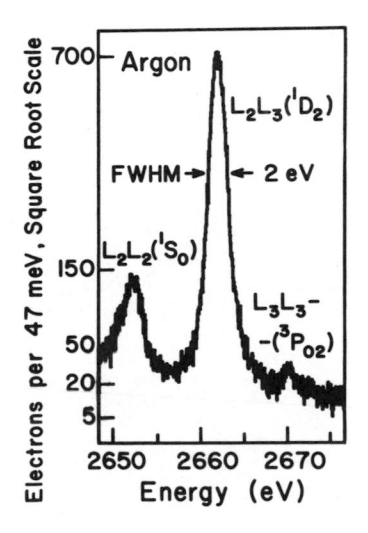

Figure 2

The KL_2L_3 ($^1D_2$) and KL_3L_3 ($^3P_{02}$) lines in the Ar K Auger spectrum.

states via the "spectator" channel could be responsible for intensity variations in this region of the spectrum.

The 4 and 2 MeV proton induced Auger spectra were carefully
fitted with a sum of Gaussian functions after subtraction of the
continuous background. There is, as stated before, no difference
in the line intensities of the electron and 4 MeV H^+ induced
spectra within the given uncertainties. However, the numbers in

Table I show a tendency for a decrease in the normal line intensities in general. This tendency becomes more strongly pronounced
for a proton projectile energy of 2 MeV. Since there is no exact
value for a branching ratio of the normal lines obtainable at this
point of the study, a rough method was used to determine the
increase of the structure due to satellite Auger transitions. For
the case of the 2MeV proton induced spectra, a relative increase of
about 20% was found after subtracting the normal line intensities
from the total Auger intensity.

As a result of this work an indication for a projectile energy
dependence for the satellite line production in the Ar K Auger
spectrum comparable to that for neon {4} was found. Different projectiles give identical Auger spectra if their velocities are high
enough. Present theories are found to be adequate to predict line
intensities in the Auger spectrum. This inference, should hold
for KLL Auger spectra for elements up to at least Z=18.

We are indebted to Dr. M. O. Krause for helpful communication.

R E F E R E N C E S

1   M. O. Krause, J. Phys. (Paris) 32 (1971) C4-67
2   H. Körber, W. Mehlhorn, Z. Phys. 191 (1966) 217
3   A. K. Edwards, M. E. Rudd, Phys. Rev. 1970 (1968) 140
4   D. Schneider, D. F. Burch, N. Stolterfoht, Proceedings of the
     VIIIth International Conference on Electronic and Atomic
     Collisions, Abstract of Papers, edited by B. Cobic and M. V.
     Kurepa (Institute of Physics, Belgrade, (1973), p. 729
5   D. L. Matthews, B. M. Johnson, J. J. Mackey, and C. F. Moore,
     Phys. Rev. Lett. 31 (1973) 1331
6   N. Stolterfoht, Z. Phys. 248, 81 (1971); Z. Phys. 248, 92
     (1971)
7   W. Mehlhorn, Z. Phys. 208, 1 (1968); O. Keski-Rahkonen and
     M. O. Krause, At. Data Nucl. Data Tables 14, 139 (1974)
8   M. O. Krause, Phys. Rev. Lett. 34 (1975) 11; Atomic Inner-
     Shell Processes (Academic Press, New York, 1975) edited by
     B. Crasemann (Dept. Phys., University of Oregon, Eugene, Or.)

9   J. J. Mackey, L. E. Smith, B. M. Johnson, C. F. Moore and
     D. L. Matthews, J. Phys. B: Atom. Molec. Phys. 7, 16 (1974)
     L 447
10   D. Burch, J. S. Risley, D. Schneider, N. Stolterfoht, H. Wie-
     man, Annual Rep. of the University of Washington, Seattle
     (1974) p. 165
11   D. L. Matthews, B. M. Johnson, J. J. Mackey, L. E. Smith, W.
     Hodge, and C. F. Moore, Phys. Rev. 1 10 (1974) 1177
12   M. H. Chen, B. Crasemann, Phys. Rev. A 8, 7 (1973)
13   F. P. Larkins (to be published 1975)
14   K. Seigbahn, C. Nordling, G. Johansen, J. Hedman, P. F. Heden,
     K. Harris, W. Gelius, T. Bergman, L. O. Werme, Y. Manne, Y.
     Baer, 1969 ESCA, Applied to free Molecules (Amsterdam: North
     Holland) p. 39

SECONDARY ELECTRON EMISSION FROM H^+ AND H_2^+ PASSAGE THROUGH THIN
CARBON FOILS

M. G. Menendez and M. M. Duncan

University of Georgia

Athens, Georgia 30602

Secondary electrons emitted with very nearly the same velocity as a fast ion, ($v_e = v_i$), have been observed in ion/gas
experiments[1], ion/foil experiments[2], and with channeled ions
through crystals[3]. Herein we report on the measurement of the
energy profiles and resulting angular distributions of these
electrons emitted near $\theta = 0°$ for H^+ and H_2^+ ions incident on
amorphous carbon foils. Figure 1 shows the experimental arrangement.

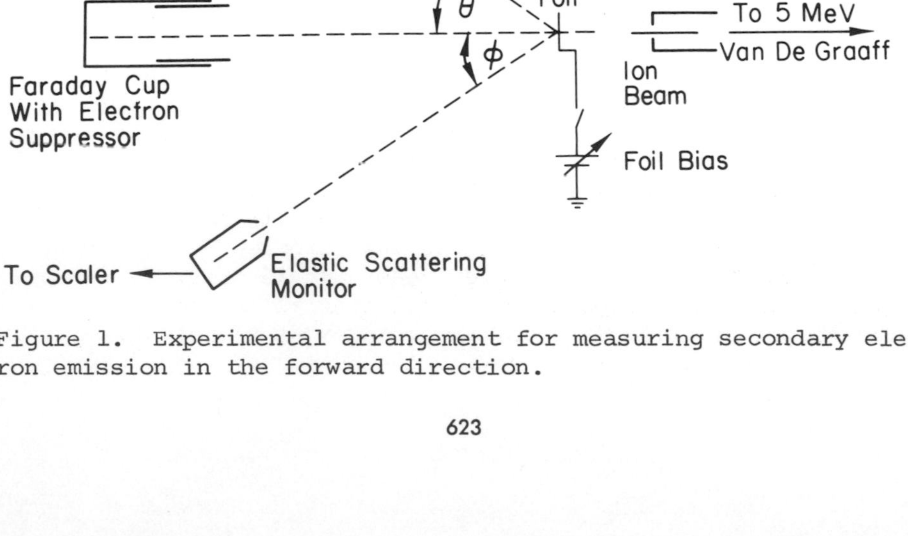

Figure 1. Experimental arrangement for measuring secondary electron emission in the forward direction.

The ion beams pass through collimators and into the chamber. To be noted is the 3mm channel completely through the analyzer. The beam that passes through this channel is measured in a suppressed Faraday cup. By rotating the analyzer about the center of the chamber the beam profile was measured both with and without a target. Incident beam diameters were less than 1mm. As the beam passed through the foils it was spread by scattering. However, the angular spread was small compared to the measured angular distribution of the electrons. Periodic beam profiles were measured to monitor the beam. If needed, adjustment was made to maintain constant beam conditions. Beam currents were typically a few nA.

The hemispherical electron analyzer[4] was calibrated with electrons from a hot tungsten filament and had an energy spread of $\Delta E/E = 0.02$ (FWHM) for the 1mm apertures used in these experiments. Standard electronics were used to obtain the electron energy distributions. The analyzer could be positioned in angle to 0.05° and subtended a cone of half angle 0.36° from the center of the chamber. A laser beam was used to align the analyzer with the center line of the chamber. Magnetic field compensation was provided by three sets of Helmholtz coils surrounding the chamber.

The pressure in the chamber was maintained at approximately 2×10^{-6} torr and target foils were bombarded by the beams for at least thirty minutes before any data were taken. Commercial carbon foils with thicknesses from 5 to 50 $\mu g/cm^2$ were used. Also, Ni foils of thickness approximately 70 $\mu g/cm^2$ were investigated. The H^+ and H_2^+ beams had equivalent proton energies between 0.175 and 1 MeV.

Since the electron analyzer blocks all or part of the scattered ion beam at all angles up to 18°, it was impossible to use current integration to normalize one run to another. In order to overcome this difficulty, a surface-barrier charged particle detector was used to count the elastic scattering events from the target.

The most dominant feature of the secondary electron spectrum at all energies greater than a few eV is the appearance of a group of electrons whose energy is centered about an energy E_i, where the velocities of the electron and the ion are the same, $v_e = v_i$. At small angles one finds many extraneous secondary electrons produced when the scattered ion beam hits the entrance apertures of the analyzer. This problem disappears at angles where there is no scattered beam. To identify this beam-aperture background, a positive bias potential was applied to the target of sufficient voltage to prevent all electrons produced at the target from leaving the region of the target.

This potential had no discernable effect on the ion beam.   Therefore, near 0° two runs were made, one with and one without bias.
After subtraction there remained, in addition to the group of
interest, a smooth continuous electron spectrum in agreement
with measurements at angles where beam-aperture electrons were
absent.   At angles near 0° the $v_e = v_i$ group rises sharply from
the smooth electron spectrum which is treated as a background
in our experiments.   To correct for the variable energy spread
of the analyzer the data in each channel are divided by a number
proportional to the electron energy of that channel.   Figure 2
shows two typical energy spectra.

    Upon integration over the $v_e = v_i$ groups one obtains the
angular distributions.   Figure 3 shows two typical distributions.
These distributions have been normalized to the same value at 0°
by fitting a smooth curve (not shown) through the data points.
They are narrow, but much wider than the angular spread of the
beam as measured by beam profiles.   All angular distributions
using 10 µg/cm² carbon foils fall to half their value at 0°
at angles $\theta_{\frac{1}{2}} = 1.5 \pm 0.5°$.   In general, the distributions from
the faster ions are somewhat narrower than those for slower ions.
Changes in beam spreading with the ion energy can account for
some of this variation in $\theta_{\frac{1}{2}}$.   The H_2^+ angular distribution seems

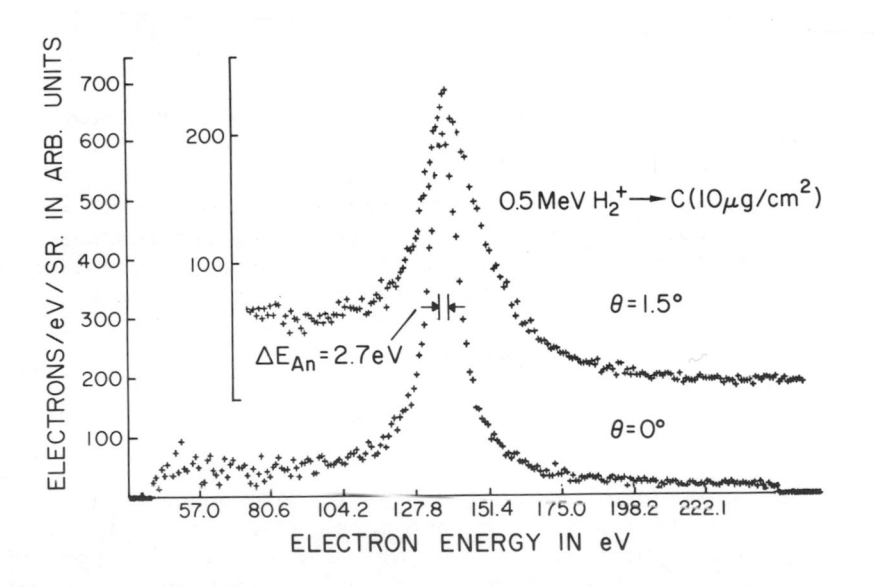

Figure 2.   Typical energy spectra corrected for analyzer resolution. $v_e = v_i$ electrons have an energy of 136eV.   The analyzer
resolution (FWHM) for that energy is shown.

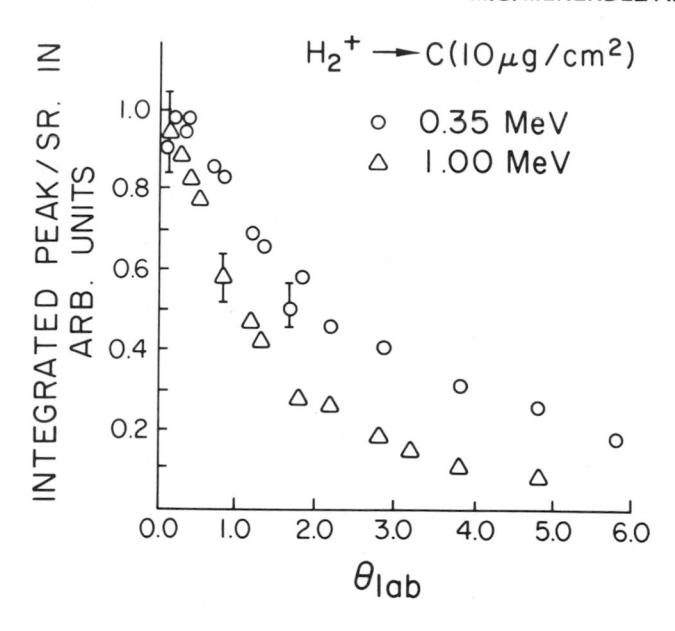

Figure 3.   Angular distributions of $v_e = v_i$ secondary electrons produced when H_2^+ ions bombard amorphous carbon foils.   Data for 0.5 and 0.7 MeV H_2^+ is also available and falls between the 0.35 and 1 MeV data shown here.   All angular distributions have been normalized to 1 at 0°.

to be narrower than the H^+ angular distributions at the same equivalent proton energy.

   The widths (FWHM) of the energy distribution of the $v_e = v_i$ group increases with ion energy for fixed angles and increases with angle for fixed ion energy.   Both dependences seem approximately linear except near 0° where beam spreading can cause a departure from linearity.   As a result of these approximately linear dependences, $FWHM/E_i$ should be the same for all ion energies for a given angle.   Figure 4 shows a small fraction of our data.   Widths, except near 0°, are the same for nickel as for carbon.

   Within experimental errors there is no dependence on the thickness of the carbon foils except those associated with beam spreading, in agreement with our experiments using He^+.   For our most extensive data, taken with H^+ and H_2^+ incident on 10 µg/cm^2 carbon targets, we found that the yield per proton at 0° decreases as E^{-3} in agreement with earlier results taken with less angular resolution[2].

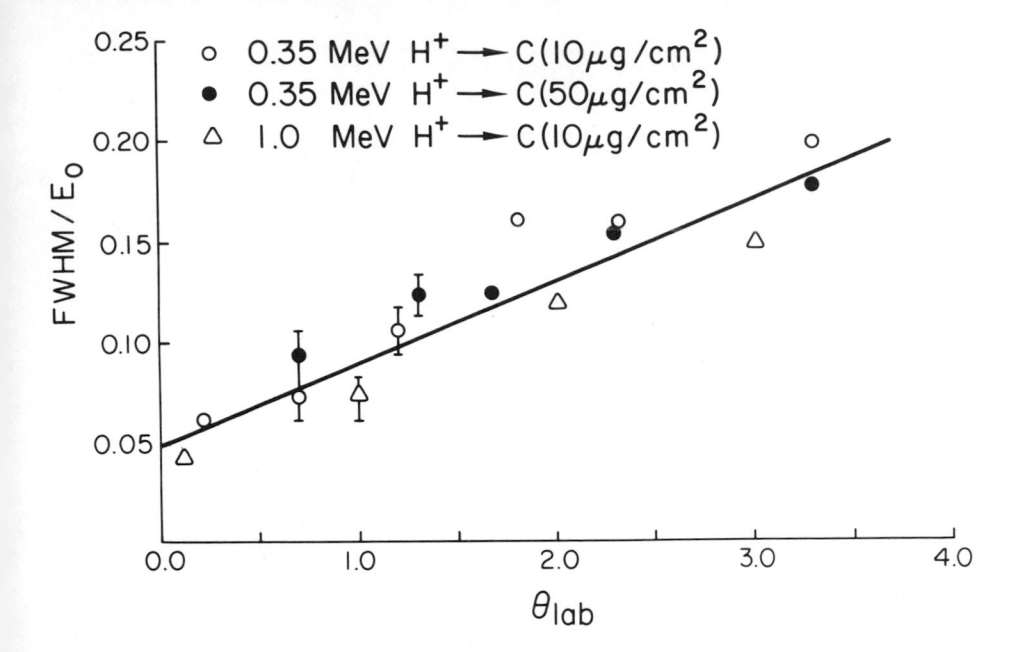

Figure 4. ΔE(FWHM)E_i for the $v_e = v_i$ group. Data is shown for two energies and two foil thicknesses. Only a fraction of the data used to develop the line is shown.

References

1. G. B. Crooks and M. E. Rudd, Phys. Rev. Lett. 25, 1599 (1970).

2. K. G. Harrison and M. W. Lucas, Phys. Lett. 33A, 142 (1970).

    S. Datz, B. R. Appleton, J. R. Biggerstaff, M. G. Menendez and C. F. Moak, Bull. Am. Phys. Soc. 18, 662 (1973).

    K. Dettman, K. G. Harrison, and M. W. Lucas, J. Phys. B., Proc. Phys. Soc., London 7, 264 (1974)

3. Datz, Appleton, Biggerstaff, Menendez and Moak, (unpublished).

4. J. A. Simpson, Rev. Sci. Instr., 35, 1698 (1964).

RELATIVE MULTIPLE IONIZATION CROSS SECTIONS OF NEON BY DIFFERENT

PROJECTILES*

C. P. Bhalla

Kansas State University

Manhattan, Kansas   66502   USA

ABSTRACT

The experimental relative x-ray intensities of neon produced in the heavy ion neon collisions are analyzed with the theoretical fluorescence yields and the statistical model of multiple ionization. The calculated relative multiple ionization cross sections of neon and average fluorescence yields are given for those cases where the experimental and the theoretical x-ray relative intensities are in agreement.

INTRODUCTION

Recently several high-resolution x-ray measurements of neon are reported for heavy ion-neon collisions (Kauffman et al. 1973, 1974, 1975; Matthews et al. 1974a, 1974b; Mowat et al. 1974). There are several well-defined broad peaks (in these K x-ray spectra of neon) which are usually identified with different degrees of multiple ionization in the L-shell, KL^n. The notation KL^n refers to a defect electronic configuration with a single K- and n-vacancies in the L-shell. Matthews et al. (1974b) and Kauffman et al. (1975) noted that the peak labeled KL^5 contains contributions from the ^4P states of KL^6 and, therefore, a corrections needs to be applied to the x-ray intensities in these two peaks.

Our calculations of multiplet x-ray energies (Bhalla 1975a) show that the peaks labeled KL^4 and KL^3 also need corrections in addition to the peaks KL^5 and KL^6 so that the relative x-ray intensities represent correctly the contributions of the various degrees of ionization as indicated by the notation, KL^n.

A detailed analysis of the experimental x-ray relative intensities to obtain the relative multiple ionization corss sections
requires a model of multiple ionization for example the statistical
model (Hansteen and Mosebekk 1972, McGuire and Richard 1973) and
the assumption of the statistical population of the spectroscopic
terms.   It is desirable to compare directly the experimental and
the theoretical x-ray intensities for the purpose of testing the
theoretical models of multiple ionization. We have presented
elsewhere (Bhalla 1975c) the results of our analysis for several
cases.

We present in this paper a consistent analysis of the x-ray
relative intensities of neon, which are produced in the heavy ion-
neon collisions, in particular for Ne^{+}-Ne collisions.

THEORETICAL PROCEDURE AND NUMERICAL RESULTS

We use the model of multiple ionization, proposed by Hansteen
and Mosebekk (1972) and also discussed in detail by McGuire and
Richard (1973), and by Hopkins et al. (1973). This model has been
recently utilized by Kauffman et al. (1975).

The ionization cross section, $\sigma_{K,2s^{\ell},2p^{m}}$, of single K-shell,
ℓ 2s-electron and m 2p-electrons is given by

$$\sigma_{K,2s^{\ell},2p^{m}} \Big/ \sigma_K = \binom{2}{\ell} P_{2s}^{\ell}(0) [1-P_{2s}(0)]^{2-\ell}$$

$$\times \binom{6}{m} P_{2p}^{m} [1-P_{2p}(0)]^{6-m} \tag{1}$$

where

$$\sigma_K = \int 2\pi b \, db \, \binom{2}{1} P_K(b)[1-P_K(b)] \quad.$$

The probabilities per electron of ionization in the 2s and 2p shells
for zero impact parameters are denoted respectively by $P_{2s}(0)$ and
$P_{2s}(0)$. We assume a statistical population of the various spectroscopic terms of an electron configuration. Since we calculate the
position and the relative intensity of all the lines from the
configurations $(1s) (2s)^{2-\ell} (2p)^{6-m}$ it is easy to compute the total
intensities in the various experimental peaks. Consequently, we
compare directly the experimental and the theoretical x-ray relative intensities.

Table I   Values of $P_{2p}(0)$, P_{2s}/P_{2p} and Average Fluorescence
Yields, ω, for Ne^{+}- Neon Collisions

Energy (KeV)	$P_{2p}(0)$	P_{2s}/P_{2p}	χ^2	$\omega \times 100$
100	0.53	0.00	4.6	2.12
200	0.53	0.00	3.2	2.12
300	0.54	0.00	2.6	2.12
400	0.53	0.002	1.7	2.13
500	0.55	0.004	1.6	2.16
600	0.55	0.00	2.7	2.13
800	0.58	0.01	0.5	2.24
1000	0.62	0.01	0.5	2.27

Table I contains the results of our analyses for Ne^+-Ne colli-sions. A comparison of the theoretical and the experimental (Matthews _et al._,1975) relative intensities is given in Table II. In these cases the search was made for the values of $P_{2p}(0)$ and $P_{2s}(0)/P_{2p}(0)$ so that a minimum value of χ^2 was obtained. The uncertainties in the experimental data in Table II were assigned by us. For ion-neon collisions in the bombarding energy range of $\simeq$ 1-2 MeV/amu, we tried first to see if there is an acceptable fit for values of $P_{2p}(0)$ and by assuming $P_{2s}(0) = P_{2s}(0)$. Table III contains these values, χ^2 and average fluorescence yields when there was acceptable fit to the data ($\chi^2 <$ 2.5). For Cl^{+7} neon col-lisions at 40 MeV, the experimental relative intensities were in reasonable agreement with the theory only when $P_{2s}(0)/P_{2p}(0)$ was taken in the range of 2.4 to 3.0 with $P_{2p}(0)$ as 0.43 to 0.38.

In most cases of stripped atoms (N, O, F) the values of χ^2 were too large.

DISCUSSION AND CONCLUSIONS

We have presented a consistent analysis of the x-ray rela-tive intensities of neon, which are produced in the heavy ion-neon collisions. The present procedure avoids any errors and ambiguities arising from the overlap of several multiplets of configurations with different number of L-shell electrons. Fur-ther, calculated relative x-ray intensities are compared directly with the experimental values to ascertain the validity of the sta-tistical model of multiple ionization. Tables I, II and III show that the statistical model is indeed valid for a large number of heavy ion-neon collision experiments.

In the cases of highly ionized, usually fully stripped, heavy ions the statistical model does not appear to be valid. One pos-sible reason could be that for bare heavy-ion projectiles there is a large probability of electron capture from the K-shell of neon by the projectile, which leads to a higher probability for double K-shell vacancies. The contributions to the KL^n x-ray spectra from these cascades do not lead to a statistical ionization dis-tribution because the dominant de-excitation occurs by the Auger process.

It should be noted that $P_{2s}(0) \simeq$ o for low energy Ne^+-Ne col-lisions. Furthermore, there are not acceptable fits to the experi-mental data for Ne^+ projectiles with energies $\leq$ 200 keV. It is also known that the molecular promotion model (Briggs 1975) is manily responsible for the ionization. Consequently, the ionization pro-cess may not statistical. Further work, however, is needed to arrive at a difinitive conclusion.

Table II Comparison of Experimental and Theoretical x-ray Relative Intensities, R_n, for Ne^+-Ne Collisions

Energy (KeV)		R_o	R_1	R_2	R_3	R_4	R_5	R_6
	Exp.	0.030(30)	0.160(16)	0.440(44)	0.310(31)	0.060(48)		
100	Th.		0.059	0.183	0.286	0.338	0.134	
100	Exp.		0.040(40)	0.170(17)	0.410(41)	0.330(33)	0.050(50)	
200	Th.		0.058	0.183	0.286	0.338	0.135	
200	Exp.		0.040(40)	0.160(16)	0.390(39)	0.340(34)	0.070(56)	
300	Th.		0.053	0.173	0.282	0.347	0.145	
300	Exp.		0.030(30)	0.170(17)	0.370(37)	0.330(33)	0.090(45)	0.010(10)
400	Th.		0.058	0.182	0.284	0.338	0.136	0.001
400	Exp.		0.030(30)	0.150(15)	0.360(36)	0.340(34)	0.110(55)	0.010(10)
500	Th.		0.048	0.161	0.274	0.356	0.158	0.003
500	Exp.		0.030(30)	0.130(13)	0.320(32)	0.370(37)	0.130(13)	0.020(20)
600	Th.		0.048	0.164	0.277	0.356	0.154	0.001
600	Exp.		0.020(20)	0.100(50)	0.280(28)	0.370(37)	0.200(20)	0.030(30)
800	Th.		0.034	0.131	0.251	0.377	0.197	0.010
800	Exp.		0.010(10)	0.080(64)	0.240(24)	0.380(38)	0.240(24)	0.050(50)
1000	Th.		0.022	0.099	0.224	0.397	0.245	0.014

Table III Values of $P_{2p}(0)$, χ^2 and Average Fluorescence Yields,
          ω, for Ion-Neon Collisions in the Projectile Energy
          Range of 1-2 MeV/amu

Projectile	Energy MeV	$P_{2p}(0)$ $(= P_{2s}(0))$	χ^2	$\omega \times 100$
C^{+3}	18	0.285	0.08	2.85
C^{+4}	18	0.305	0.09	3.16
C^{+5}	18	0.335	0.06	3.57
C^{+6}	18	0.410	1.04	5.2
N^{+4}	21	0.340	2.14	3.68
N^{+6}	21	0.395	2.5	5.03
N^{+7}	21	0.525	0.93	9.42
O^{+4}	16	0.360	2.31	3.91
O^{+4}	24	0.375	2.40	4.53
O^{+5}	24	0.390	0.50	4.66
O^{+5}	30	0.370	1.65	4.22
O^{+5}	30	0.345	0.94	3.81
O^{+6}	24	0.405	1.07	5.31
O^{+6}	35	0.350	1.92	3.91
Cl^{+11}	40	0.640	0.11	16.3
Cl^{+13}	40	0.725	1.79	22.3
Ar^{+6}	80	0.525	0.43	9.74

The relative multiple ionization cross section of neon are given by Eq (1) with the values in Table I and III, for those cases where there is an acceptable agreement between the theory and the experimental relative x-ray intensities.

It is a pleasure to thank Professor Patrick Richard for stimulating discussions and Mr. John Shellenberger for his assistance in the computations.

REFERENCES

*Work supported by the US Army Research Office, Durham, North Carolina.

Bhalla, C.P. 1975a, Phys. Rev. A 12, 122.

Bhalla, C.P. 1975b, J. Phys. B: Atom. Molec. Phys. 8 1200.

Bhalla, C.P. 1975c, J. Phys. B: Atom. Molec. Phys. (Submitted for publication).

Briggs, J. 1975, International Conference on the Physics of Electronic and Atomic Collisions (invited talk).

Hansteen, J. M. and Mosebekk, O. P., 1972, Phys. Rev. Lett. 29 1361.

Hopkins, F. Elliott, D. O., Bhalla, C. P. and Richard, P. 1973, Phys. Rev. A 8 2952.

Kauffman, R. L., Hopkins, F. F., Woods, C. W. and Richard, P. 1973 Phys. Rev. Lett. 31, 621.

Kauffman, R. L., Woods, C. W., Jamison, K. A. and Richard, P. 1974 J. Phys. B: Atom. Molec. Phys. 7 L335.

Kauffman, R. L., Woods, C. W., Jamison, K. A. and Richard, P. 1975, Phys. Rev. A 11 872.

Matthews, D. L., Johnson, B. M. and Moore, C. F., 1974a, Phys. Rev. A 10 451.

Matthews, D. L., Johnson, B. M., Hoffmann, G. W. and Moore, C. F., 1974b Phys. Lett. 49A 195.

Matthews et al., 1975, Physics Lett., 50A, 441. Matthews, D. L., (Private communication).
McGuire, J. H. and Richard, P. 1973 Phys. Rev. A 8 644.

Mowat, J. R., Laubert, R., Sellin, I. A., Kauffman, R. L., Brown, M. D, Macdonald, J. R. and Richard, P. 1974 Phys. Rev.A10, 1446.

SPECTRAL AND ELECTRON COLLISION PROPERTIES OF ATOMIC IONS

K. D. Chao[†], J. L. Dehmer[¶], U. Fano[*], M. Inokuti[¶],
S. T. Manson[†], A. Msezane[†], R. F. Reilman[†], C. E.
Theodosiou[*]

[†]Dept. of Physics, Georgia State U., Atlanta, Georgia
30303, USA [††]

[¶]Argonne National Laboratory, Argonne, Illinois 60534,
USA [¶¶]

[*]Dept. of Physics, Univ. of Chicago, Chicago, Illinois
60637, USA [**]

The aim of this study is to delineate the systematics of various properties of positive atomic ions over a wide range of nuclear charge Z and electron number N. These properties include energy levels, transition rates, electron collision cross sections, and other directly observable quantities. From a theoretical point of view, many of these properties are derived from a few basic quantities: phase shifts of partial waves $\delta_\ell(\varepsilon)$ [including both $\varepsilon>0$ and $\varepsilon<0$] and their energy derivatives, enhancement factors[1], $\alpha_\ell(\varepsilon)$, and oscillator strength distributions, $df/d\varepsilon$. It is these quantities that we calculate, mapping them extensively over Z and N. The point of view taken herein has been presented in a study of these same basic quantities for neutral atoms[1] (hereafter called FTD).

Since any property of the ions can be considered as a function of two variables, Z and N, our data can be analyzed in terms of at least three alternative pictures: *isoelectronic* (N kept constant), *isonuclear* (Z kept constant), and *isoionic* (z=Z-N kept constant). Each of these pictures brings out different aspects of the variation of a given quantity over Z and N.

[††]Supported by the U.S. Army Research Office and NSF.
[¶¶]Work performed under the auspices of U.S. ERDA.
[**]Supported by U.S. ERDA, Contract No. COO-1674-109.

The *isoelectronic* picture is the most straightforward from a computational view point and is widely used. In addition, it is directly applicable to 1/Z expansions[2]. Each of the properties approaches the known hydrogen values asymptotically[3] when plotted vs. 1/Z (1/Z→0). The *isonuclear* picture simplifies the variation of ionic properties from one order of the spectrum to the next. In particular we shall see that inner shell properties are virtually unaffected by stripping outer shell electrons. Further, the *isonuclear* picture is most suitable for certain applications. For example, if we consider a given impurity contamination in a CTR plasma, the many ions of that impurity nuclide which are relevant constitute an isonuclear sequence. This is also true for astrophysical applications[4]. The *isoionic* picture maintains a constant asymptotic potential, $(z+1)/r$, in the ion and focuses on the interplay between increasing Z and increasing N. This provides a framework for the transfer of the extensive experience with neutral atoms, z=0, gained by traditional spectroscopy and collision physics.

The calculations presented in this paper are based on single-electron Hartree-Slater (HS) wave functions[5]. The HS model is used since it is quite amenable to large-scale calculations. In addition, HS functions have been shown to be reasonably realistic for neutral atoms by FTD[1] and can be expected to be better for ions. We view this study as the beginning of a first approximation to the (Z,N) mapping of ionic properties. This initial survey attempts to elucidate gross systematics and to identify areas for deeper examination, incorporating, e.g., correlation and relativistic effects[6,7]. In this paper we present several results exemplifying the usefulness of the different pictures.

As an example of the *isoionic* picture, Fig. 1 shows the phase shift $\delta_\ell(0) = \pi\mu_\ell$ which characterizes the scattering of zero-energy electrons in the field of ions at various stages of ionization. Two striking features are observed. First, the behavior of $\delta_\ell(0)$ shows the known[1] rich structure (associated with chemical properties) for the first spectrum (neutrals). This structure still persists to a lesser extent for the second spectrum (singly ionized species), but disappears completely for all higher-order spectra. Second, in the generally smooth behavior of $\delta_\ell(0)$ for each higher-order spectrum, prominent changes of the slope appear at the same Z for all partial waves. These slope discontinuities occur whenever the total number of electrons equals 28, 10, and 2. These values correspond to the complete filling through the n=3, 2, and 1 shells, respectively, of a hydrogenic atom. Accordingly, we expect to observe a similar behavior of $\delta_\ell(0)$ for ions with 60 electrons which will the n=4 shell. This is in contrast to the neutral atom case where "closed-shell" systems occur when the electron number is 86, 54, 36, 18, 10, and 2, i.e., the noble gases.

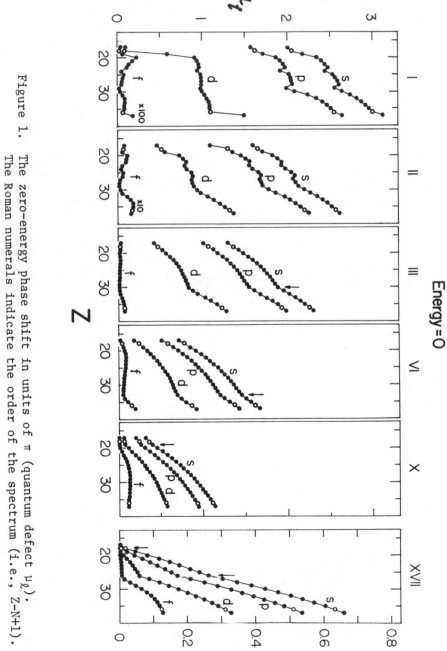

Figure 1. The zero-energy phase shift in units of π (quantum defect μ_ℓ). The Roman numerals indicate the order of the spectrum (i.e., Z-N+1).

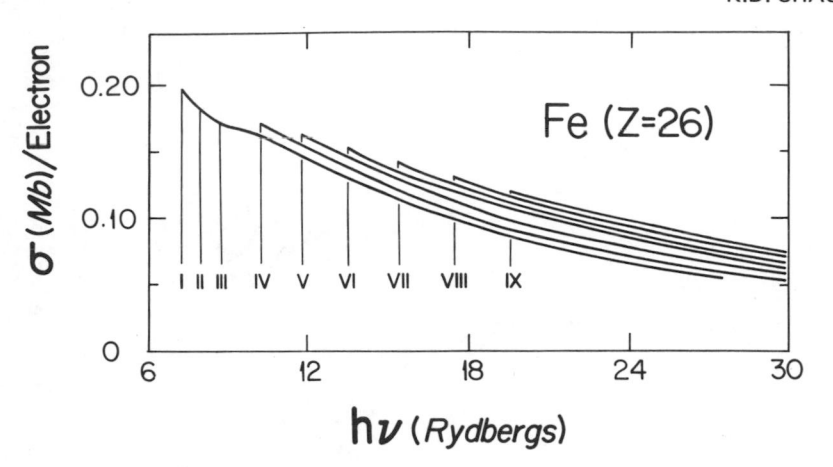

Figure 2.   The 3s subshell photoionization cross section for the
            Z=26 (Fe) *isonuclear* sequence.   The Roman numerals
            indicate the order of the spectrum (i.e., Z−N+1).

        As a second example, Fig. 2 shows photoionization cross section
(or rather, oscillator-strength distributions) for the 3s subshell
along the Z=26 (Fe) *isonuclear* sequence.   Note that the cross sections are plotted against photon energy, rather than photoelectron
energy.   Displaying the data in this fashion reveals an important
simplification: the cross sections for the neutral, singly ionized, and doubly ionized species lie on a common curve, apart from
the obvious shift of the threshold.   This is in sharp contract to
the members of the sequence of higher ionicity, which follow independent curves.   In other words, successive removal of outer 4s
electrons maintains the cross section on the common curve.   The
removal of the 3d electrons, i.e., of electrons with the same
<u>principle</u> quantum number as the 3s, however, alters the internal
dynamics enough to make the cross sections deviate significantly.
As a generalization of this result, we expect that the removal of
outer-shell electrons leave inner-shell properties largely unaffected.   A consequence of this generalization is extremely important,
namely, that a large portion of our knowledge on neutrals is directi
applicable to positive ions; in addition, the amount of calculation
necessary for the Z,N mapping of ionic properties has thus been
greatly reduced.

REFERENCES

1.   U. Fano, C. E. Theodosiou, and J. L. Dehmer, Rev. Mod. Phys.
     (to be published) and references therein.

2. D. Layzer, Ann. Phys. $\underline{8}$, 271 (1959).
3. M. W. Smith and W. L. Wiese, Ap. J. Supp. $\underline{23}$, 103 (1973).
4. B. C. Fawcett, Adv. Atom. Molec. Phys. $\underline{10}$, 223 (1974).
5. F. Herman and S. Skillman, <u>Atomic Structure Calculations</u>
    (Prentice-Hall, Englewood Cliffs, N.J., 1963).
6. H. T. Doyle, Adv. Atom. Molec. Phys. $\underline{5}$, 337 (1969).
7. L. Armstrong and S. Feneville, Adv. Atom. Molec. Phys. $\underline{10}$,
    1 (1974).

DISTINCTIVE FEATURES OF CAPTURE AND LOSS OF ELECTRONS BY EXCITED IONS

I. S.   Dmitriev

Institute of Nuclear Physics, Moscow State University, Moscow, USSR

INTRODUCTION

The present report is a review of the experimental studies carried out at the Institute of Nuclear Physics, Moscow State University, in which beams of helium- and lithium-like ions of light elements containing the metastable component were used. The required ions were obtained either in the passage of a beam of accelerated particles through a thin solid film or in the process of electron loss and capture by ions having a greater or smaller initial charge in a thin gas target. Based on the results of previous measurements of electron loss and capture cross sections some conclusions have been drawn about the corresponding cross sections for excited particles.

THE PROCEDURE FOR DETECTING METASTABLE EXCITED IONS

In our experiments designed to detect excited ions in beams of fast particles, we made use of the phenomenon of variation of the mean electron loss cross section in the process of ion excitation and the spontaneous loss of electrons by metastable ions in autoionizing states.

As a source of fast ions we used a 72 cm cyclotron, which made it possible to obtain ions of the various light elements with ion velocities from 2.6×10^8 to 12×10^8 cm/sec /1,2/. Ions of a definite charge were directed either to a flowing-gas target or to a thin celluloid film. After the passage of which there up peared particles of different

charges in the ion beam, some of which being in excited states. Separated out by the magnetic separator helium- or lithium-like ions of charge i were directed to the collision chamber to determine their mean cross sections for electron loss as a result of collisions with gas atoms and a spontaneous process of autoionization.

The relative number of ions that lost an electron $\Phi_{i,i+1}$ depends on the effective electron loss cross sections $\sigma^0_{i,i+1}$ and $\sigma^*_{i,i+1}$ for unexcited and excited particles, the relative number of excited particles in the beam, α, and on the mean lifetime of particles in excited states, τ /3/. At sufficiently low gas pressures in the collision chamber the relative number of ions that have lost one electron may be written as

$$\Phi_{i,i+1} = \sigma^0_{i,i+1}t + \alpha \exp(-R/\tau v) \cdot G(\tau, \sigma^0_{i,i+1}, \sigma^*_{i,i+1})$$

(1)

where t is the thickness of the gas layer in the collision chamber, R the distance between the target and the magnetic separator. When it is possible to neglect spontaneous decay of excited particles in the collision chamber, for example, in the case of helium-like ions, the qunatity G has the following form:

$$G = (\sigma^*_{i,i+1} - \sigma^0_{i,i+1}) \cdot t$$

(2)

Thus, if the cross sections, $\sigma^*_{i,i+1}$, and, $\sigma^0_{i,i+1}$, are known from the experimentally measured mean cross sections for electron loss, $\overline{\sigma}_{i,i+1}$, one can determine the value of α :

$$\alpha = \exp\left(-\frac{R}{\tau v}\right)\left[(\overline{\sigma}_{i,i+1}/\sigma^0_{i,i+1}) - 1\right]\left[(\sigma^*_{i,i+1}/\sigma^0_{i,i+1}) - 1\right]^{-1}$$

(3)

In experiments with lithium-like ions, i.e., when autoionizing particles were present in the ion beam, there appeared a component in the quantity $\Phi_{i,i+1}$, which was due to the spontaneous increase in the particle charge and which was independent of the gas layer thickness t in the collision chamber. In this case the function G is expressed as follows

$$G = 1 - \exp(-L/\tau v)$$

(4)

where L is the length of the collision chamber. From the value of the t-independent component of $\Delta\Phi_{i,i+1}$ obtained at different distances R it was possible to determine the lifetime τ of autoionizing particles and the relative number α . From (4) we have

$$\alpha = \Delta\Phi_{i,i+1} \cdot \exp(-R/\tau v)\left[1 - \exp(-L/\tau v)\right]^{-1}$$

(5)

Experiments on the determination of the long-lived excited component in beams of fast ions of light elements have been carried out for more than 30 ion species of eight light elements, namely ions: $Li^{1,2}$, $Be^{+}(1-3)$, $B^{+}(1-4)$, $C^{+}(1-4)$, $N^{+}(1-6)$, $O^{+}(1-6)$, $F^{+}(1-3)$, $Ne^{+}(1-7)$. For each of these ions, we investigated the t-dependence of the quantities $\Phi^{s}_{i,i+1}$ and $\Phi^{o}_{i,i+1}$, obtained in cases when solid target was placed in the path of the ion beam before the magnetic separator or when there was no target. As one can see from Fig. 1, showing the situation for B ions, for all the ions investigated the

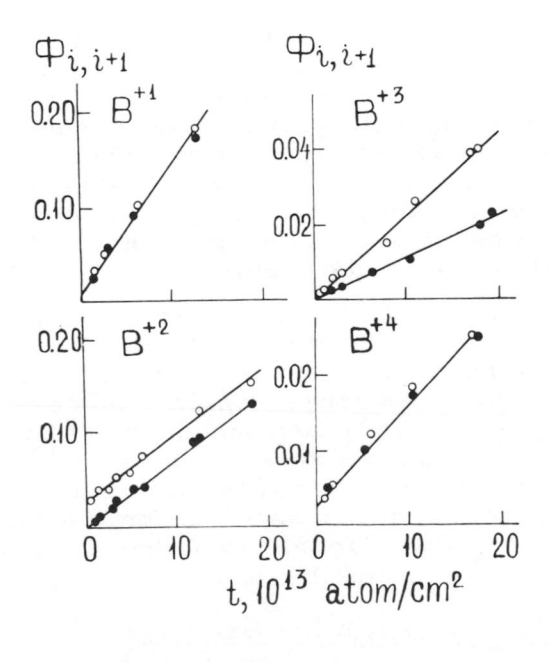

Fig. 1. The t-dependence of $\Phi_{i,i+1}$ obtained with the solid target (o) or without a target (•).

values $\Phi^{o}_{i,i+1}$ are proportional to t. The values $\Phi^{s}_{i,i+1}$ are close to the values $\Phi^{o}_{i,i+1}$ for all ions, except for helium- and lithium-like ions, i.e. for particles possessing two and three electron. For these ions, the values $\Phi^{s}_{i,i+1}$ considerably exceed the values $\Phi^{o}_{i,i+1}$. And for helium-like ions the ratio $\Phi^{s}_{i,i+1}/\Phi^{o}_{i,i+1}$ remains nearly the same with the increasing t, which points to an increase in the mean electron loss cross section. For lithium-like ions it has been found that the values $\Phi^{s}_{i,i+1}$ are markedly nonzero at $t \rightarrow 0$, which suggests the presence of autoionizing particles in the beam. A clearly defined correlation of the observed effect with the

number of ionic electrons indicates that this effect is associated with the structural peculiarities of helium- and lithium-
like ions. Thus, the measurement of the ion charge distribution after the ions have passed the collision chamber enables one to detect excited metastable particles in the beam.

ON THE CROSS SECTIONS FOR THE LOSS OF ELECTRONS BY UNEXCITED AND METASTABLE HELIUM-LIKE PARTICLES

As the experience has shown, in beams of helium-like
ions a significant fraction of particles is practically always
in the metastable states $(1s2s)^{1,3}S$. Therefore, the cross sections found in experiments with helium-like ions cannot, generally speaking, be ranked among ions in the ground state
$(1s^2)^1S$. Separately, the cross sections for electron loss by
unexcited and metastable particles were determined for the He
atoms by Gilbody /4/ who used the technique of atomic beam
attenuation in a collision chamber. This method, however, cannot be applied to multiply charged helium-like ions for which
the electron loss cross section is comparable with or even lower than that for electron capture.

The cross sections for unexcited and metastable ions
have been obtained from the results of measurements of helium-like particles formed by ionization of unexcited lithium-
like ions in a thin gas target /5/. In this case the number of
ions in the ground $(1s^2)$ and metastable $(1s2s)$ states is proportional to the cross sections for loss of 2s and 1s electrons by lithium-like ions $\sigma_{2s}(1s^2 2s)$ and $\sigma_{1s}(1s^2 2s)$.
For the ratios of the cross sections for electron loss by unexcited ions $\sigma^0_{i,i+1}$ to the experimental average cross
sections $\overline{\sigma}_{i,i+1}$ we have

$$K^0 = \frac{\sigma^0_{i,i+1}}{\overline{\sigma}_{i,i+1}} \approx \left[1 + \frac{\sigma_{1s}(1s^2 2s)}{\sigma_{1s}(1s^2)} \frac{\sigma_{2s}(1s2s)}{\sigma_{2s}(1s^2 2s)} \right]^{-1} \tag{6}$$

The values $\sigma_{1s}(1s^2)/\sigma_{1s}(1s^2 2s)$ and $\sigma_{2s}(1s2s)/\sigma_{2s}(1s^2 2s)$
are the ratios of the loss cross sections for the same elect-
rons (1s and 2s, respectively) from helium- and lithium-like
ions and depend, therefore, very weakly on the assumptions
about the values of the cross sections themselves, being
close to 1. Accordingly, the coefficient K^0 turns out to be
close to 1/2 and does not, practically, depend on either the
relationship between the loss cross section for excited and
unexcited ions or on the relative number of metastable particles in the ion beam, α_i. Thus the electron loss cross
section for unexcited helium-like ions is about half of the
values $\overline{\sigma}_{i,i+1}$ obtained in the experiment. A more exact
calculation performed in /5/ for helium-like ions with Z from

3 to 8 and velocities from 4×10^8 to 12×10^8 cm/sec leads to $K^° = 0.55 \pm 0.15$. The cross sections $\sigma^°_{i,i+1}$ obtained in He and N are close to twice the cross sections for electron loss by hydrogen-like ions having the same binding energy for an electron, I.

After the cross sections $\sigma^°_{i,i+1}$ have been found from the experimental values of $\overline{\sigma}_{i,i+1}$, it is possible to obtain also the cross sections for electron loss by metastable ions $\sigma^*_{i,i+1}$ for which the ratio $K^* = \sigma^*_{i,i+1} / \overline{\sigma}_{i,i+1}$ is

$$K^* \approx \frac{1}{\alpha_i}\left[1 + \frac{\sigma_{1s}(1s^2)}{\sigma_{1s}(1s^2 2s)} \frac{\sigma_{2s}(1s^2 2s)}{\sigma_{2s}(1s 2s)}\right]^{-1} \qquad (7)$$

The values of K^*, in contrast to those of $K^°$, are essentially dependent on the value of α_i which is given by the relation of the cross sections for the loss of 1s and 2s electrons by lithium-like ions. Therefore the error in the value of the cross section obtained will be about 1.5 times that for the cross section $\sigma^°_{i,i+1}$. For the metastable helium-like ions we have studied the cross sections $\sigma^*_{i,i+1}$ which are practically indistinguishable from those for electron loss by unexcited lithium-like ions with the same binding energy of electrons, I.

For all the cases investigated the ratio $\sigma^*_{i,i+1}/\sigma^°_{i,i+1}$ is greater than unity, increasing from ~ 2 for Li^+ ions to ~ 50 for O^{+6} (at $V = 8 \times 10^8$ cm/sec) as the values of Z increase. Thus, depending upon the way of formation of helium-like ions, the mean value of $\overline{\sigma}_{i,i+1}$ will vary over a wide range. For obtaining beams of helium-like ions with a different content of the metastable component, beams of fast hydrogen- and lithium-like ions of a certain element were incident on a thin gas target; helium-like ions were also separated out from the ion beams that had traversed solid target. As one can see from Fig. 2, the maximum value of the mean cross section was obtained in this case for ions arising from electron capture by hydrogen-like ions and the maximum value, for the same ions formed in the process of electron loss by unexcited lithium-like ions. The intermediate value of the cross section is observed for ions formed in the solid target.

For ion systems which are more complex than the helium-like ones, due to the larger number of electrons and the lower ratio of the excitation energy to the electron binding energy, the ion excitation should lead to a much smaller increase in the electron loss cross section. In particular, for lithium-like ions for a beam of particles in the $(1s2s2p)^4P$ metastable autoionizing state the value of the electron loss

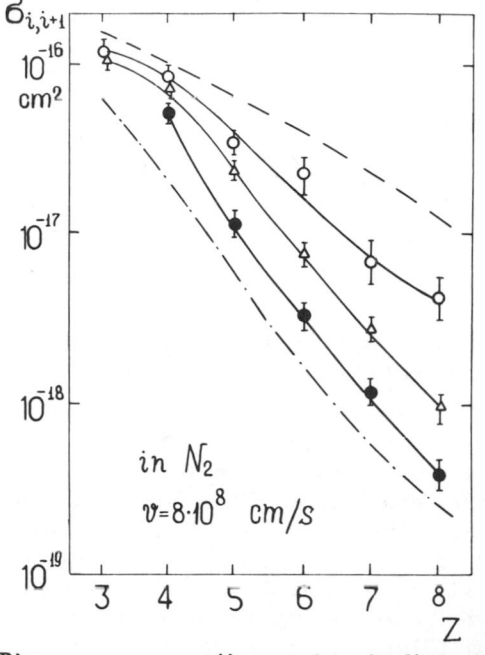

Fig.2. The cross sections for helium–like ions,
obtained in the passage through the thin gaseous nitrogen gas target by hydrogen–like ions
(○), lithium–like ions (●) and which were also
separated out from the beam of ions that passed
through the solid target (△).(—·—·—) and
(————) the values of cross sections $\sigma^{0}_{i,i+1}$
and $\sigma^{*}_{i,i+1}$, respectively.

cross section $\sigma^{*}_{i,i+1}$ should be only about twive as great
as the corresponding cross section for unexcited particles in
the (1s2s) state. Since the relative number of such excited
particles in beams of lithium–like ions formed in various ways
is also, on the average, smaller than that for helium–like particles, the mean value of the electron loss cross section for
beams of ions involving the excited component will not be increased by more than 20%, which has been borne out experimentally.

All these results refer to the range of relatively high
ion velocities, where the electron loss cross section depends
very weakly on the structure of the initial state of an ion
and is mainly determined by its binding energy. The theoretical calculations and experimental data on the ionization cross
sections for the atomic L-subshells indicate that in the range
of lower relative velocities of colliding particles the situation
may be different. In particular, the cross sections for
electrons loss from the excited states   with $\ell > 0$ may be
considerably lower than those from the unexcited S states /6,7/.

THE RELATIVE NUMBER OF METASTABLE PARTICLES IN BEAMS OF HELIUM- AND LITHIUM-LIKE IONS

For helium–like ions the values of α are minimal (3 to 10%) in ion beams produced by ionization of lithium–like particles in a thin gas target; they are maximum ($\sim$ 50%) when these ions are formed as a result of electron capture by hydrogen–like ions (See Fig.3).

In all the cases considered /10/ for lithium–like ions the maximum number of auto–ionizing particles (10–15%) was obtained in the passage of hydrogen-like particles through a thin gas target, i.e., as a result of capture of two electrons. Where autoionizing particles arose from capture of one electron by helium–like ions, the value α was 1.5 or 2 times smaller. In this case, as special experiments have shown, the autoionizing particles are formed as a result of electron capture by metastable helium–like ions. The lowest number of autoionizing lithium–like ions was obtained for particles separated out from the beam of ions that passed through a solid target.

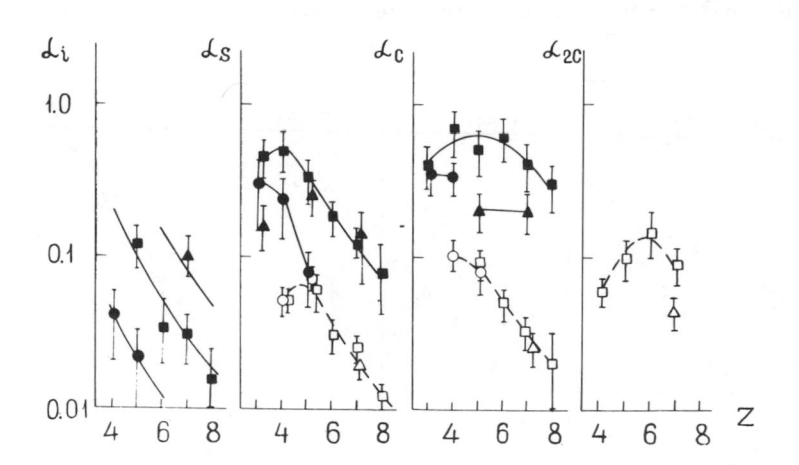

Fig.3. The relative number of metastable particles α, in beams of helium–like (solid marks) and lithium–like (open marks) ions formed in the solid target (s) as a result of loss of one electron (i), capture of one electron (c) and two electrons (2c). The ion velocities v= 4×10^8 cm/sec ($\bullet$,o), 8×10^8 cm/sec ($\blacksquare$,$\square$) and 12×10^8 cm/sec ($\blacktriangle$,$\triangle$).

Thus, from the given experimental data it follows that in single collisions of fast ions with gas atoms metastable helium–like ions in the $(1s2s)^{1,3}$ S states and autoionizing lithium–like particles in the $(1s2s2p)$ $^{4}P_{5/2}$ quartet state are most probably formed in the process of electron capture; the necessary condition for the formation of metastable lithium–like ions from the capture of one electron by helium–like ions being the presence of metastable particles in the $(1s2s)$ ^{3}S state in the ion beam.

THE CROSS SECTION FOR ELECTRON CAPTURE BY EXCITED IONS

Before considering the influence of the excitation of fast ions on the process of electron capture by these ions, we shall dwell upon the basic regularities of the process of electron capture by unexcited ions /5/. A great body of the accumulated experimental data on the electron capture cross sections $\sigma_{i,i-1}$ for ions with different electron number $N = Z-i$ show that for ions with a large charge i, for which the ratio $\eta = I(n_o)/I_v$ is greater than unity (here I(n_o) is the electron binding energy in the unfilled shell closest to the nucleus with the principal quantum number n_o, $I_v = \frac{1}{2} \cdot \mu v^2$, μ

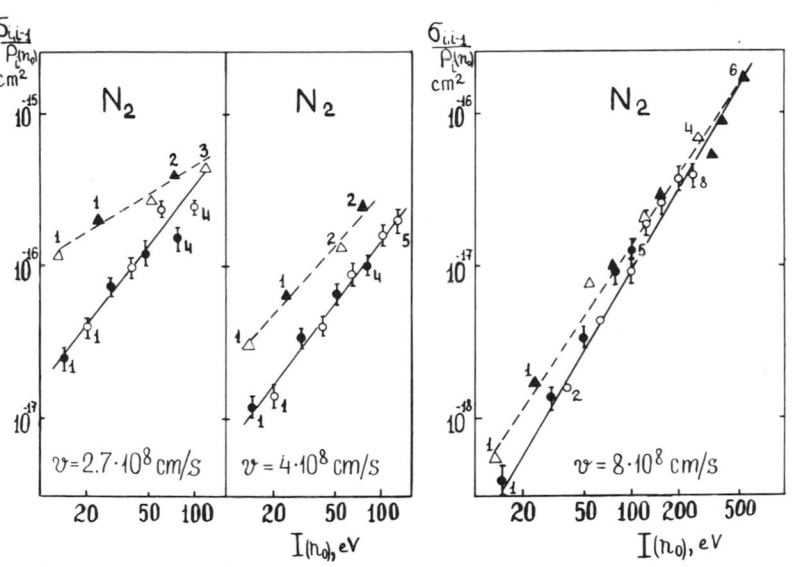

Fig. 4. The reduced electron capture cross sections $\sigma_{i,i-1}/p(n_o)$ for bare nuclei ($\triangle$) hydrogen–like ions ($\blacktriangle$), nitrogen ions ($\bullet$) and neon ions (o) as functions of I(n_o). The ionic charge i is labelled by numbers.

is the electron mass) the cross sections $\sigma_{i,i-1}$ are practically independent of the number N of electrons present in the ion, proportional to i^2. But at $\eta < 1$ they are significantly different for ions with different values N. However, the values of the cross sections divided by the number of vacancies $p(n_0)$ in the unfilled shell closest to the nucleus are practically equal at equal values of $I(n_0)$ and are independent of the number of electrons in the ion. When $V = 8 \times 10^8$ cm/sec the reduced cross sections $\sigma_{i,i-1}/p(n_0)$ for bare nuclei and hydrogen-like ions are identical within the experimental errors to the cross sections $\sigma_{i,i-1}/p(n_0)$ for ions with a partially filled L shell. However, when $V \leq 4 \times 10^8$ cm/sec as an analysis of the known electron capture cross sections at fixed values $I(n_0)$ shows, the values of the reduced cross sections decrease with increasing n_0 (see Fig. 4).

Some of these conclusions also follow from the straightforward quantum-mechanical formula of Brinkman-Kramers /9/ for the cross sections for electron capture by bare nuclei with charge Z if, when applied to particles with the electron number $N > 0$, the quantity Z/n is replaced by $[I(n)/I_0]^{1/2}$ and the quantity n^2 by $P(n)/2$ where $I(n)$ and $P(n)$ are the mean electron binding energy and the number of vacancies in the states with the principal quantum number n, respectively, and $I_0 = 13,6$ eV.

$$\sigma_{i,i-1} = \frac{2^{17}}{5} \pi a_0^2 I_0^{7/2} I_v^4 \sum_n \frac{P(n)[I(n)]^{5/2}}{\left\{I_v^2 + 2I_v[I(n)+I_c] + [I(n)-I_c]^2\right\}^5}$$

(8)

From this formula it follows that when $\eta > 1$ the electron is captured predominantly into excited states with the binding energy $I(n) \approx (1/2)I_v + I_c$ (I_c is the binding energy of the captured electrons in the target atom), therefore the electron capture cross section is determined by the ionic charge i and the ion velocity and is independent of N. In the range $\eta < 1$ the values $\sigma_{i,i-1}$ should be proportional to the sum $\sum_n p(n)[I(n)]^{5/2}$. Therefore, for unexcited helium-like particles, if the relations $I(n) = i^2 I_0/n^2$ and in particular: $I(n_0) = i^2 I_0/4$ are satisfied, the values $\sigma_{i,i-1}/p(n_0)$ should be 30% higher than those for bare nuclei with the same value $I(n_0)$. But due to the fact that for helium-like ions the real $I(n_0)$ are greater than $i^2 I_0/4$ the values $\sigma_{i,i-1}/p(n_0)$ for ions with N=1 and 2 do not differ by more than 15%.

In considering the electron capture cross sections $\sigma^*_{i,i-1}$ for metastable helium-like ions it should be kept in mind that the electron capture into states with $n \geq 2$ is

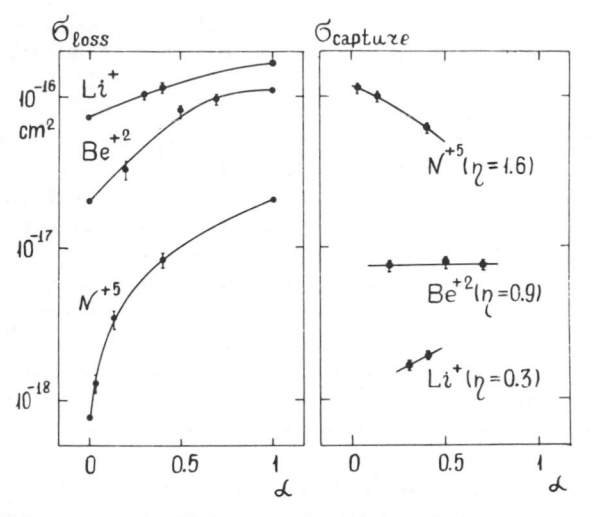

Fig.5. The electron loss and capture cross sections for helium-like ions Li^+, Be^{+2} and N^{+5} in N_2 at v= 8×10^8 cm/sec as functions of α.

accompanied by the formation of lithium-like particles with K-vacancies most of which autoionize before entering the collision chamber /3/. Therefore the experimental electron capture cross section for metastable helium-like ions should be close to the cross section for electron capture into the K shell vacancy alone. It thus appears that at $\eta \geqslant 1$, when metastable ions capture an electron mainly into states with $n \geqslant 2$, the cross sections $\sigma^*_{i,i-1}$ should be smaller than the values $\sigma^0_{i,i-1}$ for unexcited particles and at $\eta < 1$ greater than $\sigma^0_{i,i-1}$.

In order to verify these conclusions, we have measured the electron capture cross sections for Li^+, Be^{+2} and N^{+5} beams containing a considerable number of meastable particles /10/. These ions were formed both in the processes of capture of one electron by hydrogen-like particles in a thin gas target and in the passage of a beam of fast particles through a solid film. For Be^{+2} ions we have also measured the cross sections for ion beams extracted directly from the cyclotron. The cross sections for N^{+5} ions formed in the process of electron loss by lithium-like N^{+4} ions were also measured. Measurements have been made at an ion velocity of V= 8×10^8 cm/sec.

From experiment it w as found that for N^{+5} ions (for which η =1.6) the electron capture cross section σ_{54} monotonically decreases approximately by one half with increasing values α from 0.03 to 0.4. For Be^{+2} ions (for which η = 0.9) variations in the values α from 0.2 to 0.7 do not

practically lead to variations in the cross sections σ_{21}, and for Li^+ ions (for which $\eta = 0.3$) one observes a moderate increase in the cross sections σ_{10} with increasing α from 0.3 to 0.4. At the same time the cross section for electron loss from the same ions increases steadily with the increasing α (see Fig. 5).

The mechanism proposed allows one to explain a decrease in the electron capture cross sections for beams of He^+ ions /11/ and N^+ ions /12/, containing a metastable component, and a deviation from the monotonic dependence of the cross sections $\sigma_{i,i-1}$ on i for P^{+5}, Br^{+7} and I^{+7} ions /13,14/ which were obtained in the passage of ion beams through a solid film. As with helium-like ions, ions of P^{+5}, Br^{+7}, and I^{+7} have completely filled electronic shells in the ground state: a completely filled L shell for P^{+5} ions, a completely filled M shell for Br^{+7} ions and completely filled 4spd electronic subshells for I^{+7} ions. In the works by Betz /14/ it was indicated that the decrease in the electron capture cross sections for Br^{+7} and I^{+7} ions is largely connected with the residual excitation of fast particles created by autoionizing events arising from the electron capture into excited states. The experimentally observed electron capture cross sections can decrease up to one half. As a result, the capture of an electron by excited ions with charge i leads, in appropriate cases, not to the formation of particles with charge i-1 in the ion beam that has traversed the collision chamber but only to the deexcitation of ions with charge i, i.e., ultimately to a decrease in the electron capture cross sections. However, until recently it has not been clear why precisely this decrease for ions with a definite electron number N= Z-i is significantly greater than that for other particles.

Multiply charged ions of nitrogen, phosphorus, bromine and iodine, for which a decrease in the electron capture cross sections is discovered, were formed by passing ion beams through a solid or gas target located at a distance of several tens of centimeters from the collision chamber. In connection with this, excited ions having a lifetime $\tau > 10^{-7}$ sec could only be present in a large quantity in the collision chamber. Autoionization of excited ions after they have captured electrons into states with a binding energy I can occur only in those cases when the excitation energy E exceeds the value I. But when these conditions are satisfied, electron capture by an excited ion will practically always lead to autoionization. As a result, the experimentally observed electron capture cross sections for excited ions will be much smaller than those for unexcited particles with the same charge i. Thus, for the experimental electron capture cross sections to be

markedly decreased, it is necessary and sufficient that a large number of excited ions in metastable states with the lifetime $\tau > 10^{-7}$ sec and excitation energy $E > I$ (which corresponds to ~ 25 eV at $V \approx 3 \times 10^{8}$ cm/sec) should be present among ions with charge i arising from the passage of fast particles through matter.

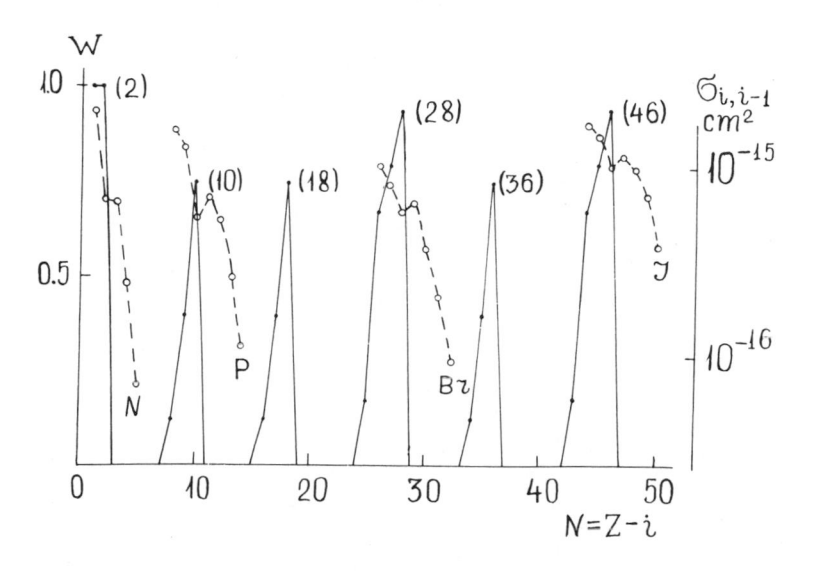

Fig. 6. Statistical weight, W, of metastable states with $E \gtrsim 25$ eV (the solid line) and electron capture cross sections in helium, $\sigma_{i, i-1}$ (the dotted line) for ions with electron numbers $N = Z-i$. The ion velocities are: 8×10^{8} cm/sec (nitrogen ions), 2.7×10^{8} cm/sec (phosphorus ions); 4.9×10^{8} cm/sec (bromine ions), and 2.7×10^{8} cm/sec (iodine ions).

As a result of our consideration of excited states of many-electron systems it is shown that metastable states have the maximum statistical weight W for ions with electronic numbers $N = 2, 10, 18, 28, 36$ and 46 i.e. for ions which form completely filled s, p and d electronic shells in the ground state (see Fig. 6). These states are similar to the metastable states of helium-like particles and decay practically only due to the admixture of states from which dipole transitions are allowed. The magnitude of admixture is generally rather small, and therefore their lifetimes τ should be large. In the process of electron capture by ions in the states just mentioned autoionizing particles are formed which are similar

to autoionizing lithium-like particles.

Thus, in addition to the decrease in the electron capture cross sections for excited ions with numbers of electrons N = 2; 10; 28 and 46, which has already been observed, a similar decrease should be expected in the case of multiply-charged ions with N= 18 and 36.

REFERENCES

1. V.S. Nikolaev, I.S. Dmitriev, Ya. A. Teplova and L.N. Fateeva, in book "Accelerators" (in Russian), Moscow, 1960, pp.90-96.
2. V.S. Nikolaev, Uspekhi Fiz. Nauk 85, 679 (1965).
3. I.S. Dmitriev, Ya. A. Teplova and V.S. Nikolaev, ZhETF, 61, 1359 (1971).
4. H.B. Gilbody, K.F. Dunn, R. Browning, C.J. Latimer, J. Phys. B3, 1105 (1970).
5. I.S. Dmitriev, V.S. Nikolaev, Yu. A. Tashaev and Ya. A. Teplova, ZhETF 67, 2047 (1974).
6. V.S. Nikolaev, V.S. Senashenko and V. Yu. Shafer, J. Phys. B6, 1779 (1973).
7. V.S. Nikolaev, V.P. Petukhov, E.A. Romanovsky, Abstracts IX ICPEAC, p.513, Seattle (1975).
8. I.S. Dmitriev, V.S. Nikolaev and Ya. A. Teplova, Abstracts VI ICPEAC Boston 460 (1969).
9. M.C. Brinkman, H.A. Kramers, Proc. Acad. Sci. Amsterdam 33, 973 (1930).
10. V.S. Nikolaev, I.S. Dmitriev, Yu. A. Tashaev, Ya. A. Teplova, Yu. A. Fainberg, J. Phys. B8, 58 (1975).
11. G.J. Lockwood, Phys. Rev. A2, 1406 (1970).
12. M. Vojovic, M. Matic, B. Cobic, Abstracts VIII ICPEAC, Belgrad, 779 (1973).
13. Yu. A. Tashaev, I.S. Dmitriev, V.S. Nikolaev, Abstracts VIII ICPEAC, Belgrad, 793 (1973).
14. H.D. Betz. Rev. Mod. Phys. 44, 465 (1972).

MULTIPLE ELECTRON LOSS CROSS SECTIONS FOR 60 MeV

I^{+10} IN SINGLE COLLISIONS WITH XENON*

L.B. Bridwell
Murray State University
Murray, Kentucky

J.A. Biggerstaff, G.D. Alton, C.M. Jones, P.D. Miller
Oak Ridge National Laboratory
Oak Ridge, Tennessee

Q. Kessel
University of Connecticut
Storrs, Connecticut

and

B.W. Wehring
University of Illinois
Urbana, Illinois

INTRODUCTION

The production of highly ionized beams of heavy ions has been a subject of intensive investigation in several laboratories during the last decade. The resulting technology provides the basis on which many high-energy heavy-ion accelerators operate. Along the path of an ion as it traverses a stripping medium, the charge state of a particular ion may fluctuate many times. The charge state distribution of such a beam of particles reaches equilibrium after traversing a certain thickness of the stripping medium. A recent survey shows

---

*Research sponsored by the Energy Research and Development Adminis-
tration under contract with Union Carbide Corporation.

that extensive information has been accumulated concerning charge state distributions of heavy ions as they emerge from both solid and gaseous targets[1]. Most of these, however, have been obtained with experimental apparatus which confines the emerging beam within a very small solid angle in the forward direction. The work by Kessel shows that the emerging charge state distributions produced in single event scattering depend strongly on the distance of closest approach[2,3]. We have previously reported a series of experiments in which absolute yields of highly charged ions were obtained from high atomic number stripping gases at target thicknesses somewhat above that of single collision processes, but somewhat less than equilibrium thickness[4]. That work was performed specifically for determining design parameters for a terminal stripper in a large tandem accelerator. The present results were obtained using the same apparatus but with a smaller target thickness to examine single scattering events and to obtain cross sections for the loss of several electrons in a single collision.

EXPERIMENTAL APPROACH

The experimental apparatus is virtually identical to that reported earlier[4]. A schematic illustration is reproduced in Fig. 1. A momentum analyzed beam of 60 MeV I^{10+} ions, produced in the ORNL tandem accelerator, was collimated prior to entering the target chamber by two apertures. The apertures were separated by 154 cm and had diameters of 3 mm and 0.5 mm. The smaller aperture was placed just before the target chamber. A surface barrier detector, used to detect particles scattered at 60° from a chemically etched annular film surrounding the first aperture, served as the beam monitor. The details of the monitoring system have been reported previously by Appleton et al[5].

A differentially pumped target cell was positioned 27 cm after the second aperture and mounted directly above a 1400 ℓ/sec diffusion pump. The target cell consisted of two circular entrance apertures 1 mm in diameter and two exit apertures 1 x 2.5 mm and 1 x 4 mm. All apertures were spaced 2 cm apart as shown in the insert to Fig. 1, permitting measurements at scattering angles through 3°. The target gas was introduced between the second and third apertures. The target gas pressure was measured by a capacitance manometer.

Charged particles scattered at a given angle were analyzed with an electrostatic charge state analyzer which has been described previously[6]. The charge state resolution was determined by a vertical collimator 1.03 mm high positioned at the entrance to the analyzer. The angular resolution was determined by a slot in a mask 4.1 mm wide directly in front of the detector which was 429 cm from the

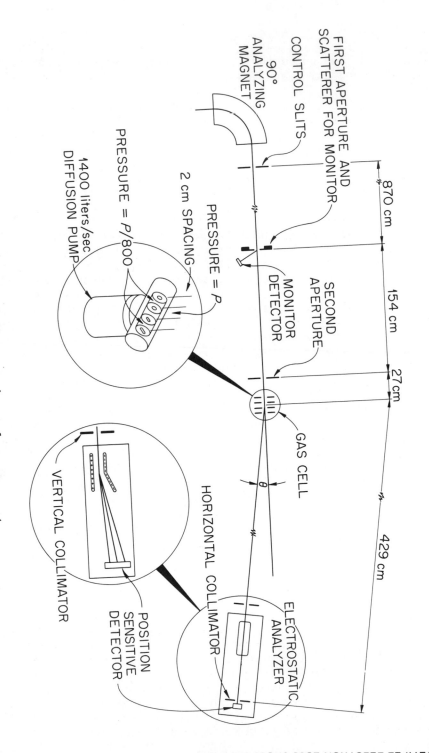

Fig. 1. The experimental arrangement.

center of the target. The resulting angular resolution was $\Delta\theta =$ .054° and the solid angle was $\Delta\Omega = 2.9 \times 10^{-7}$ ster. The monitor efficiency was measured by observing the beam intensity at 0° with no target gas. The result was 1064 ± 60 beam particles per monitor count. Pressures in the flight tube before and after the target cell were approximately 2×10^{-6} Torr throughout the measurements. Target gas pressures were (5.1 and 11.9) $\times 10^{-3}$ Torr, and the scattering angles were 0.05°, 0.10° and 0.20°.

RESULTS

Figure 2 shows the charge state distribution measured at Xe pressures of 5.1×10^{-3} and 11.9×10^{-3} Torr at a scattering angle of 0.2°. The similarity in shape suggests that these target thicknesses are essentially in the single collision region although there is some evidence of electron pick-up at the higher pressure.

Figure 3 shows the differential scattering cross section plotted against the number of electrons removed for the three scattering angles observed. For a Δq of 15, the last N shell electron must be removed in the collision. At a scattering angle of $\theta = .05°$, the range of θ actually extends from .025° to .075° so that the distance of closest approach, r_0, varies from .17 to .26 Å. Self-consistent field calculations[7,8] show that the radii of maximum radial charge density are approximately .13Å for the M shell and .42Å for the N shell in I or Xe. At the minimum distance of closest approach within the allowed range for $\theta = .05°$, the M shell electrons of the I - Xe system are just beginning to merge during the collision. This may result in the excitation or removal of one or more of the M shell electrons which then leads to ionization cascades where all of the electrons in the N shell are removed. The much higher probability of removing 10-12 electrons probably results from similar processes but with the initial electron excitation or removal occurring in the N shell itself.

At $\theta = .10°$ the range of r_0 extends from .13 to .17Å. In this case the merging of the M shells within the I - Xe system is substantial and therefore M shell electron excitations are relatively more probable as a result. Finally at $\theta = .2°$ the interpenetration of the M shells is complete with r_0 extending from .09 to .11Å. The probability of removing all N shell electrons relative to other processes then becomes greater.

At these selected scattering angles L shell interpenetration did not occur, thus very little M shell ionization was observed.

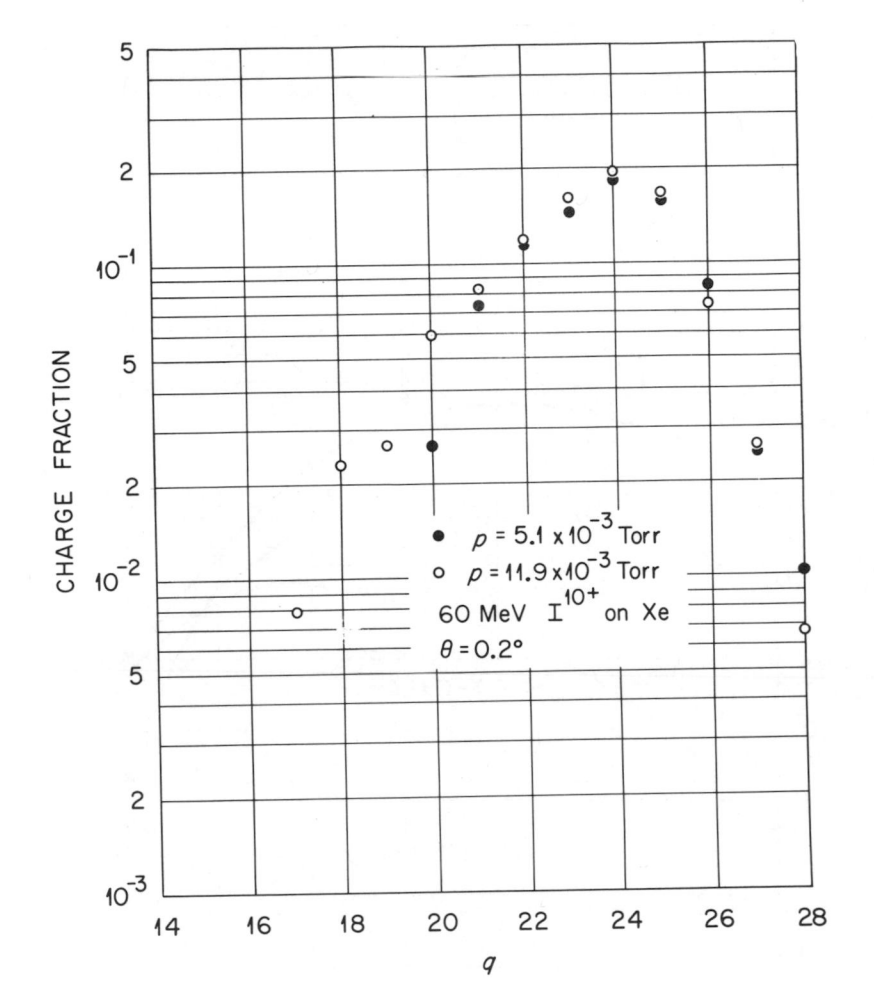

Fig. 2. Charge state distributions for 60 MeV I^{10+} on Xe at
a scattering angle of $\theta = 0.2°$ for target gas pressures
of 5.1 x 10^{-3} Torr (●) and 11.9 x 10^{-3} Torr (o).

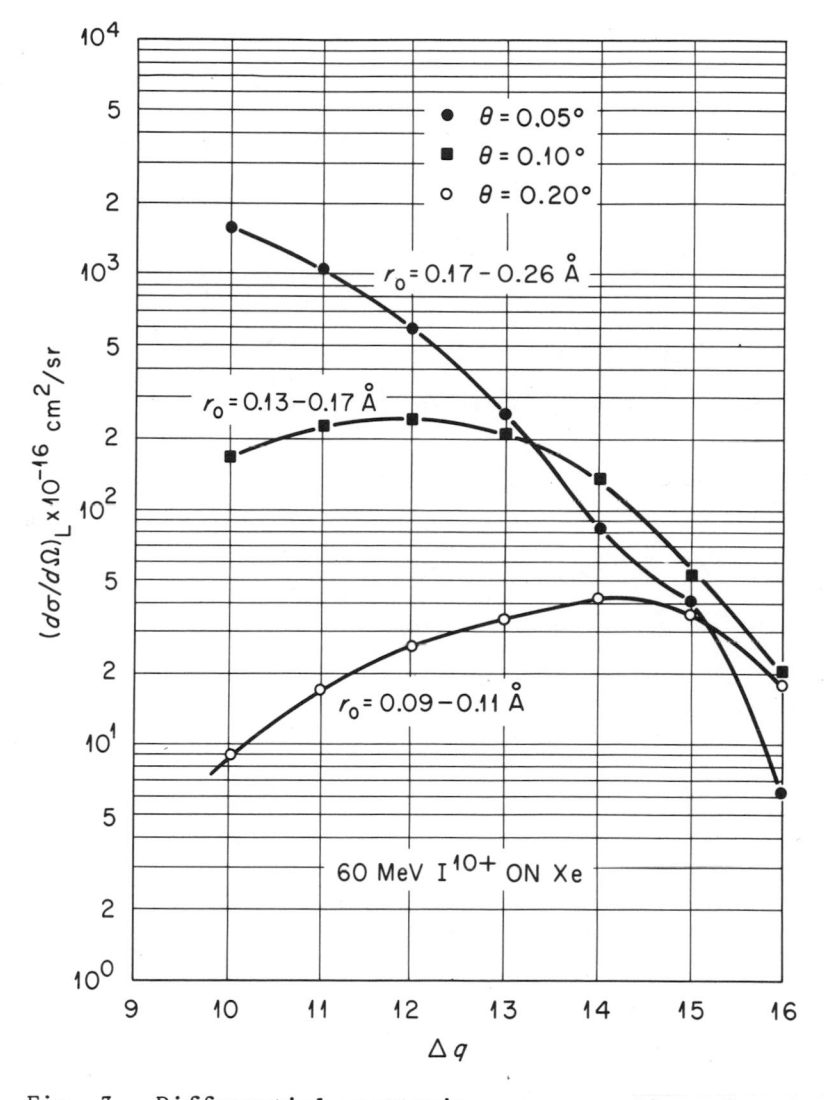

Fig. 3. Differential scattering cross sections for the removal
of Δq electrons in a single collision between 60 MeV
I^{10+} and Xe at angles of .05°, .10°, and .20°.

Δq	Total σ ($\times 10^{-18}$ cm^2)
11	16 $\pm$ 4
12	8.7 $\pm$ 0.9
13	2.8 $\pm$ 0.5
14	1.9 $\pm$ 0.3

Table 1. Total removal cross sections for Δq electrons
from 60 MeV I^{10+} when scattered by Xe gas in
single collisions.

Numerical intergration of the differential cross sections for
the removal of 11, 12, 13, and 14 electrons over the angular
range observed yields the results shown in Table 1.

CONCLUSION

From these data and those reported earlier[3,4,6] it is clear
that ultra high charge state ions are produced in single collisions with the target atoms. With the large magnitude of the
cross sections for these events it is possible to make experimentally useful beams of such ions with a relatively simple gas
cell provided that a sufficiently large acceptance solid angle is
employed.

REFERENCES

1. H. D. Betz, Rev. Mod. Phys. $\underline{44}$, 465 (1972).

2. Q. Kessel, Phys. Rev. $\underline{A2}$, 1881 (1970).

3. Robert A. Spicuzza and Quentin C. Kessel, BAPS $\underline{19}$, 782 (1974) and BAPS $\underline{20}$, 675 (1975).

4. G. D. Alton, J. A. Biggerstaff, L. B. Bridwell, C. M. Jones, Q. Kessel, P. D. Miller, C. D. Moak and B. W. Wehring, IEEE Transactions on Nuclear Science NS-$\underline{22}$, 1685 (1975).

5. B. R. Appleton, J. H. Barrett, T. S. Noggle, C. D. Moak, Radiation Effects $\underline{13}$, 171 (1972).

6. C. D. Moak, H. D. Lutz, L. B. Bridwell, L. C. Northcliff, and S. Datz, Phys. Rev. $\underline{176}$, 426 (1968).

7. F. Herman and S. Skillman, "Atomic Structure Calculations," Prentice-Hall, Inc., Englewood Cliffs, N.J. 1963.

8. J. T. Waber and D. T. Cromer, J. Chem. Phys., $\underline{42}$, 4116 (1965).

CHARGE STATES OF BACKSCATTERED He IONS

Allen Lurio and J. F. Ziegler

IBM Thomas J. Watson Research Center

Yorktown Heights, New York 10598

Previous measurements (1) have shown that charge state equilibrium is attained on passage of a He beam through even the thinnest of self-supporting foils. The fraction of He atoms (normalized to one) which emerge from the solid as neutrals, or singly, or doubly charged ions are the charge state fractions. Equilibrium occurs when an incremental change in penetration depth does not change these fractions. In order to set an upper limit on the amount of matter through which a He beam must pass to attain charge state equilibrium, we have studied the charge state fractions of a He beam backscattered from very thin surface layers on a supporting substrate. In this way one can investigate charge equilibration in thicknesses of matter of the order of 10 Å.

For these measurements we have developed a simple experimental technique (2). The basic idea of the method is shown in Figure 1. A beam of He$^+$ ions from an ion accelerator is incident on a target and those ions backscattered at 170° are counted by a surface barrier detector. In order to separate by energy ions backscattered from the surface layer from those backscattered from the substrate, we placed our layer on a substrate with atomic mass constituents much lighter than those of the surface layer. For an incident ion energy E_0, the energy E of the elastically backscattered ions is

$$E = K E_0 = E_0 \left[\frac{M_1 \cos\theta + \sqrt{M_2^2 - M_1^2 \sin^2\theta}}{M_1 + M_2} \right]^2 \qquad (1)$$

where M_1 is the projectile ion mass, M_2 the target atom mass and θ is the angle of scattering.

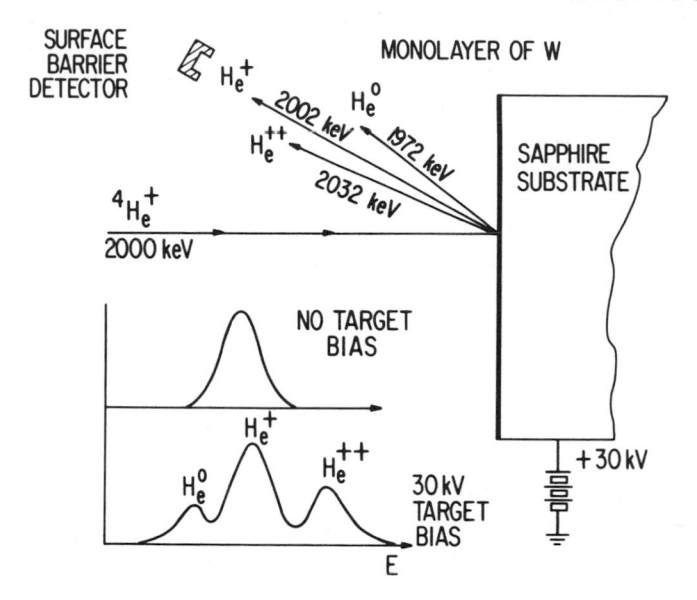

Figure 1.   Schematic diagram of experiment.   Curves show back-scattering spectra expected for zero and 30 keV bias on the target.

        The upper curve in Figure 1 shows the single isolated peak due to backscattering from the surface layer.   Its width is approximately 15 KeV and is entirely due to the detector resolution.   The backscattering from the substrate occurs at much lower energy and is not shown on the curves.   If one now applies a bias of + 30 KeV to the target, then the single peak splits into three well resolved peaks which correspond to He°, He$^+$ and He^{++} ions.   This is shown in Figure 1 and the lower curve of Figure 1.   The energy dependence of these peaks is given in the following equations.

$$E(He^\circ) = K(E_0 - eV)$$

$$E(He^+) = K(E_0 - eV) + eV \qquad\qquad (2)$$

$$E(He^{++}) = K(E_0 - eV) + 2eV,$$

where K is the recoil factor of Equation 1 and V is the target bias. Since the shape of each peak is very nearly Gaussian, we can estimate the charge state fractions by fitting our experimental peaks to the sum of three Gaussians where the area under each peak will be proportional to its charge state fraction.

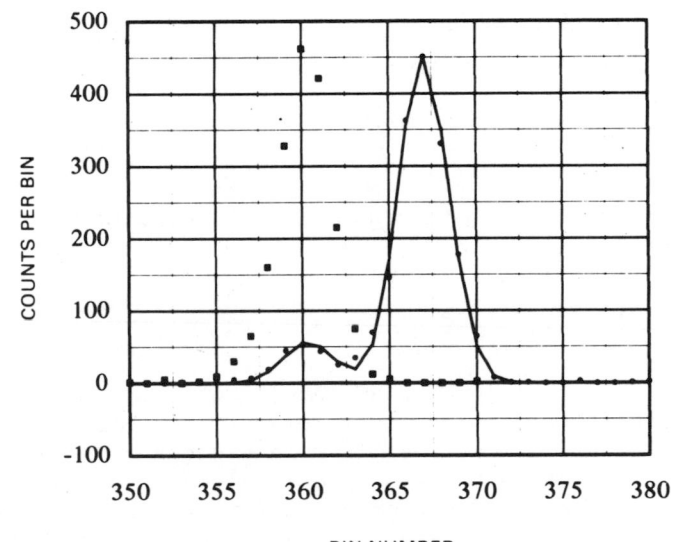

Figure 2. Experimental backscattering curve (5 keV/bin) for He$^+$ at 2 MeV incident on a W layer on a sapphire substrate. Solid squares are for zero bias, solid dots are for + 30 KeV bias and the curve is the theoretical fit to the data.

In Figures 2 and 3 we show experimental data for He$^+$ at 2 MeV and 300 keV backscattered from a target of tungsten on sapphire. The surface layer of W has an areal density of 3.3×10^{15} atoms/cm^2 which corresponds to a thickness of 6 Å. This combination of surface layer and substrate was chosen because tungsten is least

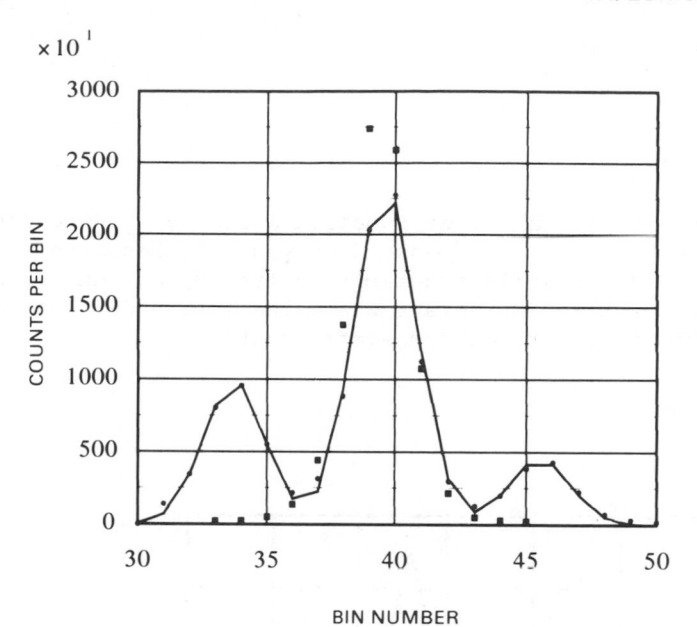

Figure 3. Experimental backscattering curve (5 KeV/bin) for He$^+$ at 300 KeV incident on a W layer on a sapphire substrate. A linear backround subtraction was applied to the data.

likely to agglomerate on a sapphire surface. Data were taken for a number of He$^+$ ion energies. The recoil factor K for W at $\theta=170°$ is .917. To see if our results were dependent on the recoil factor we also used a target of Be metal which was etched to remove oxide and other surface contaminants. Backscattering analyses showed that a natural oxide of $\sim$ 20 Å had regrown during the time between the etch and placement of the sample in the vacuum system. The recoil factor K for oxygen at 170° is .363. A typical experimental curve for He$^+$ backscattered from oxygen at 1 MeV is shown in Figure 4.

All the experimental results are plotted in Figure 5. For comparison, the solid lines drawn in Figure 5 are the charge state fractions obtained by Armstrong, Mullendorn, Harris and Marian (1) from their measurements on He$^+$ ions transmitted through thin carbon foils. Within experimental error they are the same. One important difference between these measurements and transmission measurements is that in backscattering the He ion suffers a large angle nuclear collision during which it is likely to lose its electrons. For ion

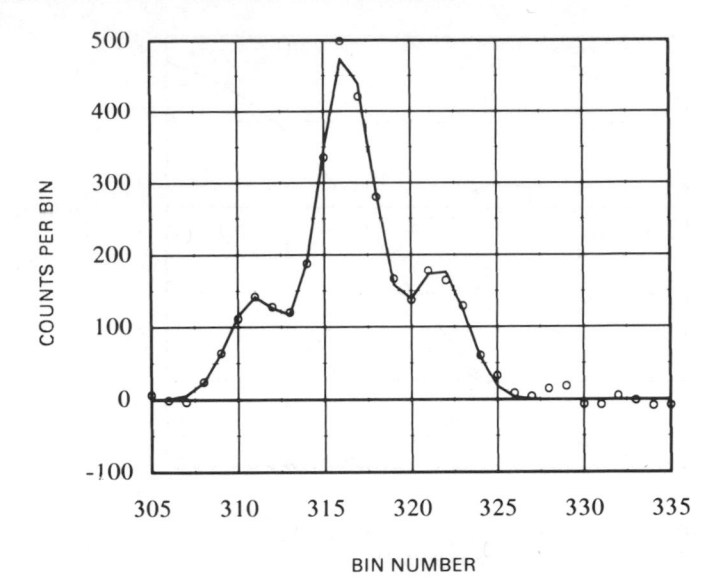

Figure 4. Experimental backscattering curve (5 keV/bin) for He[+]
at 1 MeV incident on Be O layer on Be. Open circles are for 30
KeV bias and the solid curve is the theoretical fit. A linear
backround subtraction was applied to the data before analysis.

transmission experiments on the other hand the detected ions have
undergone only a number of very small angle collisions. Hence the
backscattered ion has largely forgotten its charge state prior to
the collision. We conclude that charge state equilibrium must
occur in the 5-10 Å between the backscattering collision and the
surface or in the region just outside the surface. We have compared
our results with carbon because in our vacuum ($\sim 2 \times 10^{-6}$ torr) even
with frequent changing of the beam position on the target several
angstroms of carbon will form on the surface during the measurement.
Our conclusion is in agreement with Buck, Feldman and Wheatly (3)
that for "practical surfaces" (i.e. ones not cleaned in ultra high
vacuum) there is little evidence that anything but the last few
angstroms of the surface are important for the emerging charge state
distribution. The question whether equilibrium (4) occurs outside
or inside the surface is not resolved by these measurements.

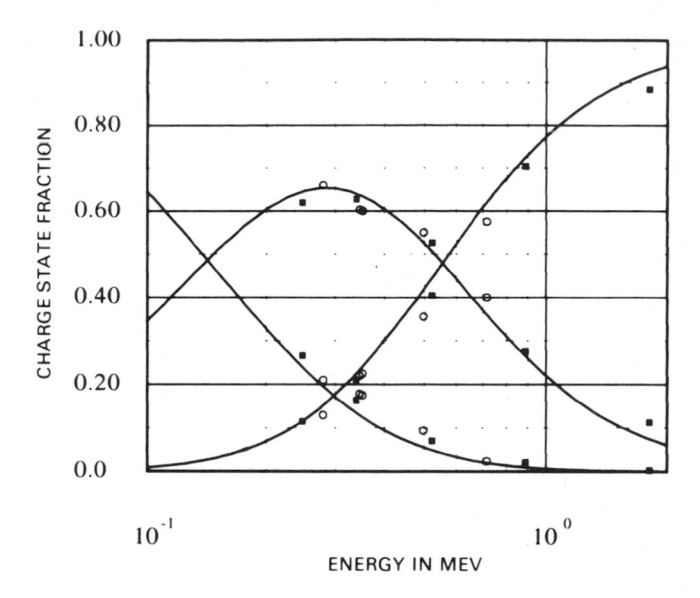

Figure 5. Measured charge state fractions. Solid squares are for He$^+$ on W. Open circles are for He$^+$ backscattered from oxide layer on the Be metal. The solid curves are results from Reference 1 for comparison. At high energies the charge state is all He^{++}, at low energies the backscattered beam is neutral.

We wish to thank Johann Keller for his collaboration in all parts of the experimental work, and J. Cuomo and J. Baglin for several helpfull suggestions.

REFERENCES

1.   J. C. Armstrong, J. V. Mullendorn, W. R. Harris and J. B. Maria Proc. Phys. Soc. **86**, 1283 (1965).

2.   John A. Davies in a private discussion suggested the idea of applying a voltage to the target in order to observe by back- scattering the <u>incident</u> charge states in the He beam coming from the accelerator.

3.   T. M. Buck, L. C. Feldman and G. H. Wheatley "Atomic Collisions in Solids" (Plenum Press, New York 1975) Vol. 1, pp. 331.

4.   J. Davidson and W. S. Bickel, Nuc. Instrum. Methods, **110**, 253 (1973).

CHARGE-STATE DISTRIBUTIONS IN SINGLE ATOMIC COLLISIONS OF 2.5 MEV N^{+i} WITH N_2 AT SMALL IMPACT PARAMETERS[*]

F. W. Martin and R. K. Cacak[†]

Department of Physics and Astronomy
University of Maryland
College Park, Maryland 20742

Abstract

Charge-state distributions have been measured for 1.5 to 4.0 MeV nitrogen ions emerging from single collisions with nitrogen gas. For impact parameters near 0.1 Angstrom, the distributions at 2.5 MeV are independent of incident ion charge. The charge-state distributions are similar to those found for multiple collisions at larger impact parameters, suggesting that the projectile obtains charge-state equilibrium in a single collision.

The fraction of particles in each state of ionization has been measured for nitrogen ions emerging from single collisions with nitrogen atoms in N_2 gas at scattering angles of approximately 0.14°, corresponding to interpenetration of the K shells, and at projectile velocities greater than the velocity of L-shell electrons in the separated atoms.

The apparatus used is nearly the same as that for studies of the projectile charge-state dependence of carbon K X rays from methane excited by megavolt C, N and O ions.[1] In brief, ions accelerated in a 3-MeV Van de Graaff are stripped to high charges in low pressures of air intentionally introduced following the energy-analyzing magnet.[2] Ions of a given charge are selected by an electrostatic analyzer and directed through the collimators labeled EA and HV in Fig. 1. Ions scattered in the molecular beam gas target pass through the movable aperture labeled XY, the horizontal coordinate of which provides definition of the scattering angle θ. Behind this aperture are electrostatic deflector plates and a cooled position-sensitive detector[3] which serve to disperse the scattered ions and measure the number in each charge state.

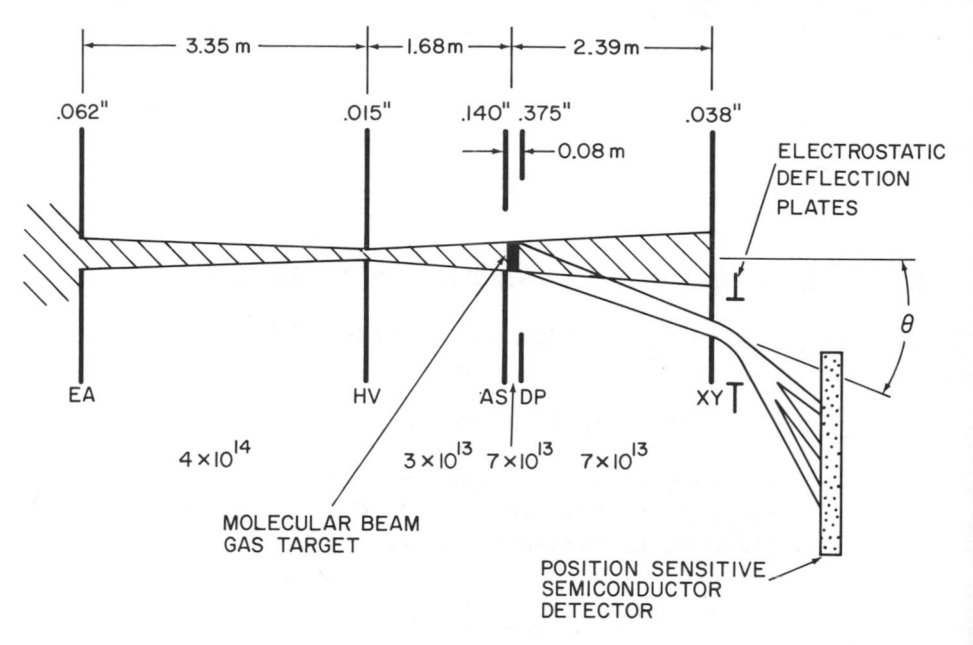

Fig. 1. Schematic diagram of apparatus. Ions of known mass, energy and charge are directed into the system of collimating apertures shown, and the charges of ions scattered at angle θ from the window-less gas target are determined. Aperture dimensions and separations are shown at the top of the figure. The numbers at the bottom of the figure are the thicknesses of residual gas; for example, between the differential pumping apertures HV and AS the residual gas amounted to 3×10^{13} molecules/cm^2 in the beam path.

The gas target consists of a molecular beam effusing from an array of glass capillary tubes in a direction perpendicular to the first ion beam. The density within the beam is proportional to the driving pressure behind the array[4] and for a pressure $\Delta h=80$ mm oil corresponds to a pressure of 1.0×10^{-2} Torr within the beam and a target thickness of about 3×10^{14} molecules/cm^2. Targets of this extreme thinness are necessary because the cross sections for charge transfer can be as large as 0.6×10^{-15} cm^2,[5] and if dependence on the incident charge is to be studied, there cannot be appreciable charge change within the target itself. In addition, there must not be charge exchange on the residual gas in the large flight paths necessary for accurate collimation. The equivalent thickness of all

the residual gas in the system, computed from dimensions and residual gas pressures measured with $\Delta h=75$ mm oil, was about 6×10^{14} molecules/cm^2, which corresponds to an incident charge purity of at least 70%. The purity of the beam was measured for 4 MeV N^{+4} ions by turning off the effuser and moving the detector to zero scattering angle, and was found to be about 95%.

Zero scattering angle was determined by measuring the left and right edges of the direct beam for $\Delta h=0$. No particles from the intense direct beam could enter the XY aperture when $\theta>1.5$ mrad. The movable aperture HV was carefully adjusted relative to the anti-scattering aperture AS so that no particles scattered on the edges of HV could reach the detector when $\theta>1.7$ mrad. As a check, the rate of scattering when the detector was at $\theta=3.4$ mrad was measured as a function of Δh for 2 MeV N^{+1} ions, and was found to be proportional to Δh above 20 mm oil. A slight non-linearity below $\Delta h=20$ is probably associated with scattering from residual gas. This measurement indicates that the results do not contain significant components associated with the direct beam or with scattering from apertures.

Pulses from the position output of the detector were amplified and recorded in a multichannel analyzer, and the fraction $\phi_j(\theta)$ of ions in emergent charge state j was determined by summing the counts in each peak and dividing by the total number of counts. No background corrections were used. Results obtained for 2.5 MeV N^{+i} ions with incident charges i between 2 and 4 are shown in Fig. 2. It is apparent that for $2<\theta<3.5$ mrad, corresponding to impact parameters between 0.08 and 0.14 Angstroms, and for $2 \leq i \leq 4$, there is no appreciable dependence of the emergent charge state $\phi_j(\theta)$ on either θ or i. Similar independence of θ and i was found for 1.5 MeV N^{+i} ions in the same angular range and for i=2,3. No dependence on θ was found for 3.0 or 4.0 MeV N^{+3} ions in the same angular range.

Two sample distributions are plotted as a function of j at constant θ in Fig. 3. The solid curves represent the distributions of this experiment. At 2.5 MeV the single solid curve for i=3 also closely approximates the data for i=2 and 4. This is in marked contrast to the strong dependence on i observed in single collisions at large impact parameters. Distributions for such collisions are calculated for the present target thickness using published values for ϕ_{ij}[6-9] and are plotted as dashed lines for i=2,4 in Fig. 3. Also shown in Fig. 3 as dotted lines are equilibrium distributions for ions which have undergone multiple small-angle collisions.[10] A remarkable feature of the distributions of this experiment is their similarity to these equilibrium distributions. In effect the scattered ions behave as if they had come to charge-state equilibrium in a single collision.

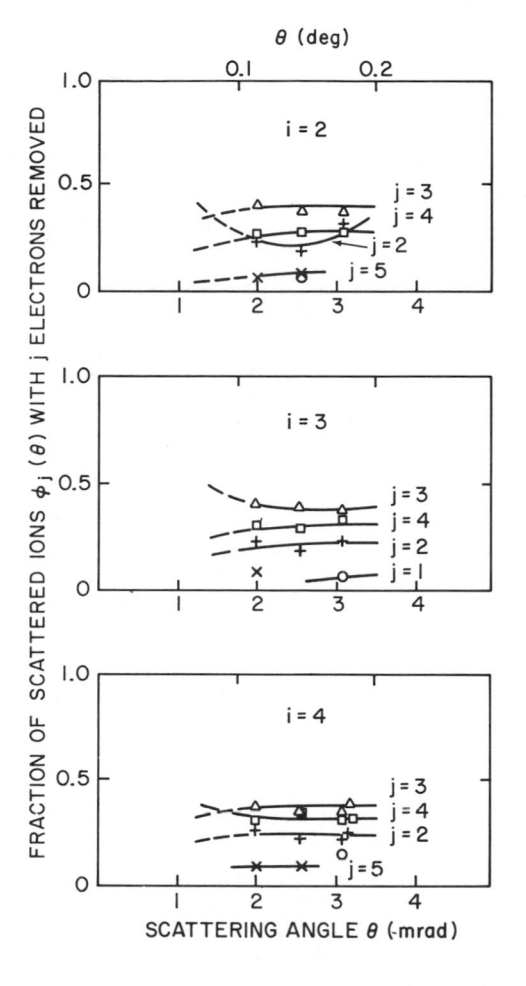

Fig. 2. Charge-state populations $\phi_j(\theta)$ as a function of scattering angle θ for 2.5 MeV nitrogen ions scattered in single collisions with nitrogen molecules at impact parameters of 0.08 to 0.14 Angstroms. The populations have very little dependence on scattering angle, and demonstrate a remarkable independence of the charge state i of the incident nitrogen ion for $2 \le i \le 4$.

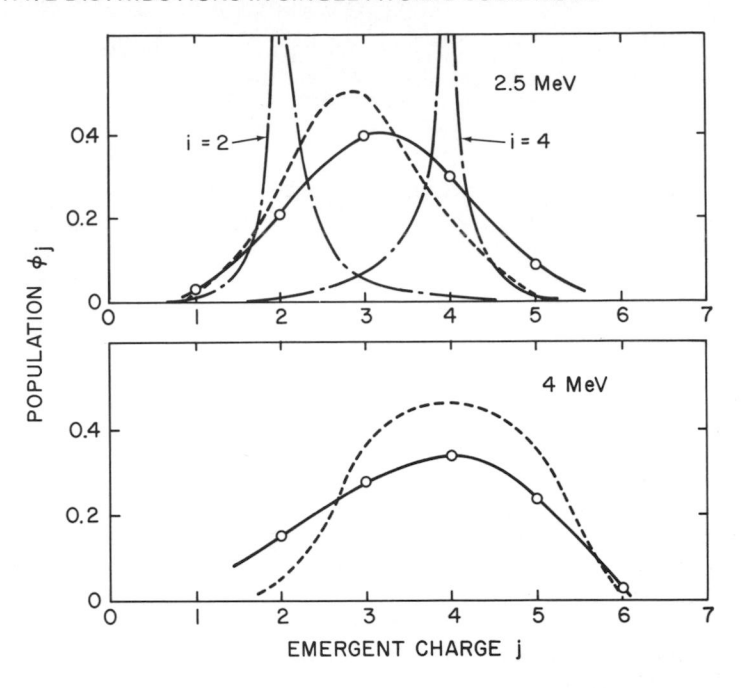

Fig. 3.   Charge state populations as a function of ion charge j.
The solid lines are drawn through the present experimental data for
2.5 and 4.0 MeV nitrogen ions incident with charge i=3 and emergent
with charge j at a scattering angle θ=1.9 mrad from single collisions
with nitrogen molecules.   As shown in Fig. 1, similar distributions
are found for i=2 and 4 at 2.5 MeV.   The dashed lines indicate the
populations after single collisions at large impact parameters for
2.8 MeV nitrogen ions incident with charge i=2 and i=4, as determined from Refs. 5 to 9.   The dotted lines indicate the equilibrium
distributions after multiple collisions at large impact parameters,
as found in Ref. 10.

        Some general comments about possible interpretations of these
results may be made.   One possible framework for interpretation is
that of the Fano-Lichten theory of transitions between molecular
energy levels formed as various shells of the projectile and target
overlap.[11]   This theory predicts promotion of K electrons into the
L shell during the collisions, followed by Auger electron emission
and consequent increase of charge on the part of the collision
partners.   The K X-ray yield for the similar system C→C up to 3 MeV
is approximately described by such a K vacancy production process,[1]
and in N→N collisions at 0.16 MeV the average charge of ions scat-

tered from collisions in which a promotion has occurred increases
by 1/2, indicating that the de-excitation of the promoted electron
by Auger emission occurs half the time on the scattered projectile
and half the time on the struck target.[12] In the present case,
there are 8 electrons to be shared among the L shells of the collision partners when a N^{+3} ion causes promotion of one K electron,
and if these are equally shared and Auger emission occurs half the
time, the charge of the resulting ions will be j=2, which is not
close to the observed values. Better agreement with the data is
obtained by assuming that two electrons are promoted.

Still higher values of the final charge j can be obtained by
considering promotion of L shell electrons. In particular, electrons from the N L shell can follow either the 3dσ molecular state
into the united atom, where they can be doubly promoted by rotational coupling to the 3dδ state leading to the M shell,[13] or they can
follow the 4fσ molecular state, which crosses a large number of
states that return to the M shell or other excited states.[14] Each
excited electron is presumed to cause Auger electron emission after
the collision is over; for example in $Ar^{+}\rightarrow Ar$ the average charge of
the system (sum of the average charges of the products) increases
slightly more than unity for each L shell vacancy.[15] However a
general feature of models of this type is a weak dependence of j
upon i, because the number of promotions occurring is proportional
to the number 10-i of electrons in the L shells of the nitrogen
atoms. In addition, the theory is not expected to be valid when
the velocity V of the projectile is greater than the velocity u of
the electrons in the shells involved, since in this case there is
not time for the molecular state functions to be formed. For the
present collisions $V/u_L \sim 2.5$. On the other hand, proper inclusion of
momentum factors can extend the range of the molecular theory,[16]
and the appearance of carbon KL^2 satellite X-ray lines for N→C collisions at 1.75 MeV can be interpreted in terms of promotions from
the L shell.[17]

An alternative framework for interpretation utilizes a description in terms of multiple capture and loss of electrons by the
heavy projectile along its trajectory inside the target atom. The
usefulness of such an interpretation is suggested by the following
numerical estimate. As the density of electrons within a N atom is
of the order of 10^{24} cm^{-3}, while the atom presents a target some
0.5×10^{-8} cm^2 in thickness, the projectile sees a target with an
areal density of 5×10^{15} electrons/cm^2 during the scattering event.
Measured cross sections for N^{+3} on He at 2.8 MeV are 3×10^{-17} cm^2
for electron loss[8] and 1×10^{-16} cm^2 for electron capture.[6] On the
basis of these numbers the probability of interaction is 0.15 for
loss and 0.5 for capture in the present scattering events. These
cross sections represent interaction probabilities averaged over
impact parameter, with most events having low probabilities and
large impact parameters; in the present small-impact parameter

collisions, the interaction probability can be expected to increase over the average value. In addition, there is a greater density of electrons than in helium. It can therefore be expected that the ion undergoes several capture or loss events in a single collision. The increase in capture cross sections and decrease in loss cross sections with ion charge then should result in a equilibrium charge, in the same way as for collisions with successive atoms.

Competition between capture and loss in single collisions has been described by Nikolaev, as it influences the loss cross section for $Mg^+ \to N$.[17] Experimental distributions suggesting equilibrium in single collisions have been found for $O^{+i} \to Kr$ collisions in which the projectile reaches approximately the K shell of the target and exits through an M shell containing 18 electrons for which $V/u_M \approx 2$.[18] A different situation exists for $I^{+i} \to Xe$ collisions at impact parameters of about 0.08 Angstroms, in which case the average charge of 22 in single collisions is much greater than the value of 8 after multiple collisions.[19] Here only the eight O-shell electrons have $V/u_O > 1$, and promotion of some of the 36 M- and N-shell electrons accounts for the high charges.[20]

Theoretical calculation of capture and loss cross sections or of high-velocity promotion probabilities for fixed impact parameters is required if either the equilibrium model or the electron promotion model is to be shown to be a correct quantitative description. An experimental distinction between the models might be made in the present case where $V/u_L > 1$ by scattering C and O ions from N_2. Since the molecular model suggests that fewer L-shell vacancies should appear in the heavy collision partner in asymmetric cases, it would predict that the average charge after single collisions of O ions at small impact parameters should be less than the equilibrium charge after multiple collisions, while in the case of C ions it should be greater.

The authors wish to acknowledge useful discussions with Drs. B. Rosner and Q. Kessel concerning the interpretation of these results.

References

*Supported in part by the National Science Foundation.
†Present address: Univ. of Colorado Medical Center C-278, 4200 East 9th Avenue; Denver, Colorado 80220

1. F. W. Martin and R. K. Cacak, submitted to Phys. Rev.; see also S. S. Choe, R. K. Cacak and F. W. Martin, Bull. Am. Phys. Soc. 19, 591 (1974).
2. F. W. Martin, Nucl. Instr. and Meth. 124, 329 (1975).

3. E. Laegsgaard, F. W. Martin and W. M. Gibson, IEEE Trans. on
    Nucl. Sci., NS-15-3, 239 (1968).

4. J. C. Johnson, A. T. Stair and J. L. Prichard, J. Appl. Phys.
    37, 1551 (1966).

5. V. S. Nikolaev, I. S. Dmitriev, L. N. Fateeva and Ya. A. Teplova,
    Soviet Phys. –JETP 13, 695 (1961) [Russian original ZETF 40,
    989 (1961)]; a worse case, σ_{43} for +4 nitrogen ions in N_2 at
    0.1 MeV/amu.

6. Ref. 5, σ_{43} at 0.2 MeV/amu.

7. V. S. Nikolaev, L. N. Fateeva. I. S. Dmitriev and Ya. A. Teplova
    Soviet Phys. –JETP 14, 67 (1962) [Russian original ZETF 41,
    89 (1961).

8. I. S. Dmitriev, V. S. Nikolaev, L. N. Fateeva and Ya. A. Teplova,
    Soviet Phys. –JETP 15, 11 (1962) [Russian original ZETF 42,
    16 (1962)].

9. I. S. Dmitriev, V. S. Nikolaev, L. N. Fateeva and Ya. A. Teplova
    Soviet Phys. –JETP 16, 259 (1963) [Russian original ZETF 43,
    259 (1962)].

10. V. S. Nikolaev, I. S. Dmitriev, L. N. Fateeva and Ya. A. Teplova
    Soviety Phys. –JETP 12, 627 (1961) [Russian original ZETF 39,
    905 (1960)].

11. M. Barat and W. Lichten, Phys. Rev. A5, 211 (1972), and refer-
    ences therein.

12. B. Fastrup, G. Hermann, Q. Kessel and A. Crone, Phys. Rev. A9,
    2518 (1974).

13. Q. C. Kessel and B. Fastrup, Case Studies in Atomic Physics 3,
    137 (1973); see page 195.

14. Ref. 13, p. 175.

15. B. Fastrup. G. Hermann and K. J. Smith, Phys. Rev. A3, 1591
    (1971).

16. W. R. Thorson, Phys. Rev. A12, XXX (1975).

17. F. W. Martin and R. K. Cacak, submitted to Phys. Rev.; see also
    Bull. Am. Phys. Soc. 20, 638 (1975).

18. V. S. Nikolaev, Soviet Phys. –Uspekhi 8, 286 (1965) [Russian
    original UFN 85, 679 (1965)].

19. B. Rosner, K. C. Chan, D. Gur and L. Shabason, Bull. Am. Phys.
    Soc. 20, 620 (1975).

20. G. D. Alton, C. M. Jones, P. D. Miller, B. Wehring,
    J. A. Biggerstaff, C. D. Moak, Q. C. Kessel and L. Bridwell,
    to be published in Nucl. Instr. and Meth.; see also Bull.
    Am. Phys. Soc. 20, 195 (1975).

STOPPING POWER FOR IONS OF INTERMEDIATE ATOMIC NUMBERS

B. W. Wehring and R. G. Bucher

University of Illinois at Urbana-Champaign

Urbana, Illinois USA

INTRODUCTION

The stopping power for energetic heavy ions has been characterized in terms of Z_1 and Z_2, the atomic numbers of the ion and stopping atom, respectively, and v, the velocity of the ion. Lindhard[1]
developed a unified theory for the stopping power of heavy ions using
the Thomas-Fermi description of the atom. Firsov[2] obtained a semi-
classical expression for energy loss in terms of the impact parameter
which was integrated over all impact parameters[3] to obtain the stopping power. The tables of Northcliffe and Schilling[4] provide a
tabulation of stopping powers theoretically extrapolated from experimental values obtained for a limited number of different heavy ions.
Experimental results have also been fitted by semiempirical formulae--
the expression of Moak and Brown[5] was fitted to experimental results for solid absorbers[6] and that of Pierce and Blann,[7] for
gaseous absorbers. None of these invesitgations, however, is capable
of describing the detailed behaviour of the stopping power for energetic ions, in particular the Z_1 and Z_2 oscillations which have been
observed for low-energy ions.

We have developed a multiparameter technique[8] to measure the
energy loss of fission fragments as a function of fragment atomic
number, Z_1. The atomic numbers were identified by detection of characteristic K x rays emitted in coincidence with fission.[9] These
x rays result from nuclear transitions in the fission fragment which
internally convert a K shell electron. The half-lives for the fragment internal-conversion decays range from a fraction of a ns to µs.

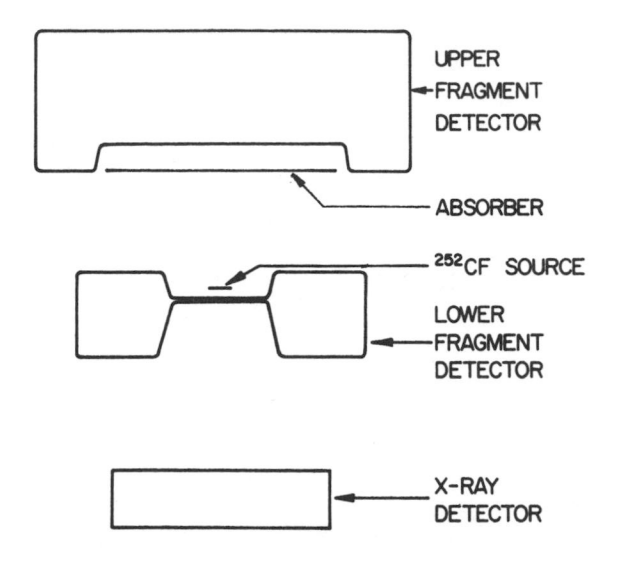

Fig. 1.   Physical arrangement of the detectors used to collect the
multiparameter data.   The sizes and spacings of the detectors are
shown in correct proportion.

EXPERIMENT

A ^{252}Cf source on a thin nickel backing was located between two
fragment detectors as is shown in Fig. 1.   An absorber holder with
a nickel absorber at the center and open annular slots around the
absorber was placed in the space between the source and the upper
fragment detector, a heavy-ion Si(Au) detector.   In this way, both
degraded and undegraded fragments were measured simultaneously in
the upper fragment detector to provide an accurate energy calibration for the degraded fragments.   From a measurement of the degra-
dation of the ^{252}Cf alpha particles, the effective thickness of the
nickel absorber was determined to be 1.35 ± 0.04 mg/cm^2.

In all events in which one fragment was detected in the upper
fragment detector, the complementary fragment was detected in the
lower fragment detector, a fully-depleted 100-μ thick Si(Au) detector, into which the source holder was recessed to minimize Dopple
shifting of the x-ray energies.   A Si(Li) x-ray detector with a resolution of 500 eV at 22 keV was located below the lower fragment
detector.   Figure 2 shows an x-ray spectrum taken in coincidence
with fission and identifies the atomic number responsible for each
Kα x-ray peak.   For the fragment energy-loss measurement, three-
parameter data, two fission-fragment pulse heights and an x-ray puls
height, were recorded event by event on magnetic tape and analyzed
off-line.

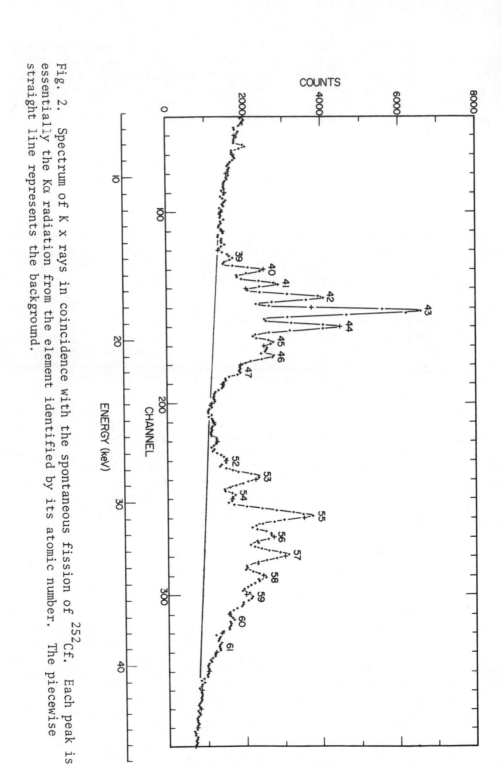

Fig. 2. Spectrum of K x rays in coincidence with the spontaneous fission of ^{252}Cf. Each peak is essentially the Kα radiation from the element identified by its atomic number. The piecewise straight line represents the background.

ANALYSIS

The three-parameter data were sorted according to the values
of the x-ray pulse height and the lower-fragment-detector pulse
height. To avoid the problem of sorting on Doppler-shifted x rays
emitted by the fragment moving toward the upper detector, only those
events were analyzed for which the x ray was emitted by the fragment
detected in the lower detector. For example, the energy loss for
fragments with Z_1 = 43 was studied by selecting three-parameter
events with an x-ray pulse height in the 55 peak of the x-ray spectrum and with the lower-fragment-detector pulse height in the heavy
peak of the fragment spectrum. The lower-fragment-detector pulse
heights were further sorted into narrower intervals in order to yield
different initial energies for the fragments under study. This is
possible because of the correlation which exists between the two complementary fission-fragment energies.

Using a Z-dependent energy calibration,[8] energy spectra which
contained both degraded and undegraded fragments were accumulated for
the various x-ray peaks ($40 \leqslant Z_1 \leqslant 45$, $53 \leqslant Z_1 \leqslant 58$) and various initial energies. Under each x-ray peak, however, there are background
events which do not belong to the atomic number assigned to the x-ray
peak (Fig. 2.). This slowly varying background in the x-ray spectrum
results primarily from prompt fission gamma rays which Compton scatter in the x-ray detector.[10] These Compton events were subtracted
from the individual fragment energy spectra by assuming that all possible fragment pairs contribute to this Compton background in proportion to their abundance. The resulting energy spectra, foreground,
background, and difference, are shown in Fig. 3 for Z_1 = 55 and an
initial fragment energy of 80 MeV. Using the most probable mass[9]
associated with this Z_1 yielded an initial velocity of 1.044 cm/ns for
this case.

The energy loss for each Z_1 initial-velocity combination was tak-
en to be equal to the difference in the mean energies of the undegraded and degraded peaks of the fragment energy spectra. In order
to compare these results to predicted results, stopping powers were
integrated over the foil thickness. The following integral was calculated numerically:

$$\Delta E = \int_0^{\rho t} \left(\frac{1}{\rho} \frac{dE}{dx} \Big|_\varepsilon + \frac{1}{\rho} \frac{dE}{dx} \Big|_\nu \right) d(\rho x)$$

where ε denotes the electronic stopping power, ν denotes the nuclear
stopping power, and (ρt) is the effective thickness of the nickel
foil. The nuclear component of the stopping power was assumed to be
that given by the Lindhard theory.[11] The contribution to the total
energy loss from nuclear stopping ranged from 0.6 to 1.0% for the
light fragments and from 2 to 3.5% for the heavy fragments.

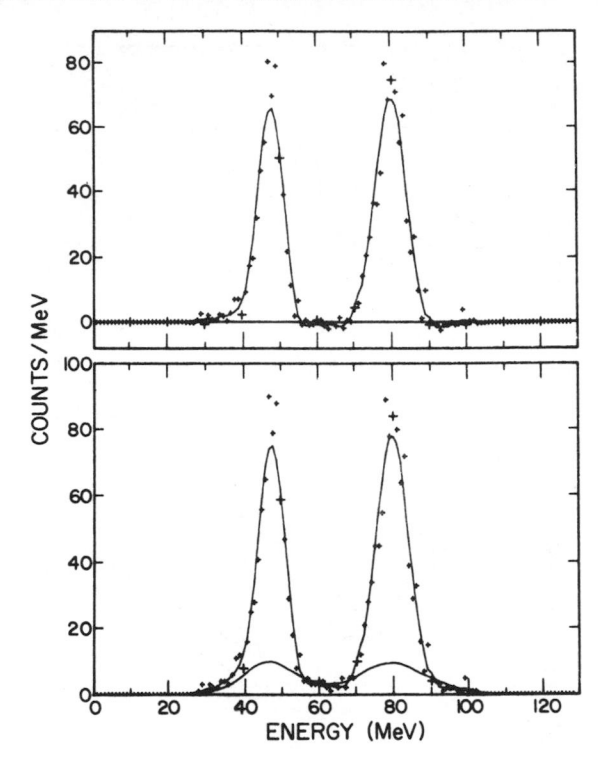

Fig. 3. Energy spectra of degraded and undegraded fragments for
Z_1 = 55 and initial fragment velocity of 1.044 cm/ns. The points (+)
and corresponding solid curve in the lower graph are the foreground
spectrum; the other solid line represents the background. The points
(+) and corresponding solid curve in the upper graph are the foreground minus the background.

RESULTS

     The measured energy losses in the nickel foil were plotted as
a function of initial velocity for the different fragment atomic
numbers, Z_1. Two of these graphs are shown in Fig. 4. The error
bars indicate the standard deviation of the uncertainty in the energy loss due to the counting statistics of the undegraded and degraded peaks. The measured results were compared to predicted energy losses using the electronic stopping theories of Linhard[1] and
Firsov,[2,3] the tabulations of Northcliffe and Schilling, [4] and
the semiempirical formula of Moak and Brown.[5] For the light fragments the predictions of Moak-Brown and Northcliffe-Schilling agree
with the results of the measurement; for the heavy fragments the

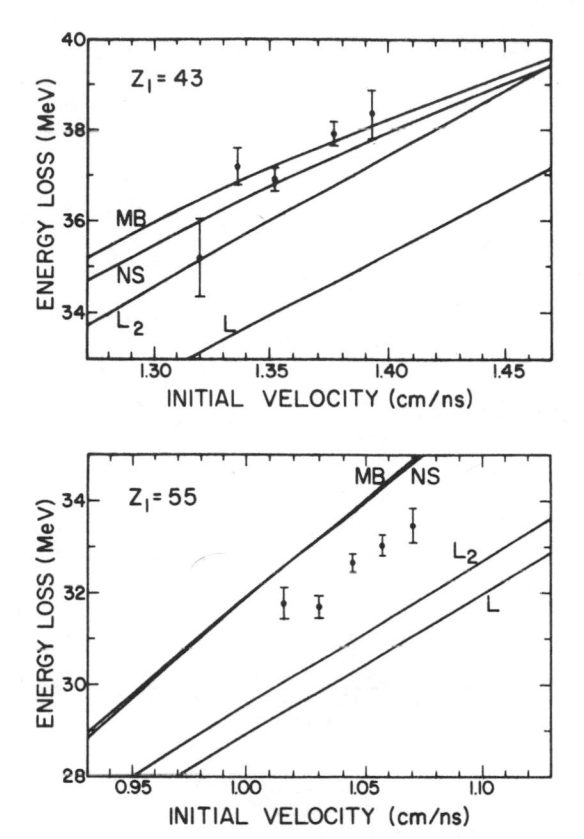

Fig. 4.   Comparison of the measured and predicted energy losses for fragment atomic numbers Z_1 = 43 and 55 in a 1.35 mg/cm^2 nickel foil. The curves are predictions of: (L) Linhard $\xi_\varepsilon = Z_1^{1/6}$; (L$_2$) Lindhar ξ_ε = 2; (NS) Northcliffe-Schilling; and (MB) Moak-Brown.

measured values fell approximately midway between the predictions of Moak-Brown and Northcliffe-Schilling, and those of Lindhard for ξ_ε = 2.

      In order to study the Z_1 dependence of the stopping power of nickel, the Moak-Brown velocity dependence[5] was used to interpolat and extrapolate the measured results to a single light-fragment ve- locity, 1.36 cm/ns, and a single heavy-fragment velocity, 1.04 cm/ns These results are shown in Fig. 5.   The Z_1 dependence predicted by Moak-Brown and Northcliffe-Schilling for the light fragments is in good agreement with the measurements.   For the heavy fragments, how-

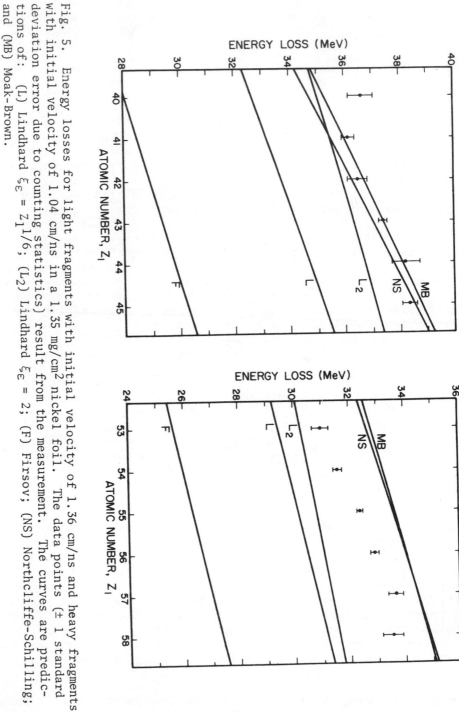

Fig. 5. Energy losses for light fragments with initial velocity of 1.36 cm/ns and heavy fragments with initial velocity of 1.04 cm/ns in a 1.35 mg/cm² nickel foil. The data points (± 1 standard deviation error due to counting statistics) result from the measurement. The curves are predictions of: (L) Lindhard $\xi_E = z_1^{1/6}$; (L₂) Lindhard $\xi_E = 2$; (F) Firsov; (NS) Northcliffe-Schilling; and (MB) Moak-Brown.

ever, none of the predictions contain a strong enough dependence on Z_1.

The agreement for the Z_1 dependence shown in the light fragments and the disagreement for the heavy fragments can be interpreted in terms of Z_1 oscillations. The percent deviation of the experimental results shown in Fig. 5 from the Moak-Brown results also shown in Fig. 5 suggests a Z_1 oscillation with a maximum around $Z_1 = 42$, a minimum around $Z_1 = 51$ and a peak-to-peak deviation on the order of 5%. This oscillation is consistent with that observed for lower energy ions.[12] It is in disagreement with the interpretation by Schmidt et al. of their secondary-electron-production measurements for fission fragments passing through thin carbon foils.[13]

REFERENCES

1. J. Lindhard and M. Scharff, Phys. Rev. 124, 128 (1961).
2. O. B. Firsov, Sov. Phys. - JETP, 9, 1076 (1959).
3. Y. A. Teplova, V. S. Nikolaev, I. S. Dmitriev, and L. N. Fateeva, Soc. Phys. - JETP, 15, 31 (1962).
4. L. C. Northcliffe and R. F. Schilling, Nucl. Data, A7, 233 (1970).
5. M. D. Brown and C. D. Moak, Phys. Rev. B 6, 90 (1972).
6. C. D. Moak and M. D. Brown, Phys. Rev. 149, 244 (1966).
7. T. E. Pierce and M. Blann, Phys. Rev. 173, 390 (1968).
8. R. G. Bucher, "An Experimental Study of Stopping Powers for Ions of Intermediate Atomic Numbers," Ph.D. Thesis, University of Illinois at Urbana-Champaign (1975).
9. W. Reisdorf, J. P. Unik, H. C. Griffin, and L. E. Glendenin, Nucl. Phys. A 177, 337 (1971).
10. C. J. Withee, "Measurement of Internal Conversion Cascading in U^{235} Fission Fragments," Ph.D. Thesis, University of Illinois, Urbana (1973).
11. J. Lindhard, V. Nielsen, and M. Scharff, Kgl. Danske Videnskab. Selskab, Matt.-Fys. Medd. 36, No. 10 (1968).
12. C. P. Bhalla, "Inelastic Atomic Collisions and the Stopping Power of Heavy Ions," Proceedings of Heavy-Ion Summer Study, CONF-720669 Oak Ridge, Tennessee (1972).
13. K.-H. Schmidt, H.-G. Clerc, H. Wohlfarth, W. Langs and W. Lehmann-Carpzov, Phys. Letters 56B, 250 (1975).

ANGULAR BEHAVIOUR OF STOPPING POWERS OF CARBON FOIL

FOR ARGON IONS BELOW 250 keV

Gilles Beauchemin and Robert Drouin

Laboratoire de l'accélérateur Van de Graaff, C.R.A.M. and
Département de chimie, Université Laval, Québec, Canada
G1K 7P4

The stopping power $\frac{dE}{dX}$ as a function of scattering angle θ is
studied for argon ions ($Z_1 = 18$) between 40 keV and 240 keV. The dependence of $\frac{dE}{dX}(\theta)$ upon foil thickness (4-14 $\mu g/cm^2$) and incident
energy has been derived using the method of least squares. The analysis also examined other elements from which the dependence on Z_1
is deduced. For argon the values of stopping power averaged over
all angles are calculated and compared with the Lindhard-Scharff-
Schiott theory.

INTRODUCTION

Numerous physical processes are simultaneously involved in beam-
foil collisions. One of these processes, the slowing down of the
ions in the carbon foil, is especially important for mean life studies and a knowledge of energy loss ΔE is required. Traditionally,
these energy losses have been determined from Northcliffe's tables
(Ref.1), from previous experimental results (Ref.2-6), or from the
Lindhard-Scharff-Schiott theory (Ref.7); sometimes they were simply
ignored. But T. Andersen (Ref.8) has indicated the need for a better
knowledge of ΔE, mainly for heavy ions at low energies.
In relation to these beam-foil problems we were interested at
Université Laval in studying the following phenomena: 1) the stopping
power dE/dX compared with the LSS theory, and 2) the energy loss
$\Delta E(\theta)$ as a function of the emerging angle θ.

THEORY

The comparison of our results with the LSS theory was our first aim. According to this theory the stopping power of an amorphous target is the sum

$$\frac{dE}{dX} = \frac{dE}{dX}_{electronic} + \frac{dE}{dX}_{nuclear} \tag{1}$$

The electronic stopping power represents inelastic processes involving atomic electrons and is said to be $kE^{1/2}$ where E is the incident energy and k a constant for a given pair of incident ion and target atom. The nuclear stopping power describes elastic processes involving nuclei and is a more complicated function $f(E, Z_1, Z_2, M_1, M_2)$. According to the theory, for argon at 40 keV, the nuclear stopping power is 80% of the total and at 240 keV, it is 45%. The total stopping power is calculated to be around 5 keV-cm^2/µg for the entire range of energies.

EXPERIMENTAL CONDITIONS

We performed experiments with argon ions at energies from 40 to 240 keV on home-made carbon foils. The thicknesses of the foils varied form 4 to 14 µg/cm^2 and were determined using the stopping power of carbon on 60 keV protons as given by Ormrod and Duckworth (Ref.3). We observed a variation in ΔE values produced by the thickening of the target due to the ion beam. This has been observed by Bickel and Buchta (Ref.9) and others. In our calculations, corrections were made taking into account this thickening.

The ion beam passes through the foil, is mass-analysed by a Browne-Buechner variable angle mass-spectrometer, and detected by a channel electron multiplier. The signal is transmitted to a strip chart recorder.

An energy spectrum is recorded on the chart. The energy spectrum presents the following characteristics: a) a Gaussian curve produced by a large number of soft collisions, and b) a tail towards lower energies produced by a small number of collisions which involve large energy losses. The most probable energy loss $\Delta\hat{E}$ is taken at the maximum of the curve. The mean energy loss $\Delta\bar{E}$ is the value averaged over the total curve. For comparison with the theory we always used, as does Högberg (Ref.10), the mean energy loss $\Delta\bar{E}$, while most other authors used the most probable energy loss $\Delta\hat{E}$(cf. Ref.3, 4,5,6,11). The $\Delta\bar{E}$ values are generally 10% higher than $\Delta\hat{E}$. In the present text we assume about the stopping power that $\frac{dE}{dX} \simeq \frac{\Delta E}{\Delta X}$.

RESULTS

Preliminary results were presented elsewhere (Ref.12), while

these are final results analysed by computer. The experiments were
made for 7 different energies from 40 to 240 keV with two foils of
different thickness for each energy, at angles from $0°$ to $40°$ in $2°$
or $3°$ steps.

Stopping Power and LSS Theory

We now compare our results for stopping powers with the LSS
theory. In Figure 1, the solid line represents the total stopping
power according to the LSS theory. The lower curve illustrates values for the stopping power in the forward direction $\frac{d\bar{E}}{dX}(0°)$ obtained
with the mean energy loss values $\Delta \bar{E}$. Our values are slightly lower
than those of Högberg (Ref.10) for $\Delta x > 6$ µg$/$cm^2.

Using the most probable energy losses ΔE would yield another
curve, 10% lower than the one shown, and closer to the results of
Ormrod et al.(Ref.4), Fastrup et al.(Ref.5) and Carriveau et al.
(Ref.11).

The curve shown here indicates a large discrepancy with the LSS
theory for the energy range studied. But the LSS theory does not
refer to any specified direction of the emerging ions. Therefore the
value predicted by the theory shoud be compared with the stopping
power averaged over all angles

$$\frac{d\bar{E}}{dX}\bigg|_{\text{all angles}} = \frac{\int \frac{d\bar{E}}{dX}(\theta)\ I(\theta)\ G(\theta)\ d\theta}{\int I(\theta)\ G(\theta)\ d\theta} \qquad (2)$$

where $\frac{d\bar{E}}{dX}(\theta)$ is the mean stopping power measured at the angle θ from
the forward direction and $I(\theta)$, the relative number of particles
emerging at angle θ. $G(\theta)$ is a geometrical factor [$G(\theta) = 2\pi R^2$
$\sin\theta \ \Delta\theta$] corresponding to the annular area obtained at distance R
from the collision centre.

Taking into account this calculation, the mean stopping powers
averaged over all angles are shown in the middle curve of Figure 1.
These integrated stopping powers are approximately 25% higher than

$\frac{d\bar{E}}{dX}(0°)$, and nearer the LSS theoretical curve. But the discrepancy
between theory and this experimental curve remains significant for
energies lower than 100 keV.

Stopping Power Angular Distribution $\frac{d\bar{E}}{dX}(\theta)$

The method. The method used in the second part of this study
is based on least squares fitting and minimization of standard de-

viations. From experimental values of energy loss drawn on a graph, we obtained by computer a least squares fitted curve. From this fitted curve, standard deviation was calculated for the set of the experimental values. We then sought to minimize this standard deviation by changing the form of the coördinates.

Thickness dependence. Figure 2 represents experimental results, with their fitted curves, of energy loss at different angles θ, for two foils at 150 keV energy. The values $\Delta\bar{E}(\theta)$ increase as a nearly parabolic function added to the value $\Delta\bar{E}(0^\circ)$.

As expected, the thinner foil has lower energy loss at 0°, but the curve is steeper for this foil and shallower with the thicker foil. A thickness dependence was found by the minimization method when the abscissa is expressed in terms of $\theta/\Delta X$. ΔX is the thickness of the foil. It could be seen either as $\Delta X(0^\circ)$, or $\Delta X(\theta)$. $\Delta X(0^\circ)$ is the actual thickness of the foil and is the path-length of ions emerging at 0°. $\Delta X(\theta) = \frac{1}{2}[\Delta X(0^\circ)(1+\sec\theta)]$ is the path-length of an ion colliding in the middle of the target and emerging at angle θ.

Minimum deviations were found (cf. Table 1a) for two sets of coördinates. The first set has as ordinate the change in stopping

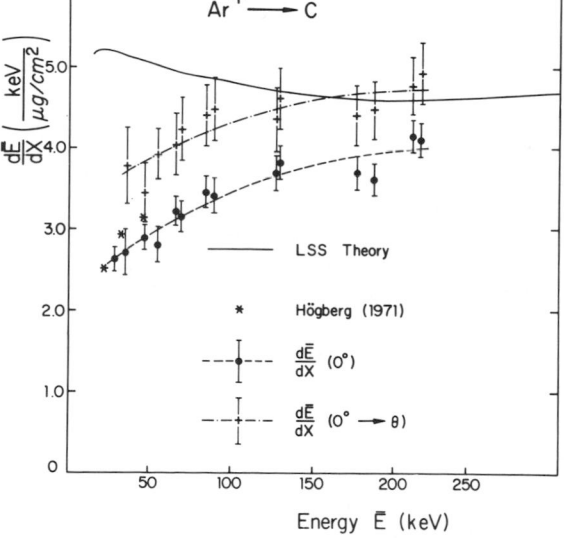

Figure 1.   Stopping Power

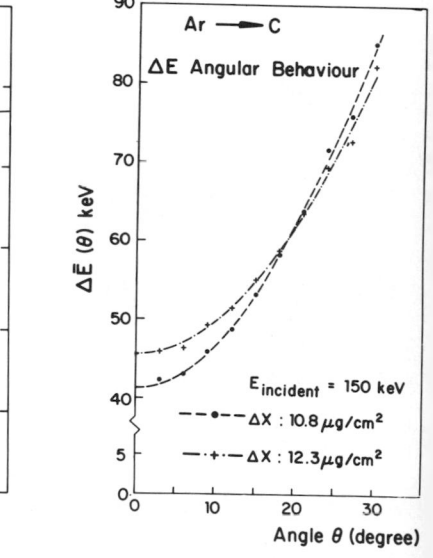

Figure 2.   Angular Behavi
of ΔE

power from $0°$ to angle θ, while the second set expresses the <u>change in the ratio</u> of stopping power between $0°$ and θ. The least squares fitting gives 8.6% as deviation averaged over the seven energies for the first set, and 6.8% for the second set. Thickness dependence was mentioned by Högberg (Ref.10) for a coördinate system proportional to our first one.

Figure 3 illustrates a typical curve of this thickness dependence with two foils at 80 keV energy. The standard deviation is 7.5%. The fitted curve is parabolic for lower values of abscissae and linear for higher values, as mentioned by Högberg (Ref.10).

<u>Energy dependence</u>. Calculations were made to study the energy dependence. Assuming the thickness dependence, if the energy is varied, the curves are steeper for higher energies and shallower for lower energies. An energy dependence was found when abscissae were expressed in terms of $\dfrac{\theta(\bar{E})^p}{\Delta X}$, where $\theta/\Delta X$ contains the previous thickness dependence. $\bar{E}$ is the mean energy in the foil $[\bar{E} = E_{inc} - \Delta E/2)$, and p an adjustable parameter. Table 1b presents the best results obtained. Minimum deviations were found for 2 sets of coördinates. The first set has 11.5% deviation for p = .95, the second 9.2% for p = .77.

Figure 4 presents the fitted curve for the first set of coördinates of Table 1b. Nearly 170 experimental values are shown. Values for argon on carbon from Högberg (Ref.10) have the same trend and are slightly higher than ours. They were obtained with 2 foils, one energy, and for angles $0°$ to $15°$. As in previous figure, the present solid curve is parabolic for lower values of abscissae and linear for higher values. Figure 5 presents the fitted curve for the second set of coördinates of Table 1b. This curve too has the same feature as the previous ones.

	a) <u>Thickness-dependence</u>		b) <u>Energy-dependence</u>	
Ordinates	Abscissae	% deviation	Abscissae	% deviation
$\left\|\dfrac{\Delta\bar{E}}{\Delta X}(\theta) - \dfrac{\Delta\bar{E}}{\Delta X}(0°)\right\|$	$\dfrac{\theta}{\Delta X(0°)}$	8.6%	$\dfrac{\theta(\bar{E})^{.95}}{\Delta X(0°)}$	11.5%
$\dfrac{\Delta\bar{E}(\theta)}{\Delta\bar{E}(0°)} - 1$	$\dfrac{\theta}{\Delta X(\theta)}$	6.8%	$\dfrac{\theta(\bar{E})^{.77}}{\Delta X(\theta)}$	9.2%

<u>Table 1</u> Forms of coördinates giving minimum deviation from the fitted curve.

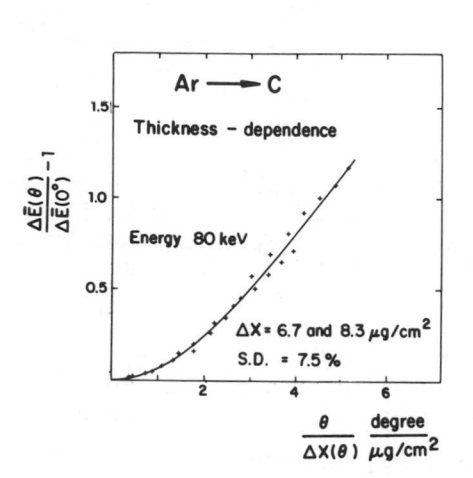

Figure 3.   Thickness—Dependence

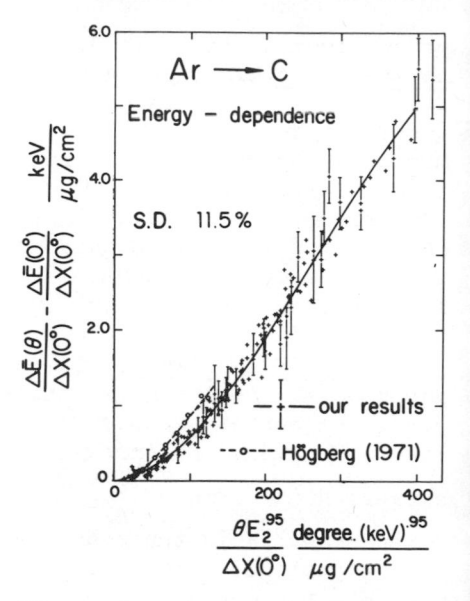

Figure 4.   Energy-Dependence (1

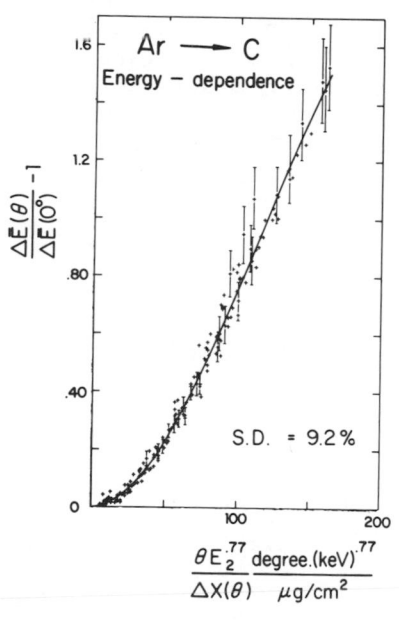

Figure 5.   Energy—Dependence (2)

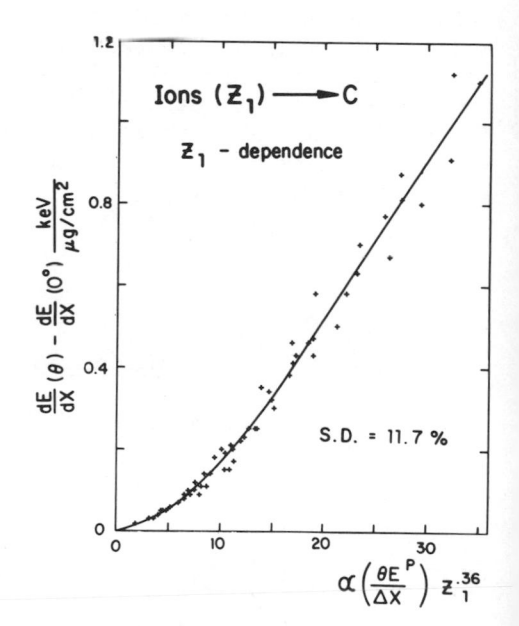

Figure 6. Z_1-Dependence

Z_1-dependence. With the same method, we tried to extend our analysis to other Z_1-dependence. We used experimental results figured in Högberg (Ref.10), obtained at 46 keV with 5 different ions; in his figure 3 the abscissa is expressed in terms of $\theta/\theta_{1/2}$. Assuming $\theta_{1/2} = k \, Z_1^{0.75} \, (\frac{\Delta X}{E^p})$ (Ref.14), we found the best fit with abscissae expressed in terms proportional to $(\frac{\theta E^p}{\Delta X}) Z_1^{0.36}$; the deviation from the fitted curve is 11.7%. The curve has the same parabolic form for low values of abscissae and is linear for high values (figure 6).

According to V. Neilsen (see Ref.10), a $1/r^2$ interaction would give a parabolic function for the curve. Skoog and Högberg (Ref.14) have tried to reproduce the experimental curves of $\frac{dE}{dX}(\theta)$ by computer simulation.

The application of these results to beam-foil spectroscopy is in progress.

CONCLUSION

According to our analysis we can say

a) about the stopping power, that the discrepancy between the LSS theory and experiments is significantly smaller if we take into account the stopping power integrated over all angles. However, the discrepancy remains significant for E < 100 keV.

b) that experimental "universal" curves can be drawn describing the angular distribution of the stopping power for carbon foils of various thicknesses and for argon in a wide range of energies.

c) that this last conclusion can be extended also to various Z_1-values of incident ions.

REFERENCES

1) L.C. Northcliffe, Ann. Rev. Nucl. Sci. 13, 67(1963).
    L.C. Northcliffe and R.F. Schilling, Nuclear Data Tables, A7 233(1970).
2) D.I. Porat and K. Ramavataram, Proc. Phys. Soc. 77, 97(1960); 78, 1135(1961).
3) J.H. Ormrod and H.E. Duckworth, Can. J. Phys. 41, 1424(1963)
4) J.H. Ormrod, J.R. Macdonald, H.E. Duckworth, Can. J. Phys. 43, 275(1965).
5) B. Fastrup, P. Hvelplund and C.A. Sautter, Matt.-Fys. Medd. Dan. Vid. Selsk. 35, no. 10 (1966).
6) P. Hvelplund and B. Fastrup, Phys. Rev. 165, 408(1968).

7) J. Lindhard, M. Scharff and H.E. Schiott, Matt-Fys. Medd. Dan.
      Vid. Selsk. 33, no. 14 (1963)
8) T. Andersen, Nucl. Instr. Meth. 110, 35(1973).
9) W.S. Bickel and R. Buchta, Physica Scripta 9, 148(1974).
10) G. Högberg, Phys. Stat. Sol. 48b, 829(1971).
11) G.W. Carriveau, G. Beauchemin, E.J. Knystautas, E.H. Pinnington
      and R. Drouin, Phys. Lett. 46A, 291(1973).
12) G. Beauchemin and R. Drouin, in Proceeding of the Third Confe-
      rence on application of small accelerators (1974) vol. 1,
      (U.S. Energy Research and Development Administration.) Conf.
      741040 P1, p. 336.
13) G. Högberg, H. Norden, H.G. Berry, Nucl. Instr. Meth. 90, 283
      (1970).
14) R. Skoog and G. Högberg, Rad. Eff. 22, 277(1974).

VACUUM ULTRAVIOLET EMISSION SPECTRA FROM keV ENERGY RARE GAS ION-ATOM COLLISIONS[*]

W. W. Smith,[†] D. A. Gilbert and C. W. Peterson

University of Connecticut

Storrs, Connecticut 06268

1. INTRODUCTION

Beam-gas spectra have been observed under single-collision conditions in the 500-1100 Å wavelength range for several rare-gas ion-atom collision combinations at energies from 8-30 keV. We report here a preliminary determination of the absolute cross sections for the most prominent spectral features observed, based on normalization to the data of deHeer et al. (1) for He^+ + Ar collisions at 10 keV. Similar spectra have been observed for some of these collisions at energies up to 10 keV by Isler (2). We report here observations for He^+, Ne^+, Ar^+ + Ar collisions, as well as spectra from Ar^+ + Ne, Ar^+ + He, and Ne^+ + Ne collisions. Simple explanations of the prominence of many of the observed spectral features are possible on the basis of the electron-promotion model (3). Comparison of these spectra gives considerable physical insight into the nature of the excitation process in asymmetric as well as symmetric collisions in the low keV energy range. This may have applications, for example, to excitation processes produced by particles in the solar wind or the solar corona. The high resolution obtainable spectroscopically provides a useful supplement to previous studies (4) of these same ion-atom systems by inelastic energy loss measurements in differential scattering.

---

[*]Research supported in part by a grant from AROD.

[†]On sabbatic leave as a Visiting Fellow, Joint Institute for Laboratory Astrophysics, National Bureau of Standards and University of Colorado, Boulder, CO 80302.

The independent-electron promotion model (3) has been extensively used to study inner-shell excitation processes in symmetric
ion-atom collisions like Ar^+ + Ar. It has also recently been applied to outer shell electronic excitation (4) where its validity
must be examined more critically. This model is based on the Born-
Oppenheimer separation of the electronic and nuclear motions when
the nuclei move together slowly compared to characteristic electron velocities, forming a quasimolecule during the collision. A
study of the diabatic (3) correlation of the molecular orbitals of
this quasimolecule as a function of internuclear distance, from
the separated to the united atom limit, can give semiquantitative
insight as to which excitations will be most important as a result
of the collision. Such a correlation diagram for Ar^+ + Ar is
shown in Figure 1.

The Pauli principle causes "promotion" of the MO energies
(Fig. 1) at small internuclear distances since there are enough
electrons in the molecular system to overfill the atomic orbitals
of the "united-atom" formed when the two nuclei are very close together. This promotion to higher atomic orbitals in the united
atom limit induces crossings or degeneracies between the various
MO energies at certain internuclear distances R. Near a crossing
of two such diabatic curves (of the same molecular symmetry σ, Π,
etc.), certain residual interactions and dynamic couplings (due to
the nuclear motion) can cause transitions from one diabatic state

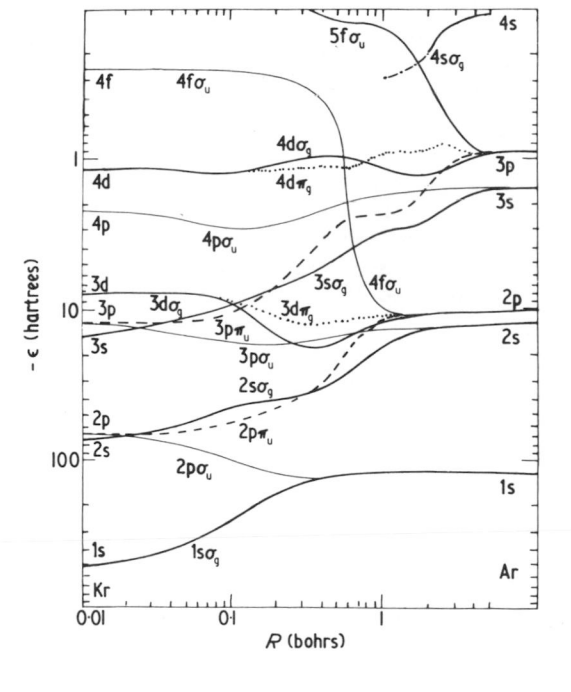

Fig. 1. Ab initio
diabatic molecular-
orbital (MO) energy
level diagram for
Ar_2^+, showing corre-
lations vs. the
internuclear dis-
tance R. [From V.
Sidis et al. (4).]

to another with high probability. Rapid rotation of the internuclear axis (particularly at the smallest values of R encountered
in the collision) can cause transitions from one symmetry to
another (σ-Π, Π-δ, etc.) (5). The particular excited states produced depend very selectively on the MO curve crossings encountered
in the course of the collision from R = ∞ to the distance of closest approach R_o, as well as on the collision energy.

2. EXPERIMENTAL TECHNIQUE

Ion beams of the rare gases (~50–100 nA) are produced in a 30
keV electrostatic accelerator (6) and transmitted through collimating apertures into a differentially-pumped collision chamber
containing the target gas at a pressure ~2–3 × 10^{-4} Torr. The
target pressure is kept low enough so that charge-exchange neutralization of the ion beams in the collision chamber is no more
than 5% and so that the UV emission is approximately linear in
the target gas pressure.

A McPherson (Model 235) Seya–Namioka 0.5 meter vacuum monochromator is coupled without windows to the collision chamber.
The spectra are taken with a B & L 1200 line/mm gold replica
grating blazed at 700 Å and a channel electron multiplier is used
in a photon-counting mode as a UV detector. The manufacturer's
stated quantum efficiency for the solar-blind electron multiplier
is 22% at 584 Å, dropping off rapidly above 900 Å, but providing a
good detector in the 500–1100 Å wavelength range of interest here.

The spectra are taken automatically in digital form by slowly
scanning the spectrometer wavelength and simultaneously recording
the photon counts in the memory of a multiscaler. With a 1 mm entrance and exit slit, ~50–100 counts are accumulated in a one minute period at the center of a bright line. Dispersion of the instrument is 16.6 Å/mm and slit widths in the range 0.3 mm to 1 mm
provide the best compromise between resolution and signal counting
rate.

The relative emission cross sections, Q, for a given spectral
line are determined from the formula

$$Q = [s/(Ipd^2)] \sum_i (C_i - B_i) \quad ,$$

where the summation is over the counts minus background $(C_i - B_i)$
for each of the channels in the line, s is the scanning speed, d
the spectrometer slit width, p the target chamber pressure, and I
the ion beam current (7). The background signal B_i is primarily
due to beam-produced light in the collision chamber with no target
gas, plus a small component (a few counts per minute) due to detector dark count. The absolute emission cross sections σ are

then determined from $Q = k(\lambda)\sigma$, where $k(\lambda)$, the system spectral
sensitivity, is determined empirically for each spectral line of
interest by normalizing our He$^+$ + Ar data at 10 keV to the cross
sections obtained for each line at the same energy by deHeer <u>et al</u>.
(1). The absolute cross sections were stated in (1) to be determined to an accuracy of better than 25% at 537 and 1216 Å, with
possible 100% uncertainties in the interpolated values at intermediate wavelengths. No cascade corrections have been applied to
the emission cross sections to get true excitation cross sections
because the cascading transitions (mostly in the visible) were not
observed in this UV experiment. Estimates of these cascade contributions, which in some cases can be large, are given in (1).

3. RESULTS

3.1 Ar$^+$ + Ar Spectra

Emission spectra from Ar$^+$ + Ar collisions in the energy range
8-30 keV are similar to those from He$^+$ + Ar collisions at lower
energy (2). The most prominent transitions observed in the 500-
1100 Å range involve the excitation of the 4s and perhaps 3d orbitals in Ar$^+$, the 4s and 3d orbitals of neutral Ar, and the production of Ar$^+$ ions with 3s vacancies (920 and 932 Å lines).

Figure 2 shows the Ar$^+$ + Ar spectrum at 8.56 keV. The prominent features A-E are identified in Table 1. In addition, several weak features are evident between 830 and 850 Å, possibly due

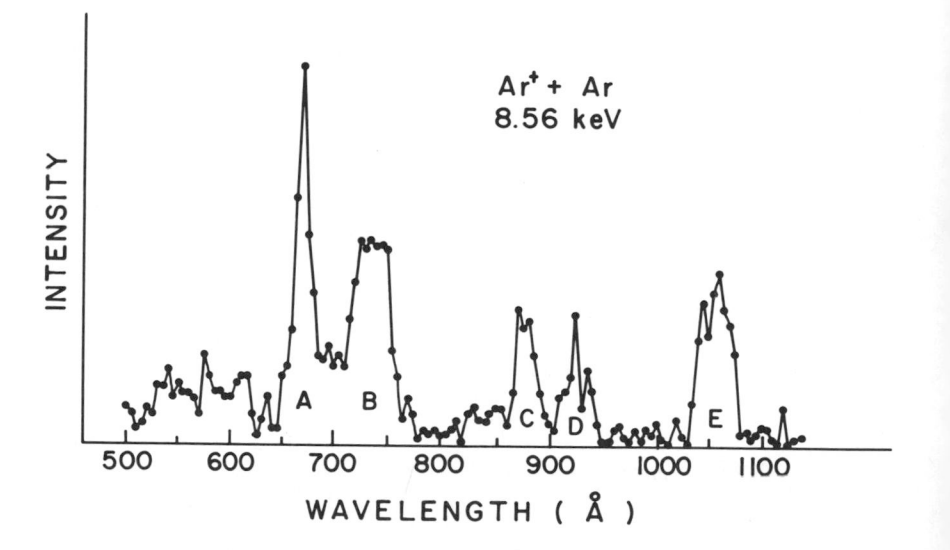

Figure 2. Ar$^+$ + Ar single collision spectra at 8.56 keV.
           (Resolution $\simeq$ 17 Å.)

Table 1. Identification of Prominent Lines in Ar$^+$ + Ar Spectra

Feature	Observed Wavelength (Å)	Probable Identification
A	670	Ar$^+$: 4s^1 ^{2}D(672,673,679 Å) 3d ^{2}D(662,664 Å)
B	720–750	Ar$^+$: 4s ^{2}P(718,723,726,731 Å) 4s ^{4}P(multiplet average 743 Å)
C	870–880$^+$	Ar$^\circ$: 3d(867,876,894 Å (weak)) also 5s(870,880 Å)?
D	920,935	Ar$^+$: 3s3p^6 ^{2}S(920,932 Å)
E	1045,1065	Ar$^\circ$: 4s(1048,1067) Resonance lines

to 4d excitation of Ar$^\circ$. Reproducible but unresolved features are evident between 500 and 600 Å, probably due to excitation of Rydberg states of Ar$^+$: 3d, 4s, 4d, 5s and 6s. Excitation of the 4p state of Ar$^\circ$ and Ar$^+$ has been seen in these collisions by energy loss spectroscopy (4) but these transitions are in the visible. The 4s emission, however, reflects the 4p excitation as a cascade contribution. Weak features near 537, 578 and 695 Å may be due to 3d excitation in Ar^{++}.

The estimated absolute cross sections for Features A-E (Table 2) are based on normalization at 10 keV to the He$^+$ + Ar data of (1), as described previously. Most of these cross sections have sharp onsets and some structure below 1 keV, reach a peak at ~2 keV and tend to rise again slightly somewhere between 10 and 30 keV (8).

The largest cross section observed (Table 2) is for the Ar$^\circ$ resonance lines (Feature E), involving 3p^6 → ^{3}P^5(^{2}P)4s excitation. The MO diagram (Figure 1) provides two obvious pathways for this

Table 2. Emission Cross Sections (10^{-17}cm^2 units) in Ar$^+$ + Ar Collisions (errors are reproducibility errors)

Spectral Feature	k(λ) (10^{17})	Beam Energy (keV)				
		10	13.3	16.7	20	30
A	6.07	6.4	6.6	6.3	7.4±0.5	7.5±0.6
B	27.02	1.2	1.4	1.6	1.5±0.3	1.4±0.1
C	14.0	0.81	1.04	1.08	1.54±0.12	1.29±0.22
D	9.06	1.10	1.21	1.05	1.39±0.12	1.34±0.51
E	3.5	7.2	9.2	--	9.3±2.0	11.3±0.6?

excitation. One of these is the crossing between the $5f\sigma_u$ MO curve
(correlating to the 3p at R=∞) and the $5p\sigma_u$ curve (not shown) which
correlates to the 4s orbital of Ar at R=∞. This crossing should
occur at $R_c \sim 0.4$-0.5 bohrs, leading to a maximum cross section
$\pi R_c^2 \sim 1.8 \times 10^{-17}$cm^2). Since the observed cross section ($\sim 10^{-16}$
cm^2) is larger than this one must look for another mechanism: a
radial two-electron excitation process is possible in the vicinity
of the crossing of the opposite-parity $5f\sigma_u$ and $4s\sigma_g$ MO curves near
R=2 bohrs (see Brenot et al., Ref. 4). The observed sharp onset of
the Ar° 4s excitation at ≲100 eV (8) supports this hypothesis of
radial coupling at a far-out curve crossing. We also note that
the Massey criterion predicts the peak in the cross section for
Ar° 4s excitation to occur at $\sim$185 keV, using the separated atom
excitation energy.

A plausible explanation can also be found in the MO model for
the production of 3s vacancies (Feature D: 920 and 932 Å lines of
Ar$^+$). There is a pronounced crossing of the $4f\sigma_u$ and $4p\sigma_u$ MO
curves at $R_c \sim 0.6$ bohr, leading to a maximum cross section
$\pi R_c^2 \sim 3 \times 10^{-17}$cm^2. A radial one-electron transition at this
point is possible from the filled $4p\sigma_u$ MO (correlating to the 3s of
Ar) to the partially empty $4f\sigma_u$ MO (correlating to the 2p of Ar)
during the final phases of the collision. The absolute cross section for producing 2p ($4f\sigma$) vacancies in Ar$^+$ + Ar collisions has
been measured experimentally [see Ref. (5)] and found to rise
sharply to a plateau value of 2.5×10^{-17}cm^2 at $\sim$20 keV. This is
greater by a factor $\sim$2 than the measured cross section for Feature
D, which also rises sharply between 1 and 10 keV (1) and appears to
level off at $\sim$20 keV. The proposed mechanism implies correlated
inner-shell and outer-shell excitations and must await further experimental verification.

The excitation of Feature C (Ar° 3d) as well as other weak
features attributable to 3d excitation of Ar^{++} can also be related
to the partial emptying of the $4f\sigma$ MO: the 3d orbital can be excited by rotational coupling (9) with the nearly degenerate $4f\pi$ MO
out to fairly large R. Both the peaking of the 3d cross section
(Feature C) at about 20 keV and its magnitude appear reasonable on
the basis of this rotational coupling mechanism. -- The energy dependences of Features C and D are similar enough to suggest a common origin: we suggest it may be the depletion of the ($4f\sigma$)2p MO.

3.2 Ne$^+$ + Ar and Ar$^+$ + Ne Spectra

Representative spectra from these two collision combinations
are compared in Figure 3 (a and b). Note the relative weakness of
Feature D (Ar$^+$ $3s^{-1}$) in the Ar$^+$ + Ne case. In the Ne$^+$ + Ar collision, the cross section for Feature D (920+932 Å) is $2.5 \pm 0.1 \times 10^{-17}$
cm^2 at 10 keV, rising nearly linearly and then leveling off to
$\sim 7 \times 10^{-17}$ cm^2 near 30 keV. The only explanation presently at hand

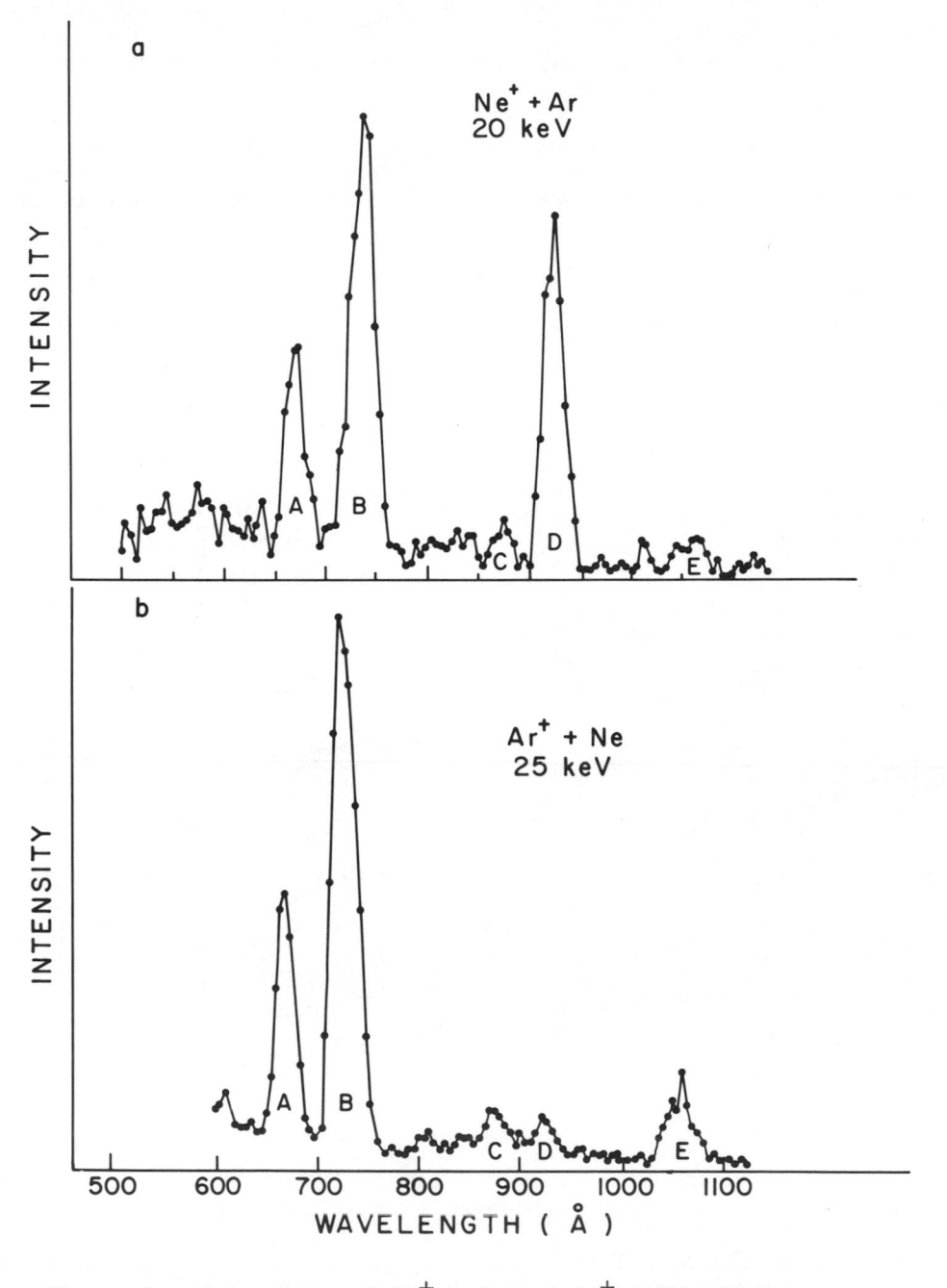

Figure 3.   Comparison of Ne$^+$ + Ar and Ar$^+$ + Ne spectra.
(Resolution ≃ 17 Å.)

for the difference in the two cases is that there is a crossing of the $4f\sigma_u$ MO (correlating to Ne$^+$ 2p) and the $3s\sigma_g$ MO (correlating to Ar 3s) at large R (~1 a.u.) (10). This can depopulate the Ar 3s orbital via a transfer to the Ne$^+$ (2p) in the Ne$^+$ + Ar case only. The striking contrast between these two collision cases, together with the large magnitude of the 920 and 932 Å cross sections in the Ne$^+$ + Ar case, provide strong indications that MO selection rules hold even for quite asymmetric collisions and for outer-shell electrons.

3.3 He$^+$ + Ar and Ar$^+$ + He Spectra

Figure 4 shows the approximate absolute cross sections for the four most prominent spectral features in the He$^+$ + Ar case. Note that Feature C (Ar° 3d) is not prominent in this spectrum. The

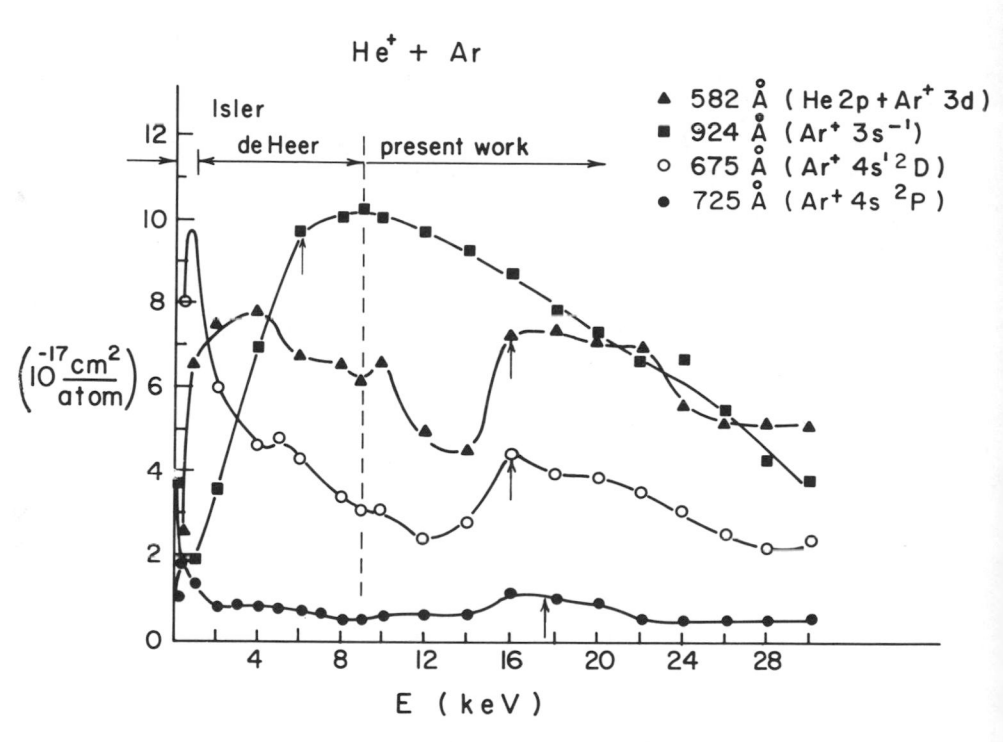

Figure 4. Cross sections for brightest features (~17 Å resolution) in He$^+$ + Ar excitation (500–1000 Å). Normalization is based on Ref. 1 (deHeer et al.) and some low energy data of Isler (2) are shown. The scale of the curve labeled 924 Å should be multiplied by 3/2 to include both the 924 and 932 Å cross sections. Vertical arrows indicate the energy of the predicted maximum cross section, based on the Massey criterion, with atomic-state energy defects and an impact parameter of 1.4 bohrs.

largest cross section is for depopulation of the 3s orbital of Ar (Feature D, 920+932 Å of Ar II), but the peak occurs at much lower energy than in the Ne^+ + Ar case. The $HeAr^+$ correlation diagram is similar to Fig. 6 of (10); inspection of this diagram suggests that the electron transfer occurs through radial coupling at large R between the filled $3s\sigma_g$ (Ar 3s) and the half vacant $3d\sigma_g$ (He 1s) MO's. We note that the Ar^+ 4s' ^2D excitation is always greater than the Ar^+ 4s ^2P. Feature C (Ar° 3d excitation) is much weaker than the other main features in this case. For He^+ + Ar the $4f\sigma$ orbital is initially empty and Ar° (3d) excitation requires the relatively inefficient $(3d\sigma)$ He 1s -- $(3d\pi)$ Ar 3d rotational transition at small R.

In the Ar^+ + He spectrum at 10 and 20 keV (not shown) the features between 500 and 600 Å (Ar^+ Rydberg states) and Features A and B (Table 1 -- Ar^+ 3d and 4s) are prominent while D and E are weak. The apparent cross section for Feature A at 10 keV is $\sim 3 \times 10^{-17} cm^2$. Feature D is weak even at the highest energies observed, presumably because in the Ar^+ + He channel there is no vacancy in the $3d\sigma_g$ MO initially so the Ar 3s electron cannot transfer into it. Feature E (Ar° 4s) is weak in both the He^+ + Ar and the Ar^+ + He cases, probably because the 4s of Ar correlates with the $4s\sigma$ MO, which lies above and does not cross any of the initially-filled MO's of the $HeAr^+$ system.

3.4 Ne^+ + Ne Spectra

The most prominent lines observed in the 500-1100 Å range were the 626 and 630 Å lines of Ne I ($2p^5$ 4s) and the somewhat stronger Ne I ($2p^5$ 3s) lines at 736 and 743 Å. The energy dependences were different, with the 3s excitation falling to a minimum at ~ 20 keV from a maximum below 10 keV. Isler (8) gives the 736+743 Å cross section as $2.5 \times 10^{-17} cm^2$ at 8 keV. The 4s excitation has a local maximum at ~ 15 keV in the 10-30 keV range. There is some indication of weak Ne I 3d excitation (619 Å) in our spectra. -- The MO model provides an efficient $4f\sigma_u$ (Ne 2p) to $4p\sigma_u$ (Ne 3s) one-electron radial coupling to explain the 3s excitation (see Figure 1). The 4s excitation should be weaker since it would require a two-electron $4f\sigma_u$-$4s\sigma_g$ excitation to conserve parity.

4. CONCLUSION

Our limited experience with low keV energy rare-gas ion atom collisions shows that there is considerable selectivity in the states that are excited. This is particularly evident in asymmetric collisions such as Ar^+ + Ne and He^+ + Ar. Cross sections for the most prominent vacuum ultraviolet lines, corresponding to n=4 excitation of the atom or ion, are between 10^{-17} and $10^{-16} cm^2$ at 10 keV. Molecular-orbital correlation diagrams and energy level curves as

a function of internuclear distance make it possible to estimate
which excitations will predominate, even in the case of outer-shell
excitations. MO curve crossing radii are useful in setting geometrical upper limits to the cross sections; these limits are in
fact reached in a number of cases. Some of the data are suggestive
of correlated collisional production or quenching of 2p vacancies
with excitation of specific outer-shell orbitals (11).

We acknowledge with thanks the technical assistance and encouragement of David A. Clark in setting up this experiment.

References

1. F. J. deHeer et al., Physica 41, 588 (1969).
2. R. C. Isler, Phys. Rev. A 10, 117 (1974).
3. U. Fano and W. Lichten, Phys. Rev. Letters 14, 627 (1965); W.
    Lichten, Phys. Rev. 164, 131 (1967); M. Barat and W. Lichten,
    Phys. Rev. A 6, 211 (1972).
4. For example, J. C. Brenot et al., Phys. Rev. A 11, 1245, 1933
    (1975); F. J. Eriksen, S. M. Fernandez, A. B. Bray and E.
    Pollack, Phys. Rev. A 11, 1239 (1975); V. Sidis, M. Barat and
    D. Dhuicq, J. Phys. B 8, 474 (1975).
5. These phenomena have been reviewed, for example, in W.W. Smith
    and Q. C. Kessel, Ch. 7 of New Uses of Ion Accelerators, J.
    Ziegler (ed.), Plenum Press, N.Y. (1975).
6. The accelerator was originally built for low-energy inner-
    shell excitation studies: G. Thomson, P. Laudieri and E.
    Everhart, Phys. Rev. A 1, 1439 (1970).
7. The slit width d appears twice in the denominator since the
    observation time $\Delta t \propto d/s$.
8. R. C. Isler and L. E. Murray, in Electronic and Atomic Colli-
    sions (IX ICPEAC Abstracts), J.S. Risley and R. Geballe, Eds.,
    Univ. of Washington Press, Seattle 1975, page 609; and R.C.
    Isler, private communication.
9. This is analogous to the $1s\sigma$-$2p\pi$ rotational coupling process
    discussed by Bates and Williams for H_2^+, Proc. Phys. Soc.
    (London) 83, 425 (1964); and by Briggs and Macek for Ne_2^+,
    J. Phys. B 5, 579 (1972). See also R. J. Fortner, Phys. Rev.
    A 10, 2218 (1974).
10. M. Barat and W. Lichten, ibid. Ref. (3), Figure 5.
11. See also R. S. Thoe and W. W. Smith, Phys. Rev. Letters 30,
    525 (1973).

CALIBRATION OF SPECTROMETER DETECTION EFFICIENCY IN THE ULTRAVIOLET[*]

WARD WHALING

California Institute of Technology

Pasadena, California 91125

ABSTRACT

Methods of calibrating the relative detection efficiency of spectrometer systems are compared with emphasis on spectrometers to be used with the beam-foil light source at wavelengths greater than 1100 Å. Capabilities and limitations of tungsten and deuterium standard lamps, of known atomic and molecular branching ratios, and of cascade radiation are discussed. Transitions between Rydberg levels excited in the beam-foil source provide many additional line pairs for the atomic branching ratio method of calibration.

I. INTRODUCTION

One advantage of the beam-foil time-of-flight method over other methods of measuring atomic lifetimes is that it is readily applicable to ions of any charge. Beam-foil practitioners have been quick to exploit this capability and the beam-foil literature reports lifetimes in many multiply-charged ions. However, if we consider the use of this lifetime information to provide transition probabilities, we find results for only a few transitions from levels with simple decay schemes for which the relative transition probability or branching ratio can be calculated. The systematic conversion of lifetimes into transition probabilities through

---

[*]Supported in part by the Office of Naval Research [N00014-67-A-0094-0022] and the National Science Foundation [MPS71-02670 A05].

laboratory measurement of branching ratios has been limited to
neutral atoms and to low-lying levels of singly-charged ions,
levels with decay branches $\gtrsim$ 2500 Å. This limitation on the use
of BFS lifetimes is imposed by the difficulty in comparing the intensity of branches in the visible spectrum with lines in the vacuum
ultraviolet. We have investigated various ways of overcoming this
obstacle and this paper is a report of that investigation.

Many different methods have been used for calibrating detection
systems in the VUV, an immediate indication that no single method
is completely satisfactory or superior to all others. We cannot do
justice to all of the various methods and refer the reader to recent
reviews of the subject (1,2) for a discussion of the many other
methods that we do not treat. We focus our attention on methods
that appear to be most useful to beam-foil spectroscopists who wish
to obtain, with minimum effort and expense, a calibration of their
detection efficiency that matches the accuracy of their lifetime
measurements. Relative detection efficiency as a function of wavelength is sufficient for branching ratio measurements and we do not
consider absolute efficiency calibration. Although many of the
methods we discuss are applicable to any wavelength, our particular
interest has been for wavelengths longer than the LiF cut-off near
1100 Å which is near the lower limit of normal-incidence multiple
reflection spectrometers.

II.   STANDARD LAMPS

To calibrate a detection system one would like to have a
source that radiates a known number of photons per second per cm^2
per steradian per unit wavelength interval for all wavelengths of
interest. The tungsten ribbon standard lamp meets this requirement
for $2500 \leq \lambda \leq 60,000$ Å. These lamps are available commercially (3)
and are easy to use for $\lambda \gtrsim 3000$ Å. Between 2500 Å and 12,000 Å
the lamp output increases by five orders of magnitude and some care
is required to reject scattered photons when observing the small
output below 3000 Å. Narrow-pass interference filters can be used
to eliminate unwanted photons; their transmission should be measured
in the geometry in which they will be used by substituting a line
source at the lamp position. If photon counting is used, the opposite problem of reducing the counting rate at the red wavelengths
can be met with neutral density filters. Again, the filter transmission and that of any filter combinations should be measured in
the exact geometry to be used in the calibration procedure.

The manufacturer claims an accuracy of ± 3% to 10%, with the
larger uncertainty at the shortest wavelengths. Our comparison of
two lamps indicated a maximum discrepancy of 12%, again at the
shortest wavelengths. This accuracy is adequate for measuring

branching ratios since level lifetimes are rarely known with better precision. More precise calibration of the tungsten ribbon lamp down to 2250 Å can be obtained from the National Bureau of Standards (4). Buckley (5) has investigated the use of tungsten lamps down to 1500 Å.

The calibration supplied by the manufacturer is guaranteed for 50 hours of lamp use. It is convenient to acquire two lamps, one to be used only briefly to check the calibration of the second lamp which is used for routine calibrations. After 60 hours of use, the output of one such lamp was found to have increased by 10% to 4% over the range 3000 Å to 7500 Å.

For wavelengths short of 2500 Å the deuterium low pressure discharge lamp is a commercially available continuum source. The commercial lamps are now available calibrated (3) over the range 1800 Å ≤ λ ≤ 4000 Å, with an accuracy of ± 10%. The National Bureau of Standards can calibrate (4) these lamps against the wall-stabilized hydrogen arc (6) down to 1670 Å. There are reports (7) of rapid decrease in lamp output with time or with use although the reason for this deterioration is not understood. Since the decrease in output is reported to be independent of wavelength, it may not be a serious problem if one is interested only in relative sensitivity calibrations. The deuterium lamp is many orders of magnitude brighter than the tungsten lamp for λ < 2500 Å, but the small area of the source (~ 0.3 mm in diameter) requires some care in positioning and alignment.

III. BRANCHING RATIO METHOD

Since we intend to use the calibrated spectrometer to measure unknown branching ratios, the use of known branching ratios to carry out the calibration is particularly attractive since it allows the use of the same geometry and procedures in the calibration procedure and in the measurement of unknown lines. In figure 1 are represented two decay branches from upper level u. The photon intensity of branch λ_{u1} is $I(\lambda_{u1}) = A_{u1} N_u$ where A_{u1} is the transition probability for the transition from level u to level 1, and N_u is the population of level u. The signal detected at wavelength λ_{u1} is $S(\lambda_{u1}) = Eff(\lambda_{u1}) I(\lambda_{u1})$ where $Eff(\lambda_{u1})$ is the spectrometer detection efficiency, including all geometrical factors, at the wavelength λ_{u1}. The ratio of the two signals is then

$$\frac{S(\lambda_{u1})}{S(\lambda_{u2})} = \frac{Eff(\lambda_{u1})}{Eff(\lambda_{u2})} \frac{A_{u1}}{A_{u2}} . \tag{1}$$

The left-hand side of equation (1) is the measured quantity in both

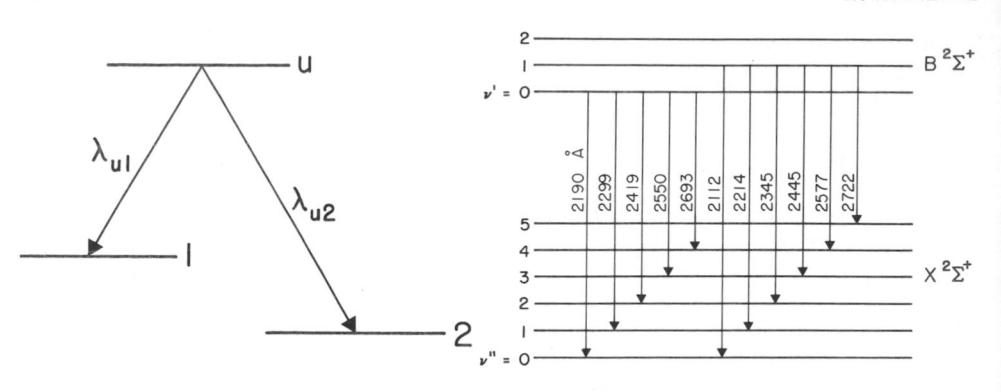

Figure 1.   Two transitions from a common upper level that can be used to calibrate a detection system if the branching ratios are known.

Figure 2.   Electronic vibrational energy level diagram for a few strong members of two v'' progressions in the first negative system of CO^+.

the spectrometer calibration and the branching ratio measurement. If the two lines have known transition probabilities, the ratio measurement determines the relative detection efficiency of the spectrometer system at the two wavelengths. Once the relative efficiency has been established for the wavelengths of interest, then a similar measurement of the intensity ratio for unknown lines is used to measure their relative transition probabilities, A_{u1}/A_{u2}, and the branching ratio, $BR_{u1} = (1 + A_{u2}/A_{u1})^{-1}$, and the individual transition probability $A_{u2} = BR_{u2}/\tau_u$. By using identical experimental conditions for the calibration and the unknown branching measurement, geometrical factors, window transmissions, etc., cancel each other and one measures the ratio of unknown transition probabilities directly in terms of known transition probabilities.

The branching ratio method has been discussed by Hinnov and Hofmann (8), by Boland et al. (9), and by Griffin and McWhirter (10). These authors suggest useful line pairs in H, He^+, and several 3-electron ions that are listed in Table 1. The required transition probabilities are given by Bethe and Salpeter (11) for the hydrogenic lines, and for the remaining line pairs the transition probabilities have been compiled by Wiese et al. (12). Hinnov and Hofmann describe a simple hollow cathode source for the H and He^+ lines, and they consider the likelihood that the levels are populated in this source in statistical equilibrium, a necessary condition since the degenerate levels of the one electron system makes it impossible to separate transitions from individual levels.

Table 1. Atomic lines for branching ratio calibration. The lines
at the top of the table are from Boland (9). The lines below the
dashed line are transitions between Rydberg levels excited in the
BF spectrum. Wavelengths in parentheses are transitions that have
not yet been observed and are only approximate since no polarization shift has been included. The transition probabilities listed
for the Rydberg transitions are computed from the matrix elements
of Karzas and Latter (13). Rydberg levels are designated by the
quantum numbers n_ℓ.

Transition	$\lambda(\text{Å})$	$A\ (\text{sec}^{-1})$	Observer
H $3 \to 1$	1025	5.57×10^7	
$3 \to 2$	6562	4.41×10^7	
$He^+\ 4 \to 1$	243	0.204×10^9	
$4 \to 2$	1215	1.35×10^9	
$4 \to 3$	4685	1.44×10^9	
$He^+\ 5 \to 1$	237	6.60×10^7	
$5 \to 3$	3203	3.52×10^7	
$C^{3+}\ 3p \to 2s$	312	4.56×10^{10}	
$3p \to 3s$	5801	3.19×10^7	
$N^{4+}\ 3p \to 2s$	209	1.20×10^{10}	
$3p(P_{3/2}) \to 3s$	5801	4.15×10^7	
$O^{5+}\ 3p \to 2s$	150	2.59×10^{10}	
$3p(P_{3/2}) \to 3s$	3811	5.10×10^7	
- - - - - - -	- - - - -	- - - - - - -	- - - - -
$Cl^{6+}\ 7_4 \to 6_3$	2501	9.0×10^8	Bashkin et al.
$7_4 \to 5_3$	944	13.1×10^8	(14)
$7_4 \to 4_3$	(442)	15.4×10^8	
$Cl^{6+}\ 8_4 \to 7_3$	3854	3.70×10^8	Bhardwaj et al.
$8_4 \to 6_3$	(1530)	5.42×10^8	(15)
$8_4 \to 5_3$	(763)	7.51×10^8	
$8_4 \to 4_3$	(398)	8.71×10^8	
$Fe^{4+}\ 7_5 \to 6_4$	4904	3.35×10^8	Lennard et al.
$7_5 \to 5_4$	(1861)	3.17×10^8	(16)
$Mn^{6+}\ 8_6 \to 7_5$	3874	6.75×10^8	This paper.
$8_6 \to 6_5$	(1530)	5.21×10^8	see fig. (4).
$B^{2+}\ 5_3 \to 4_2$	4487	2.10×10^8	Bromander (17)
$5_3 \to 3_2$	(1424)	3.64×10^8	

The beam-foil light source provides many additional line pairs that extend the range of the branching-ratio method. The one-electron ions excited in the BF source are not useful since it is known (18) that the population of levels of different ℓ is not statistical. However, transitions between Rydberg levels, which are the most conspicuous feature of heavy-ion BF spectra, offer many useful possibilities. Bromander (17) has shown that the branching ratios between Rydberg levels closely approximate those calculated for the hydrogen atom. The total transition probability for Rydberg level decay is also thought to follow that for the hydrogen atom, scaled by the factor (core charge)^4, but for our purposes hydrogenic branching ratios are sufficient. The Rydberg transitions have distinct advantages over hydrogenic transitions:
(i) The polarization of the heavy-ion core removes the ℓ-degeneracy and transitions from levels of the same n and different ℓ can be distinguished. One thereby avoids the question of statistical equilibrium.
(ii) Line pairs covering a wide range of wavelengths are available. A few possibilities are appended at the bottom of table 1 but this list is by no means complete. Lennard et al. (16) have tabulated the $\Delta n = -1, -2, -3$ hydrogenic transitions in ions with charge up to 14+. Their calculated wavelengths do not include the polarization shift, which can amount to several tens of angstroms, but they indicate the range of wavelengths covered by line pairs from a common upper Rydberg level.

With arbitrary ion charge and n-value, almost any wavelength range can be spanned but practical considerations restrict the choice. The ion charges produced in the BF source are limited by the beam energy and ion species available. No limit has been clearly established for the range of n-values populated, although Hallin (19) did not observe transitions from levels with n = ≥ 15. Very large values of n and ℓ may not be useful since the polarization shift may become too small to permit clean separation of transitions from particular n_ℓ upper levels. For low values of n, there is evidence for coupling to the core other than simple polarization, at least for some core configurations. Whether this coupling, discussed further in Section IV, affects the branching ratios is not known and transitions to levels exhibiting such coupling should be avoided.

(iii) The BF source is not afflicted by self-absorption which can distort measurements of ground state transitions in the plasma source.
(iv) The BF source is, of course, ideally suited to calibrate spectrometers which are to be used for branching measurements on levels excited in the BF source.

A weakness of atomic branching ratios for calibration is that

the calibration lines are few and far between, even when augmented by lines from the beam-foil source. To increase the density of available lines, molecular spectra have been used; more specifically the vibrational progressions that have a common upper vibrational level and a sequence of lower vibrational levels. Such a progression gives rise to a series of lines with roughly equal spacing. Figure 2 illustrates two progressions in the first negative band of CO^+. The stronger transitions which are illustrated provide eleven calibration points in the wavelength interval $2112 \leq \lambda \leq 2722$ Å.

For a transition between upper vibrational level v' and lower level v'' the transition probability $A_{v'v''}$ is proportional to

$$A_{v'v''} \sim E_\gamma^3 \, R_e^2 \, q_{v'v''}$$

where E_γ is the photon energy, R_e is the electronic transition moment, and $q_{v'v''}$ is the Franck-Condon factor for the transition. The molecular analogue to equation (1) becomes

$$\frac{S(\lambda_{v'v''_1})}{S(\lambda_{v'v''_2})} = \frac{Eff(\lambda_{v'v''_1})}{Eff(\lambda_{v'v''_2})} \frac{R_e(v'v''_1)}{R_e(v'v''_2)} \frac{q_{v'v''_1}}{q_{v'v''_2}} . \qquad (2)$$

The Franck-Condon factors $q_{v'v''}$ have been calculated for many spectral bands. The good agreement between values calculated by different authors using different model potentials gives one some confidence in their accuracy, at least for the bands that have been extensively studied.

The ratio of electronic transition moments appearing in equation (2) was set equal to unity by the first users of molecular branching for calibration on the assumption that the electronic transition moment, defined as $R_e = \int \Psi_e'^* \, \tilde{M}_e \, \Psi_e'' \, dv_e$ where Ψ_e is the electronic wave function and $\tilde{M}_e$ is the dipole operator, should vary but little from one lower vibrational level to the next since all of the lower vibrational levels involve the same electronic level. More recent papers have attempted to estimate the variation of R_e from one line to the next, usually by the r-centroid approximation introduced by Fraser and Jarmain (20). This approximation has been criticized (21) and defended (22), and the current confusion regarding the variation of the electronic transition moment, summarized in a recent review by Kelmsdal (23), must be acknowledged as a serious weakness of molecular branching ratios for calibration. Klemsdal cites an example of an N_2 band for which R_e varies by a factor of 5 for different members of the band. Although the variation over a particular progression with common upper level is expected to be smaller, the difficulty of estimating R_e with confidence

restricts the usefulness of molecular branches for calibration purposes.

Mumma (24) has reviewed the various molecular bands that have been used for calibration in the wavelength region $1000 \le \lambda \le 3000$ Å, listing wavelengths and relative line intensities. The $B^1\Sigma_u^+ - X^1\Sigma_g^+$ Lyman band in H_2 and HD cover the range 1062-1660 Å and seems especially useful for calibrations between the lower limit of the deuterium lamp and the cut-off of LiF windows near 1100 Å. Becker et al. (25) have described the use of these hydrogen bands to calibrate a spectrometer system using transition probabilities calculated by Allison and Dalgarno (26) for which an accuracy of $\pm$ 6% is claimed.

Because molecular spectra are so rich in lines, many of the experiments reviewed by Mumma make use of resonance fluorescence or electron beam bombardment to excite selectively particular upper levels of interest. To maintain the symmetry between calibration procedure and branching ratio measurement, we have used a hollow-cathode discharge for both purposes. Using a graphite cathode with discharge in a He + CO_2 mixture to excite the CO^+ first negative band, we find no difficulty in picking out the desired lines from the many other bands excited in this source.

Each 'line' in the progression is actually a rotational band. With the reduced line density that can be achieved with selective excitation, the spectrometer resolution may be reduced to include the entire band. In the absence of selective excitation, we have taken a particular line in the rotational band, i.e., particular upper and lower rotational levels, as representative of the intensity of the entire band. The same rotational levels are used for each member of the progression. Fairly good resolution is required since the spacing between members of the rotational bands in the case of the CO^+ first negative band is $\sim$ 0.25 Å.

IV.   CASCADE RADIATION

If level a decays to level b which decays to level c, and level b is fed by no transitions other than those from level a, then the number of λ_{ab} photons emitted from the source is equal to the number of λ_{bc} photons. If the lifetime of the level a is $\tau_a \gg \tau_b$, then the level populations N(t) approach after long times $N_a(t)/\tau_a \to N_b(t)/\tau_b$, or $I_b \to I_a$, where I is the photon intensity emitted in the decay of the subscript level.

The Rydberg levels of maximum angular momentum that are excited in the beam-foil source decay through a succession of $\Delta\ell = -1$ transitions, and this cascade decay chain produces a series of

lines covering a broad wavelength range. The transitions originating from levels with $\ell = n - 1$ are always far stronger than lines from levels of lower ℓ (cf. figure 4) which have more than one decay channel. Under certain conditions this cascade radiation produces lines of approximately equal intensity which may be used for calibration.

The Rydberg levels have the attractive feature that all transition probabilities are known. Within the accuracy of experimental measurements, both the total transition probability and the individual transition probabilities (17) are given by the hydrogenic values scaled by a factor Q_c^4 where Q_c is the core charge. If the initial population of the Rydberg levels at the foil were known, one could calculate the subsequent intensity of all of the lines as a function of time or position. Unfortunately, the evidence concerning the initial population is incomplete and unclear. There is general agreement that the initial population decreases with increasing n, perhaps as fast as n^{-3} (ref. 18). However, the evidence regarding the ℓ-dependence is puzzling: For hydrogen (18), helium (27), and B^{2+} (12) the initial population decreases with increasing ℓ; for high-Z ions the initial population increases with ℓ, and even more rapidly than $2\ell + 1$ (28). Nor is it known whether there is a cut-off at high n or ℓ. Thus, it is not possible to calculate the expected intensities. On the other hand, $\Delta n = \Delta \ell = -1$ transitions are the most conspicuous features of the beam-foil spectrum of heavy ions at the visible wavelengths. These transitions between levels of n = 5 to 10 or more must eventually cascade down through a series of transitions. Are there conditions under which these cascade lines have equal intensity?

If it were possible to excite selectively a Rydberg level of maximum angular momentum $\ell = n - 1$, the level would decay by cascading down the set of levels of maximum angular momentum, which we will call the yrast levels, borrowing a term from nuclear physics. All of the lines emitted in the cascade would have equal photon intensity. The BF source offers the possibility of making observations at selected times after excitation, and one possibility for selectively observing only the longest-lived levels of maximum n and ℓ is to wait until the shorter-lived levels have decayed by making the observations at some distance from the foil. In figure 3 we show the most probable decay branches for some levels with n = 14. We have chosen this particular value of n because there is evidence that it may represent a maximum value: Hallin et al. (19) could not find transitions from levels with n ≥ 15 in their C1 and O spectra at beam energies up to 42 MeV although they saw essentially all of the expected transitions from levels with n ≤ 14.

As can be seen from figure 3, the level lifetimes increase with increasing n and ℓ and it is interesting to ask how the longest-lived levels decay. The level 14_{13} decays only along the yrast line

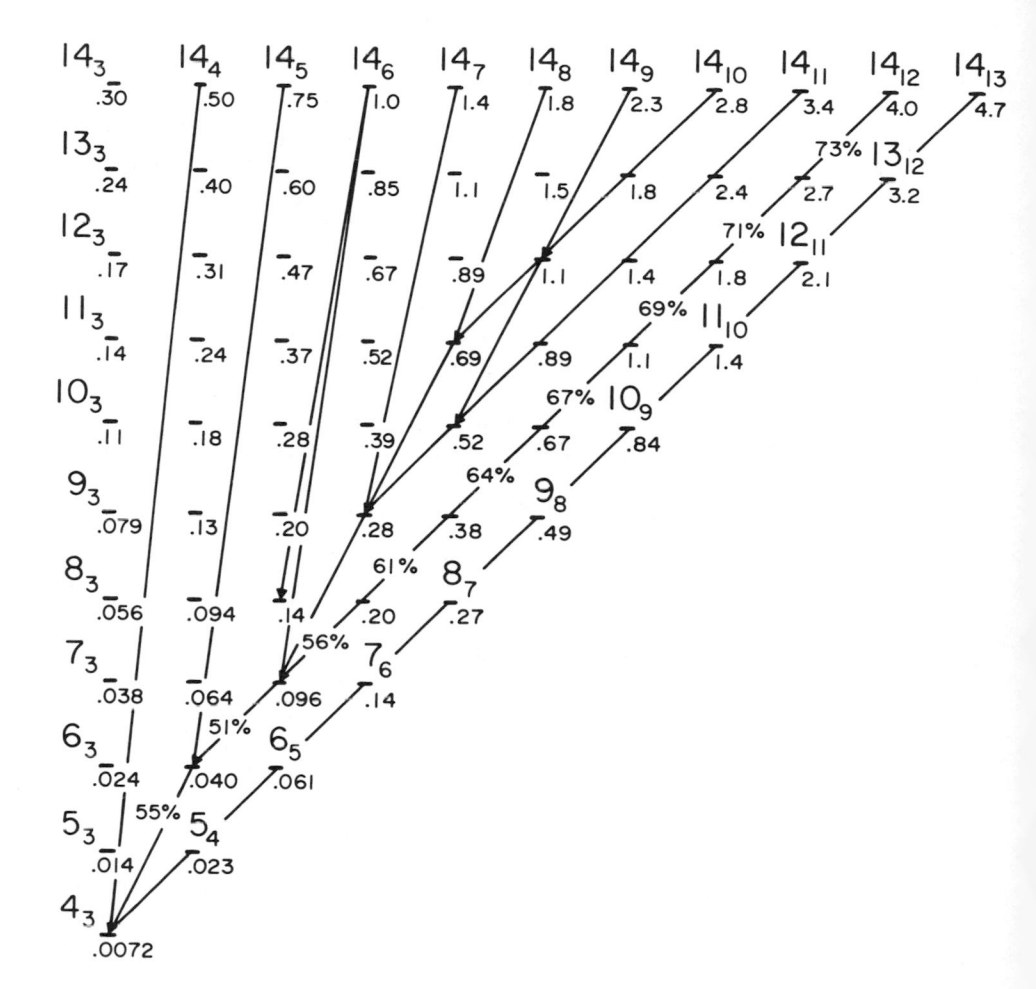

Figure 3.   Schematic representation of Rydberg levels.   Energy
separations are not to scale and polarization shifts are not shown.
Levels are designated by the quantum numbers n_ℓ.   Numbers under the
levels are mean lifetimes in units of 10^{-5} seconds for core charge
of one; for other ions the lifetime is reduced by the fourth power
of the core charge.   Lifetimes are from Ref. (11) or calculated
from the matrix elements of Ref. (13).   Most probable decay paths
are shown for some levels with n = 14, and for the level 14_{12} the
branching ratios along the most probable decay path are labeled.

of maximum angular momentum. The level 14_{12} decays predominantly
along the line parallel to the yrast line until it reaches 6_4,
whence it decays to 4_3. This same decay parallel to the yrast line
is characteristic of all of the longest-lived states and arises
from the large overlap between wave functions with the same value
of the radial quantum number $n - \ell - 1$, hence the same number of
radial nodes, large overlap, and large matrix element. Only at
small values of n does the E_γ^3 term in the transition probability
overcome the matrix element and push the transition to the lowest
level accessible under the selection rule $\Delta\ell = \pm 1$. This same
property of the wave functions is responsible for the negligible
probability of transitions with $\Delta\ell = +1$.

From an examination of the decay paths in figure 3, one might
conclude that cascading along the yrast line will be dominant only
for levels with $n \leq 4$. This is a crippling restriction since the
transitions $4_3 \rightarrow 3_2$, $3_2 \rightarrow 2_1$ are widely spaced in wavelength, and
the level $n = 2$ is likely to interact with the core. However, if
one supposes an initial population of 1000 in the 14_{13} state and
asks how many of these eventually populate the 6_4 state, we find
that the number of cascades through the 6_4 level is only 27, while
973 go through the yrast level 5_4. Only 3% of the initial population follows the most probable transition path, while the total
following one or another of the alternate decay paths leading to
the level 5_4 amount to 97%. Similarly, 95% of the initial population of 14_{13} goes through yrast level 6_5 and 90% goes through 7_6.
Thus it seems possible that useful cascades may begin several steps
further up the diagram at a value of n large enough to permit
several cascade lines of equal intensity which might serve as useful calibration lines. In the absence of better information concerning the initial populations, one cannot draw stronger or more
specific conclusions.

We have looked at this problem experimentally by observing
cascade lines with a calibrated spectrometer to see where, or if,
the subsequent cascade lines approach equality. An initial examination of several transitions in various ionization states of mangan-
ese showed that, with the exception of transitions in Mn^{6+}, all of
the lines had greater than instrumental width or displayed complex
structure that cannot be explained in terms of polarization of the
core. This structure has been reported earlier by Lennard and
Cocke (28) who suggest that there are interactions between the
outer electron and the core other than simple polarization. Because of the uncertainty of interpreting the experimental results
when the lines are complex, we have limited our observations to
the Mn^{6+} ion. The core, Mn^{7+}, is isoelectronic with argon and the
closed 18-electron structure appears to interact with the outer
electron only by virtue of the polarization.

Figure 4 shows the lines that we have examined. The spectrum shown is taken 2 mm downstream from the foil at 90° to a 1.4 MeV Mn beam excited in a 10 μgm/cm² carbon foil. The spectrum has been put on a relative intensity scale by dividing the observed photon yield by the detection efficiency as determined with a tungsten ribbon standard lamp.

In the spectrum shown at 2 mm from the foil, the photon intensity ratio for the adjacent cascade steps $\lambda 2514/\lambda 3883 = 1.6$. At 25 mm from the foil, this ratio has dropped to 1.3, and at 50 mm it is 1.14, in accord with our expectation that at the greater distances one observes predominantly cascading that originates from the longer-lived states of large n and ℓ. One would expect the next step in the cascade sequence, $6_5 \rightarrow 5_4$ at 1522 Å, to be stronger than $\lambda 2514$, but by a factor less than 1.14 at 50 mm from the foil. One can adopt the value 1.07 ± 0.07 as a conservative estimate of the photon intensity ratio $\lambda 1522/\lambda 2514$ and thereby obtain a calibration point at 1522 Å. A factor closer to unity can be achieved by moving further downstream but at the cost of reduced intensity. Because of the strong cascading, the intensity of these lines in the yrast cascade decreases more slowly than the calculated lifetime would indicate: the intensity of the two lines at 3883 Å and 2514 Å decreases by only a factor of four as one moves from 2 mm to 50 mm.

IV.   SUMMARY

Standard lamps are the most satisfactory method for calibrating spectrometer detection efficiency for wavelengths greater than 1670 Å. To extend the calibration down to 1100 Å, the molecular branching ratio method offers the best density of calibration points but some care is required in selecting a band system of known intensity. For those spectrometers to be used with the beam-foil light source, the beam-foil source itself provides a number of scattered calibration points that make use of transitions between Rydberg levels. These calibration points should be useful in checking the calibration of the spectrometer in exactly the same geometry to be used for measurements of lines from the beam foil source.

ACKNOWLEDGMENTS

It is a pleasure to express my thanks to Geo. M. Lawrence and Wm. R. Ott for their help with questions that arose during the preparation of this paper, and to the students who have taken part in this investigation, notably J. W. Hugg, Jr., who carried out the CO^+ branching ratio calibration, and Craig Lage who assisted with the cascade intensity measurements.

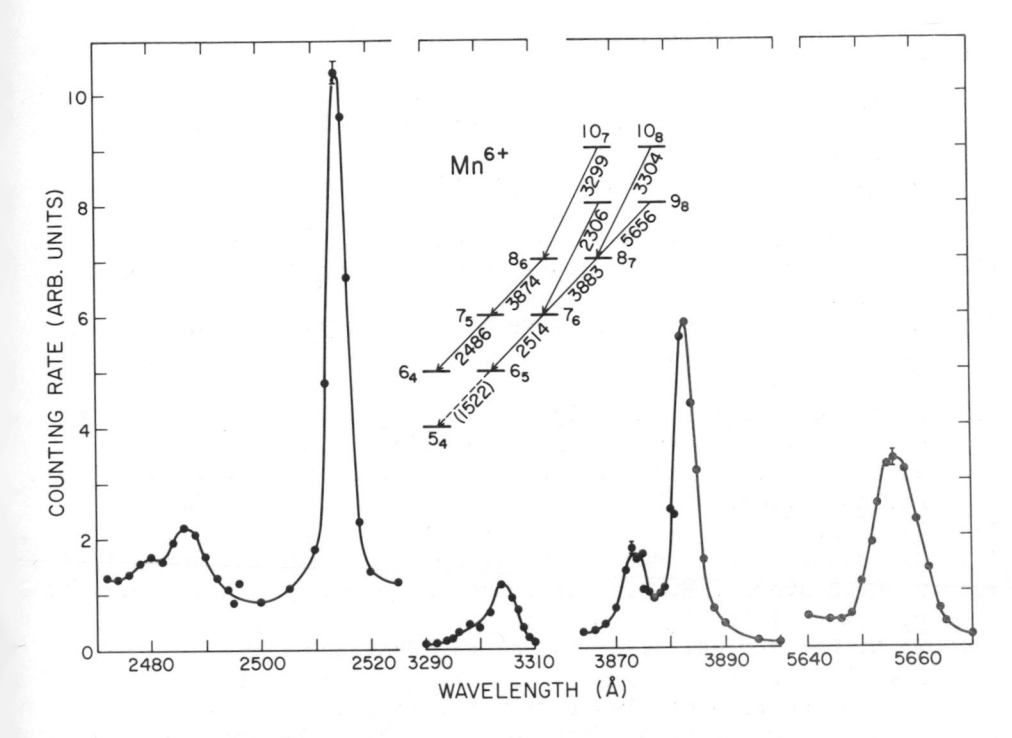

Figure 4. Cascade lines between Rydberg levels in Mn^{6+} observed at 90° and 2 mm. downstream from a 10μgm/cm^2 carbon foil. Incident Mn^+ beam energy is 1.40 MeV. The observed counting rate has been divided by the monochromator detection efficiency to obtain the relative counting rate that is plotted in the figure. The monochromator passband is 2.5 Å for $\lambda < 5000$ Å, 5 Å for $\lambda > 5000$ Å. $\lambda 2306$ was observed but not plotted since the monochromator efficiency is not known at that wavelength. $\lambda 1522$ was not observed.

REFERENCES

(1)   R. W. P. McWhirter, in New Techniques in Space Astronomy, ed.
      by F. Labuhn and R. Lüst (D. Reidel Publishing Co., Dordrecht,
      Holland, 1971).

(2)   Calibration Methods in the UV and X-Ray Regions. Report No.
      ESRO SP-33 (Published by the European Space Research Agency,
      1968).

(3)   Optronic Laboratory, Inc., 7676 Fenton St., Silver Springs,
      MD 20910.

(4)   N.B.S. Measurement Users Bulletin No. 4 (A Supplement to N.B.S.
      Special Publication 250, 1970 Edition), p. 5.9, February, 1973.

(5)   J. L. Buckley, Appl. Opt. 10, 1114 (1971).

(6)   W. R. Ott and W. L. Wiese, Optical Eng. 12, 86 (1973).

(7)   E. Pitz, Appl. Opt. 8, 255 (1969).

(8)   E. Hinnov and F. W. Hofmann, J. Opt. Soc. Am. 53, 1259 (1963).

(9)   B. C. Boland, T. J. L. Jones, and R. W. P. McWhirter, p. 31
      in Ref. (2).

(10) W. G. Griffin and R. W. P. McWhirter, in Optical Instruments
      and Techniques, ed. by K. J. Habell (John Wiley, New York,
      1963) p. 14.

(11) H. A. Bethe and E. E. Salpeter, Handbuch der Physik XXXV, ed.
      by S. Flügge (Springer Verlag, Berlin, 1957) p. 352.

(12) W. L. Wiese, M. W. Smith, and B. M. Glennon, Atomic Transition
      Probabilities, NSRDS-NBS 4 (U.S. Nat. Bur. Stand. 1966).

(13) W. J. Karzas and R. Latter, Astrophys. J. Suppl. 6, 167 (1961).

(14) S. Bashkin and I. Martinson, J. Opt. Soc. Am. 61, 1686 (1961).

(15) S. N. Bhardwaj, H. G. Berry, and T. Mossberg, Physica Scripta
      9, 331 (1974).

(16) W. N. Lennard, R. M. Sills, and W. Whaling, Phys. Rev. 6A,
      884 (1972).

(17) J. Bromander, Nucl. Instr. & Meth. 110, 11 (1973).

(18) H. H. Bukow, H. V. Buttlar, D. Haas, P. H. Heckman, M. Holl,
      W. Schlagheck, D. Schurmann, R. Tielert, and R. Woodruff,
      Nucl. Instr. & Meth. 110, 89 (1973).

(19) R. Hallin, J. Lindskog, A. Marelius, J. Pihl, and R. Sjödin,
      Physica Scripta 8, 209 (1973).

(20) P. A. Fraser and W. R. Jarmain, Proc. Phys. Soc. 66A, 1145
      (1953); P. A. Fraser, Can. J. Phys. 32, 515 (1954).

(21) T. C. James, J. Mol. Spect. 20, 77 (1960).

(22) J. Drake and R. W. Nicholls, Chem. Phys. Lett. 3, 457 (1969).

(23) H. Klemsdal, J. Quant. Spect. Rad. Transfer 13, 517 (1973).

(24) M. J. Mumma, J. Opt. Soc. Am. 62, 1459 (1972).

(25) K. H. Becker, E. H. Fink, and A. C. Allison, J. Opt. Soc. Am.
      61, 495 (1971).

(26) A. C. Allison and A. Dalgarno, Atomic Data 1, 289 (1970).

(27) J. A. Jordon, G. S. Bakken, and R. E. Yager, J. Opt. Soc. Am.
      57, 530 (1967).

(28) W. N. Lennard and C. L. Cocke, Nucl. Inst. & Meth. 110, 137
      (1973).

A HIGH-INTENSITY METHOD FOR BEAM-FOIL SPECTROSCOPY, WITH RETAINED

SPATIAL RESOLUTION ALONG THE BEAM

Karl-Erik Bergkvist

Research Institute for Physics

Stockholm, Sweden

1.   INTRODUCTION

In all spectroscopic work high intensity and high resolution in wave length are desirable features. In beam-foil spectroscopy - because of the time information given by the coordinate along the beam - one in addition requires also a good spatial resolution along the beam. There are, ultimately, conflicts involved between the three requirements. We wish to report here a new method which allows a material increase in intensity without unduly setting aside the two other requirements. In a preliminary test of the method an intensity gain by a factor of 70 has been obtained; there are prospects for even higher gains.

A central feature of the new method is the deletion of the entrance slit in the spectrometer. In this respect our work owes considerably to an earlier work by Stoner and Leavitt[1]. These authors pointed out that a significant intensity gain in beam-foil spectroscopy can be achieved without loss in resolution in wave length by removing the entrance slit in the spectrometer and replacing it by a suitably positioned cylinder lens. In this earlier work the possibility to combine such a gain in intensity with a reasonable retention of the spatial resolution along the beam does not seem to have been recognized. We show below that the situation studied by Stoner and Leavitt is actually a special case in a more general formulation of the problem. From this latter it appears that both higher intensity and better spatial resolution along the beam can be achieved than previously assumed.

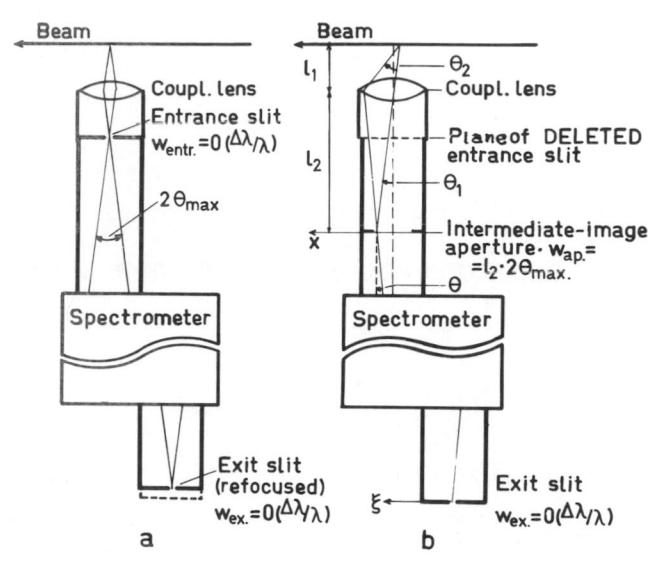

FIGURE 1.   *This figure shows schematically (a) the conventional*
arrangement of ion-beam spectrometer in beam-foil
spectroscopy and (b) the arrangement as studied in the
present work.

2.   BASIC PRINCIPLES

Consider Fig. 1(a), illustrating schematically the conventional
layout of ion beam and spectrometer in beam-foil spectroscopy. The
ion beam is assumed to have a negligible cross section and a well-
defined velocity. The beam direction and the dispersion at the
spectrometer exit slit are assumed to be in the plane of the figure.
Light emitted from an atom in the beam is focused on the spectrometer entrance slit by the indicated coupling lens. Because of the
velocity of the ions, the light reaching the entrance slit will be
Doppler shifted, with an amount varying with its direction of emission from the beam. It is possible, and common practice, to com-
pensate for this Doppler spread by introducing a suitable amount of
astigmatic focusing in the spectrometer (as indicated in the figure)
Let Ω be the entrance solid angle of the spectrometer, and $w_{entr.}$
the width of the entrance slit. With the coupling lens placed
roughly intermediately between the beam and the entrance slit (as
conventionally done) the effective radiating beam length will be
roughly equal to $w_{entr.}$ and the effective solid angle at the beam
roughly equal to Ω. The intensity I_{old} observed for a given spectral line with the arrangement of Fig. 1(a) can therefore be written

$$I_{old} = c \cdot w_{entr} \cdot \Omega, \tag{1}$$

where c is a factor containing all dependence of the intensity not
contained in the other two factors. The intensity is proportional
to $w_{entr.}$ and hence necessarily decreases when one proceeds to a
higher resolution $\Delta \lambda / \lambda$.

Now consider the arrangement in Fig. 1(b). We first estimate
its intensity capabilities. The entrance slit of the spectrometer
has been deleted. The coupling lens images a point of emission in
the beam in an intermediate-image plane, positioned some distance
inwards in the spectrometer. An aperture in this plane, of width
w_{ap}, determines the beam length seen by the arrangement. If $2\theta_{max}$
is the maximum acceptance angle of the spectrometer aperture in the
plane of the figure, and l_2 the distance between the coupling lens
and the intermediate image, the aperture in the intermediate-image
plane can be opened to a value $l_2 \cdot 2\theta_{max}$ before the spectrometer
finite aperture starts significantly to restrict the light entering
the spectrometer. If l_1 is the distance between the coupling lens
and the beam, the utilized beam length then is $l_1 2\theta_{max}$. Assuming
the same spectrometer as in case (a), the effective solid angle of
the arrangement in (b) is $(l_2/l_1)^2 \Omega$, provided that the coupling lens
has an adequately large opening ratio. Instead of the expression
(1) we now have for the intensity , I_{new}, in the new arrangement:

$$I_{new} = c \cdot l_1 \cdot 2\theta_{max} (l_2/l_1)^2 \Omega \qquad (2)$$

where c is the same factor as in eq. (1). For the intensity ratio
I_{new}/I_{old} we therefore get

$$\frac{I_{new}}{I_{old}} = \frac{l_1 \cdot 2\theta_{max}}{w_{entr.}} \cdot (\frac{l_2}{l_1})^2 . \qquad (3)$$

Now consider the imaging at the detector slit of the new arrangement. Let x and ξ be coordinates in the intermediate-image plane
and the detector-slit plane, respectively, as indicated in Fig. 1(b).
Let θ be the entrance angle (in the plane of the figure) of a ray
entering the spectrometer from a point x in the intermediate-image
plane, and let $\delta \lambda$ be the Doppler shift pertaining to the ray in
question. The resolution determining image coordinate ξ can then
be written, to lowest order,

$$\xi = M \cdot x + A \cdot \theta + D \cdot \delta \lambda , \qquad (4)$$

where M is the magnification, A the astigmatism and D the dispersion
of the spectrometer, all entities referred to an unshifted wave
length λ_o and to an object plane defined by the intermediate-image
aperture.

The variables x, θ and $\delta \lambda$ in eq. (4) depend on the variables
θ_1 and θ_2, defining, respectively, the point of emission along the

beam and the relevant direction of the ray within the entrance aperture of the coupling lens (cf. Fig. 1(b)). For not too large values of the variables involved we have the following relations:

$$x = l_2 \theta_1, \tag{5}$$

$$\theta = \theta_1 - \frac{l_1}{l_2} \theta_2 \tag{6}$$

$$\delta\lambda = \lambda_o \frac{v}{c} (\theta_1 + \theta_2), \tag{7}$$

where v/c is the ion velocity in units of the velocity of ligth. Using (5)-(3) in (4), the vanishing of ξ amounts to

$$Ml_2\theta_1 + A'(\theta_1 - \frac{l_1}{l_2} \theta_2) + D\lambda_o \frac{v}{c} (\theta_1 + \theta_2) \equiv 0. \tag{8}$$

For the indentity (8) to hold we must clearly have a simultaneous satisfication of the two equations

$$Ml_2 + A + D\lambda_o \frac{v}{c} = 0 \tag{9}$$

$$-A \frac{l_1}{l_2} + D\lambda_o \frac{v}{c} = 0 \tag{10}$$

It turns out that the two eqs. (9) and (10) can indeed be simultaneously satisfied. Although the set-up in Fig. 1(b) does not contain any entrance slit, it can therefore be arranged that a given spectral line is brought to focus within a narrow exit slit in the spectrometer.

    A full discussion of eqs. (9) and (10) requires somewhat more space than available in this conference report. We just state the central implications of the equations:
(i) Irrespective of the ratio l_2/l_1 the appropriate focal length f of the coupling lens is given by

$$f = \left| \frac{D \lambda_o \frac{v}{c}}{M} \right|, \tag{11}$$

i.e., the focal length is directly proportional to the magnitude of the dispersion and to the magnitude of v/c, and inversely proportional to the magnitude of the magnification.
(ii) Irrespective of the ratio l_2/l_1 the requirement on the astigmatism parameter A will be satisfied if the coupling lens is mounted at a distance equal to its focal length f in front of the entrance slit plane of the spectrometer (as adjusted for no astigmatism).
(iii) For a given ratio l_2/l_1 the appropriate position of the intermediate -image aperture is f · (l_2/l_1) inwards in the spectrometer from the plane of the original entrance slit. The appropriate (maximum) width of the aperture is $w_{ap} = f \cdot (l_2/l_1 + 1) \cdot 2\theta_{max}$.

3.   NUMERICAL EXAMPLE

Let us now estimate numerically, in a concrete example, the intensity gain (3) and the associated spatial resolution along the beam. The result depends on λ_0, on the data of the spectrometer, and on the value of v/c for the ions. Let us assume a spectrometer with a dispersion of 0.2 mm/Å, referred to $\lambda_0 = 5 \cdot 10^3$ Å, and an opening ratio of 1:12 (i.e. $2\theta_{max} = 1/12$). The latter figure implies a diffraction line width of about 0.01 mm and, thus, $w_{entr.} \approx 0.01$ mm in eq. (3). To make the intensity ratio (3) as large as possible, the value of l_1 should be made as small as possible. This means giving l_1 a value close to f, the focal length of the coupling lens. Assuming v/c = 10^{-2} and the magnification M equal to unity, we get, from eq. (11), f = 10 mm. The value of l_2 should, according to eq. (3), be made as large as possible. The theoretical upper limit on l_2 - within the present first-order treatment - is set by the condition that the opening ratio of the coupling lens must exceed that of the spectrometer by roughly a factor l_2/l_1. Let ur choose $l_2/l_1 = 6$ (thus implying a minimum opening ratio of 1:2 for the coupling lens). Eq. (3) then gives

$$\frac{I_{new}}{I_{old}} = 3 \cdot 10^3 \tag{12}$$

The spatial resolution width along the beam is given by $l_1 \, 2\theta_{max}$. With the above data we hence have:

Spatial resolution along beam = 0.8 mm.   (13)

The result (13) illustrates the important property of the arrangement of Fig. 1(b) of utilizing a beam length which is not infinitesimal, in the sense of the arrangement of Fig. 1(a), but at the same time not large enough so as to really destroy the spatial resolution along the beam. If a better resolution along the beam is required than given by the result (13), this is easily accomplished by restricting the width of the intermediate-image aperture. The sacrifice is a proportional loss in intensity. This may well be acceptable in view of the large factor (12).

Of the over-all factor of $3 \cdot 10^3$ in (12) a factor of around 80 derives from the fact that the method of Fig. 1(b) utilizes a beam portion which is $l_1 2\theta_{max}/w_{entr.}$ longer than the conventional method. A factor of 36 stems from assumed ratio $l_2/l_1 = 6$ and is the magnification of the solid angle of the arrangement as compared to the bare spectrometer.

It should be emphasized that the estimate (12) refers strictly to the somewhat idealized model for our treatment, and to the particular data involved. In reality higher-order effects, not con-

tained in the present first-order treatment, may become of importance. Also, finite beam cross section and spread in direction and
magnitude of the ion velocity will ultimately impose limitations.
Clearly, though, the result (12) speaks promisingly about the basic
possibilities of an arrangement like that in Fig. 1(b).

At this point it is appropriate to relate the present analysis
to that of ref. 1. The method of ref. 1 implies replacing our coupling lens in Fig. 1(b) with a cylinder lens, dispensing with the
defining aperture in the intermediate-image plane, and choosing l_1
such that $l_1 \gg f$. With this choice our result (12) for the intensity gain is replaced by a six times smaller figure. As to the spatial resolution width this depends on the precise value of l_1. For,
e.g., $l_1 = 5 f$ the result (13) is replaced by a spatial resolution
width about 25 times larger.

4.   PRACTICAL DEMONSTRATION OF THE NEW METHOD

To test experimentally the basic ideas presented we have recorded the He$^+$ 4686 Å line both in the conventional manner of Fig.1(a)
and with the new method of Fig. 1(b). The spectrometer used was of
the type of Seya and Namioka, with some minor modifications to suit
the present particular application. In the setting used the magnification M was measured to be M = 0.65, while the dispersion D was
measured to be 0.10 mm/Å. With a focal length f = 7.4 mm for an
available coupling lens these data by means of eq. (11) gave a value
$v/c = 1.11 \cdot 10^{-2}$ as the appropriate ion velocity, corresponding to
an ion energy of 225 keV for He. For the ratio l_2/l_1 a value equal
to 6 was chosen, implying that the ion beam should pass the coupling
lens at a distance $l_1 \approx 8.0$ mm. The cross section of the ion beam
was defined by an aperture 1 mm in diameter mounted in front of the
foil. The beam portion focused on by the arrangement was about 10
mm downstream from the foil. Fig. 2(a) shows the 4686 Å line as recorded with the new method, Fig. 2(b) as recorded with the method
of Fig. 1(a). The beam current is the same in both cases. The intensity with the new method is about 70 times higher than with the
conventional method. A very minor increase in line width (f.w.h.m.),
from 2.8 Å to 3.0 Å, can be traced in the recording with the high-
intensity method.

Fig. 2(c) illustrates the spatial resolution along the beam
with the new method. A kind of knife edge was mounted close to the
beam and moved along the beam while the peak intensity of the line
of Fig. 2(a) was observed. As seen, the peak counting rate of the
line first remains unaffected of the movement of the knife edge and
then, over a distance of about 0.3 mm, drops to the background.

The results exhibited in Fig. 2 are in satisfactory agreement
with the already mentioned data and the values $w_{entr.} \approx 0.17$ mm,
$2\theta_{max} \approx 0.05$ involved.

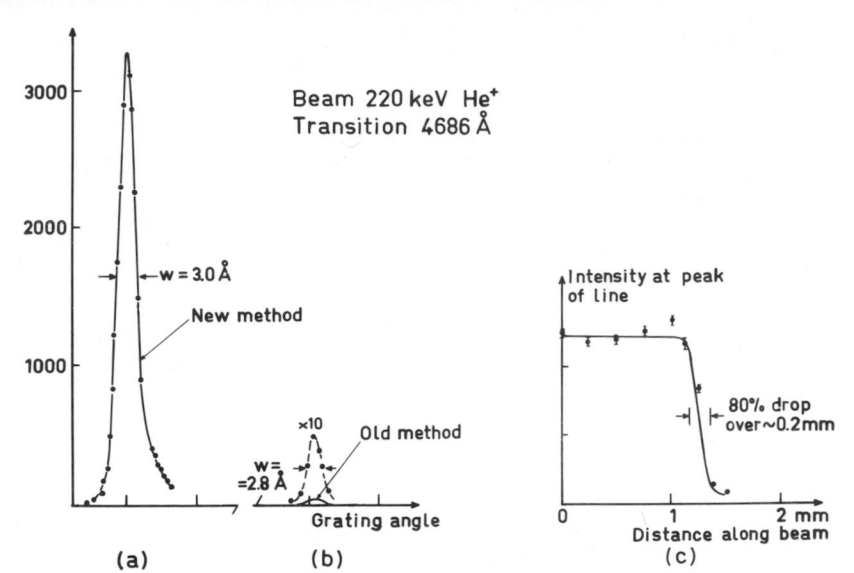

FIGURE 2.   *Recordings of the He$^+$ 4686 Å line with (a) the high-intensity method of the present work and (b) the conventional method. Fig. 2(c) serves to illustrate the spatial resolution along the beam with the high-intensity method.*

5.   CONCLUDING REMARKS

The difference between the experimentally observed intensity gain, a factor of 70, and the much higher theoretical figure (12) largely stems from the very different values of w_{entr} involved in the two cases. In the experimental investigation the comparison refers to an entrance slit width of $\sim$ 0.17 mm in the conventional recording. As employed the spectrometer resolution actually did not motivate an entrance slit width much smaller than this value. Experiments are now being planned to employ a spectrometer of higher quality, allowing a test of the new method at higher resolution than the 2.8 - 3.0 Å in Fig. 2. The superiority of the new method should then be expected to be more pronounced than the obtained factor of 70. Any statement as to how closely one may actually approach the theoretical intensity gain (12) must await until some experience is gained from such higher-resolution experiments.

ACKNOWLEDGEMENTS

     The author acknowledges informative discussions with Professor
John O. Stoner and valuable cooperation with Mr. Leif Liljeby in the
test recordings of Fig. 2.

REFERENCE

1.   J.O. Stoner and J.A. Leavitt, Optica Acta 20, (1973) 435

NONADIABATIC SPIN TRANSITIONS:

A POSSIBLE SOURCE OF POLARIZED ELECTRONS

R. D. Hight[*] and R. T. Robiscoe

Department of Physics

Montana State University, Bozeman, Montana 59715

We have developed a model for the nonadiabatic passage of an oriented beam of atoms through an inhomogeneous magnetic field. The passage is completely characterized by a single parameter called the adiabaticity parameter α, the ratio of the spin Larmor frequency to the field rotation rate. Using our model, we have derived differential equations for the amplitudes describing the time behavior for a spin one atom. We have solved these equations using S-matrix theory and numerical integration. Using the $F = 1$ hyperfine state of the $2S_{1/2}$ level in atomic hydrogen, we have confirmed our model in part. The theory suggests a rather novel method for producing a beam of atomic hydrogen polarized in any one of the three $F = 1$ hyperfine states of the $2S_{1/2}$ level. With photo-ionization, this atomic beam could produce a beam of 100% polarized electrons or protons.

MODEL

Consider the entry of an atom with oriented spin at a velocity v into a spacially varying magnetic field. As the atom traverses the field region, it will experience a time varying field. If the field varies slowly enough, the spin will simply follow the field and remain aligned with it as predicted by the adiabatic theorem. However, if the variation is rapid enough, the spin will not be able to follow the field and therefore become anti-aligned with it. The adiabaticity parameter α, defined as the ratio of the Larmor frequency to the field rotation rate, is used to determine whether or not a transition or spin flip will occur. When the field rotates slowly, $\alpha \gg 1$, no spin flips are expected. On the other hand, if the field rotates quickly enough, $\alpha \ll 1$, spin flips occur. A more

727

detailed discussion has been given by Robiscoe[1] using a spin 1/2 as an example.

Our model assumes a cylindrically symmetric magnetic field whose axial component reverses direction linearily in a finite region, rotation of direction by π radians, and remains a constant, purely axial field elsewhere. The adiabaticity parameter for this field is $\alpha = \alpha_0/\sin^3 \theta$, where α_0 is a constant proportional to the field strength and θ is the polar angle of the field direction with respect to the cylindrical axis of the field. Using time dependent perturbation theory, we have solved Schrodinger's equation for a spin one system with the help of W. R. Thorson[2]. Denoting $a_i(\pi)$ as the time dependent amplitude for the three $M_F = 1,0,-1$ states with $i = 1,2,3$ respectively, we have for the final state amplitudes

$$\begin{pmatrix} a_1(\pi) \\ a_2(\pi) \\ a_3(\pi) \end{pmatrix} = \begin{pmatrix} s_{13} & -s_{12}e^{i(\lambda_\pi+\phi)} & s_{11}e^{2i(\lambda_\pi+\phi)} \\ s_{12}e^{-i(\lambda_\pi+\phi)} & -s_{22} & -s_{12}e^{i(\lambda_\pi+\phi)} \\ s_{11}e^{-2i(\lambda_\pi+\phi)} & s_{12}e^{-i(\lambda_\pi+\phi)} & s_{13} \end{pmatrix} \begin{pmatrix} a_1(0) \\ a_2(0) \\ a_3(0) \end{pmatrix}$$

where $a_i(0)$ are the initial state amplitudes. The phase λ_π is the integrated adiabaticity parameter $\int_0^{\pi/2}\alpha(\theta)d\theta$ and ϕ is a complicated function which is approximately given by $-(\alpha_0\rho^2 + 1.2)^{-3/2}$, where ρ is a dimensionless parameter that is determined by the distance off axis of the atom. The elements s_{ij} are related to s_{13} by $s_{11} = 1 - s_{13}$, $s_{12} = \{2s_{13}(1 - s_{13})\}^{\frac{1}{2}}$, and $s_{22} = (1 - 2s_{13}^2)^{\frac{1}{2}}$. The element s_{13} is simply $\exp(-\pi\alpha_0\rho^2/4)$. By substituting in the appropriate initial amplitudes, we may determine the population of each magnetic substate as a function of applied magnetic field strength.

EXPERIMENT

For our spin one atom, we used the $F = 1$ hyperfine state of the metastable $2 S_{\frac{1}{2}}$ level in atomic hydrogen. The beam of atomic hydrogen was produced by thermal dissociation and excited by electron impact. At thermal energies, the metastable state $(2S_{\frac{1}{2}})$ of atomic hydrogen is the only excited state remaining after several milli- meters of travel due to its long lifetime (~1/7 sec). Since all states in the $2S_{\frac{1}{2}}$ level are approximately equally populated after electron excitation, all but the $M_F = +1,0$, $F = 1$ states were removed by stark quenching. The atom then traveled through our magnetic field region before being detected. By stark quenching after the magnetic field, we were able to measure the $M_F = -1$ state as a decrease in beam signal. The magnetic field was generated by two oppositely wound solenoids butted together. The measured adiabaticity parameter was extremely close to the calculated α, less than one percent difference.

Due to the finite cross section of the beam and the velocity distribution of the atoms, the theoretical amplitude for the $M_F = -1$ state had to be averaged. The velocity distribution was measured by time-of-flight techniques and the averaging was done numerically. The finite cross-section averaging was performed also by numerical integration assuming uniform cross sectional density. The maximum radius of the beam was allowed to vary to fit the experimental data. The results are shown in the Figure. The theoretical fit is sufficient to confirm our model in part. To complete the experiment, we plan to determine the populations of the $M_F = +1,0$ states as a function of field strength.

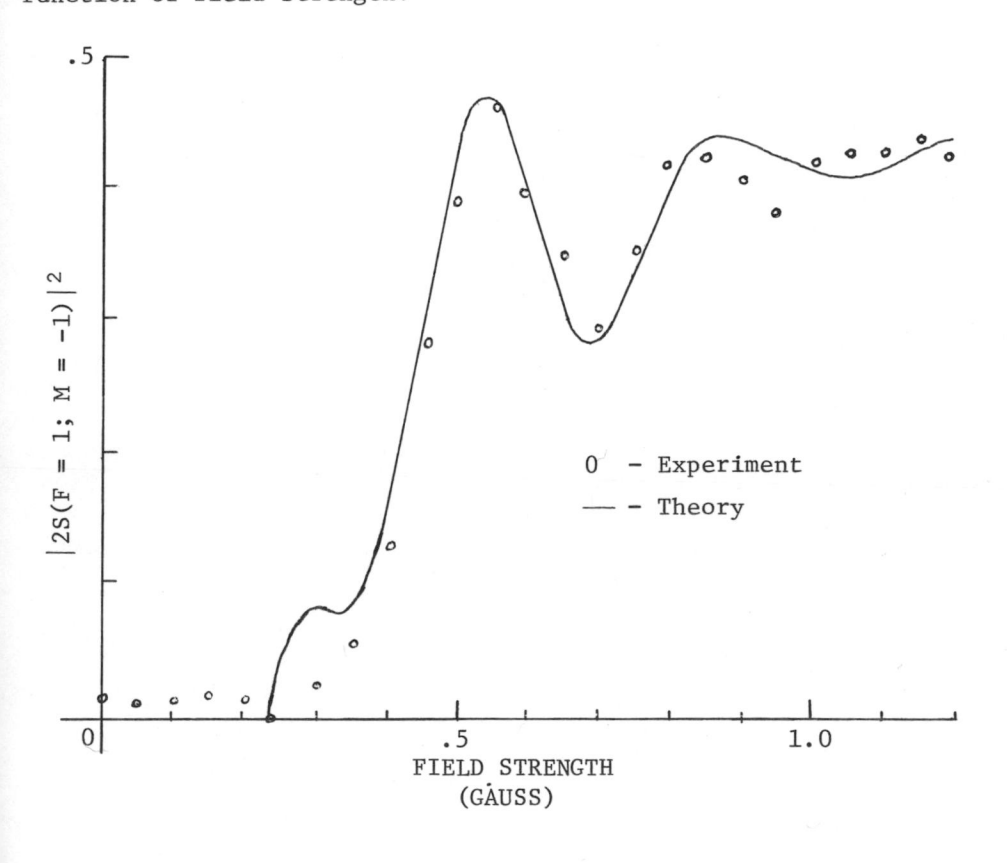

POLARIZED BEAM

The theory predicts that at large fields, $\alpha \gg 1$, the population of the $M_F = -1$ state becomes 0.5 while the $M_F = +1$ state is drained to zero. This leaves the $M_F = 0$ state with a relative population of 0.5. By stark quenching, we may remove the $M_F = -1$ state leaving the $M_F = 0$ state the only excited state in the beam. Or, we may bend the $M_F = -1$ state out of the beam adiabaticly. Photo-ionization with circular polarized light should give a beam

of either polarized electrons or protons.

*Present address: Department of Physics and Astronomy, University of Toledo, Toledo, Ohio 43606

[1] R. T. Robiscoe, AJP, Vol. 39/2, p 146 (Feb. 1971).

[2] W. R. Thorson, Private communication.

ALIGNMENT AND ORIENTATION

PRODUCTION MEASUREMENT AND CONVERSION

M. LOMBARDI

Laboratoire de Spectrométrie Physique, Université Scientifique et Médicale de Grenoble, B.P. 53 - 38041 Grenoble-Cédex - France

Since the experimental proof given one year ago by Berry, Curtis, Ellis and Schectmann (1) that, according to a suggestion made by Fano and Macek (2) and by Ellis (3), the light emitted in a beam tilted foil experiment can be partially elliptically polarized, beam foil physicists have paid much attention to the ways of producing, measuring and describing elliptical polarization. The theoretical framework needed, and some experimental tricks have been established some years ago by people working in the optical pumping field, who produced elliptical polarization in a genuine manner by using resonance excitation of atoms by elliptical light. The purpose of this paper is first to review the more useful methods of description of the state of the atom, with emphasis on physical signification and geometrical representation of the various theoretical quantities introduced, and their connection with measurements of light polarization, second to review some non trivial methods of producing orientation, with also emphasis on symmetry considerations and physical meaning.

1 - Description and measurement of orientation and alignment

1.1. - Definition and geometrical representation.

The concepts of orientation and alignment where first loosely defined in situations of cylindrical symmetry along O_z as follows. A level is said oriented if the populations of $+m$ and $-m$ sublevels

are different. It is said aligned if the populations of the couple
of ±m levels are equal but the populations of levels with different
$|m|$ are unequal. The first situation arises in a level excited by
light circularly polarized along O_z and the second arises in a level excited by light linearly polarized along O_z or by an electron
or ion beam directed along O_z.

More precisely the orientation along O_z of a level of total momentum J is the first moment of the population distribution :

$$O_z = \sum m\ p_m\ = <J_z> \tag{1}$$

where p_m is the population of sublevel $|J,m>$.

The alignment along O_z is the difference between the second moment
of the distribution and its mean :

$$A_z = \sum (m^2 - \overline{m^2})\ p_m = \sum (m^2 - \frac{\sum m'^2}{2J+1})\ p_m$$

$$= \sum (m^2 - \frac{1}{3} J\ (J+1))\ p_m$$

as can be verified directly by induction, or

$$A_z = < J_z^2 - \frac{1}{3} \vec{J}^2 > \tag{2}$$

i.e. the difference between $<J_z^2>$ and its mean value $\frac{1}{3} <\vec{J}^2>$.

In more general situations, one is lead to use density matrix elements σ_{mm} to generalise the populations p_m and to study orientation
and alignment along any u direction. The basic problem to solve is :
how many independant informations can one collect from this general
study, and which are they ? The answer is very simple in the case
of orientation. Since

$$<J_u> = u_x <J_x> + u_y <J_y> + u_z <J_z> \tag{3}$$

where u_x, u_y, u_z are the direction cosines of the unitary $\vec{u}$ vector,
there are three independant orientations parameters, the three
components of the orientation vector $<\vec{J}>$.

Of paramount importance for the second part of this review is the
fact that the $\vec{J}$ vector is not polar but axial, so that its correct
geometrical representation is a sense of rotation along a straight
line, not an arrow (Fig. 1). The situation is only slightly more
complicated for alignment. Introducing the alignment tensor (4)

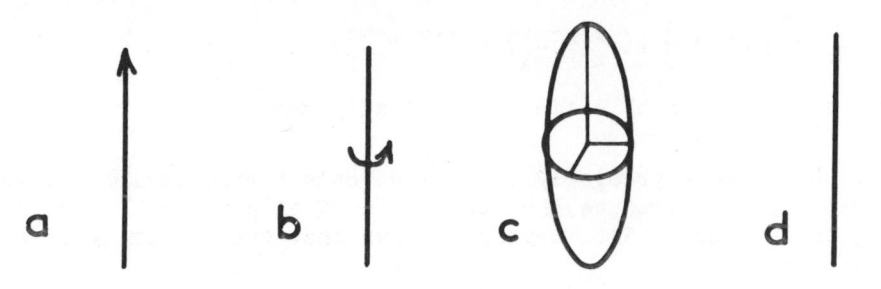

Fig. 1 : Uncorrect (a) and correct (b) representation of orientation. Full (c) and simplified (d) representation of alignment.

$$M^{(2)} = \begin{vmatrix} <J_x^2 - \frac{1}{3}\vec{J}^2> & \frac{1}{2}<J_x J_y + J_y J_x> & \frac{1}{2}<J_x J_z + J_z J_x> \\ \frac{1}{2}<J_x J_y + J_y J_x> & <J_y^2 - \frac{1}{3}\vec{J}^2> & \frac{1}{2}<J_y J_z + J_z J_y> \\ \frac{1}{2} J_x J_z + J_z J_x & \frac{1}{2}<J_y J_z + J_z J_y> & <J_z^2 - \frac{1}{3}\vec{J}^2> \end{vmatrix}$$

one verify that

$$<J_u^2 - \frac{1}{3}\vec{J}^2> = \sum u_i M_{ij}^{(2)} u_j = \vec{u} \overset{\leftrightarrow}{M}^{(2)} \vec{u}$$

Since $M^{(2)}$ is symmetric and traceless, there are only five independant parameters. One possible choice is the following which, within
factors varying from author to author constitute the Fano's "standard real" components of alignment (4)

$$M_{0+}^{(2)} = <J_z^2 - \frac{1}{3}\vec{J}^2> \tag{4}$$

$$M_{1+}^{(2)} = <J_x J_z + J_z J_x> = <J_{x/z}^2 - \frac{1}{3}\vec{J}^2> - <\vec{J}_{x\backslash z}^2 - \frac{1}{3}\vec{J}^2>$$

$$M_{1-}^{(2)} = <J_y J_z + J_z J_y> = <J_{y/z}^2 - \frac{1}{3}\vec{J}^2> - <\vec{J}_{y\backslash z}^2 - \frac{1}{3}\vec{J}^2>$$

$$M_{2+}^{(2)} = <J_x^2 - J_y^2> \qquad = <J_x^2 - \frac{1}{3}\vec{J}^2> - <J_y^2 - \frac{1}{3}\vec{J}^2>$$

$$M_{2-}^{(2)} = <J_x J_y + J_y J_x> = <J_{x/y}^2 - \frac{1}{3}\vec{J}^2> - <J_{x\backslash y}^2 - \frac{1}{3}\vec{J}^2>$$

where $J_{x/y}$ and $J_{x\backslash y}$ mean components of $\vec{J}$ along directions at + and
$-45°$ of the axis in the xOy plane. The four last standard real ali-
gnment components are then merely differences of alignments. Since

$$J_z^2 - \frac{1}{3} J^2 = \frac{1}{3} \left((J_z^2 - J_x^2) + (J_z^2 - J_y^2) \right) ,$$

$M_{0+}^{(2)}$ is also a sum of differences of alignments.

To visualise more precisely the alignment's signification, it is interesting to introduce the "ellipsoïd of alignments" $\mathcal{E}_{AL}$ of Laloe, Leduc and Minguzzi (5). They have shown that the extremity of a vector of length.

$$r = \frac{1}{\sqrt{J_u^2}} \tag{5}$$

lies on an ellipsoïd $\mathcal{E}_{AL}$ which lies partly inside, partly outside the sphere of radius $\sqrt{3/J(J+1)}$. In the frame OXYZ of principal axis of $\mathcal{E}_{AL}$, $M_{1+}^{(2)} = M_{1-}^{(2)} = M_{2-}^{(2)} = 0$. If the ellipsoïd has a symmetry of revolution along OZ, one has furthermore $M_{2+}^{(2)} = 0$. When some of the $M^{(2)}$ components are zero, one can deduce some particularities of $\mathcal{E}_{AL}$ by using the remark that the last four components are merely differences of alignment, so that the cancellation of one of these $M^{(2)}$ components implies equality of alignment along two particular directions. But in general one uses the opposite way : from the symmetry of the problem one deduces the position of some axis of the ellipsoïd and then the cancellation of some $M_i^{(2)}$. For example if the problem is symmetric on the plane xOz, $<J_{x/y}^2 - \frac{1}{3} J^2>$ and $<J_{x\backslash y}^2 - \frac{1}{3} J^2>$ which are two mirror alignments in this plane are equal and then $M_{2-}^{(2)} = 0$. Fig. 1 gives a complete and a simplified (valid only for revolution invariant cases) representation of alignment deduced from this discussion.

1.2. Description of the state of polarization of a light beam.

This is a classical description that can be found in standard text books (6). If the plane perpendicular to the beam is referred to two axis OX and OY and since the signal detected is quandratic in the electric fields, one defines a polarization matrix

$$\pi(t) = \begin{vmatrix} <|E_x|^2> & <E_x E_y^*> \\ <E_y E_x^*> & <|E_y|^2> \end{vmatrix}$$

The brackets < > denote averaging over the response time of the photodetector. To relate easily the polarization of the light emitted to the orientation and alignment standard real components defined in section (1), it is interesting to introduce as in (1) the Stroke

parameters defined as

$$I = |E_x|^2 + |E_y|^2 \quad = \quad I_x + I_y \qquad (6)$$

$$M = |E_x|^2 - |E_y|^2 \quad = \quad I_x - I_y$$

$$C = 2\text{Re} \, (E_x \, E_y^*) \quad = \quad I_{+45°} - I_{-45°}$$

$$S = 2\text{Im} \, (E_x \, E_y^*) \quad = \quad I_{\sigma+} - I_{\sigma-}$$

The third column indicates how they are related to measurements with linear (for the first three) or circular polarizers. This justify their names of "total intensity", "preferred O_x linear polarization", "preferred +45° linear polarization" and "preferred right circular polarization".

1.3. Relation between polarization of the light emitted and orientation and alignment of the emitting level.

The light emitted is given by the formula

$$I_\lambda = C \, T_2 \, [\rho(\vec{e}_\lambda^* \cdot \vec{d}) \, P_f \, (\vec{e}_\lambda \cdot \vec{D})] = C \, <(\vec{e}_\lambda^* \cdot \vec{d}) \, P_f \, (\vec{e}_\lambda \cdot \vec{d})> \qquad (7)$$

where P_f is the projector upon the lower state of the observed transition, ρ the density matrix, $\vec{e}_\lambda$ the polarization vector, $\vec{d}$ the dipole moment and C a constant.

First if one uses a linear polarizer along the O_z direction,

$$I_z = C \, <d_z \, P_f \, d_z>$$

$$= C \, <(d_z \, P_f \, d_z - \tfrac{1}{3} \vec{d} \, P_f \, \vec{d})> + \tfrac{1}{3} \, C \, <\vec{d} \, P_f \, \vec{d}>$$

Owing to the Wigner Eckart theorem

$$I_z = C^{(2)} <J_z^2 - \tfrac{1}{3} \vec{J}^2> + C^{(0)} <\vec{J}^2> \qquad (8)$$

is the sum of a constant and of a term proportional to the alignment in the O_z direction. Moreover since

$$(J_x^2 - \tfrac{1}{3} \vec{J}^2) + (J_y^2 - \tfrac{1}{3} \vec{J}^2) = - (J_z^2 - \tfrac{1}{3} J^2)$$

the total intensity emitted along O_z is $3 \, C^{(0)} <\vec{J}^2> - I_z$.

On the other hand the difference of right and left circular polari-

zations around the direction 0_z, whose polarization vectors e_λ are $(1/\sqrt{2}, i/\sqrt{2}, 0)$ and $(1/\sqrt{2}, -i/\sqrt{2}, 0)$, is

$$I_{\sigma+} - I_{\sigma-} = i <d_x P_f d_y - d_y P_f d_x>$$
$$= C^{(1)} <J_z> \tag{9}$$

owing again to the Wigner Eckart theorem.

Since all polarization measurements can be experessed by linear combination of the components of the Stoke's vector (6), they are all linear combinations of the alignment and orientation parameters defined in section (1). Explicit expressions have been given by various authors (7)(5)(2). We shall only quote here that if the light propagates along the 0_z direction, the Stoke's parameters I, M, C, S are equal to $2 C^{(0)} - C^{(2)} M_{0+}^{(2)}$, $C^{(2)} M_{2+}^{(2)}$, $C^{(2)} M_{2-}^{(2)}$, $C^{(1)} M_0^{(1)}$, which is a direct consequence of equations (1), (4), (8) and (9).

1.4. Measurement.

We shall in this section only discuss some aspects which are related to the fact that all orientation or alignment parameters are proportional to differences of polarizations either linear for alignment or circular for orientation. In experiments in which high light levels enable continuous (non photon counting) methods of detection, this has lead to a wide use of synchronous detectors following rotating polarizers. One very interesting device described by Pavlovic and Laloe (8) is the following : a $\lambda/4$ plate rotates before a fixed polarizer (not fixed plate and rotating polarizer). This device was designed originally to have alternatively $\sigma+$ and $\sigma-$ detection when the axis of the $\lambda/4$ plate are interchanged, which results in an orientation signal at twice the frequency f of rotation of the plate But a full analysis has shown several interesting features (8). First the signal at frequency 2f depends only on orientation even if the plate is not exactly $\lambda/4$. Only the detection efficiency decreases as sin ϕ, ϕ being the dephasing angle of the plate. Since sin ϕ is stationnary in the vicinity of $\phi = 90°$, this enables one to use only one simple commercial plastic sheet quarter wave at $\lambda = 5600$ Å to measure orientation between 4000 and 7000 Å with a loss of detection efficiency of at most 20 % and with great rejection of superimposed alignment signals. This rejection ratio has been measured to be nearly 1000 if one uses a glazebrook prism as polarizer (9) (but only a few percent with polaroïds). Another unusual feature of this device is that at frequency 4f, twice the orientation frequency, there is a signal proportional to the alignment. With a two phase synchronous detector set to 4f, one extract

the M and C Stoke's components and simultaneously a 2f synchronous
detector extract the orientation S. The only drawback of this method of measuring alignments is that the yield is only one half of
the rotating polaroïd method, but this is partially compensated by
the possibility of simultaneous measurements which increases the
measurement times. Otherwise since the polaroïd behind the $\lambda/4$
plate is fixed, one has no problem with the parasitic polarization
of the monochromator. It is necessary to digitize this system to
apply it to photon counting techniques of beam foil spectroscopy,
but it would give more meaninful results that method using long
time counting with fixed polarization (1) because it limits problems
due to variation of polarization due to foil ageing.

Another widely used method consists, instead of rotating a polarizer
between a fixed atom and a detector, to rotate the atom by a magnetic field while keeping fixed the polarizer and detector. This is
possible only in an isolated spin state since otherwise the effect
of magnetic field is much more complicated than a simple rotation
(see part 2). This method, first introduced in optical pumping by
Bell and Bloom (10), has already been applied to beam foil spectroscopy. A detailed study and geometrical interpretation of its use to
determine orientation and alignment parameters in a general polarization case has been given by Laloe et al (5). We extract from their
paper the physical explanation of the alignment ω and 2ω beats which
have caused trouble recently to some beam foil spectroscopists (11).
A substantially equivalent explanation was given earlier by Nedelec
(12), using the simplified representation of alignment given in
Fig. 1. If one chooses the axis of quantization along the magnetic
field, $<J_z>$ is time invariant and the time evolution of $<J_\pm>$ is

$$<e^{i \omega J_z t} \ J_\pm \ e^{-i \omega J_z t}> = <e^{\pm i \omega t} \ J_\pm>$$

so that the frequency ω corresponds according to equations (3) and
(4) to $M^{(2)}_{1\pm}$ (and $M^{(1)}_{1\pm}$), and the frequency 2ω to $M^{(2)}_{2\pm}$. To show this
geometrically choose first a state with $M^{(2)}_{2\pm} \neq 0$ and $M^{(2)}_{1\pm} = 0$ which
will give pure 2ω modulations. In this case O_z is one of the principal axis of the ellipsoïd of alignments which rotates about it.
The two extreme situations depicted in Fig. 2a correspond to the
cases in which the major and minor axis of the ellipsoïd situated
in the xOy plane are on the O_x direction, the observer being in the
O_y direction. Since every half turn the ellipsoïd becomes identical
to itself, the frequency observed is 2ω. One sees immediately that
as a function of the angle Ψ between O_z and the polarizer direction,
the signal is maximum at $\Psi = 0$, zero at $\Psi = \pi/2$ and one can show (5)
that it varies as $\sin^2\Psi$. Its amplitude is deduced from the difference between the two axis of the ellipse which lie in the xOy plane,
i.e. $M^{(2)}_{2+} = <J^2_x - J^2_y>$ when M_{2-} is zero as in the tilted foil confi-

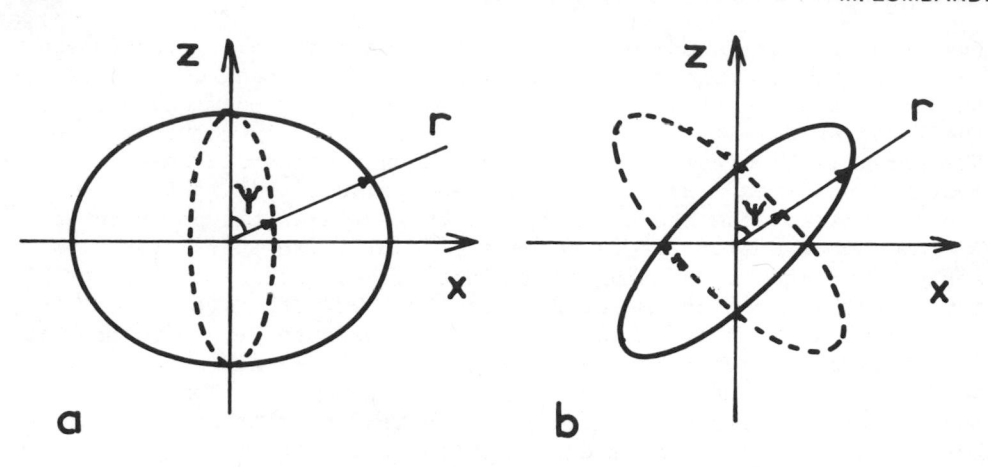

Fig. 2 : Rotation of alignment.

a) $M^{(2)}_{\pm 2} \neq 0$, $M^{(2)}_{\pm 1} = 0$

b) $M^{(2)}_{\pm 1} \neq 0$, $M^{(2)}_{\pm 2} = 0$

guration with magnetic field parallel to the beam. If conversely $M^{(2)}_{1\pm} \neq 0$ and $M^{(2)}_{2\pm} = 0$, one axis of the ellipse is at 45° of the O_z axis (in the xOz plane if only $M^{(2)}_{1\pm} \neq 0$) and its section by the xOy plane is a sphere. In this case after half a turn one go from the full line to the dotted line situation of Fig. 2b and it is only after a full turn that the ellipsoïd coïncides with itself. The frequency of the modulation is then ω. Its amplitude is zero if $\Psi = 0$ of $\pi/2$ and maximum at $\Psi = \pm \pi/4$ and it varies with Ψ as $\sin 2\Psi$ (5). An analogous argument shows that orientation, being a vector, can coïncide with itself only after a full turn so that its frequency is always ω. One can with this method easily visualise also what happens when one varies the direction of the magnetic field.

2 - Creation of orientation.

We describe in this section some non trivial methods of creating orientation. By non trivial we mean here other than exciting an atom with circularly polarized light. The essential point is to note that orientation means a net rotation of the atom around some axis (Fig. 1) and we will focus our discussion on symmetry considerations and illustrative examples.

2.1. Methods using magnetic fields.

The magnetic field is a good candidate to make orientation since it
has by itself the correct symmetry, being an axial vector with a
sense of rotation defined by the magnetizing current. However it is
well known that a magnetic field acting on an isolated aligned level does not orient it, but induces a Larmor precession of the alignment. The question is then : is this an unlucky peculiar circums-
tance or is there any fundamental principle which can distinguish
in any situation when there can be creation of orientation and when
there can be only rotation of the alignment ? The answer is that
the fundamental principle involved is time reserval T. Indeed $\vec{J}$ is
odd under T so that the Hamiltonian γ $\vec{H}.\vec{J}$ (and even more general
Hamiltonians representing inhomogeneous magnetic field) and the
part of the density matrix which describes orientation are odd under T, whereas a purely aligned density matrix is even under T.
The equation of motion of the density matrix :

$$\frac{d\rho}{dt} = -\frac{i}{\hbar} [\mathcal{H},\rho]$$

shows that if at time t the density matrix is aligned
($\rho(t) = K \rho(t) K^+$, with K operator of time reserval (13)), at time
t + dt

$$K \rho(t+dt) K^+ = K (\rho(t) - \frac{i}{\hbar} [\mathcal{H},\rho(t)] \, dt) \, K^+$$

$$= K \rho(t) K^+ + \frac{i}{\hbar} [K \mathcal{H} K^+, K \rho(t) K^+] \, dt,$$

since K i K^+ = -i because of the antilinearity of K. It follows
that if K $\mathcal{H}$ K^+ = - $\mathcal{H}$, K $\rho(t+dt)$ K^+ = $\rho(t+dt)$ so that the atomic
state remains aligned. The conclusion is that to create orientation
in an aligned (or isotropic) state, there must be a T invariant
Hamiltonian involved. We shall give now a few examples of "cooperation" in which the sense of rotation is defined by the axial
character of H whereas an other, T invariant, Hamiltonian, which by
itself does not define a sense of rotation, couple states of opposite parity under K.

The first is the creation of a beam of oriented particules in a
Stern-Gerlach apparatus. One needs a slit to separate one particular m_J beam, and it is the Hamiltonian representing the action of
this slit which is the needed T invariant Hamiltonian.

A less obvious example is the experiment made by Lehmann (14). He
excites an atomic level with a "broad line" beam of unpolarized
light directed along the axis O_z of the field H which creates an
alignment $M^{(2)}$ of the excited J level. If this level has an hyperfine structure a $\vec{I}.\vec{J}$ of the same order as the magnetic energy

$\gamma \vec{J}.\vec{H}$, there appears during the lifetime of the excited level an orientation along O_z both of the electronic $\vec{J}$ and of the nuclear $\vec{I}$ moments. These orientations are opposite since the projection $F_z = I_z + J_z$ is conserved because the total system is invariant by rotation around O_z. Fig. 3 shows the resulting nuclear orientation when the polarization of the pumping beam is either $\sigma+$, $\sigma-$ (ordinary optical pumping or σ (unpolarized light). This phenomenon shows that the effect on spin $\vec{I}$ of the spin $\vec{J}$ through a "magnetic" hyperfine interaction a $\vec{I}.\vec{J}$ is far different from the effect of a magnetic field which would be unable to orient it. This is one of the reasons why in N.M.R. studies (15) the theory of relaxation which treats the effect of the lattice as the effect of a fluctuating magnetic field is unable to explain how is reached the thermal equilibrium, which implies an orientation along the field due to the Boltzmann factor on populations.

An other experiment made more recently by the same group (16) is the orientation of a molecular level by combination of magnetic and natural predissociations. The magnetic predissociation is induced by coupling the observed state $|1>$ to an other, dissociative, state $|2>$ by the Hamiltonian $\gamma (L_z + 2 S_z) H$. The probability of magnetic dissociation is, according to Fermi's golden rule, given by the square of the matrix element $|<1|(L_z + 2 S_z)|2> \gamma H|^2$. The lifetime of the two $\pm$ m levels is then the same and there is no induced orientation. But if there is a matrix element of the natural dissociation Hamiltonian $\mathcal{H}_N$ between the same two levels, the probability of dissociation is given by :

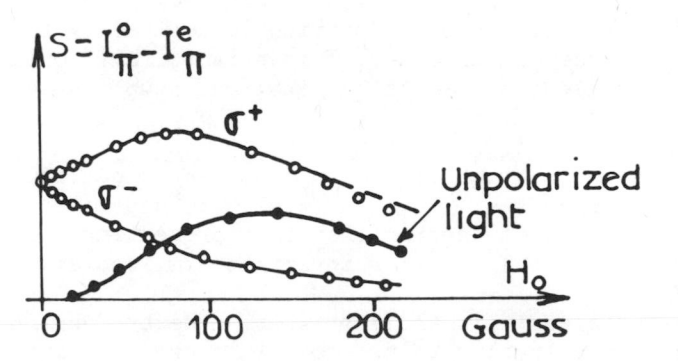

Fig. 3 : Creation of orientation by hyperfine structure anticrossing

$$|<1|\gamma\ (L_z + 2\ S_z)\ H + \mathcal{H}_N\ |2>|^2 = |<1|\gamma\ (L_z + 2\ S_z)\ H\ |2>|^2$$
$$+ |<1|\mathcal{H}_N\ |2>|^2 + <1|\gamma\ (L_z + 2\ S_z)\ H\ |2><1|\mathcal{H}_N\ |2> + C_0\ C_0$$

The cross product term, linear in J_z, gives different dissociative lifetimes to the ± m sublevels an then a net circular polarization of the light emitted : Fig. 4. Here also the orientation is maximum when the two "collaborating" Hamiltonians are equal. The amplitude of this maximum gives the ratio of dissociative to radiative lifetimes. Since the two Hamiltonians must connect the same two levels, one cannot replace the natural predissociation by an electric predissociation induced by a $-\vec{d}.\vec{E}$ Hamiltonian which connects states of opposite parity.

An example of combined electric/magnetic field creation of orientation is the method used by Lamb (17) to select the $m_S = + 1/2$ spin state of the $S_{1/2}$ metastable level of H by coupling the $S_{1/2,-1/2}$ state to the $P_{1/2,+1/2}$ state by an electric field when an applied magnetic field makes them cross. The difference with the previous case is that the useful part of the magnetic field is not its off-diagonal matrix element between the two states (which vanish) but its diagonal matrix element within them.

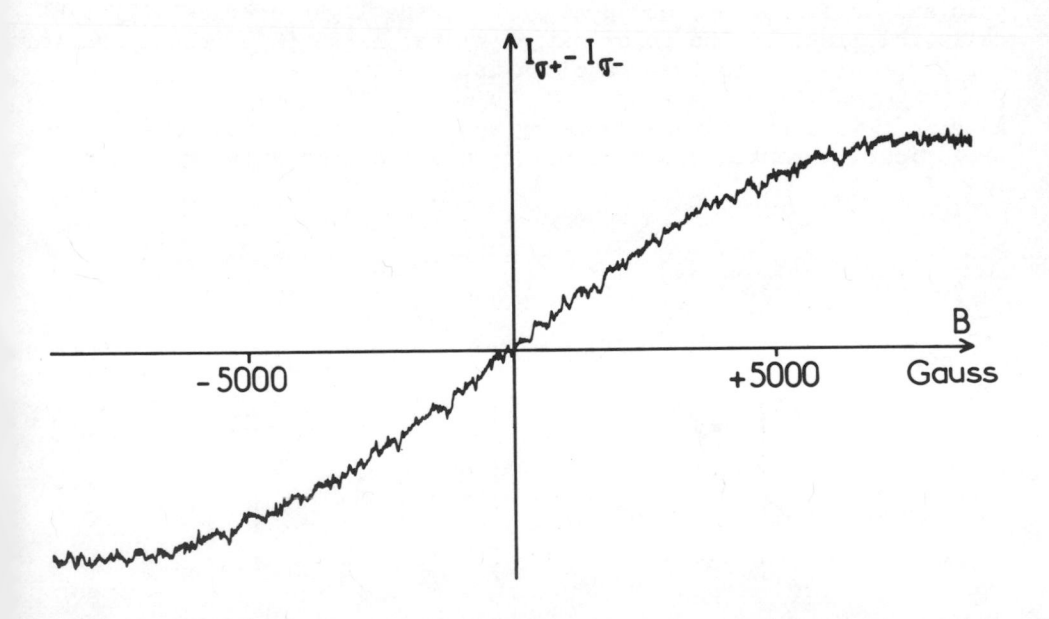

Fig. 4 : Orientation by interference of natural and magnetic predissociation in A_2^+ laser excited I_2 (B $^3\Pi_{o+u}$ R(77) 40→2)

2.2. Orientation as combination of two alignments.

Since one T invariant Hamiltonian is needed in any case to create orientation, the next step is to try to eliminate the magnetic field. One is then faced with the problem of finding the way by which the sense of rotation defining the orientation is defined. We shall first study the case in which an aligned atom is submitted to an interaction having also the symmetry of alignment. These two alignments will be supposed to have the symmetry of revolution, since the more general case adds nothing new, and will be represented by a straight line, according to Fig. 1d. This means that these two alignments are invariant by rotation around their axis , inversion and time reserval. The orientation cannot have components in the plane defined by the two alignments since the system is invariant by reflexion on this plane while such orientation components would change sign in this reflexion. However the two senses of rotation around the perpendicular O_z to this plane which bring alignment 1 onto alignment 2 (Fig. 5) are clearly unequivalent unless the two alignments are either parallel or perpendicular. In those latter cases, reflexion in the plane perpendicular to any of the two alignments leaves the whole system invariant whereas it changes the sign of the orientation along O_z.

The first realisation of this symmetry scheme is the usual birefringent plate. If one sends a linearly polarized beam of light (the first alignment) on a birefringent plate (whose crystallographic axis defines the second alignment, the light emerges elliptically polarized if the polarization of the light is neither parallel nor perpendicular to the plate's axis.

An atomic version of this symmetry scheme is the action of an electric field, through second order Stark effect, on an aligned atom.

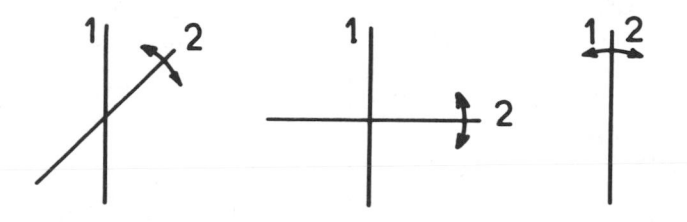

Fig. 5 : Orientation by combination of two alignments.

The Hamiltonian is in this case :

$$\mathcal{H} = \alpha \, E^2 \, (J_z^2 - \frac{1}{3} J^2)$$

The creation of orientation in an aligned level is prooved by the following operator equation :

$$\frac{d}{dt} <J_y> = - \frac{i}{h} <[J_y, \mathcal{H}]> = \alpha \, E^2 <J_x J_z + J_z J_x>$$

$$= \alpha \, E^2 \, (<J_{x/z}^2 - \frac{1}{3} \vec{J}^2> - <J_{x\backslash z}^2 - \frac{1}{3} \vec{J}^2 \,)$$

This shows that if the ellipse which is the intersection of $\mathcal{E}_{AL}$ with the xOz has not O_x and O_z as principal axis, i.e. if the alignment lies somewhere in between O_x and O_z, there appears an orientation along O_y.

This phenomenon was studied by Giroud and Lombardi (18) in a discharge where the alignment was produced by electron bombardment and the electric field by the external capacitors which sustained the discharge. It has been studied more recently by Nouh et al (19) in a beam (Na$^+$, Li$^+$, 100 keV) – gas (He 4^1D_2) experiment designed to measure the higher multipoles of the atomic state after excitation. In both cases however the electric field was, for experimental convenience, parallel or perpendicular to the beam induced alignment. One then used a magnetic field to rotate the alignment so that orientation could appears. Thus these experiments belong in fact to the preceding paragraph. One example of recent experimental curve is given in Fig. 6. The complicated shape of the signal is

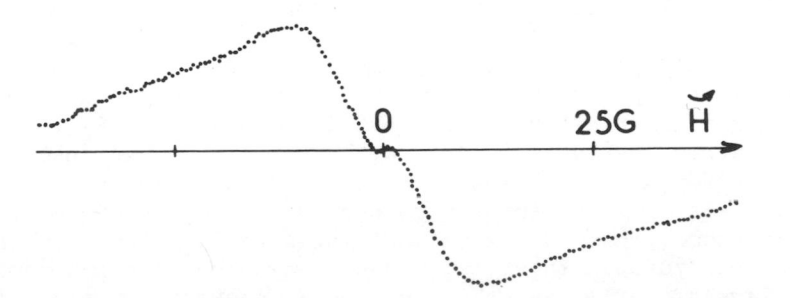

Fig. 6 : Orientation of the He 4^1D_2 level excited by a Na$^+$ (70 keV) beam : E = 250 V/cm $\perp$ beam $\perp$ H // observator.

due to the fact that one studies a J=2 level and is used to measure the k=4 multipole moment of the excitation.

Another recent experiment has been made in which the electric field, instead of being macroscopic, is the electric microfield equivalent to the interaction of two colliding atoms (19). This electric field is rapidly varying in direction and in strength during the collision but if the direction of the relative velocity of the two atoms is not distributed at random there remain a net average direction of the electric microfield. However, due to this bad averaging, the measured orientation is only 3 % of the original alignment so that much care had to be taken in order to avoid possible spurious effects. In this case the non randomness of the direction of collision was due to the recoil velocity of the He (4^1D_2) atoms excited by a 70 keV Na^+ beam. In this case however, the direction of alignment induced by excitation and the mean direction of the electric microfield are both the direction of the beam so that one needs also a magnetic field to rotate the alignment and create orientation. The experimental curves (Fig. 7) show first an increase of the orientation signal with pressure and then a decrease due to depolarization by collision.

An other interesting application of this principle was made by Dupont-Roc (20). He used the optical frequency a.c. electric field of an off-resonant pumping beam to induce orientation in the fundamental level of ^{201}Hg aligned by an other resonant pumping beam. The orientation was monitored by a third weak resonant probe beam. It is displayed Fig. 8 as a function of the intensity of the off resonant beam and of the angle between the polarization of pump and off resonant beams.

2.3. Orientation as combination of two vectors.

The next step is to study what happens if the two alignments are replaced by polar vectors (nothing new happens if only one is replaced) In this case the vector product of these two polar vectors defines an axial vector which is maximum when they are perpendicular, to the contrary to the preceding case in which it was zero. One experiment which uses this symmetry was suggested in reference (21). When an atom is excited by an ionic beam or by a foil, he has just after the excitation a non zero value of the dipole moment $<\vec{d}>$, i.e. the optical electron is preferably before or behind the nucleus. For ordinary atoms, this $<\vec{d}>$ value oscillates rapidly and is averaged out. But on hydrogenoïd atoms, the $<\vec{d}>$ value between two opposite parity levels oscillates only slowly and, under the action of a first order Stark Hamiltonian $-\vec{d}.\vec{E}$ one gets creation of orientation through :

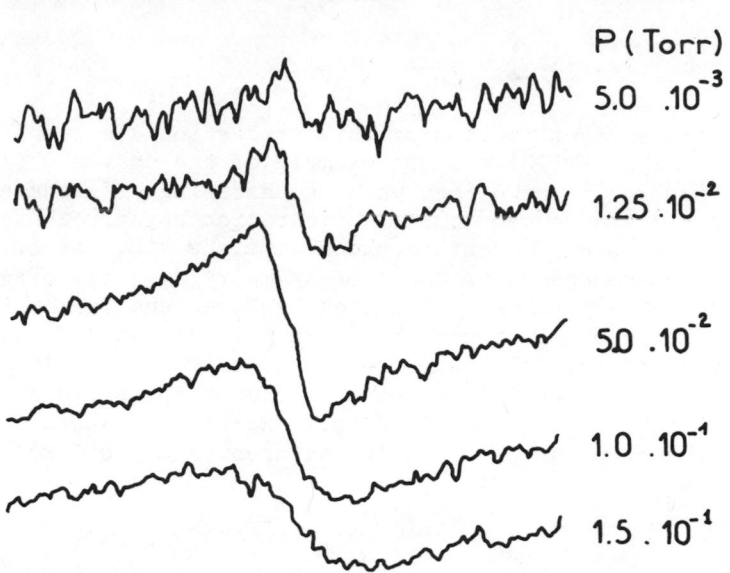

P (Torr)

5.0 .10^{-3}

1.25 .10^{-2}

5.0 .10^{-2}

1.0 .10^{-1}

1.5 . 10^{-1}

Fig. 7 : Orientation by anisotropic collision.

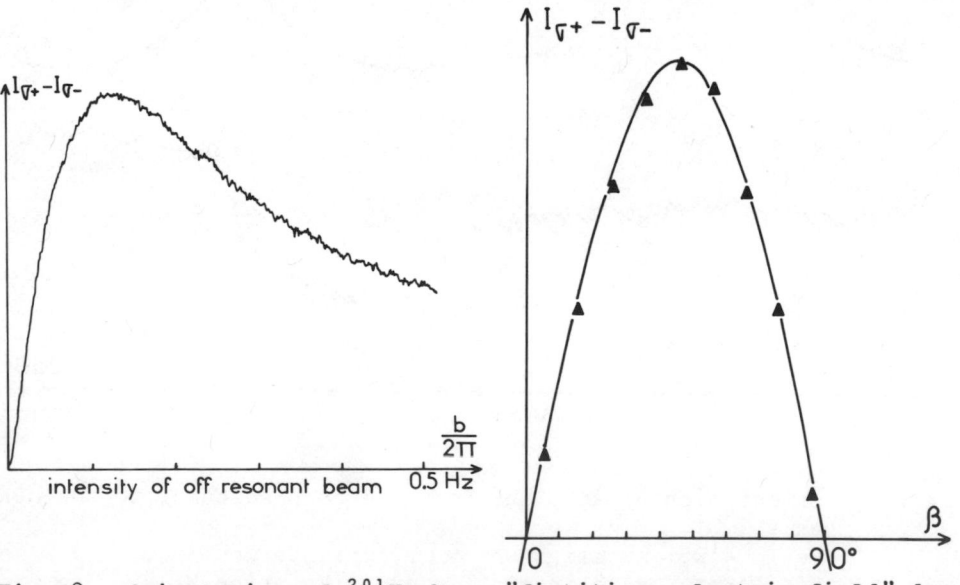

$I_{\sigma+} - I_{\sigma-}$

$I_{\sigma+} - I_{\sigma-}$

$\dfrac{b}{2\pi}$

intensity of off resonant beam   0.5 Hz

β

0   90°

Fig. 8 : Orientation of ^{201}Hg by a "fictitions electric field" due
to an off resonant pumping beam.

$$\frac{d\vec{J}}{dt} = -\frac{i}{h} [\vec{J}, -\vec{d}.\vec{E}] = \vec{d}_\Lambda \vec{E}$$

The experimental verification of this prediction has been made by
Giroud and Lombardi (22) and one example of the curves obtained is
given in Fig. 9. It shows that when the exciting Na$^+$ beam and the
electric field are perpendicular, orientation appears in the hydrogenoïd He$^+$ n=4 level but not on the neutral He 4^1D_2 level. Other
examples of phenomena using the same symmetry are the coïncidence
electron photon experiments discussed by Macek and Jaecks (23),
and polarization of electrons which are diffused at a fixed angle
by an atomic (or molecular or even solid) target (24). It has been
suggested (25) than this phenomenon is also responsible of the circular polarization in beam tilted foil experiments which show great
values of the circular polarization at grazing angles (26).

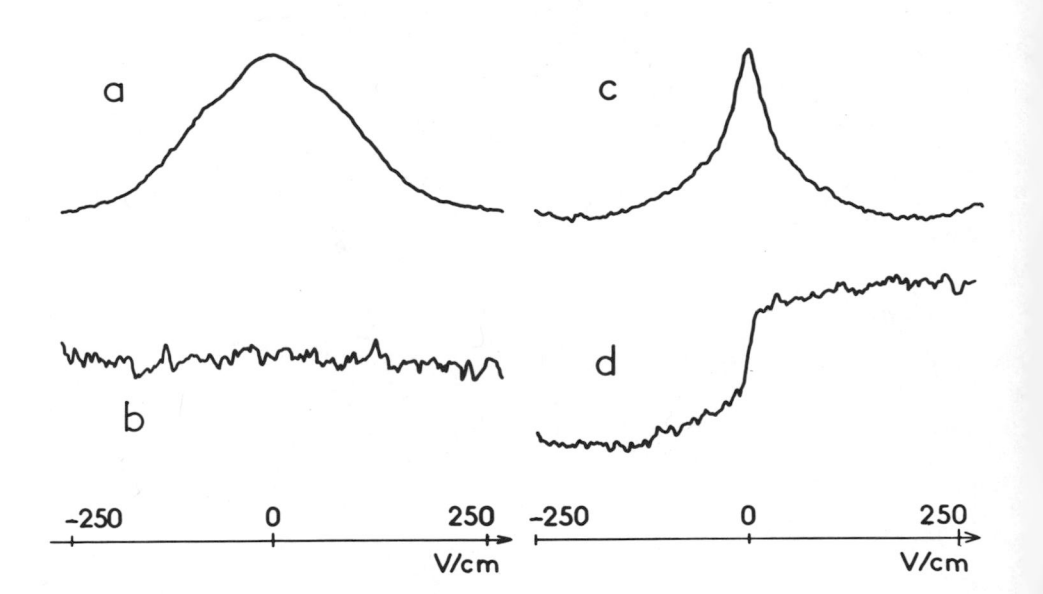

Fig. 9 : Orientation by an electric field perpendicular to the beam
         axis
         He I 4^1D_2 a) alignment   b) orientation × 16
         He II n=4   c) alignment   d) orientation × 4

(1) H.G. BERRY, L.J. CURTIS, D.G. ELLIS and R.M. SCHECTMAN,
     Phys. Rev. Lett. 32, 751 (1974)

(2) U. FANO and J. MACEK, Rev. Mod. Phys. 45, 553 (1973)

(3) D.G. ELLIS, J.O.S.A. 63, 1232 (1973)

(4) U. FANO, J. Math. Phys. 1, 417 (1960)

(5) F. LALOE, M. LEDUC and P. MINGUZZI, J. Phys. 30, 277 (1969)

(6) M. BORN and E. WOLF, Principle of Optics, Pergamon Press, 1959
     L. LANDAU and E. LIFCHITZ, Théorie du champ, Ed. de la Paix
     Moscou

(7) M.I. D'YAKONOV, Sov. Phys. J.E.T.P., 20, 1484 (1965)
                     Opt. Spectrosc. 19, 372 (1965)

(8) M. PAVLOVIC and F. LALOE, J. Phys. 31, 173 (1970)

(9) E. CHAMOUN, M. CARRE, M. GAILLARD, M. LOMBARDI, to be published
     E. CHAMOUN, Thesis, Grenoble 1973

(10) W.E. BELL, A.L. BLOOM, Phys. Rev. 107, 1559 (1957)

(11) D.A. CHURCH, W. KOLBE, M.C. MICHEL and T. HADEISHI, Phys. Rev.
     Lett. 33, 565 (1974)
     D.A. CHURCH, M.C. MICHEL and W. KOLBE, Phys. Rev. Lett. 34,
     1140 (1975)

(12) O. NEDELEC, Thèse, Grenoble, 1966

(13) A. MESSIAH, Mécanique Quantique, Dunod, 1962

(14) J.C. LEHMANN, Phys. Rev. 156, 153 (1969)

(15) A. ABRAGAM, The Principles of Nuclear Magnetism

(16) M. BROYER, J. VIGUE, J.C. LEHMANN, Chem. Phys. Lett. 22, 313
     (1973)

(17) W.E. LAMB, Phys. Rev. 85, 259 (1952)

(18) M. LOMBARDI and M. GIROUD, C.R. Acad. Sci. B 266, 60 (1968)
     M. LOMBARDI, J. Phys. 30, 631 (1969)

(19) NOUCH, E. CHAMOUN, M. LOMBARDI, M. GAILLARD, to be published

(20) C. COHEN-TANNOUDJI, J. DUPONT-ROC, Opt. Comm. 1, 84 (1969)

(21) M. LOMBARDI, M. GIROUD and J. MACEK, Phys. Rev. A 11, 1114 (1975)

(22) M. GIROUD, M. LOMBARDI, to be published

(23) J. MACEK and D.H. JAECKS, Phys. Rev. A 4, 2288 (1971)

(24) J. KESSLER, Rev. Mod. Phys. 41, 3 (1971)

(25) M. LOMBARDI, to be published

(26) H.J. ANDRÄ, Phys. Lett., to be published

QUANTUM BEATS   IN THE ELECTRIC FIELD QUENCHING OF METASTABLE
HYDROGEN*

G.W.F. Drake$^+$ and A. van Wijngaarden

Department of Physics, University of Windsor

Windsor, Ontario, Canada.   N9B 3P4

The observation of quantum beats in the Ly-α radiation of
atomic hydrogen is one of the fundamental experiments made possible
by the development of fast atomic beam technology.   Accurate
experimental results for this simple, well-understood atomic system
provide an important testing ground for time-dependent theories
applied to more complicated systems.   In this paper, we discuss the
Ly-α 'Stark beats' produced by the coherent decay of the perturbed
2s and 2p states following the rapid switching-on of an electric
field.   A detailed comparison is made between the experimental
intensity pattern and that obtained from a priori theoretical
calculations.

The quantum beat pattern is dominated by large oscillations
near the Stark shifted $2s_{1/2}$ - $2p_{1/2}$ frequency ($\sim$1000 MHz).   In
addition, the envelope of the oscillations is modulated by the
hyperfine structure of the $2s_{1/2}$ state ($\sim$180 MHz) and each peak is
structured by rapid oscillations near the $2s_{1/2}$ - $2p_{3/2}$ frequency
($\sim$10,000 MHz).   These are not single frequencies, but rather groups
of frequencies arising from the Stark and hyperfine splittings
of the states.   The effects have been observed previously by
several authors[1-4] using the beam-foil excitation technique and
the qualitative features of the beat pattern explained (for a
review, see Andrä[5]), but few precise comparisons with theory are
available.   One of the difficulties with beam-foil measurements
is that the initial state amplitudes as the beam emerges from the
foil are poorly known and are often adjusted to fit the data.
Since this adjustment alters the phase relationships among the
various frequency components, the comparison with theory becomes
somewhat arbitrary.   In addition, our calculations show that the
rapid fine-structure oscillations are of largest amplitude when

749

observed in a direction parallel to the applied field. This
implies that the field should be perpendicular to the beam. It
has been found that this geometry is difficult to produce in beam-
foil experiments without large distortions of the field near the
foil which destroy the beats.[2,6,7] Magnetic fields could be used,
but this introduces further complications of interpretation due to
the additional Zeeman splittings of the states.[3]

In the present experiment, a beam of hydrogen atoms in the
$2s_{1/2}$ state is prepared by passing a mono-energetic proton beam
through a gas cell containing H_2 and allowing sufficient time
for all other states to decay to the ground state. After
traversing a small prequenching field to remove the remaining
protons, the collimated beam enters the quenching cell shown in
Fig. 1.   The cell is basically a parallel plate capacitor with
narrow slots in the plates to allow observation of the light
emitted parallel to the field. The 1.2 mm diameter beam enters
through a small hole 1.5 mm in diameter in the plastic end plate.

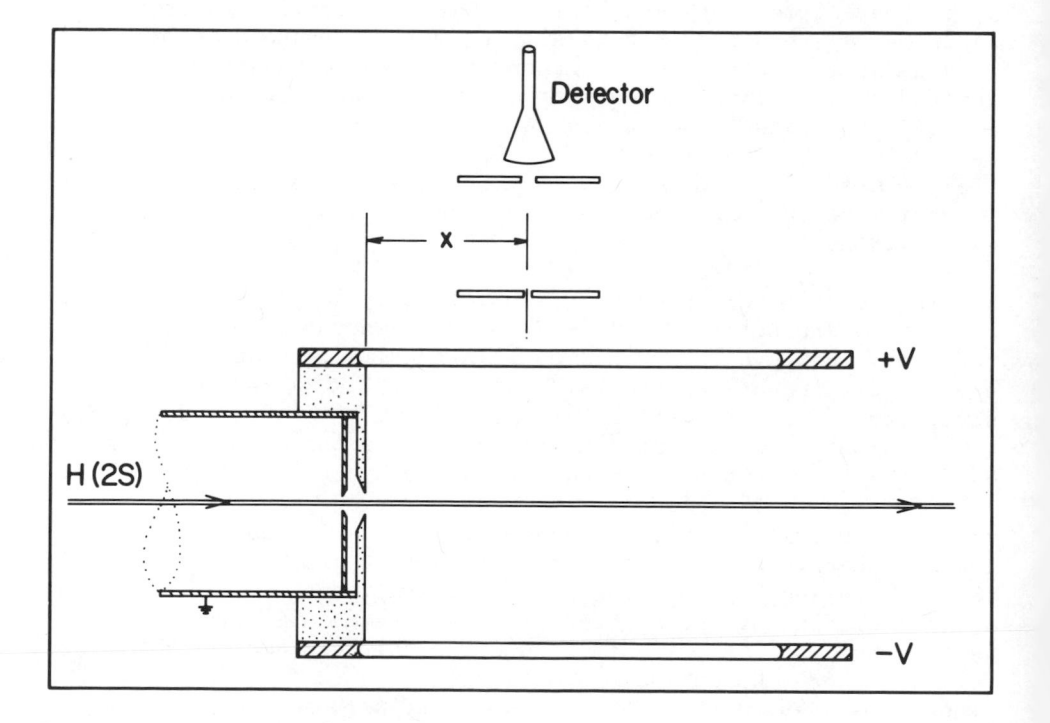

Fig. 1.   Diagram of the apparatus

The fringing field along the beam axis is greatly reduced by vacuum depositing a thin carbon film on the inside of the end plate such that it makes electrical contact with the parallel capacitor plates and provides a resistance of a few MΩ between them. Except for the perturbing effect of the hole, the current flowing through the film produces the same electrical potential at each point on the end plate as in the free space inside the capacitor. The electric field just inside the end plate is therefore almost the same as if the capacitor did not terminate. With this arrangement and a 120 keV beam (velocity ≃ 4.8 x 10^8 cm/sec), a uniform field perpendicular to the beam is switched on sufficiently fast to observe the 10,000 MHz fine-structure oscillations.

The beats were observed by translating the slit system and detector shown in Fig. 1 parallel to the beam. The slit widths 0.2 mm and 0.15 mm respectively for the inner and outer slits were

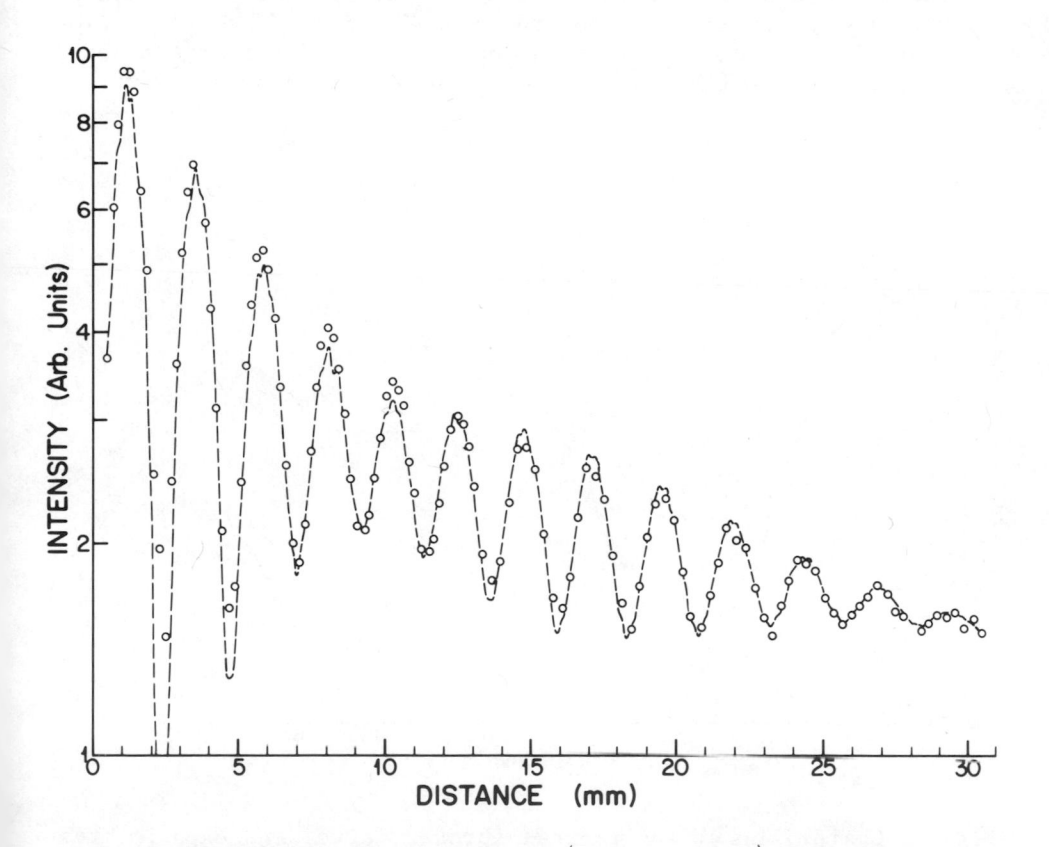

Fig. 2. Comparison of theoretical (broken curve) and experimental (circles) quantum beats for a beam energy of 47 keV and an electric field strength of 167 V/cm.

chosen to minimize Fresnel diffraction effects, which become
important at these slit widths. The inner and outer slits were
mounted 34 mm and 118 mm respectively from the beam axis. A
constant effective counting time was determined by using a second
detector to monitor the radiation from a fixed section of the beam.

The theoretical intensity can be written as a function of the
polarization vector $\hat{e}$ of the emitted photon in the form[8],[9]

$$I(\hat{e},t) = |\hat{e}\cdot\hat{F}|^2|A(t)|^2 + |\hat{e} \times \hat{F}|^2|A'(t)|^2$$

where $\hat{F}$ is a unit vector pointing in the electric field direction.
Only the $|\hat{e} \times \hat{F}|^2$ term contributes in our experiment since the
direction of observation is parallel to the field direction.
Taking $\hat{F}$ as the quantization axis, then the amplitudes $A(t)$ and
$A'(t)$ arise from transitions with $\Delta M = 0$ and ± 1 respectively.
They were calculated non-perturbatively by evaluating the time-
dependent Green's function connecting the states before and after
entering the field as described in refs. 8 and 9. The calculation
includes all sixteen $2s_{1/2}$, $2p_{1/2}$ and $2p_{3/2}$ hyperfine states and
assumes that initially, the beam contains an incoherent mixture

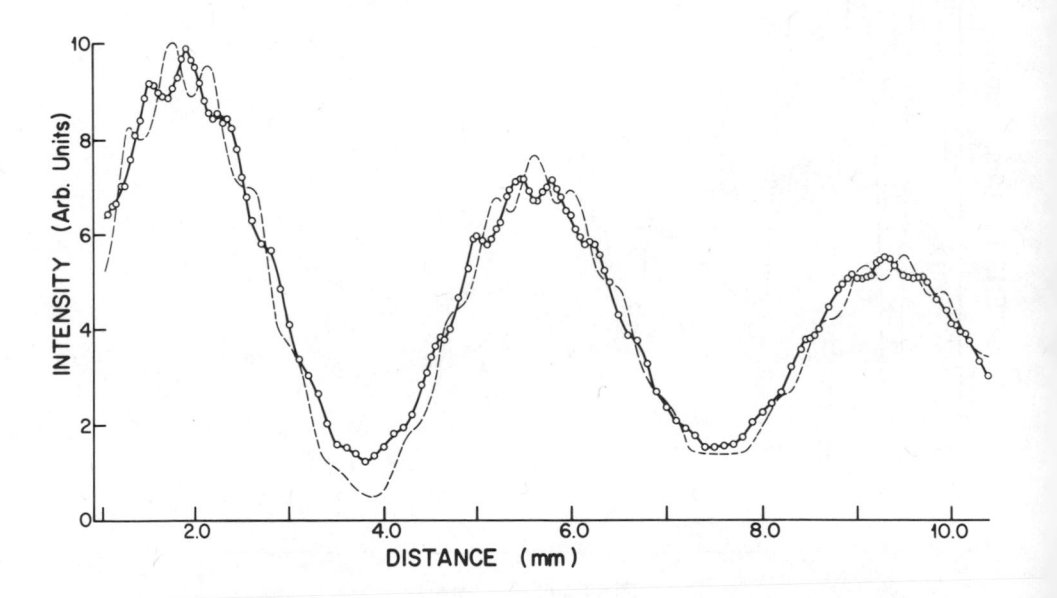

Fig. 3. Comparison of theoretical (broken curve) and experimental
        (circles) quantum beats for a beam energy of 120 keV and
        an electric field strength of 167 V/cm. The solid curve
        represents an average of the experimental data.

of the four $2s_{1/2}$ hyperfine states with equal statistical weights. This is a reasonable assumption since the metastable atoms are formed by charge exchange in a gas cell a long distance upstream from the quenching region. The theory remains valid for all field strengths, provided that the field is not so strong that mixing with states of higher n becomes important. The field is assumed to be switched on suddenly.

The experimental points corrected for ~3% noise at the peaks are compared with the theoretical curves in Figs. 2 and 3. The slit function of the detector corrected for Fresnel diffraction is folded into the theoretical curves. Each experimental point contains a minimum of 2000 counts obtained in an average counting time of 100 sec. The statistical error is roughly the size of the points. The field strength for both runs is 167 V/cm, but the beam energy in Fig. 2 is lower than in Fig. 3 so that more oscillations can be seen within the travel range of the detector. In both figures, the only adjustable parameters are an overall vertical scaling factor and a small (<0.2 mm) horizontal shift of origin. The horizontal shift is consistent with the uncertainty in the absolute location of the beam entrance hole along the beam axis. Fig. 2 shows clearly the modulation of the $2s_{1/2}$-$2p_{1/2}$ oscillations by hyperfine structure. On this scale, the agreement with theory is excellent.

Fig. 3 shows in higher resolution the fine-structure on the first three peaks. Although these results are still preliminary, the most significant difference between theory and experiment is the phase shift of the rapid oscillations. This is thought to arise from the finite time over which the field is switched on, and a possible over-shoot in field strength as the beam passes through the entrance hole. The phase shift is noticeably affected by altering the detailed design of the end plate. For example, the plastic end plate should be made thin (0.7 mm) near the entrance hole, and coated on the outside surface with a grounded conducting layer. The phase shift changed slightly depending on whether or not the outer layer made electrical contact with the inner carbon layer around the periphery of the hole. Also, a recent run with an entirely different end plate yielded results which did not contain a phase shift. Further work on this problem is in progress.

In summary, we have developed a method for switching on rapidly a uniform electric field perpendicular to the beam. Except for the phase shift of the rapid fine-structure oscillations, the results provide a precise verification of the theory of Stark beats in a case where the initial state amplitudes are accurately known. Phase shifts can be produced either by local field irregularities or by altering the initial state amplitudes. The presence of a phase shift in this experiment emphasizes the need

for care in the interpretation of beam-foil data where the initial
state amplitudes are not known.

* Research supported by the National Research Council of Canada.
+ Alfred P. Sloan Foundation Fellow.

REFERENCES

1.   I.A. Sellin, C.D. Moak, P.M. Griffin and J.A. Biggerstaff,
        Phys. Rev. 188, 217 (1969); I.A. Sellin, J.R. Mowat,
        R.S. Peterson, P.M. Griffin, R. Laubert and H.H.
        Hazelton, Phys. Rev. Letters, 31, 1335 (1973).
2.   M.J. Alguard and C.W. Drake, Phys. Rev. A 8, 27 (1973).
3.   H.J. Andrä, P. Dobberstein, A. Gaupp and W. Wittmann, Nucl.
        Instr. and Meth. 110, 301 (1973).
4.   A. Gaupp, H.J. Andrä and J. Macek, Phys. Rev. Letters, 32,
        268 (1974).
5.   H.J. Andrä, Physica Scripta, 9, 257 (1974).
6.   E.H. Pinnington, H.G. Berry, J. Desesquelles and J.L. Subtil,
        Nucl. Instr. and Meth. 110 , 315 (1973).
7.   H.J. Andrä, Phys. Rev. A 2, 2200 (1970).
8.   G.W.F. Drake and R.B. Grimley, Phys. Rev. A 11, 1614 (1975).
9.   G.W.F. Drake, P.S. Farago and A. van Wijngaarden, Phys.
        Rev. A 11, 1621 (1975).

THE SURFACE INTERACTION IN BEAM FOIL SPECTROSCOPY

H. G. Berry

Department of Physics, University of Chicago, Chicago, Ill. 60637

and Argonne National Laboratory, Argonne, Ill. 60439
and
L. J. Curtis, D. G. Ellis, R. D. Hight, and R. M. Schectman

Department of Physics and Astronomy, University of Toledo, Toledo,

Ohio 43606

Abstract

We review the measurements of the changes in light polarization in the beam-foil source when the foil tilt angle is varied. Comparisons are made with theories of the final surface interaction.

1. Introduction

The passage of fast heavy ions through solids has yet to be described in terms of an accurate theoretical model. Some progress has recently been made on the understanding of the effective charge states of the moving ions, both experimentally[1] and theoretically,[2] but the states of binding of the outer shell electrons remain essentially unknown. In particular, possible variations of such effective charge and excitation states with differing solids have not been measured or predicted. Following the discovery of atomic alignment in the beam-foil light source[3] through the observations of linearly polarized quantum beats in field-free radiative decays, it was natural to investigate whether alignment measurements could give information on the states of excitation of heavy ions in solids. We discuss below some of the progress made in this direction since the last beam-foil conference.

755

The first necessary step is to describe the state of atomic alignment produced in the beam-foil excitation mechanism in terms of the (observed) light emitted in the radiative decay of the excited state. Ellis,[4] and Fano and Macek[5] have thus related the excited state density matrix or state multipole moments to the polarization and angular distribution of the emitted radiation. We briefly discuss how symmetry conditions of the excitation can be used to predict the various possible polarizations in the emitted light.

For a spherically symmetric source, the radiation is emitted isotropically and is unpolarized, and the source can be described by a single parameter — the number of excited atoms N. In quantum mechanical terms, since no direction is specified, all different angular momentum sub-levels are equivalent and we have statistical populations.

In Fig. 1, we show successive reductions in the symmetry of the excited light source which lead to the need for more parameters to adequately describe the source, which in turn affect the emitted radiation. In Fig. 1(b) is a cylindrically symmetric source, such as electron beam excitation of a gas. The $\hat{z}$-axis now differs from the $\hat{x}$ and $\hat{y}$ directions, and a second parameter, the alignment, is introduced. The cross-sections to sub-levels of different $|m_L|$ may now be different, and Percival and Seaton[6] have related these cross-sections to the fractional linear polarization of the emitted light. For example, for a $^1P \rightarrow {}^1S$ transition, the z-axis as direction of quantization, then the fractional linear polarization of light observed perpendicular to the z-axis is

$$P_L = \frac{I^{\parallel} - I^{\perp}}{I^{\parallel} + I^{\perp}} = \frac{\sigma(m = \pm 1) - \sigma(m = 0)}{\sigma(m = \pm 1) + \sigma(m = 0)} \cdots \cdots \cdots . \tag{1}$$

Until two years ago, the beam-foil light source was considered to be such a source, describable in terms of two parameters (N, P_L) for each excited state. One additional important quality is that all excitations occur at a time $t=0$ on the z-axis defined to better than 10^{-14} sec, which gives rise to the zero field quantum beats, an example of which is shown in Fig. 2. Basically, we have produced a state which is not an energy eigenstate of the free atom Hamiltonian.

However, the beam-foil interaction at $z=0$ also depends on the direction of the beam velocity $\vec{v}$. Thus, Eck[7] pointed out that the interaction need not be invariant under reflection in the x-y plane of the foil. He proposed a simple test of comparing the Lyman α decays of $n=2$ hydrogen in an electric field parallel and anti-parallel to the $+z$-axis. A phase shift of the electric-field induced Lamb shift quantum beats between $2s_{\frac{1}{2}}$ and $2p_{\frac{1}{2}}$ indicating that a

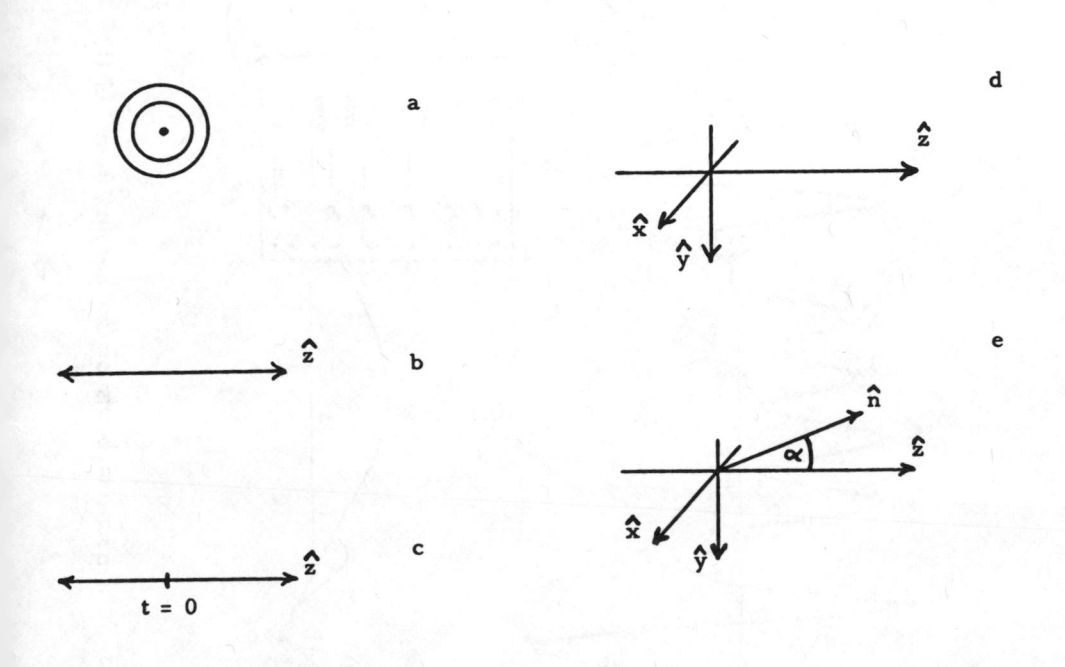

Fig. 1. Excitation source symmetries. a - spherical symmetry, b - cylindrical symmetry, c - excitation at t = 0, d - reflection assymmetry in x-y plane, e - loss of cylindrical symmetry.

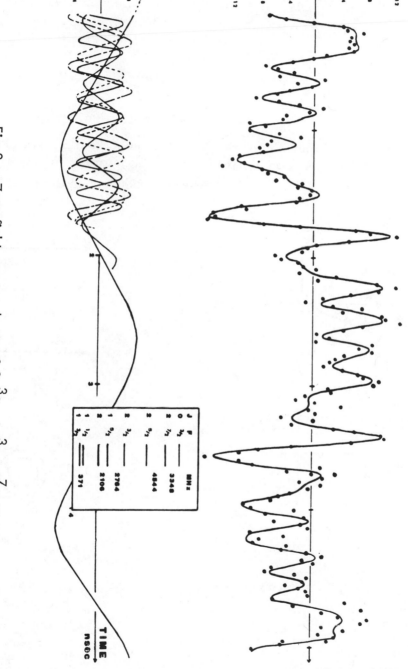

Fig. 2. Zero field quantum beats of 3s ^{3}S – 4p ^{3}P in ^{7}Li II.

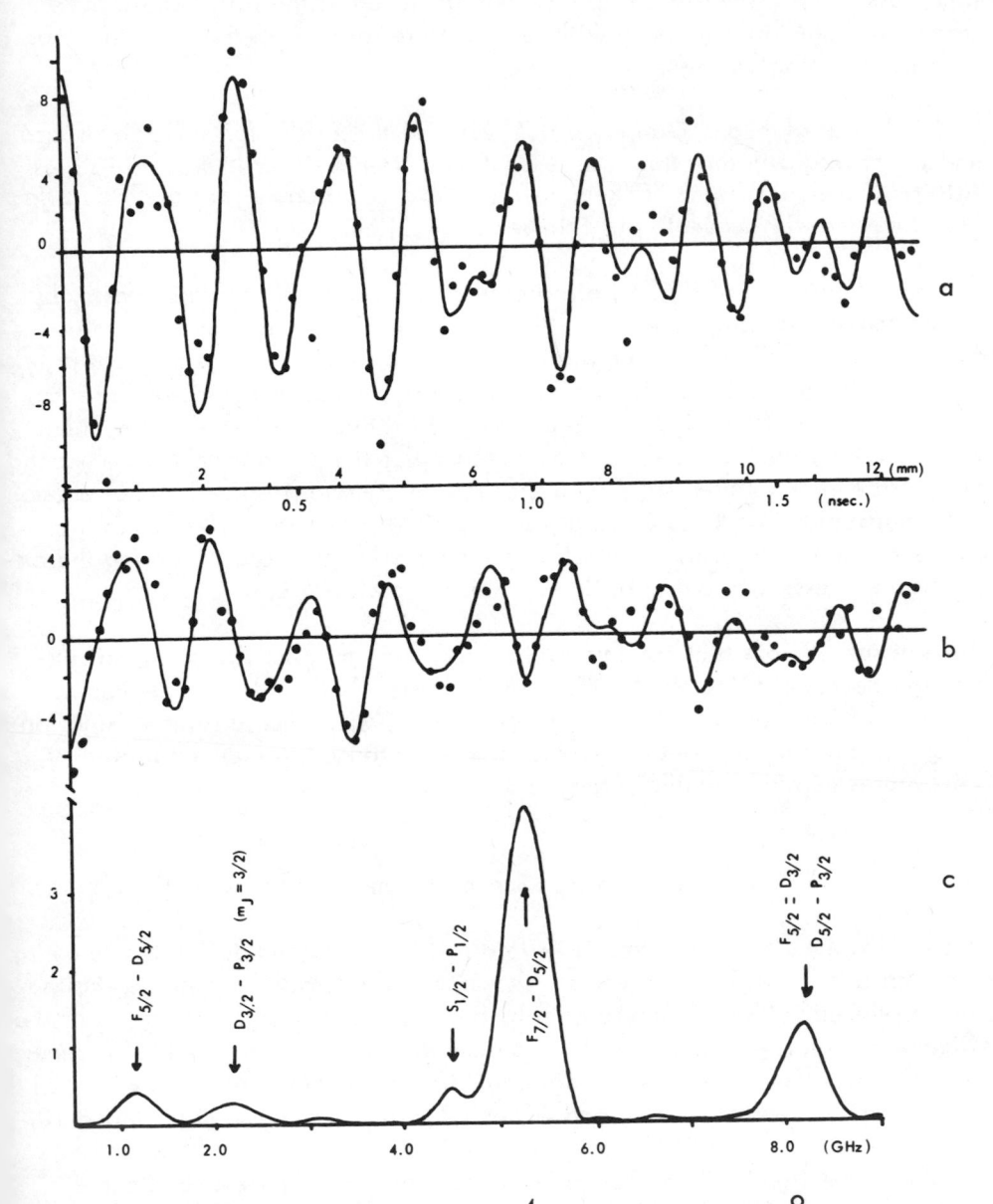

Fig. 3. Modulations in the decay of ^{4}He II n=3–4, 4686 Å in electric fields of $\pm$ 465 volts/cm in parts a and b (the solid lines are non-linear least-squares fits to sums of cosines), and the fourier transform of the difference curve in c. The modulation frequencies correspond to $m_J = \frac{1}{2}$ stark-shifted energy separations, except where noted.

superposition of these two states has been produced at the foil. Thus, the state is of non-definite parity with respect to reflections in the x-y plane, as well as not being an energy eigenstate.

Sellin et al.[8] and Gaupp et al.[9] verified this effect for n=2 of hydrogen, and we have shown that this is a general phenomenon[10] for hydrogenic states with observations of such "Eck-beats" in n=2,3 of HI and n=3,4 of ^{4}He II. In Fig. 3 we show an example of "Eck-beats" in ^{4}He II.

In the last part of Fig. 1 we indicate a further loss of source symmetry by tilting the foil so that its normal $\hat{n}$ is at an angle α to the beam-axis. An axial vector $\hat{n} \times \vec{v}$ can now be defined which corresponds to the possible pro-duction of atomic orientation, and consequently, circular polarized light may be emitted. Additionally, the atomic alignment becomes a three-component vector. Thus, a minimum of four parameters plus the population N are needed to describe each excited state. Farther source asymmetries need higher order state multipoles (see Refs. 4 and 5), but we shall here limit ourselves to the above examples where no external electromagnetic fields are disturbing the ex-citation process; such a description is then complete.

We should note that the last two examples assume that the beam-foil ex-citation depends on the final surface of the foil. Hence, these experiments are useful primarily to study this surface interaction. Hopefully, the isolation of surface interaction effects will also lead to information concerning the ionic states within the bulk of the foil.

2. The Tilted Foil Stokes Parameter Technique

In Fig. 4 we show the standard tilted foil geometry and define the relevant direction axes. We have previously shown[11] that we may describe an excited state produced in the foil interaction by a density matrix $\rho_{m\,m'}$, or ρ_q^k or the alignment vector $\underset{\sim}{A}^c$ and orientation parameter O^c. The three sets of param-eters are linearly related — m,m' are angular momentum projections, k,q$\leq$|k| are irreducible tensor indices, and $\underset{\sim}{A}^c$ and O^c are defined in Refs. 5 and 10.

Also, the light emitted from any excited state may be described com-pletely in terms of the four Stokes parameters I,M,C and S. Thus, with re-spect to a set of axes ξ, η, ζ with ζ along the observation direction and, in our specific case, ξ denoting the "parallel" direction and η the "perpendic-ular" direction, the Stokes parameters are I, the total intensity, equal to the sum of the components of plane polarized light $I^{||} + I^{\perp}$. M is the difference $I^{||} - I^{\perp}$, while C is the difference in the two plane polarized components ro-tated at 45° to $I^{||}$ and $I^{\perp}$, i.e., $C = I^{45°} - I^{135°}$. S denotes the net

circularly polarized light, $S = I_{r.h.} - I_{l.h.}$. The four parameters thus completely specify the polarization ellipse of the emitted light.[12]

The Stokes parameters are linearly related to all of the three sets of excited state parameters introduced above. Thus, for example,

$$M(\theta) = \sum_{k,q} a_{kq}(\theta) \cdot \rho_q^k \quad \cdots \cdots \qquad (2)$$

where the coupling coefficients a_{kq} depend upon the transition being observed
and the angle of observation. The a_{kq}' for a $^1P \rightarrow {}^1S$ transition are given in
Ref. 10, and they can, in general, be derived from Refs. 4 or 5. The tensor
component ρ_0^0 describes the total number of excited atoms, while $\rho_q^{k=1}$
(or O^c) describes the atomic orientation, directly proportional to the circular
polarization fraction S, and $\rho_q^{k=2}$ (or A^c) describes the atomic alignment.
For electric dipole emission without external fields, only tensor components of
$k \leqslant 2$ can be measured through this technique, while ρ itself may have un-
determined tensor components of rank $k > 2$.

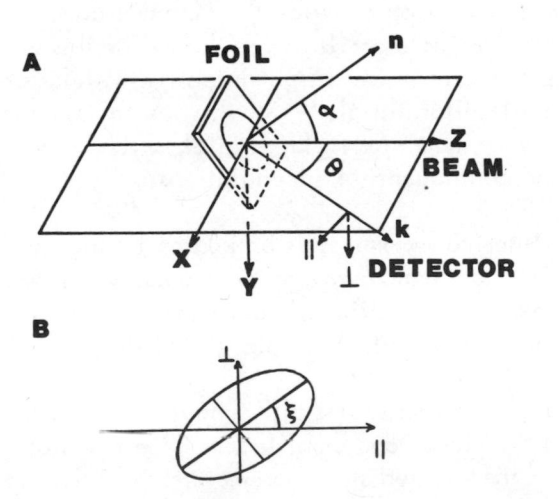

Fig. 4. A – Viewing geometry. The light vector k is in the $\hat{x}$-$\hat{z}$ plane,
perpendicular to the $\hat{n}$-$\hat{y}$-$\hat{z}$ plane at an angle θ to the z axis.
B – The polarization ellipse.

The Stokes parameters are measured by observing the light at a particular angle θ through a fixed retardation plate and a rotating polarizer. Instrumental polarization can be eliminated by the introduction of a Hanle depolarizer[13] immediately after the polarizer. The retardation plate may be removed to measure the linear polarization parameters.

Thin carbon foils were either mounted in holders of various tilt angles, or mounted on a rotatable x-axis (see Fig. 4) which allowed a continuous variation of tilt angle α. The rotation of the polarizer was controlled by an on-line ASI computer in the Argonne experiments [14] or a motor drive control system in the Toledo experiments[14]. Light collected at each step was normalized either to Faraday cup current, or to a total light yield monitor observing only the transition of interest.

The first measurements using this Stokes parameter technique[11,14,15] indicated a very large surface effect in the beam-foil excitation. That is, the changes in M, C, and S with foil tilt angle α were large compared with their values at $\alpha=0$, and the asymmetric surface interaction produced circular polarization fractions of up to 25%.

3. Magnetic Field Quantum Beat Measurements of Asymmetry Parameters

For the cylindrically symmetric $\alpha=0$ beam foil source, the excited state may be aligned relative to the beam z-axis. Thus, a perpendicular magnetic field will induce a precession of twice the Larmor frequency ω_L of the classical damped electric-dipole oscillator as the excited ion moves downbeam. The theory of these magnetic field light intensity modulations has been discussed in detail by Gaillard et al.[16] for the case of cylindrical symmetry. It should be noted that a magnetic field parallel to the beam axis will produce no precession, and consequently no modulations.

When the cylindrical symmetry is broken by tilting the foil, a parallel magnetic field will induce modulations of frequency ω_L when the q=1 (for k=1,2) terms of the density matrix ρ_q^k are non-zero, while the perpendicular magnetic field will induce both $2\omega_L$ and ω_L modulations.

Church et al.[17,18] and Liu et al.[19] have observed such magnetic field modulations, and the phases and amplitudes of the modulations have been described in terms of the excited state parameters.[5,18] Hence, both the Stokes parameter measurements, and the magnetic field quantum beats lead to the same experimental results — the alignment and orientation parameters of the source.

4. The Structure of Unresolved Multiplets

The first verification of atomic alignment in the beam–foil source was the observation[3] of quantum beats from unresolved multiplet structures. We can expect that the other excited state asymmetry parameters can also be determined through similar observations using filted foil excitation. Ellis[4] has developed the general theory, and shown that: (1) atomic orientation, $\rho_q^{k=1}$, can be measured through observations of the time-modulation of the fractional circular polarization, (2) the atomic alignment, $\rho_q^{k=2}$, describes the time-modulations of the fractional linear polarization, and (3) the relative beat amplitudes, in the case of multiple frequencies, should remain unchanged, for a given type of polarization, as the foil tilt angle is varied.

We have verified[20] these results for the hyperfine structure quantum beats of $3s^3S - 3p^3P$ in ^{14}N IV, which are shown in Figs. 5 and 6.

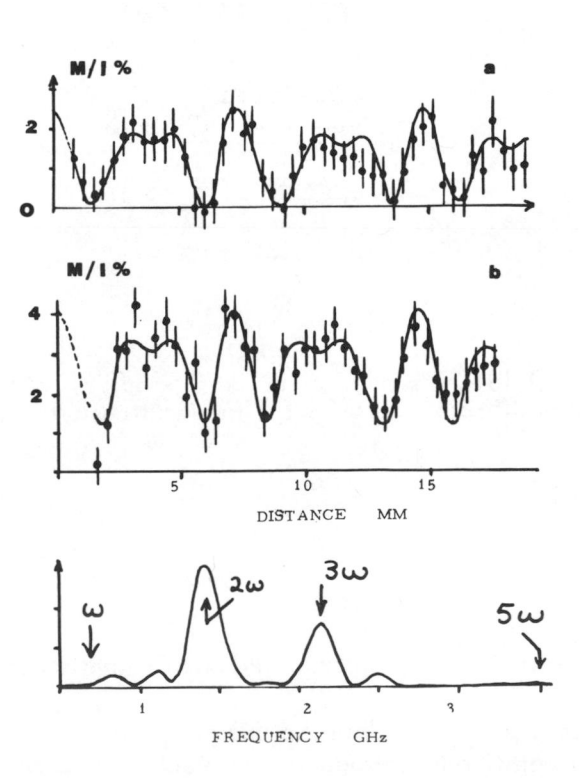

Fig. 5. Zero field quantum beats of ^{14}N IV $3s\,^3S - 3p\,^3P$, 3480 Å in linearly polarized light. a – 0° foil, b – 45° foil, c – fourier transform of b.

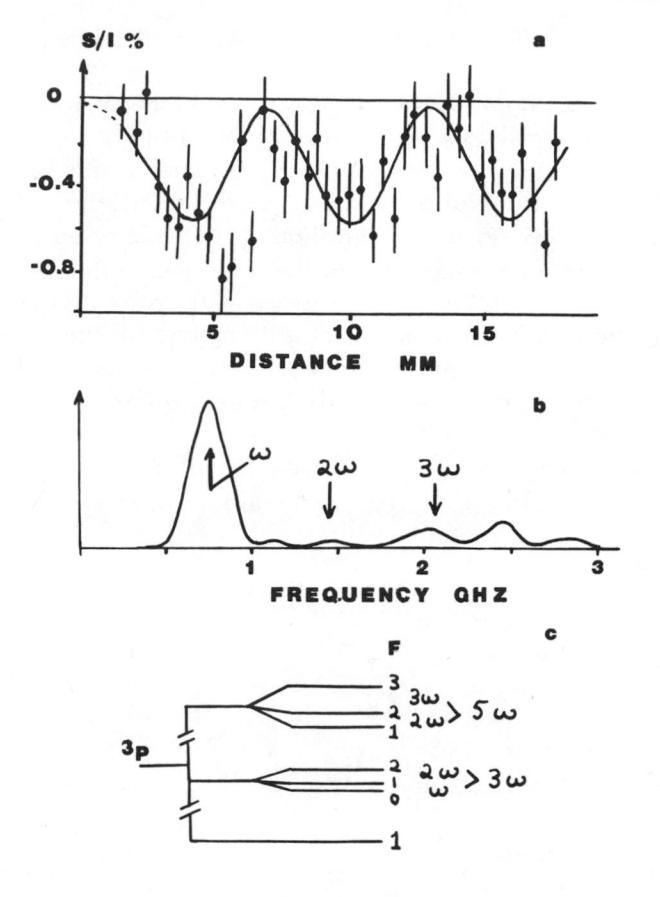

Fig. 6. a – zero field beats of ^{14}N IV 3s ^{3}S – 3p ^{3}P, 3480 Å, in circular polarized light, foil tilted at 45°, b – fourier transform, c – energy levels.

5. Theories of the Surface Interaction, and Comparisons with Experiment

Eck[21] has attempted to explain the initial results[11] of surface induced alignment and orientation by introducing an electric field perpendicular to the foil surface. This electric field removes the degeneracy in $|m_L|$ and transfers the alignment produced from excitation in the bulk into a coherence between states of different m_L. This is similar to the work of Lombardi[22] who has

shown that external electric fields skewed to an aligned excited state can produce orientation. However, Eck's theory, in particular for ^1P states, predicts that the total polarization $f_p = \sqrt{\{(M/I)^2+(S/I)^2+(C/I)^2\}}$ will be independent of foil tilt angle α. We have already shown[23] that f_p changes significantly with α and that Eck's simple theory must be modified.

Band[23] has included excited state production processes at the foil surface.
He considers the moving ion to have an "active" electron which may be excited through interactions with the foil electrons and also by the surface potential barrier as it leaves the foil. It is essentially the interference between
these two terms which gives rise to the orientation and alignment of the excited state. He obtains the following equations for the Stokes parameters of
light emitted in a $^1S-{}^1P$ transition

$$\frac{S}{I} = -E \sin 2\alpha \sin \left(\frac{\omega}{v \cos \alpha} \right) \qquad (3)$$

$$\frac{M}{I} = -E + F^2 \cos 2\alpha + 2E \sin^2 2\alpha \ \sin^2 \left(\frac{\omega}{2v \cos \alpha} \right) \qquad (4)$$

$$\frac{C}{I} = F^2 \sin 2\alpha - E \sin 4\alpha \ \sin^2 \left(\frac{\omega}{2v \cos \alpha} \right) \qquad (5)$$

where E, F^2 and ω are constants of the surface, and v is the beam velocity.
These equations are very similar to those of Eck,[21] but now the total polarization fraction f_p varies with α.

In Fig. 7, we compare Band's theory with our experimental data for
$2s{}^1S-3p{}^1P$ in He I at 246 keV beam energy. The results show partial agreement but definite discrepancies appear.

Herman[24] has calculated the change in the perpendicular foil excitation
matrix, $\rho(\alpha=0)$, for non-zero tilt angle α. He adds the contributions due to
collisions between the moving ion and those surface atoms within its forward
hemisphere as it leaves the surface. This introduces an addition due to the
surface atoms on one side and a subtraction due to the lack of surface atoms on
the other side of the moving ion. He then shows that the rank one density matrix components ($\rho k=1$, 0^C, A_{1+}^C) should vary as tan α, while the
rank two components ($\rho^{k=2}$ or A_{2+}^C) should remain fixed.

Thus, M/I should be constant, as is clearly not the case for the
transition shown in Fig. 7. The $k=1$ component, S/I does not vary as strongly
as tan α. However, our published results[23] for Ne III 2866 Å, $3s'{}^1D-3p'{}^1F$,
do show a reasonable agreement for M/I and S/I.

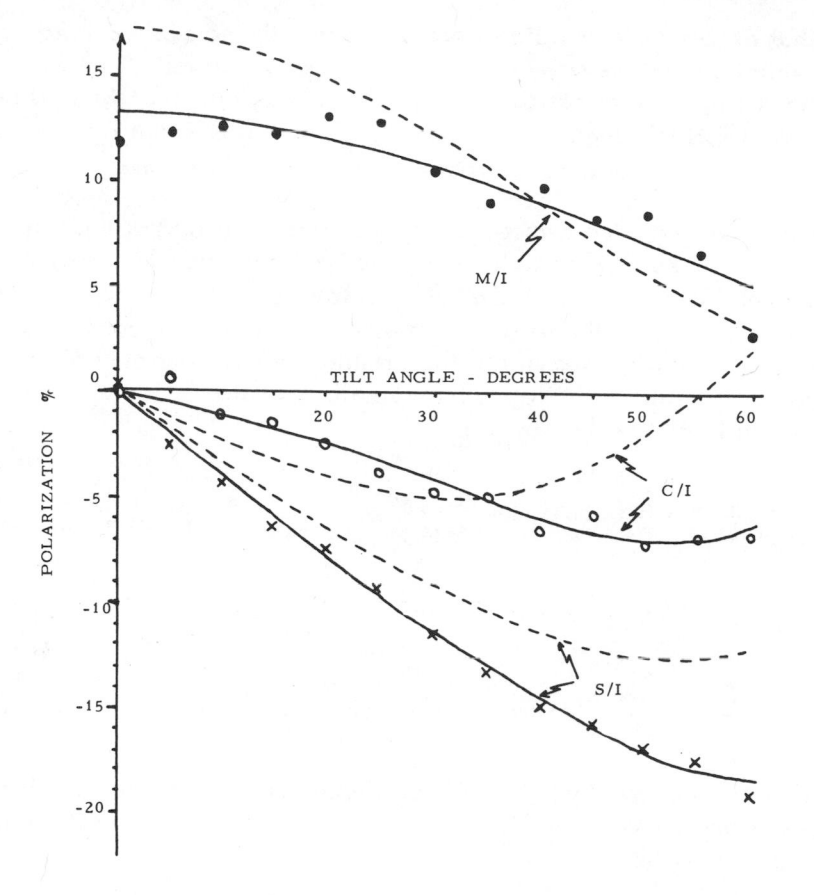

Fig. 7. Stokes parameters for ^{4}He I 2s ^{1}S – 3p ^{1}P, 5016 Å, ($\theta = 90°$, E = 246 keV) fitted to the Band theory – see Eqns. 3–5. The solid lines are independent fits to each Stokes parameter. The dashed lines are simultaneous fits with E = –0.140, F^2 =+0.0328, and $\omega/v = 0.73$.

The theories of Eck and Band assume that a state of well-defined parity is produced in the foil interaction. Consequently, the electric field interaction is a second order perturbation (present through the strong surface fields) and the Stokes parameters are functions of 2α, 4α, etc. A first order interaction should introduce terms proportional to α, 3α, etc., and we have noted earlier[20] that substitution of α/2 for α would indeed give much better agreement with experiment.

Lombardi has pointed out that the production of non–definite parity states in the beam–foil process[26] allows such first order Stark effect processes to oc-cur at the foil surface. He has derived expressions[27] for the Stokes parameters of a $^1S - ^1P$ transition assuming s–p mixing. These can be expressed as expansions of $\alpha, 2\alpha, 3\alpha$, etc. Thus,

$$I = I_0 + I_1 \cos\alpha + I_2 \cos 2\alpha \tag{6}$$

$$M = M_0 + M_1 \cos\alpha + M_2 \cos 2\alpha + M_3 \cos 3\alpha + M_4 \cos 4\alpha \tag{7}$$

$$C = C_1 \sin\alpha + C_2 \sin 2\alpha + C_3 \sin 3\alpha + C_4 \sin 4\alpha \tag{8}$$

$$S = S_1 \sin\alpha + S_2 \sin 2\alpha \tag{9}$$

where $I_i (i=0-2)$, $M_i (i=0-4)$, $C_i (i=1-4)$, and $S_i (i=1,2)$ are functions of the various s and p density matrix elements. It should be noted that $M_i = C_i$ for all i except there is no C_0, and presumably p–d mixing, etc. would increase the number of terms in the expansions. For the case of no s–p mixing all α and 3α terms disappear, reproducing Eck's results with only a phase change at the surface, and Band's results on including surface excitation.

Unfortunately, this more general theory contains a large number of parameters (the many density matrix components) and also these parameters should vary with tilt angle α, since the surface interaction time changes as $1/v \cos\alpha$. This last variation is explicitly included in the result of Eck and Band — see Eqs. (3–5).

However, an analysis of our data for Ne III, 2866Å, in Fig. 8 shows an excellent fit to Lombardi's theory with only a small number of parameters for tilt angles between 0° and 80°. Less precise data for the ^{4}He I $2p^1P-4d^1D$ transition at 4922Å in Fig. 9 also show good agreement with the theory. The $1/v \cos\alpha$ dependence has been neglected, unlike the fit to Band's theory shown in Fig. 7.

A fit to the experimental data using the Lombardi theory gives estimates of various density matrix components. Thus, for the Ne III 2866Å transition, we find $I_2 = M_4 = S_2 = 0$. These parameters are proportional to $\{\sigma(m = 0) - \sigma(m = 1)\}$ and hence indicate that $\sigma(m = 0) = \sigma(m = 1)$. However, we should than have $M_0 = 0$ which is certainly not true. An explanation may be that p–d mixing which should be as strong as s–p mixing has been neglected.

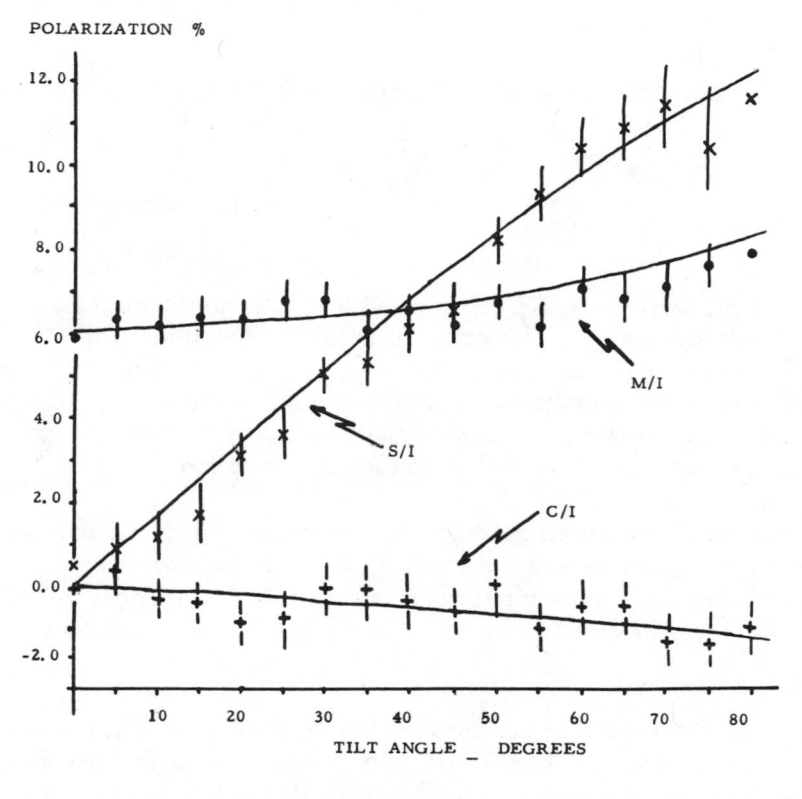

Fig. 8. Stokes parameters for Ne III, 2866 Å, 3s' ^{1}D – 3p' ^{1}F ($\theta = 90°$, E = 1 MeV), fitted to the Lombardi theory giving

$$I = 1 + 0.299 \cos\alpha$$
$$M = 0.090 - 0.012 \cos\alpha$$
$$C = -0.012 \sin\alpha$$
$$S = 0.130 \sin\alpha$$

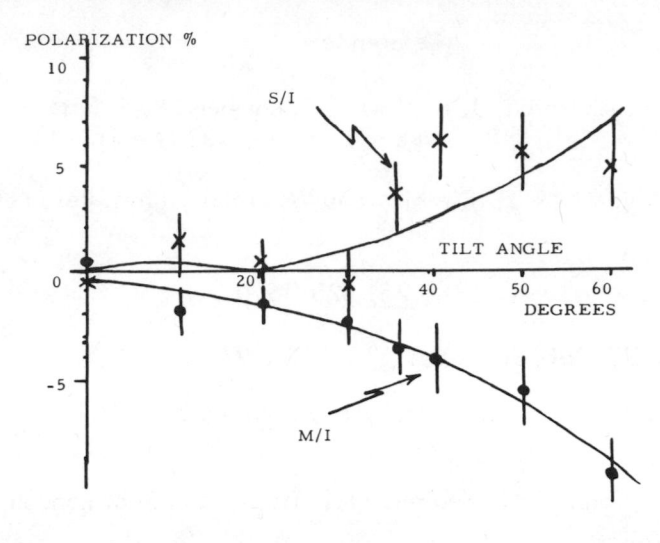

Fig.9. Stokes parameters for ^{4}He I 2p ^{1}P - 4d ^{1}D, 4922 Å (θ = 90°, E = 325 keV), fitted to Lombardi parameters I = 1, M = -0.167 + 0.162 cosα C = 0, S = 0.16 sinα - 0.08 sin2α.

6. Conclusions

The theories of Eck, Band and Lombardi are all based on an interaction between a surface electric field and the moving ions. Their treatments differ in the types of excited states produced. The most general, that of Lombardi, seems to best agree with experiment in predicting the variations of the Stokes parameters with the foil tilt angle. Thus, within the limitations of only a few experimental data and the large number of fitted parameters of the theory, the surface interaction appears to be understood.

All experiments to date have taken place in relatively dirty vacuum conditions (about 10^{-6} torr), and detailed calculations of the types of excited states produced by particular foil materials will be useful when the experiments are repeated with clean surfaces in ultra-high vacuum. Such experiments are in progress.

We acknowledge with thanks the experimental help of Jean Desesquelles, Jerry Gabrielse, Tim Gay and Chuck Batson at Argonne National Laboratory and the University of Chicago. We had helpful discussions of the theories with Yehuda Band, Tom Eck, Roger Herman, and Murray Peshkin.

References

1. S. Datz, B. R. Appleton, J. R. Mowat, R. Lambert, R. S. Peterson, R. S. Thoe, and I. A. Sellin, Phys. Rev. Lett. $\underline{33}$, 733 (1974).

2. V. N. Neelavathi, R. H. Ritchie, and W. Brandt, Phys. Rev. Lett. $\underline{33}$, 302 (1974).

3. H. J. Andrä, Phys. Rev. Lett. $\underline{25}$, 325 (1970).

4. D. G. Ellis, J. Opt. Soc. Am. $\underline{63}$, 1232 (1973).

5. U. Fano and J. Macek, Rev. Mod. Phys. $\underline{45}$, 553 (1973).

6. I. C. Percival and M. J. Seaton, Phil. Trans. Roy. Soc. London $\underline{A\,251}$, 113 (1958).

7. T. G. Eck, Phys. Rev. Lett. $\underline{31}$, 270 (1973).

8. I. A. Sellin, J. R. Mowat, R. S. Peterson, P. M. Griffin, R. Lambert, and H. H. Hazelton, Phys. Rev. Lett. $\underline{31}$, 1335 (1973).

9. A. Gaupp, H. J. Andrä, and J. Macek, Phys. Rev. Lett. $\underline{32}$, 268 (1974).

10. H. G. Berry, L. J. Curtis, D. G. Ellis, and R. M. Schectman, Proc. of the Stirling Conference on Electron and Photon Collisions with Atoms, July 1974, to be published, and R. M. Schectman et al., Proceedings of this conference.

11. H. G. Berry, L. J. Curtis, D. G. Ellis, and R. M. Schectman, Phys. Rev. Lett. $\underline{32}$, 751 (1974).

12. See, for example, D. Clarke and J.F. Grainger, Polarized Light and Optical Measurements (Pergamon Press, New York, 1971), Sec. 1.3.3. The Stokes parameters are defined in terms of the electric vectors in two arbitrary perpendicular transverse directions: $I = |E_{\parallel}|^2 + |E_{\perp}|^2$, $M = |E_{\parallel}|^2 - |E_{\perp}|^2$, $C = 2\,\mathrm{Re}(E_{\parallel}\,E_{\perp}^{*})$, $S = 2\ \ (E_{\parallel}\,E_{\perp}^{*})$.

13. W. Hanle, Z. Instrumentenk. $\underline{51}$, 488 (1931).

14. H. G. Berry, L. J. Curtis, and R. M. Schectman, Phys. Rev. Lett. $\underline{34}$, 509 (1975).

15. H. G. Berry, S. N. Bhardwaj, L. J. Curtis, and R. M. Schectman, Phys. Lett. 50A, 59 (1974).

16. M. Gaillard, M. Carré, H. G. Berry, and M. Lombardi, Nucl. Inst. Meths. 110, 273 (1973), and M. Gaillard, Thesè de Doctorat, 1974.

17. D. A. Church, W. Kolbe, M. C. Michel, and T. Hadeishi, Phys. Rev. Lett. 33, 565 (1974).

18. D. A. Church, M. C. Michel, and W. Kolbe, Phys. Rev. Lett. 34, 1140 (1975).

19. C. H. Liu, S. Bashkin, and D. A. Church, Phys. Rev. Lett. 33, 993 (1974).

20. H. G. Berry, L. J. Curtis, D. G. Ellis, and R. M. Schectman, Phys. Rev. Lett. 35, 274 (1975).

21. T. G. Eck, Phys. Rev. Lett. 33, 1055 (1974).

22. M. Lombardi and M. Giroud, C. R. Acad. Sc. (Paris) B266, 60 (1968), and M. Lombardi, J. Phys. 30, 631 (1969).

23. Y. Band, private communication, and Phys. Rev. Lett.,to be published.

24. R. M. Herman, private communication, and Phys. Rev. Lett., to be published.

25. For example, 2s-2p excitation in hyrdrogen, Ref. 7-9, and in other hydrogenic states, Ref. 10.

26. M. Lombardi, private communication, and these conference proceedings.

Work supported in part by NSF and ERDA.

THE EFFECT OF TILTED FOIL EXCITATION ON THE SPATIAL DECAY OF THE 3^{3}P STATES OF ^{4}He I IN AN APPLIED MAGNETIC FIELD

J. D. Silver and L. C. McIntyre, Jr.[+]

University of Oxford, Nuclear Physics Laboratory,

Keble Road, Oxford OX1 3RH, U.K.

[+]On sabbatical leave from the University of Arizona

Tucson, Arizona, U.S.A.

ABSTRACT

Quantum beats involving coherent decay of Zeeman levels in the 3^{3}P states of HeI differing by m = $\pm$ 1 have been observed in plane polarised light. These beats do not appear with perpendicular foil excitation and indicate the presence of density matrix components $\rho^2_{\pm1}$ in the decaying atomic state following tilted foil excitation.

1. INTRODUCTION

It has recently been discovered that the use of tilted foils in beam-foil spectroscopy can lead to orientation of the excited electronic levels studied (Berry 1975). A possible origin of this effect has been given in terms of a large electric field of short range perpendicular to the foil surface Eck (1974), Lewis (1975). We report here measurements of magnetic quantum beats between Zeeman sub-levels in neutral helium atoms produced and excited from He$^+$ beams using tilted thin carbon foils.

2. EXPERIMENTAL

Part of our experimental set-up is shown in fig. 1. The He$^+$ beam from a small Van de Graaff accelerator enters from the right. The beam is focussed by a quadrupole electrostatic lens to a spot of diameter $\sim$1 mm in the region of the carbon foil, and its

position defined with collimators.   The carbon foils so far used
have been typically 5 $\mu g/cm^2$, and amorphous, being made by electron
beam evaporation of carbon on to detergent coated glass backing
plates.   The rotating foil holder is arranged so that, when the
plane of the foil is directly imaged in the slit, the axis of foil
rotation is the optic axis of the viewing optics; the slit may also
be rotated about this axis.   Spatial quantum beats (Andrä 1974) may
thus be observed for varying foil inclinations θ.

The optical system consists of 2 lenses set to image the beam
at the slit and provide parallel illumination for the linear
polarizer and interference filter.   In our experiment a 3888 $\overset{o}{A}$
interference filter is used to pass only the $1s2s\,^3S-1s3p\,^3P_{0,1,2}$
transition of ^4HeI to the photomultiplier.   A magnetic field of up
to 68 gauss can be applied perpendicular to the beam direction as
shown in fig. 1, using a pair of Helmholtz coils fed from a current
stabilized power supply.   The foil can be translated parallel to the
beam axis by means of a stepping motor drive, in integral multiples
of a 5 micron step.   The beam charge is integrated in a suppressed
Faraday cup which travels on the foil drive bar, so that the distance between Faraday cup and foil remains constant.   Light
intensities are measured by photon counting, and decay curves showing quantum beats are obtained by integrating the number of photons
collected by the photomultiplier for fixed beam charge at each of a
number of foil positions, typically 100, 1 mm apart.   The digital
decay curves are counted into a multichannel analyzer in multiscale
mode in the standard way, then each decay curve is read directly
into a PDP 10 computer where it may be edited, and then analyzed to
find its spectral composition by a Fourier transform programme.

3. RESULTS

All our experiments so far have involved observations of quantum beats among Zeeman substates in the previously well studied
level 3^3P of ^4HeI in the $2^3S-3^3P_{0,1,2}$ 3888 $\overset{o}{A}$ decay.   We have studied
(i) the relative amplitudes of different magnetic quantum beat fre-
quencies in an applied small magnetic field as a function of θ and
(ii) the breaking of the $\Delta m = 0, \pm 2$ selection rule for quantum beat
frequencies when the cylindrical symmetry of the atomic system is
destroyed.   A physical picture of the origin of this selection rule
may be obtained by consideration of section 3.2 of Andra (1974).
The selection rule is also implicit in the tensor operator approach
of Gaillard (1973).   In the normal beam-foil set-up with foil plane
perpendicular to the beam, only alignment of the atomic states may
be produced.   In terms of Fano and Macek's definitions (Fano and
Macek 1973), the only observable alignment parameter is A_0^{col}; in the
alternative density matrix formalism, ρ_0^2 and $\rho_{\pm 2}^2$ are the only non-
zero state multipoles.   If the cylindrical symmetry of the excitation is destroyed, then the excited electronic states may be

oriented as well as aligned. The alignment also may take a more complex form, so that parameters O_{1-}^{col} and A_{1+}^{col} (Fano and Macek 1973 section II.1) or equivalently $\rho_{\pm1}^1$, $\rho_{\pm1}^2$ become observable.

Part of the Zeeman term diagram for the $3^3P_{1,2}$ state is shown in fig. 2 in a field of 53 gauss. The 3^3P_0 state is not shown; it plays no part in our measurements since it is not split by a magnetic field, and gives rise to no observable quantum beats in our set-up due to our deliberately poor time resolution. The Zeeman splittings are given in the weak field approximation; an exact solution of the Zeeman Hamiltonian shows that the small frequency errors thus introduced are not significant for our purposes.

We present here results obtained for foils tilted at 45°, showing clear evidence of the beats corresponding to coherent decay of states. Typical experimental decay curves are shown in fig. 3a, c and e, and corresponding Fourier analyses of the beat frequencies in b, d and f. a shows part of the decay curve in zero field; the modulation is due to "quantum beats" between the J = 1 and J = 2 states, the frequency being 658.55 MHz (Wieder 1957). The incident He$^+$ beam energy was 350 keV. A Fourier transform of the modulated decay curve, after subtraction of a slowly varying polynomial background, is shown in fig. 3b, and we note that only a single frequency is present; the small side bands arise because of the nature of the input data (after allowance for energy loss in the foil, the frequency is 660 $\pm$ 20 MHz, in good agreement with the known value). In this case, the foil was perpendicular to the beam axis (i.e. $\theta = 0$) and the linear polarizer was set with its axis parallel to the beam axis. Fig. 3c shows a decay curve taken under similar conditions to a, save that now a magnetic field of 53 gauss has been applied perpendicular to the beam axis. This lifts the degeneracy of the Zeeman sub-levels, giving rise to the magnetic quantum beats B($\Delta m = 2$, $\Delta J = 0$), C($\Delta m = 2$, $\Delta J = 1$), E($\Delta m = 0$, $\Delta J = 1$) and G($\Delta m = 2$, $\Delta J = 1$), where the labels B, C etc. refer to fig. 2. Observations of this type have previously been made by Andrä (1970) and more exhaustively by Zgainski (1973). Fig. 3d shows a Fourier transform of c with the observed beat frequencies. Table I presents a typical set of results for the measured frequencies.

Fig. 3e shows an experimental decay curve taken under the same conditions as in b save that the foil and slit were inclined at 45° to the beam axis, and the linear polarizer's axis was set at 45° to the beam axis. We notice that the modulations look quite different, and the Fourier transform, f, shows that   (i) beats have appeared corresponding to $\Delta m = \pm 1$, $\Delta J = 0,1$ and (ii) the relative intensities of $\Delta m = 0,2$ beats have changed. Table II presents a typical set of frequencies for such conditions, taken from fig. 3b.

CONCLUSION

We may use our measurements to draw some conclusions concerning the nature of the excited atomic state when the He atoms leave the foil surface. We shall use the density matrix to describe the excited state, and we employ the customary spherical tensor expansion (Carrington 1972)

$$\rho = \rho_q^k \, T_q^k$$

where $^J T_q^k$ are matrices whose components are given by

$$\langle JM \mid ^J T_q^k \mid JM' \rangle = (-1)^{J-M} (2k + 1)^{\frac{1}{2}} \begin{pmatrix} J & k & J \\ -M & 1 & M \end{pmatrix}$$

The advantage of this method of expressing ρ is that certain initial properties follow immediately from symmetry considerations. In particular $q \equiv 0$ for all coefficients ρ_q^k in the case of axially symmetric excitation, where the axis of quantization is the beam axis. The intensity and frequency of the quantum beats observed in our experiment is given by

$$I = \text{Tr} \, (D. \rho)$$

The development of the optical monitoring operator D in a spherical tensor basis (Gaillard 1973)

$$D = \sum_{q,k} \, ^J D_q^k \, ^J T_q^k$$

leads to the useful simplification that each multipole parameter ρ_q^k of the density matrix is sampled by the corresponding moment D_q^k of the monitoring operator.

In our geometry, and using linearly polarised light, the only non-zero components of D_q^k are for k = 0 or 2, and q ranging from +2 to −2. Thus we are not sensitive to the ρ_q^1 components generally referred to as the orientation parameters; the beats A, D, and F which are the signature of the tilted foil effect in our experiment arise from non-zero components $\rho_{\pm 1}^2$ in the atomic density matrix and an analysis of our data shows that these components, which would be rigorously zero in the absence of a tilted foil perturbation, are in fact present.

We have used the expressions of Zgainski (1973) to calculate for our case relative quantum beat amplitudes for the 3 $\Delta m = \pm 2$ beats B : C : G and the 3 $\Delta m = \pm 1$ beats A : D : F, and we obtain theoretical ratios 1 : 0.38 : 0.31 and 1 : 0.45 : 0.18 respectively. The experimental results 1 : 0.9 : 0.1, and

TABLE I

Quantum Beat	Measured ν MHz	Theoretical* ν_T MHz	$\nu_T - \nu$ MHz
B $\Delta m = 2 \Delta J = 0$	$\cdot 222 \pm 15$	223	1
C $\Delta m = 2 \Delta J = 1$	433 ± 15**	436	
E $\Delta m = 0 \Delta J = 1$	658.55**		
G $\Delta m = 2 \Delta J = 1$	880 ± 15	882	
E−C	226	223	−3
G−E	221	223	2

TABLE II

Quantum Beat	Measured ν MHz	Theoretical*** ν_T MHz	$\nu_T - \nu$ MHz
A $\Delta m = 1 \Delta J = 0$	115 ± 15	112	−3
B $\Delta m = 2 \Delta J = 0$	224 ± 15	224	0
C $\Delta m = 2 \Delta J = 1$	461 ± 20	435	
D $\Delta m = 1 \Delta J = 1$	564 ± 20	547	
E $\Delta m = 0 \Delta J = 1$	658.55**		
F $\Delta m = 1 \Delta J = 1$	790 ± 20	771	
G $\Delta m = 2 \Delta J = 1$	889 ± 20	883	
F−D	226	224	−2
G−C	428	448	20

*
The theoretical values depend on an accurate knowledge of the magnetic field; since a gauss meter accurate to 1 gauss was not available, the field was deduced from the frequencies themselves, and the theoretical values are given only to indicate consistency between B, (E−C) and (G−E). This consistency is surprisingly good. The average $m_J = 2$ splitting frequency derived in this way (223 MHz) is also in agreement with that calculated from the Helmholtz coils calibration constant (218 ± 5 MHz).

**
In order to eliminate systematic errors due to uncertainties in beam velocity, the frequency scale is normalized to E=658.55 MHz.

***
As for the results of Table I, the field was deduced from an average for the magnetic quantum beats. The underlined combinations serve as a best test of the theory, since E is of very low amplitude under this combination of foil tilt and field. With so many frequencies present, the error in finding peak position grows. The spurious peaks between B and C and C and D seem to arise from noise present in the experimental data. They are not reproducible in frequency or intensity from decay to decay.

Fig. 1.   Experimental set up

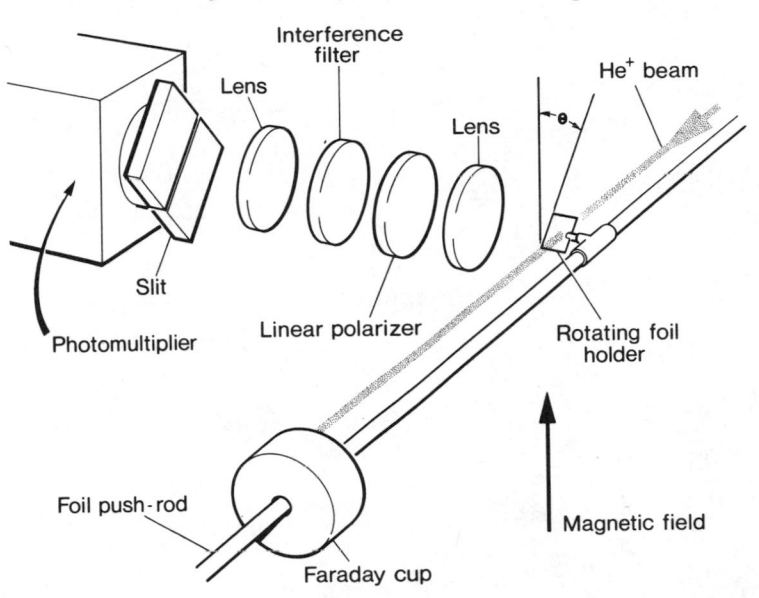

Fig. 2.   Zeeman splitting of $3^3P_{1,2}$ of ^{4}HeI in a magnetic
field of 53 gauss showing observed quantum beat
frequencies

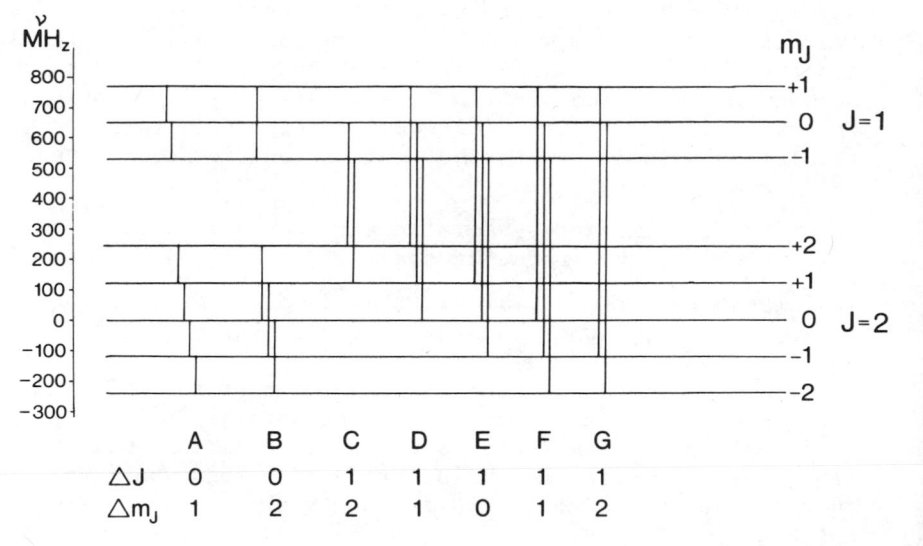

Fig. 3   Observed quantum beats

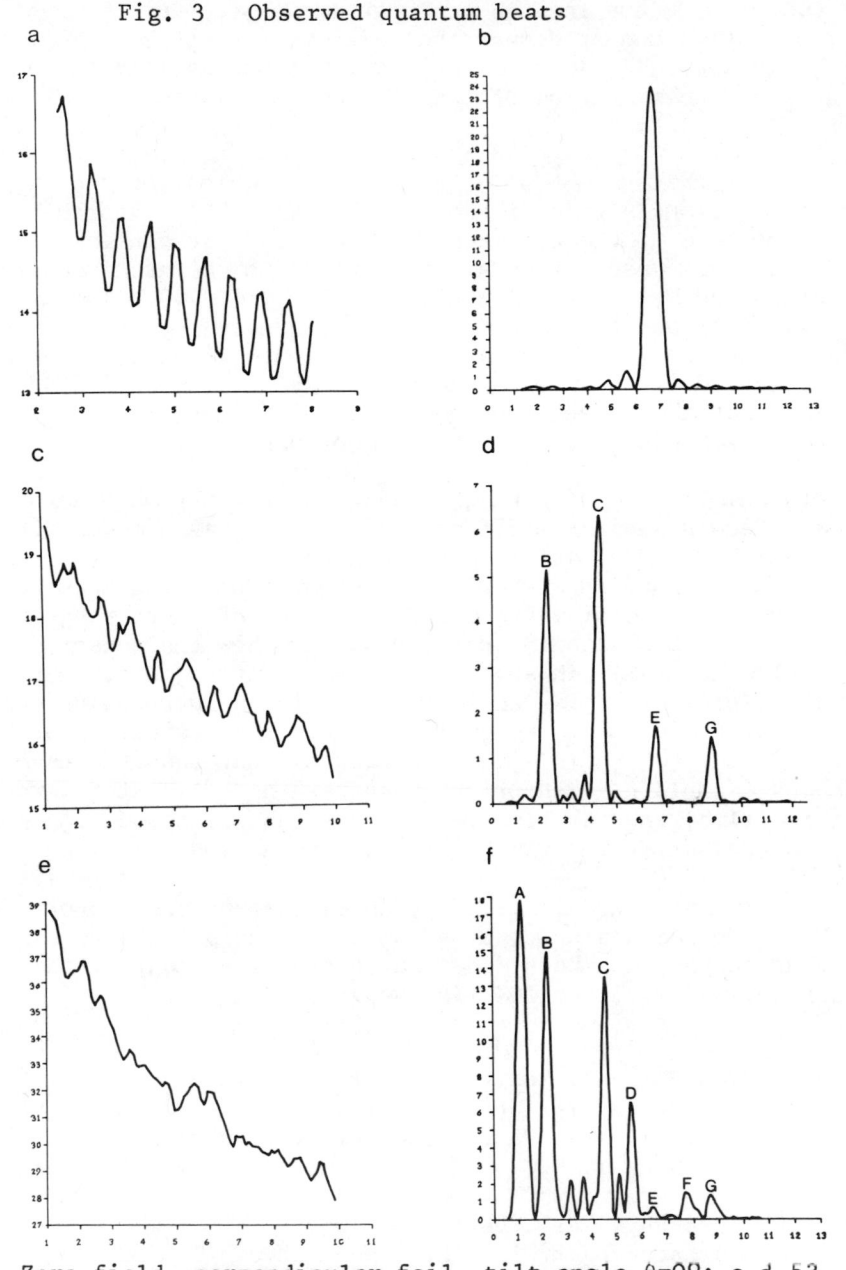

a,b, Zero field, perpendicular foil, tilt angle θ=0°; c,d,53
gauss, perpendicular foil, θ=0°; e,f,53 gauss, tilted foil
θ=45°. For a,c and e the abscissae are in cm, and the ordinates,
counts in units of 10^4. For b,d and f, the abscissae are
frequency in units of 100 MHz, and the ordinate scales are
arbitrary.

1 : 0.36 : 0.08 are in such poor agreement with our theoretical ratios that we cannot deduce the relative magnitude of $\rho^2_{\pm1}$ to $\rho^2_{0,\pm2}$ in the cases we have observed, but we may make the general point that the magnitude of these parameters is comparable at a foil tilt of 45°.

An alternative method of defining collision induced alignment and orientation is that of Fano and Macek (1973). It can be shown that their use of the parameter A^{col} and O^{col} is equivalent to the density matrix description given above, with the necessary relations between A and O, and the ρ^k_q given by Berry (1974). The magnetic quantum beats observed at the Larmor frequency in linearly polarised light by Church (1975) are a manifestation of the same phenomenon as that observed in our work; from equation 6 of Church (1975) it can be seen that the presence of beats at the Larmor frequency depends on the presence of a non-zero alignment parameter A^{col}_{1+}.

In summary, we have observed for the first time quantum beats between Zeeman substates differing by m = ± 1 as a modulation of the spatial decay of a foil excited atom. Our observations are also the first reported in any triplet system. The effect can be explained in terms of large electric fields of short range near the foil surface; Lewis and Silver (1975) show how the tensor components $\rho^2_{\pm1}$ arise naturally in such a treatment. It is possible that such surface fields also play an important role in introducing coherence between states of opposite parity in foil excited states of hydrogenic systems (Gaupp 1974). Work such as that reported here, when extended, should give information concerning solid state properties of thin films, and may also lead to a method for producing polarised excited nuclear states via hyperfine interaction.

We wish to acknowledge great encouragement and support from Dr. M. A. Grace, and unstinting and expert help from the ONP teaching course, on whose accelerator the experiments were performed, and from the OEG machine group.

REFERENCES

Andra, H.J. 1970 Nucl. Instr. and Methods 90 343.
Andra, H.J. 1974 Physica Scripta 9 257.
Berry, H.G. et al. 1975 Phys. Rev. Letts. 34 509.
Church, D.A., Michel, M.C. and Kolbe, W. 1975 Phys. Rev. Letts. 34 1140.
Eck, T.G. 1974 Phys. Rev. Letts. 33 1055.
Fano, U. and Macek, J.H. 1973 Rev. Mod. Phys. 45 553.
Gaillard, M. et al. 1973 Nucl. Instr. and Methods 110 273.
Gaupp, A. et al. 1974 Phys. Rev. Letts. 32 268.
Lewis, E.L. and Silver, J.D. 1975 to be published in J.Phys.B.
Wieder, I. and Lamb, W.E. 1957 Phys. Rev. 107 125.
Zgainski, A. 1973 Thesis, University of Lyon, unpublished.
Carrington, C. 1971 J.Phys.B,4,1222

THEORETICAL ASPECTS OF BEAM-FOIL COLLISIONS

Joseph Macek

Behlen Laboratory of Physics
The University of Nebraska
Lincoln, NE 68588

The title of this contribution "Theoretical Aspects of Beam Foil Collisions" suggests an absence of a reliable theory and emphasizes that we at best have an understanding of only some isolated aspects of the beam-foil collision process. We cannot begin to predict the population of specific atomic levels, since we have no comprehensive picture of the collision process. Rather only some general principles are available to guide our reasoning. I will focus my attention on one aspect of some current interest, namely the alignment and orientation of atomic states produced by beam-foil collisions, and the relation of this anisotropy to various general considerations. These considerations are the likely symmetry properties of the collision geometry, the presumed dominance of the electrostatic interaction in the excitation processes and the influence of cascades in populating atomic states.

Consider a typical beam-foil collision. The ion beam strikes a carbon foil whose normal is parallel to the beam. Decay radiation is observed downstream from the foil. One can measure the time evolution of the radiation pattern, its angular distribution, or the polarization of the radiation. The relation between these various measurements is particularly simple if the source has cylindrical symmetry and if the nuclear spin, or in the case of light atoms, electronic spin, is unimportant in the initial excitation process. All of these assumed properties need close examination since they are by no means obviously or universally true.

The assumption of cylindrical symmetry is the most basic one, since it tells us that magnetic substates are populated incoherently, or stated another way, that the charge distribution of beam-foil excited atoms is cylindrically symmetric. This assumption has some plausible basis in view of the nature of carbon foils. They are usually prepared by sputtering on to a glass substrate and nothing in the sputtering process picks out a particular axis other than the normal to the foil. In addition the foil undergoes further random sputtering and impurity build-up under normal use. Thus when the normal is coincident with the beam axis the beam-foil source is cylindrically symmetric.

Such a source can exhibit no orientation but it might exhibit alignment. Correspondingly, the radiation can exhibit no circular polarization but might exhibit linear polarization.

Explicitly one has for the intensity of the decay radiation detected by a linear polarizer[1]

$$I = K \{1 + A_0^{col} \tfrac{1}{4} h^{(2)}(j_i j_f)[1 - 3\cos^2\theta + 3\sin^2\theta \cos 2\psi]\}e^{-t/\tau}$$

$$(1)$$

where K is a constant incorporating the cross section for populating the level, θ equals the angle between the detector axis and the beam axis and ψ specifies the setting of a linear polarization analyzer ($\psi = 0$ corresponds to the axis of linear polarization analyzer lying in the plane of the detector axis and the ion beam axis), $h^{(2)}(j_i j_f)$ is a ratio of six-j symbols depending upon the angular momentum of the upper j_i and lower j_f levels.

The single non-zero component of the alignment tensor A_0 is given by[1]

$$A_0^{col}(t) = <(i'|\exp(iHt/\hbar)\,(3J_z^2 - J^2)\exp(-iHt/\hbar)|i)>/j_i(j_i + 1)$$

$$(2)$$

where the notation $<(i|0|i)>$ indicates an average of the operator 0 over the various substates of the initial level i. H is a Hamiltonian operator to be discussed later, and J is a general angular momentum vector. It could represent electronic, orbital, spin or total angular momentum depending upon circumstances.

Eq. (1) appears to agree with experiment, but it has not been extensively checked. If the source has a lower degree of symmetry as happens when the normal to the foil is tilted away from the beam axis, other components of the alignment tensor and an orientation vector defined as the mean value of J enter into the expression for I. Berry and co-workers[2] measured the alignment and orientation parameters of the 3^1P level in He excited by collisions with tilted foils at 130 keV incident beam energy. They found a non-zero orientation $0_{1-}^{col} = <Jy>$ which increased with increasing foil tilt angle (Table I). At zero-tilt angle 0_{1-}^{col} vanishes as do all components of the alignment tensor except A_o^{col} and possibly A_{2+}^{col}. This is one of the few experiments that check the cylindrical symmetry of the typical beam-foil geometry. Although the results at zero tilt angle are not definitive, it appears that foils are at least approximately cylindrically symmetric about the normal.

The second general aspect of beam-foil collisions is Perciual-Seaton hypothesis,[3] or the hypothesis that the hyperfine interaction and, in light atoms the fine structure play only a trivial role in the initial excitation process. Consider light ions for example. Since we suppose that initially the beam-foil system has no net electronic spin, and since we assume that spin-orbit forces play a much less significant role in the excitation than do the electrostatic forces, the states that are actually excited can have no spin anisotropy. On the other hand the spin-orbit interaction is significant during the decay of the excited state, since it splits the upper level into closely spaced sublevels. This interaction perturbs the initial alignment in that source of the alignment is transferred to the electronic spin degrees of freedom. This exchange of alignment is expressed through

TABLE I. Alignment and Orientation Parameters for the He3^1P Level Excited at 130 keV Beam Energy. Data from from Berry, et. al.[2]

Tilt Angle	A_o^{col}	A_{2+}^{col}	A_{1+}^{col}	D_{1-}^{col}
0°	-0.090(36)	0.016(9)		
20°	-0.081	0.012	-0.024(7)	-0.013(11)
30°	-0.072	0.008	-0.021	-0.038
45°	-0.054	-0.002	-0.040	-0.040

the factors $\exp(iHt/h)$ in Eq. (2), where H represents the spin-orbit interaction. The time dependence can be extracted from Eq. (2) as was done by Adler.[5] One finds

$$A_0^{col}(t) = G^{(2)}(t)A_0^{col}(0) \tag{3}$$

where

$$G^{(k)}(t) = \sum_{JJ'} \frac{(2J'+1)(2J+1)}{(2S+1)} \{_{LLS}^{J'Jk}\}^2 \cos \omega_{JJ'}t \tag{4}$$

and $$A_0^{col}(0) = <(i|3L_z^2 - L^2|i)>$$

Note here that A_0^{col} is constructed from the components of $\vec{L}$ rather than $\vec{J}$ since the electrostatic forces operative in the collision affect only the orbital motion.

To interpret Eq. (4) consider an alignment tensor constructed from the components of J. This tensor is constant in time but if we recall $\vec{J} = \vec{L} + \vec{S}$ we see that this constant is the sum of three terms, a term representing orbital alignment, spin alignment and a joint alignment of spin and orbital angular momentum, i.e.,

$$J_z^2 - J^2 = (3L_z^2 - L^2) + (3L_zS_z - L\cdot S) + (3S_z^2 - S^2).$$

Each term oscillates but the sum remains constant, consequently the total alignment is conserved, but is exchanged between various internal degrees of freedom. Time resolved measurements of the light intensity exhibit this oscillation since the electromagnetic transition depends only on the orbital degrees of freedom. Note that each term in $G^{(k)}(t)$ starts with an initial phase of 0° and then decreases reflecting the initial flow of alignment from orbital to spin degrees of freedom. The Percival-Seaton hypothesis can then be checked by determining the initial phase of the oscillations. Burns and Hancock[5] measured the initial phase of the oscillations for the $3^3P \rightarrow 2^3S$ transitions in HeI 200 keV incident He$^+$ energies. They found the initial phase to be 0°(+3.5°, - 5.2°) in very good agreement with the data. They also examined the light at $\theta = 0°$ with a linear polarization analyzer set at $\psi = 54.8°$. Then the coefficient of A_0^{col} vanishes and one should expect no oscillations. To a

good approximation none are found (see Fig. 1), but some
slight residual oscillation may be present. The slight
residual oscillation could indicate that the beam-foil
geometry is only approximately cylindrically symmetric.

Wittman, et. al.[6] made extensive measurements of the
oscillations in the 3^3P state of He resolving both the fast
ω_{20} and the slow ω_{21} components. Their decay curves are
shown in Fig. 2. Agreement between fitted curves and
experiment confirm the relative beat amplitudes given by
Eq. (4), and thus provide further confirmation of the
Percival-Seaton hypothesis. This measurement is a refined
version of the experiment by Andra[7] which first established
that foil collisions produce aligned atoms.

These and similar data by many workers establish that
different magnetic substates are excited incoherently
(cylindrical symmetry) and that levels with the same magnetic
quantum number with the Z axis along the beam axis, but
different total angular momentum J are excited <u>coherently</u>.

An interesting situation occurs in hydrogen where
radiation from levels with different orbital angular momentum
lie sufficiently close for these oscillations to be observed.
Thus for example, Burns and Hancock[8] discovered none zero
Sd coherence in their measurements on the n = 3 levels in
atomic hydrogen. In contrast to fine structure oscillations,
the oscillations related to $S_{\frac{1}{2}} - D_{3/2}$, $1/2$ splittings start
with non -zero initial phase.

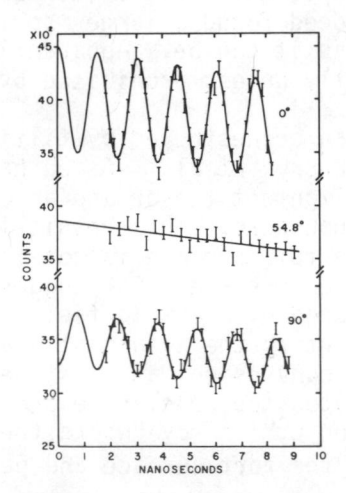

Fig. 1.   Radiation from the 3^3P level of HeI vs. time for beam-
          foil excited states. Data of Burns and Hancock.[5]

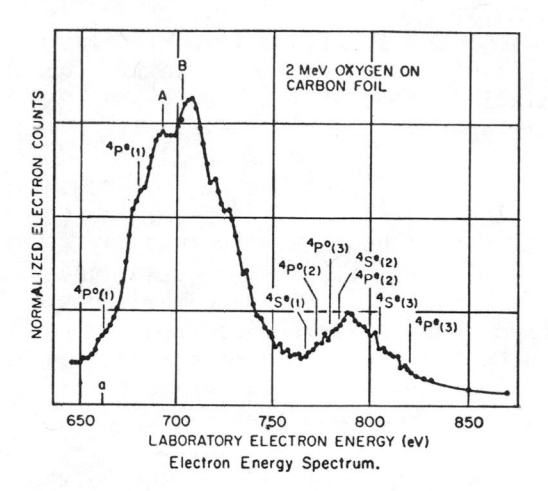

Electron Energy Spectrum.

Fig. 2.   Electrons ejected near the foil by 2 MeV oxygen ions.
          Data from Pegg, et. al.[13]

Eck[9]pointed out that the coherence between levels with
different orbital angular momentum, in particular $2S_{\frac{1}{2}}$ - $2P_{\frac{1}{2}}$
coherence could be examined observing the corresponding
electric field induced oscillations with the electric field
parallel and antiparallel to the beam velocity $\vec{v}$.   The
difference in these two signals measures $_{sp}$ the density
matrix element connecting the s and p states with $m_1 = 0$.
Sellin and later Gaupp, et. al., carried out the suggested
measurements and indeed found a large $_{sp}$ term.   Indeed $_{sp}$
is almost as large as it can be.   Apparently hydrogenic states
are excited coherently unless prohibited by symmetry.

The symmetry can be destroyed by tilting the foil as in
the experiments by Berry, et. al.[2]   As we have seen, additional
components of the alignment tensor and an orientation vector
component are then non-zero.   This result is extremely
surprising since the foil surface is not thought to be well
defined on a microscopic scale.   Eck suggested that the
orientation is produced by electric fields at the surface
of the foil which perturb the alignment via the quadratic
stork effect.   Such conversion was first demonstrated by
Lombardi.   For P states the relevant expression for the
predicted orientation $O^{col}_{1-} = \langle Jy \rangle$ where the y axis is along
an axis parallel to the fort surface and perpendicular to the
ion beam is

$$O^{col}_{1-} = A^{col}_{0}(0)\tfrac{1}{2} \sin x \sin 2\alpha \qquad (5)$$

where α is the tilt angle and x is an unknown phase related to the Stark splitting of the $M_L = 0$ and $M_L = 1$ eigenstates in an electric field. According to the simple version of this theory discussed by Eck, X is a constant, thus O_{1-}^{col} would maximize at $\alpha = 45°$. Later data by Berry and co-workers,[10] and by Church and co-workers[11] show that this prediction does not hold. O_{1-}^{col} increases with increasing α for all α between 0° and 90°. Lombardi[12] suggests that x in Eq. (5) should vary with α owing to the increasing time that atoms spend near the foil where the electric field is presumed to exist, and that the linear Stark effect could also play a role. Both of these effects give rise to an orientation which increases with α for all α.

At present the origins of the observed orientation are not clear, however it seems likely that electric fields near the surface play a significant, but as yet undertermined role.

Thus far I have mentioned nothing about the origins of this observed alignment, except to note that it is not ruled out by symmetry. We do not have a model which will enable us to calculate the alignment. We cannot even say qualitatively which specific levels are likely to show alignment. We know from experiment that the alignment is usually small, of the order of .05 for neutral atoms and singly ionized species, but is nearly zero for highly ionized species.

In the last part of my talk I will speculate on the relevant factors determining the magnitude of the alignment. I submit that the relevant question is not "Why are beam-foil excited species aligned?" rather it is "Why is the alignment so small, and why don't all levels show substantial alignment for some range of the bombardment energy?" A moderate degree alignment is expected in collision excitation since the component of the net force on an atomic electron parallel to the incident beam direction is in general rather different from the component perpendicular to the beam. Multiple collisions only enhance this net anisotropy, yet with the exception of the 2p states of atomic hydrogen at 1 MeV the observed alignment for optical or ultraviolet transitions is usually small. What has happened to the (assumed) initial alignment?

Two answers suggest themselves. When the atoms leave the foil the initial alignment may be perturbed by electric fields near the surface. If these microscopic fields have a distribution of directions experimental averaging over the directions tends to reduce the anisotropy. In contrast this

effect is absent in conventional ion-molecule collisions
since the electric field of a residual ion is always along
the final direction of the internuclear axis at large separations. Then the field cannot mix levels with different M_L,
and hence does not alter the alignment via the quadratic
Stark effect. Thus it is possible that in low Z ions the
alignment is reduced by fields near the surface.

A second possibility, especially for the more highly
ionized species relates to cascade population. Very briefly
if a lower level is populated by cascades from higher levels
the alignment of the lower level is reduced compared to the
initial alignment of the upper level. We can see why this is
true by examining the expression Eq. (2) for the alignment.

$$A_o^{col}(t) = <(i'|\exp(iHt/h)(3J_z^2-J^2)\exp(-iHt/h)|i)>/j_i(j_i+1).$$

Normally $\vec{J}$ and the eigenstates i refer to only the atomic
component of the entire system of atom plus radiation field.
If we include the radiation field in our definition so that
$\vec{J} = \vec{J}_{atom} + \vec{J}_{yield}$ then $\vec{J}$ commutes with H and the total
alignment is a constant in time, although it is exchanged
between different components of the system. In any decay
this total alignment is conserved, but the atomic alignment
alone is not. Some of it is carried away by the radiation
field. Note that this transfer of alignment is irreversible,
in contrast to the transfer of alignment between spin and
space components described by the Adler formula Eq. (4).
We have then the general rule that radiation carries away
anisotropy. Exactly how much anisotropy is carried away
depends upon the specific transition.

This rule applies to any type of radiation, in particular
it applies to Auger and autoionizing transitions. Here Auger
electrons carry away anisotropy. That Auger cascades are
important is suggested by two similar experiments where
electrons emanating from the region near the foil surface
we examined. Fig. 2 shows electrons from the foil region
emitted by a 2 MeV oxygen beam taken by Pegg, et. al.[13]
Several Auger peaks appear in the spectrum. Fig. 3 shows
similar data for 300 keV Li^+ taken by Bruch, et. al.[14] in the
Berlin group. Again autoionization lines are prominent.
These transitions occur promptly, within a few hundred
augstroms of the foil surface, thus they occur before optical
transitions are observed. Quite possibly the states of
highly ionized species are populated by a sequence that

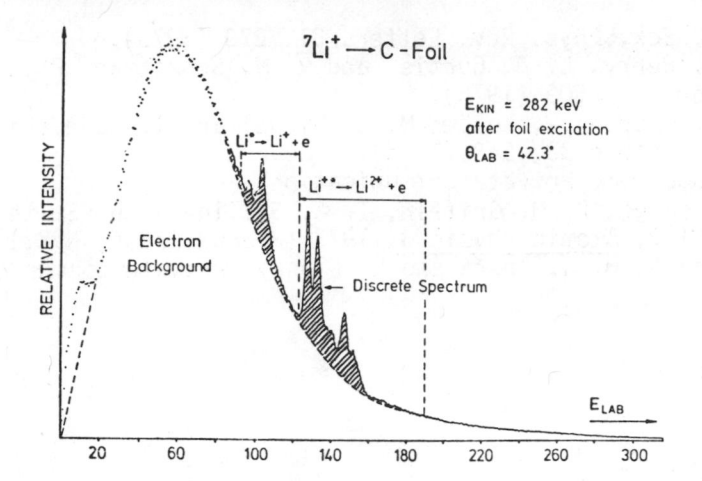

Fig. 3. Electrons ejected near the foil by 282 keV Li^+ ions. Data from Bruch et al.[14]

incorporates a few Auger transitions. Much of the initial foil-induced anisotropy would then be carried away by the Auger electrons leaving little anisotropy in the optically decaying levels.

Very little information is presently available on the anisotropy of Auger electrons, even for atoms excited in gas collisions. For purposes of understanding the alignment of beam-foil excited species, more information on the influence of the Auger effect is desirable.

REFERENCES

1. U. Fano and J. H. Macek, Rev. Mod. Phys. 45, 553 (1973).
2. H. G. Berry, L. J. Curtis, D. G. Ellis and R. M. Schectman, Phys. Rev. Letters 32, 751 (1974).
3. I. C. Percival and M. J. Seaton, Philos. Trans. R. Soc. London. A251, 113 (1958).
4. K. Adler, Helv. Phys. Acta 25, 235 (1952).
5. Donal J. Burns and Walter M. Hancock, J. Opt. Soc. 63, 241 (1973).
6. W. Wittmann, K. Tillman, H. J. Andra and P. Dobberstein, Z. Physik 257, 299 (1972).
7. H. J. Andra, Phys. Rev. Letters 25, 325 (1970).
8. D. J. Burns and W. H. Hancock, Phys. Rev. Letters 27, 370 (1971).

9. T. G. Eck, Phys. Rev. Letters 31, 270 (1973).
10. H. G. Berry, L. J. Curtis, and R. M. Schectman, Phys. Rev. Letters 34, 509 (1975).
11. D. A. Church, W. Kolbe, M. C. Michel and T. Hadeishi, Phys. Rev. Letters 33, 565 (1974).
12. M. Lombardi, Private communication.
13. D. J. Pegg, P. M. Griffin, I. A. Sellin, W. W. Smith and B. Donnally, Atomic Physic 3, 1973 (Plenum Press, N.Y.) p. 327.
14. R. Bruch, H. J. Andra and L. Lipsky, Private communication.

HYPERFINE-STRUCTURE MEASUREMENTS IN CARBON-13

J.L. Subtil, P. Ceyzeriat, J. Desesquelles and M. Druetta

Laboratoire de Spectrométrie Ionique et
   Moleculaire (associé au C.N.R.S.)
Université de Lyon I, Campus de La Doua
69621 Villeurbanne, France

As pointed out by several authors, beam-foil spectroscopy is a powerful method for obtaining fine-structure and hyperfine-structure data. This method is particularly valuable for multiply ionized atoms where other high resolution techniques fail. It has been precedently used to study several transitions in Be III , N III and N IV . We report herein an experiment on the isotope 13 of Carbon, performed with a 2.5 MeV Van de Graaff accelerator. The first results obtained with two transitions of C II and C III will be given.

The energy of the beam has been chosen using the beam-foil study of Carbon by M.C. Buchet-Poulizac, which indicates that the charge distribution has a maximum at 0.6 MeV for C II and at 1.2 MeV for C III. The intense spectral line at 3876 Å in C II has been studied at beam energies between 0.7 and 1 MeV. This line corresponds to the transition $3d\,^4F^o$ - $4f\,^4G$, and the short lifetime ($\sim$ 3 ns) of the $4f\,^4G$ term limits measurements to low frequencies. From the configuration of the higher energy term investigated, $1s^2 2s2p4f$, it is expected that the hyperfine-structure splitting is mainly due to the Fermi-contact term for the 2s electron. However, we find that contributions of the 2p electron must be included to explain the beat pattern.

The other transition investigated at 4650 Å has been studied at several energies between 1.3 and 2 MeV. This is a triplet transition $3s\,^3S$ - $3p\,^3P^o$. The configuration of the upper term is $1s^2 2s3p$. Again, the 2s electron will give a large Fermi-contact contribution to the hyperfine-structure splitting, but the contribution of the 3p electron will be reduced and may be of the order of the core contribution. The same transition has been previously studied in the isoelectronic ^{14}N IV ion.[2,3]

Theory

Fine-structure for each level of the transition investigated
has been obtained by spectroscopic measurements. The spectrum of
singly-ionized carbon has been studied by means of a condensed
hollow-cathode discharge by S. Glad[4]. K. Bockasten[5] has investigated
CIII excited in a sliding vacuum spark. The fine-structure data
are indicated in Fig. 2 and Fig. 5 which show the level scheme. The
hyperfine-structure splittings are a consequence of the nuclear
spin 1/2 in ^{13}C. The hamiltonian describing the hyperfine interactions reduces to the magnetic dipole operator. The electric
quadrupole interaction does not exist (by symmetry).

- Hyperfine-structure splitting in C II.

Since the fine-structure is very large compared with the
hyperfine structure, only diagonal elements must be considered, in
which case

$$\langle((((1_a 1_b)S_{ab}L_{ab},1_c)S,L)J,I)FM_F | H | ((((1_a 1_b)S_{ab}L_{ab},1_c)S,L)J,I)F,M_F\rangle$$

$$= \langle((((2s2p)^3 P^o, 4f)^4 G)J,I)FM_F | H | ((((2s2p)^3 P^o, 4f)^4 G)J,I)FM_F\rangle$$

where

$$H = \frac{2\beta_n \mu_n}{I} \sum_{i=1}^{3} \frac{\hat{N_i} \cdot \hat{I}}{r_i^3}$$

Expressing the interaction as a tensor operator and performing the
radial integration, which is noted symbolicaly $\langle r^{-3}\rangle$ and $\langle \delta(r)r^{-3}\rangle$,
one obtains

$$\frac{\hat{N_i}}{r_i^3} = [\hat{\ell}_i - \sqrt{10}(\hat{s}.c^{(2)})_i^{(1)}]\langle r_i^{-3}\rangle + \delta(\ell_i,0)\frac{8\pi}{3}\hat{s}_i \langle \delta(r_i)r_i^{-3}\rangle$$

J and I are assumed to be good quantum numbers, thus

$$\langle(\gamma JI)FM_F| \sum_i \frac{\hat{N_i}}{r_i^3}.\hat{I} |(\gamma JI)FM_F\rangle =$$

$$(-1)^{J+I+F} \begin{Bmatrix} J & I & F \\ I & J & 1 \end{Bmatrix} \sqrt{I(I+1)(2I+1)} \langle\gamma J\| \sum_{i=1}^{3} \frac{\hat{N_i}}{r_i^3}\|\gamma J\rangle$$

Each reduced matrix element is further reduced using the well-known
Racah method to obtain an expression involving the products of L-
and S-dependant reduced matrix elements.

- Hyperfine-structure splitting in C III.

The triplet term under study in C III can be considered, in
first approximation, as a two-electron system if the polarization
of the core is neglected. The formulation of the hyperfine-structure
in intermediate coupling by Breit and Wills[6] can be used without
relativistic corrections since they are very small[7] in carbon.

Magnetic dipole matrix elements for an s-ℓ configuration in intermediate coupling are given, e.g., by Lurio et al.[8].

- Decay curves

Several authors have derived expressions for the modulated decay curves on the basis of various assumptions concerning the nature of the excited states produced by the foil excitation process.

The intensity expression for L-S coupling in the absence of cascades can be derived using the density matrix formalism developed in the basis of irreducible tensorial operators[9]. Using the standard beam-foil technique (foil perpendicular to the incident beam), the intensity of the light emitted after the foil is given by

$$I(t) = \exp(-\Gamma t) \sum_{k=0,2} \sum_q \phi_q^k \, \sigma_q^k \, \frac{(-1)^{L_0+L+1+k}(2L+1)}{(2S+1)(2J+1)} \begin{Bmatrix} 1 & L & L_0 \\ L & 1 & k \end{Bmatrix} \sum_{FF'J} M_{FF'J}^k$$

with

$$M_{FF'J}^k = (2F+1)(2F'+1)(2J+1)^2 \begin{Bmatrix} J & J & k \\ L & L & S \end{Bmatrix}^2 \begin{Bmatrix} F & F' & k \\ J & J & 1 \end{Bmatrix}^2 \exp[-i\Omega_{(FF')J}t]$$

where ϕ_q^k and σ_q^k are the irreducible tensor components of the polarization detection operator and of the excitation matrix respectively. Axial symmetry about the beam axis allows only $q = 0$. L_0 denotes the quantum number of the lower energy level of the transition. No mixing between fine- and hyperfine-structure is assumed. $(\hbar)\Omega_{(FF')J}$ is the energy separation between the two levels of quantum numbers F, F', having the same J.

Experiment

Singly charged carbon ions were produced with the 2.5 MeV Van de Graaff accelerator of the Laboratory, from CO gas introduced into an RF ion source. Calibration of the beam energy was obtained by Al (p,γ) reactions. The targets were 10 ± 2 $\mu g/cm^2$ carbon foils. Most of the carbon impurities arising from the diffusion pump oil were eliminated by a cold trap placed near the target. This arrangement limited carbon deposition on the target foils and enabled to increase the time of the recordings. The foil was moved along the beam axis by a stepping motor, and a spectrometer detected polarized light at 90° to the beam axis, at various distances from the target. Normalization of the signal was achieved either by a light pipe collecting radiation in the visible spectral region at a fixed position relative to the foil or by the current obtained from the Faraday cup.

Results

- Modulations in C II.

The beat pattern of Fig. 1 is obtained by substracting decay

$$3d^4F^\circ - 4f^4G \quad C\ II$$

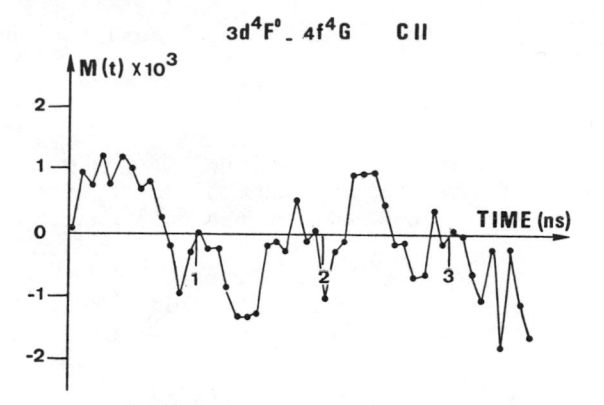

Fig. 1. Beat pattern in C II for the
transition 3d ^{4}F$^\circ$ - 4f ^{4}G (3876 Å)

curves of two different polarization directions[10] (parallel and
perpendicular to the beam axis). The fractionnal alignment is
computed from

$$T(t) = (I_{\parallel} - I_{\perp})/(I_{\parallel} + 2I_{\perp})$$

where the instrumental polarization is included. T(t) is then a sum
of cosines, the frequencies of which yield the hyperfine-structure
splittings. The theoretical expression of the modulated part of the
signal M(t), which is proportional to T(t), is deduced from the
equations above

$$M(t) = 15.29 \cos\Omega_{(3-2)5/2}t + 27.84 \cos\Omega_{(3-4)7/2}t$$
$$+ 26.0 \cos \Omega_{(4-5)9/2}t + 30.9 \cos\Omega_{(5-6)3/2}t$$

The experimental frequencies given in Table I are obtained by a fit
of the experimental curve, using this theoretical expression.

In the computation of the energy splittings, the influence of
the contact term (a_s) and of the 2p electron (a_{2p}) are included
(Table 1). The 4f contribution is negligible.

Table 1

Ω (F-F')	J	Theoretical expression	Experimental results
2-3	5/2	$-1.2856\ a_s + 1.4081\ a_{2p}$	5340 ± 200 MHz
3-4	7/2	$-0.00998 a_s + 1.24185\ a_{2p}$	460 ± 30
4-5	9/2	$0.2862\ a_s + 1.0874\ a_{2p}$	1720 ± 60
5-6	11/2	$0.5564\ a_s + 0.7980\ a_{2p}$	2750 ± 150

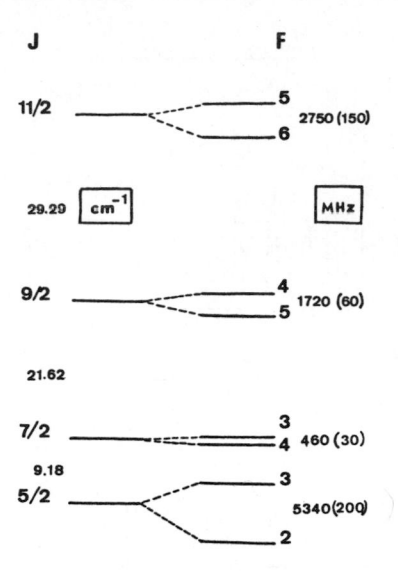

Fig. 2. Fine- and hyperfine-structure level scheme of the
4f ^{4}G term in C II. Fine structure splittings are
measured in cm^{-1} and hyperfine-structure in MHz.

The fine- and hyperfine-structure level scheme of the 4f ^{4}G
term is represented in Fig. 2. The weakness of the coefficient of
a_s for J = 7/2 allows the determination of the constant a_{2p}. From
the experimental results, the values of a_s and a_{2p} are

$$a_s = 4480 \pm 200 \text{ MHz} \qquad a_{2p} = 406 \pm 25 \text{ MHz}$$

The one-electron parameters are consistent with all measured
frequencies. Further measurements are being undertaken to reduce
the error in the determination of a_s and a_{2p}.

- Modulations in C III

The previous technique employed to obtain the modulation is not
valid for C III due to a mixing with a transition of C IV. The data
were least-squares fitted with two exponential and two cosine
functions and a background. Fig. 3 shows the modulated part of the
signal. The total expression, depending on ten parameters, gives a
χ^2 of 1.45. Fig. 4 shows the Fourier transform of the beat pattern.
As is clearly seen on the level scheme of the fine- and hyperfine-
structure (Fig. 5), there are only two frequencies. The horizontal
dashed line corresponds to the position of the P_1 level in pure
Russel-Saunders coupling.

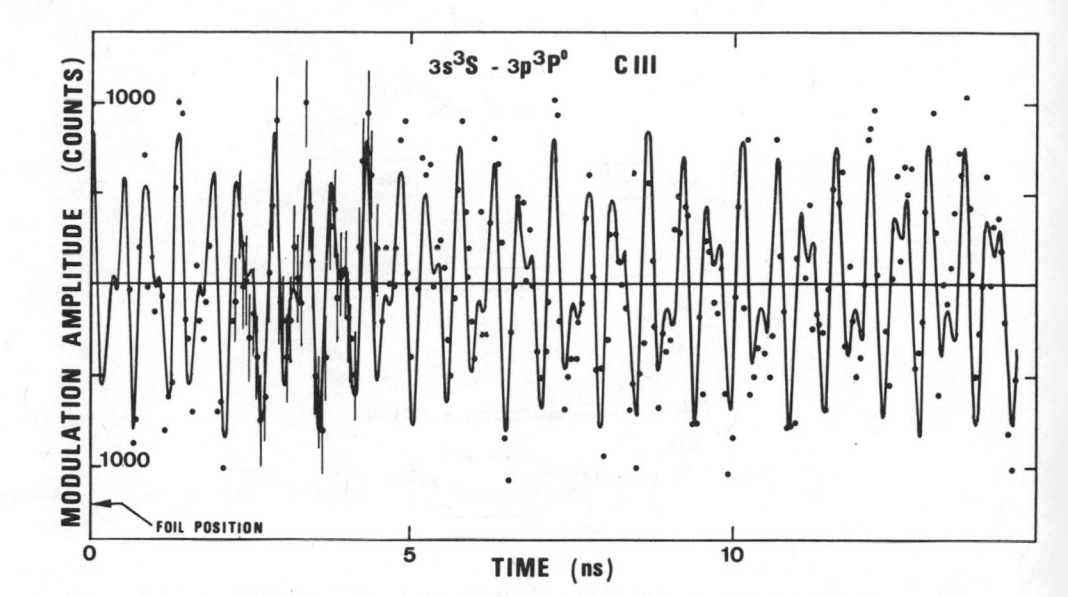

Fig.3. Beat pattern in C III. The modulation is obtained after
substraction of two exponential functions and a background. The
solide curve is a least-square fit with a sum of two cosines.

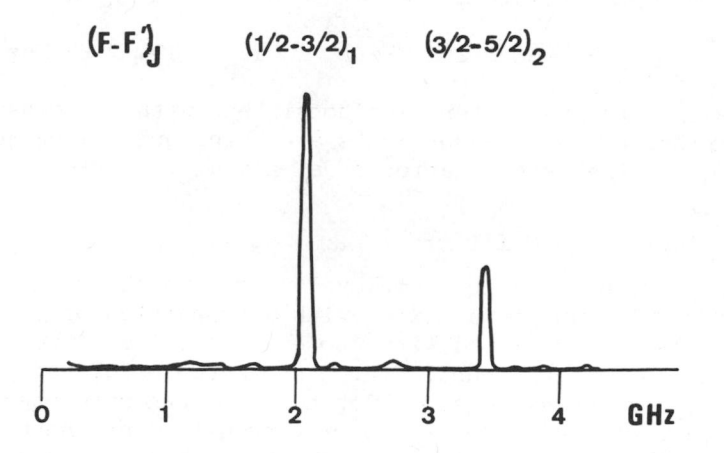

Fig.4. Fourier transform of the modulation obtained
in C III. The intensity of the higher frequency
peak is reduced due to instrumental effect.

The zero-field energy levels are given by

$$W_F = A(J) \, K/2$$

where $A(J)$ is the magnetic-dipole hyperfine-interaction constant and

$$K = F \, (F + 1) - I \, (I + 1) - J \, (J + 1)$$

The energy separations between the J levels are then

$$\Omega_{(3/2-5/2)2} = 5/2 \, A(^3P_2) \qquad \Omega_{(1/2-3/2)1} = 3/2 \, A(^3P_1)$$

The experimentally determined values of A are

$$A(^3P_2) = 1376 \pm 15 \text{ MHz} \qquad A(^3P_1) = 1363 \pm 15 \text{ MHz}$$

Furthermore, the ratio of the hyperfine constants is

$$A(^3P_2)/ \, A(^3P_1) = 1.010$$

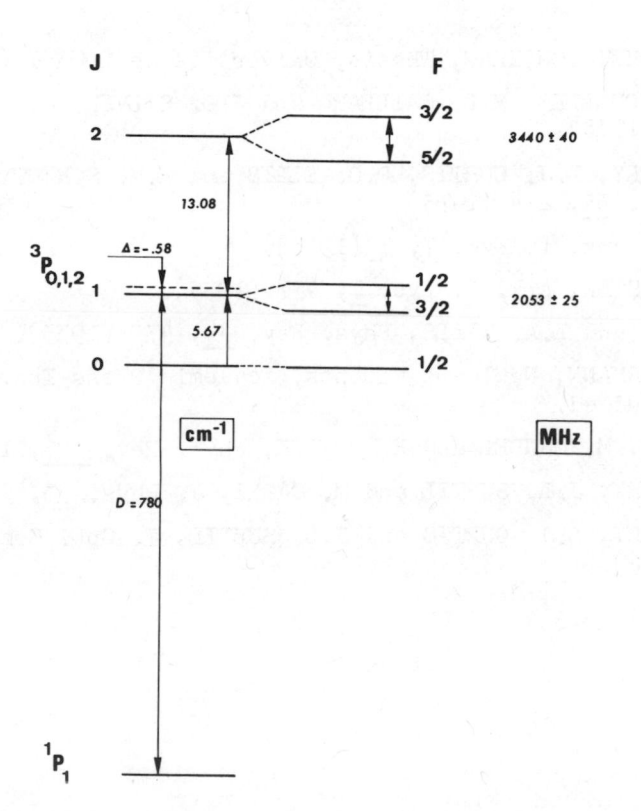

Fig. 5. Level scheme of the fine- and hyperfine-structure of the $3p \; ^3P^\circ$ term in C III. Fine-structure splittings are given in cm^{-1} and hyperfine-structure splittings in MHz. The dashed line represents the position of the P_1 term in pure Russel-Saunders coupling. Δ and D characterize the intermediate coupling.

Following the formulation of Lurio et al.[8], calculation of the theoretical energy levels was performed, but gives a result which is not in agreement with the experimental ratio. In order to improve the calculations, one must consider the corrections due to core polarization.

Conclusions

Hyperfine-structure measurements in ^{13}C II and ^{13}C III can be obtained using the zero-field quantum-beat technique.

One-electron parameters have been determined in C II and further work with increased accuracy is in preparation.

The measurements in C III are more precise and show that higher corrections are necessary to explain the hyperfine-structure splittings.

References

1. M.C. BUCHET-POULIZAC, Thesis, Université de Lyon I (1974).

2. J. DESESQUELLES, M.L. GAILLARD and J.D. SILVER, J. of Phys. B (August 1975).

3. H.G. BERRY, L.J. CURTIS, D.G. ELLIS and R.M. SCHECTMAN, Phys. Rev. Let. 35, 274 (1975).

4. S. GLAD, Ark. f. Fys. 7, 7 (1953).

5. K. BOCKASTEN, Ark. f. Fys. 7, 457 (1955).

6. G. BREIT and L.A. WILLS, Phys. Rev. 44, 470 (1933).

7. H. KOPFERMANN, Nuclears Moments, Academic Press Inc., New York (1958) 2nd ed.

8. A. LURIO, M. MANDEL and R. NOVICK, Phys. Rev. 126, 1758 (1962).

9. H.G. BERRY, J.L. SUBTIL and M. CARRE, J. Phys. 33, 947 (1972).

10. H.G. BERRY, L.J. CURTIS and J.L. SUBTIL, J. Opt. Soc. Am. 62, 771 (1972).

QUANTUM BEATS IN Hα AND Hβ AFTER BEAM-FOIL EXCITATION

A. Denis, J. Desesquelles, M. Druetta, and M. Dufay

Laboratoire de Spectrométrie Ionique et
    Moleculaire (associé au C.N.R.S.)
Université de Lyon I, Campus de La Doua
69621 Villeurbanne, France

Since the first zero-field quantum beat experiments of Bashkin and Beauchemin[1], many such studies have been undertaken with light atoms or ions. A general bibliography on the subject may be found in the compilations of Bashkin[2], Martinson and Gaupp[3] and Andrä[4]. With hydrogen projectiles, measurements have been performed principally on the 2P[5,6] and 3P[7] levels. A study of the beats observed with the Hβ transition has been carried out by Burns and Hancock[8], who have established, in addition to the fine structure beats, beats between levels with $\Delta\ell = 2$. Such beats have been predicted by Macek[9] in an extension of the theory of Percival and Seaton[10]. We have extended the experimental study of Hα and Hβ to measure the excitation cross sections of the sublevels, to confirm the SD coherence and to precisely determine phases and amplitudes. Such measurements are compared with Eck's theoretical model of beam-foil interaction.

Experimental arrangement

A 250 KeV H$^+$ ion beam from a 2 MV Van de Graaff accelerator impinges upon a carbon foil (10 $\mu g/cm^2$) moved by a stepping motor. The light emitted perpendicularly to the beam is detected by a spectrometer-photomultiplier combination. The optical detection system is composed of a fused silica lens, with a vertical slit as a diaphragm, and a linear polarizer. The axis of polarization may be set either parallel to or perpendicular to the beam direction.

The target chamber is shielded from the magnetic field of the earth to ~ 0.01 Gauss in the transverse direction, which corresponds to a motional electric field of 7 V/m.

To determine precisely the phases of the modulation, it is necessary in this experiment to know the zero position of the foil.

For this purpose, we have placed a vertical slit 4 mm upstream from the foil. A removable auxiliary foil at the entrance of the chamber is used to excite the particles of the beam which are then observed through this slit (Fig. 1).

For each position of the foil, the photomultiplier signal is amplified and analyzed by a multichannel analyzer. The integration time is determined by the ion beam current. The decay curves are recorded for the two positions of the polarizer on a length of ~ 10 cm corresponding to ~ 15 ns. At least 12 experimental points are obtained per period for the greatest frequency, with steps of 200 or 300 μm. The instrumental polarization has been taken into account.

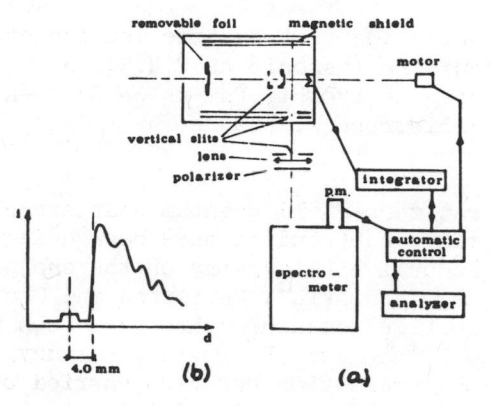

Fig. 1. Experimental set-up (a), and record with the slit upstream from the foil, for determination of the exact origin (b).

Analytical expression of the signal.

For a given quantum number n, the treatment of the density matrix expressed in the $|LM_L\rangle$ basis is classical (Macek and Jaecks[12]) Furthermore, in this case, because of the L degeneracy of the hydrogen levels, it is necessary to include the different transition probabilities from the S, P and D states. As a true coherence between states of different L may occur, the whole density matrix of a level with a given n must be treated.

In the hypothesis of a pure electrostatic interaction[10] between the hydrogen atoms and the carbon target, the density matrix may be expressed in a simple manner in the decoupled $|LM_L\rangle |SM_S\rangle$ basis. In the $|LM_L\rangle$ basis, the form of the density matrix is defined by the cylindrical symmetry of the excitation[12,13]. The cylindrical symmetry does not exclude the existence of coherence excitation terms such as

$$^{LL'}\sigma(0) = \langle LM_L|\sigma(0)|L'M_L\rangle$$

but we will limit ourselves to $^{LL'}\sigma(0)$ elements with L and L' of the same parity and which alone are coupled by the monitoring operator. Thus, at time t = 0, the density matrix may be written in the $|LM_L\rangle$ basis with the cross-sections σ_{So}, $\sigma_{Po,1}$, $\sigma_{Do,1,2}$ as diagonal terms and the coherence term σ_{SD},

By coupling in the $|JM_J\rangle$ basis, we obtain the matrix of the Table I. The evolution equation of the density matrix is

$$i\hbar d\sigma(t)/dt = H\sigma(t) - \sigma(t)H^+$$

in which $H = H_O + H_D$, where H_O is the hydrogen atom hamiltonian, and H_D the phenomenologic non-hermitic hamiltonian expressing the level de-excitation. The $\sigma(t)$ matrix is deduced from $\sigma(0)$ in this way

$$\langle LSJM_J|\sigma(t)|LSJM_J\rangle = \exp(-\Gamma_L t)\langle LSJM_J|\sigma(0)|LSJM_J\rangle$$

$$\langle LSJM_J|\sigma(t)|LSJ'M_J\rangle = \exp[-\Gamma_L t + it(\omega_{LJ}-\omega_{LJ'},)]\langle LSJM_J|\sigma(0)|LSJ'M_J,\rangle$$

$$\langle LSJM_J|\sigma(t)|L'SJ'M_J\rangle = \exp[-(\Gamma_L+\Gamma_{L'},)\tfrac{t}{2}+it(\omega_{LJ}-\omega_{L'J})]\langle LSJM_J|\sigma(0)|L'SJ'M_J\rangle$$

where Γ_L is the decay constant of the L state, and $\hbar\omega_{LJ}$ the energy of the LJ state.

The intensity of the light emitted with a polarization vector $\vec{e}_\lambda$ in a given direction is

$$I(\vec{e}_\lambda) = \text{Tr}\ (\sigma.\mathbf{M})$$

where M is the monitoring operator

$$M = \sum_\mu \langle m|\vec{e}_\lambda.\vec{P}|\mu\rangle\langle\mu|\vec{e}_\lambda^*.\vec{P}|m'\rangle$$

For an observation direction perpendicular to the beam axis and for polarization direction parallel ($I_{||}$) or perpendicular ($I_\perp$) to the beam axis respectively, we obtain the following expressions

$$I_{||} - I_\perp = \tfrac{1}{3}(\sigma_{Po}-\sigma_{P1})[1+2\cos(\omega_{P3/2}-\omega_{P1/2})t]\exp(-\Gamma_P t)A_{nP\ \to\ 2S}$$

$$+ \tfrac{1}{50}(\sigma_{Do}+\sigma_{D1}-2\sigma_{D2})[19+6\cos(\omega_{D5/2}-\omega_{D3/2})t]\exp(-\Gamma_D t)A_{nD\ \to\ 2P}$$

$$+\tfrac{\sqrt{2}}{5}|\sigma_{SD}|\{2\cos[(\omega_{D3/2}-\omega_{S1/2})t-\varphi] + 3\cos[(\omega_{D5/2}-\omega_{S1/2})t-\varphi]\}$$

$$\times\ \exp[-(\Gamma_S+\Gamma_D)/2.t](A_{nS\ \to\ 2P}A_{nD\ \to\ 2P})^{1/2}$$

$$I_{||} + 2I_\perp = \sigma_{So}\exp(-\Gamma_S t)A_{nS\ \to\ 2P} + (\sigma_{Po}+2\sigma_{P1})\exp(-\Gamma_P t)A_{nP\ \to\ 2S}$$

$$+ (\sigma_{Do}+2\sigma_{D1}+2\sigma_{D2})\exp(-\Gamma_D t)A_{nD\ \to\ 2P}$$

Table 1. The density matrix in the $|\,J\,M_J\,\rangle$ basis.

The density matrix is written in the $|\,J\,M_J\,\rangle$ basis. Columns and rows are labelled by J (= $S_{1/2}$, $P_{3/2}$, $P_{1/2}$, $D_{5/2}$, $D_{3/2}$) and M_J.

Row/column M_J values:
- $S_{1/2}$: $\tfrac12,\ -\tfrac12$
- $P_{3/2}$: $\tfrac32,\ \tfrac12,\ -\tfrac12,\ -\tfrac32$
- $P_{1/2}$: $\tfrac12,\ -\tfrac12$
- $D_{5/2}$: $\tfrac52,\ \tfrac32,\ \tfrac12,\ -\tfrac12,\ -\tfrac32,\ -\tfrac52$
- $D_{3/2}$: $\tfrac32,\ \tfrac12,\ -\tfrac12,\ -\tfrac32$

$S_{1/2}$ block (columns $S_{1/2}$):

$$\begin{pmatrix} \dfrac{\sigma_{S_0}}{2} & 0 \\[4pt] 0 & \dfrac{\sigma_{S_0}}{2} \end{pmatrix}$$

$P_{3/2}$–$P_{1/2}$ block (entries read from the table):

$$\frac{\sigma_{P_1}}{2}, \qquad \frac16\big(\sigma_{P_1}+2\sigma_{P_0}\big), \qquad \frac{\sqrt{2}}{6}\big(\sigma_{P_1}-\sigma_{P_0}\big), \qquad \frac16\big(\sigma_{P_0}+2\sigma_{P_1}\big), \qquad \frac{\sigma_{P_1}}{2}$$

$S_{1/2}$–D cross terms:

$$\sqrt{\frac{3}{5}}\,\frac{\sigma_{SD}}{2}, \qquad \sqrt{\frac{2}{5}}\,\frac{\sigma_{SD}}{2}, \qquad -\sqrt{\frac{2}{5}}\,\frac{\sigma_{SD}}{2}, \qquad \left(\sqrt{\frac{3}{5}}\,\frac{\sigma_{SD}}{2}\right)^{\!*}, \qquad \left(\sqrt{\frac{2}{5}}\,\frac{\sigma_{SD}}{2}\right)^{\!*}$$

$D_{5/2}$ diagonal block:

$$\frac{\sigma_{D_2}}{2}, \qquad \frac{1}{10}\big(\sigma_{D_2}+4\sigma_{D_1}\big), \qquad \frac{1}{10}\big(2\sigma_{D_1}+3\sigma_{D_2}\big), \qquad \frac{1}{10}\big(2\sigma_{D_2}+3\sigma_{D_1}\big), \qquad \frac{1}{10}\big(\sigma_{D_1}+4\sigma_{D_2}\big), \qquad \frac{\sigma_{D_2}}{2}$$

$D_{5/2}$–$D_{3/2}$ cross terms:

$$\frac{\sqrt{6}}{10}\big(\sigma_{D_2}-\sigma_{D_1}\big), \qquad \frac{2}{5}\big(\sigma_{D_1}-\sigma_{D_2}\big)$$

$D_{3/2}$ diagonal block:

$$\frac{1}{10}\big(\sigma_{D_1}+4\sigma_{D_2}\big), \qquad \frac{1}{10}\big(3\sigma_{D_1}+2\sigma_{D_2}\big), \qquad \frac{1}{10}\big(3\sigma_{D_2}+2\sigma_{D_1}\big), \qquad \frac{1}{10}\big(\sigma_{D_2}+4\sigma_{D_1}\big)$$

(All remaining matrix elements not listed above are 0.)

$I_{\parallel} + 2I_{\perp}$ is the total light intensity.

The hyperfine structure is very weak compared with the fine structure for the P and D levels. For the S levels, the hyperfine structure is 52.7 MHz for n = 3 and 22 MHz for n = 4. These frequencies are not negligible with respect to the observed frequencies. We have thus expressed the density matrix in the $|FM_F\rangle$ basis to include the S hyperfine structure in the SD coherence term, and we obtain for the SD beats

$$I_{\parallel} - I_{\perp} = \frac{\sqrt{2}}{20}|\sigma_{SD}|\{3[3\cos(\omega_{D5/2 - SF=1}t - \varphi) + \cos(\omega_{D5/2 - SF=0}t - \varphi)]$$
$$+ 2[3\cos(\omega_{D3/2 - SF=1}t - \varphi) + \cos(\omega_{D3/2 - SF=0}t - \varphi)]\}$$
$$\times \exp[-(\Gamma_S + \Gamma_D)/2 \cdot t](A_{nS \to 2P} A_{nD \to 2P})^{1/2}$$

Treatment of the signal

A Fourier transform analysis clearly confirms the existence of SD coherence. Fig. 2 shows for Hα and Hβ , in addition to the

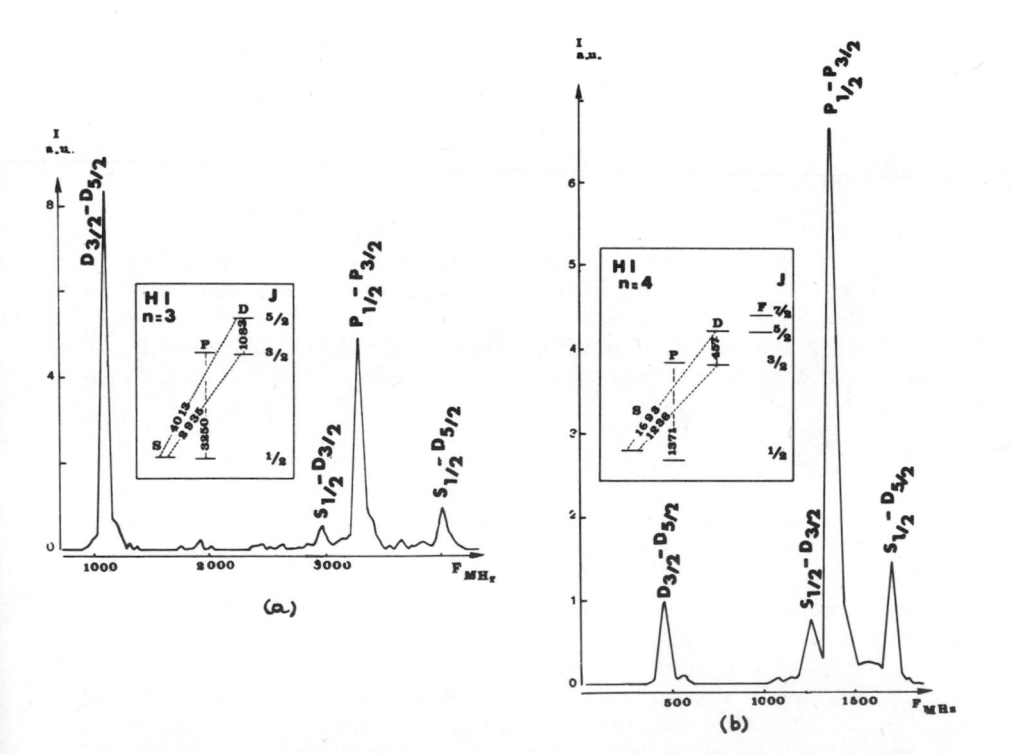

Fig. 2. Fourier transform of $I_{\parallel} - I_{\perp}$, and fine structure diagram of the upper level for Hα (a) and Hβ (b). Frequencies in MHz.

$P_{1/2}$-$P_{3/2}$ and $D_{3/2}$-$D_{5/2}$ frequencies, frequencies corresponding to $S_{1/2}$-$D_{3/2}$ and $S_{1/2}$-$D_{5/2}$. As the $S_{1/2}$-$P_{1/2}$ frequency does not appear on the spectra, it seems that the residual motional electric field does not mix these levels.

In order to determine the amplitudes and phases of the beats, data of $I_{\parallel}$ - $I_{\perp}$ have been treated with a computer program which employs a search procedure to minimize χ^2. The fitting function was

$$F(t) = A_1 \exp(-\Gamma_1 t)\left\{1 + 2\cos[(\omega_{P3/2} - \omega_{P1/2})t - \varphi_1]\right\}$$
$$+ A_2 \exp(-\Gamma_2 t)\left\{19 + 6\cos[(\omega_{D5/2} - \omega_{D3/2})t - \varphi_2]\right\}$$
$$+ A_3 \exp(-\Gamma_3 t)\left\{2\cos[(\omega_{D3/2} - \omega_{S1/2})t - \varphi_3] + 3\cos[(\omega_{D5/2} - \omega_{S1/2})t - \varphi_3]\right\} + A_4$$

with eleven parameters (beam velocity, $A_{i=1..4}$, $\Gamma_{j=1..3}$, $\varphi_{k=1..3}$).

Another function $F'(t)$ has been also used to take into account the S hyperfine structure for the SD beats.

For the $I_{\parallel} + 2I_{\perp}$ data, we have used an expression with four variable coefficients $B_{i=1..4}$

$$G(t) = B_1 \exp(-\Gamma_S t) + B_2 \exp(-\Gamma_P t) + B_3 \exp(-\Gamma_D t) + B_4$$

The best fits give a χ^2 of 0.82 for H_α and 1.09 for H_β. Fig. 3 and 4 show the results.

Results

The total cross sections σ_{So}, $\sigma_{Po} + 2\sigma_{P1}$, $\sigma_{Do} + 2\sigma_{D1} + 2\sigma_{D2}$ have been determined by the $I_{\parallel} + 2I_{\perp}$ analysis. However, due to the short decay length of the beam, corresponding to 13 ns, and to the long lifetime of the S state (τ_{3S}=158 ns, τ_{4S}=227 ns), the σ_{So} cross section values are not accurate. This may explain the discrepancy between our results and those of Bukow et al.[14] and Alguard et al.[15,16].

The σ_{Po}-σ_{P1}, σ_{Do}+σ_{D1}-$2\sigma_{D2}$ and $|\sigma_{SD}|$ coefficients are determined from the $I_{\parallel}$ - $I_{\perp}$ analysis. The results are summarized in Table II. For n = 3, the $(\sigma_{Po}-\sigma_{P1})/(\sigma_{Po}+\sigma_{P1})$ ratio ($\sim$ -0.1) is in good agreement with the value found by Lynch et al.[7] for Lyman β . We remark that the intensity at the foil for the $P_{3/2}$-$P_{1/2}$ and $D_{5/2}$-$D_{3/2}$ beats reaches a minimum when the polarizer transmission axis is parallel to the beam. Consequently, the sub-levels $M_L = 0$ are less populated. These results extend to n = 3 and n = 4 the study of the n = 2 level of the hydrogen atom [5,6].

Unlike Burns and Hancock[8], who found initial phases of 5.02 radians for the fine structure beats, we found these phases equal to zero, in agreement with the theory, except for H_β, for which the phase of the $D_{5/2}$-$D_{3/2}$ beats is $\sim$ 0.1 radian.

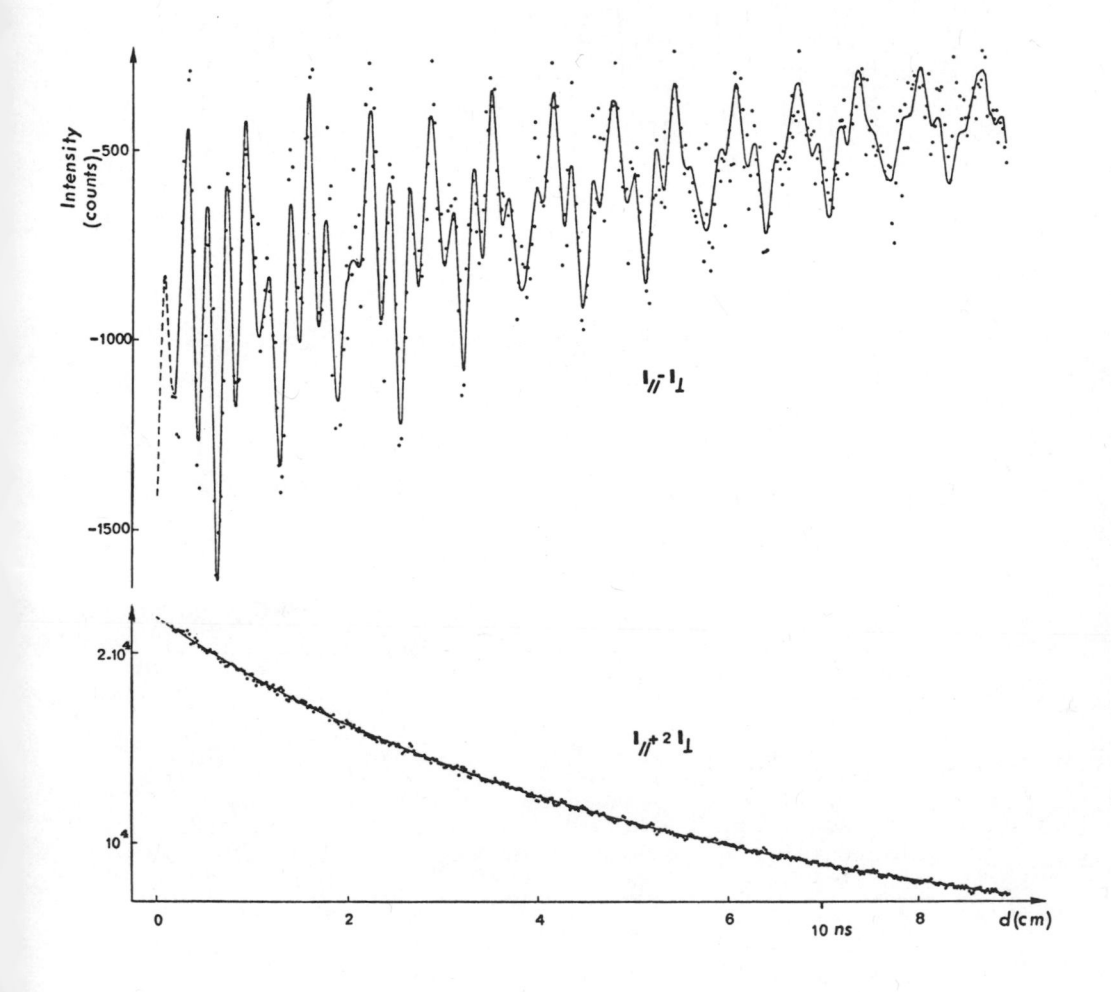

Fig. 3. Experimental and fitted curves obtained
with $I_{||} - I_{\perp}$ (top) and $I_{||} + 2I_{\perp}$ (bottom) for H_{α}

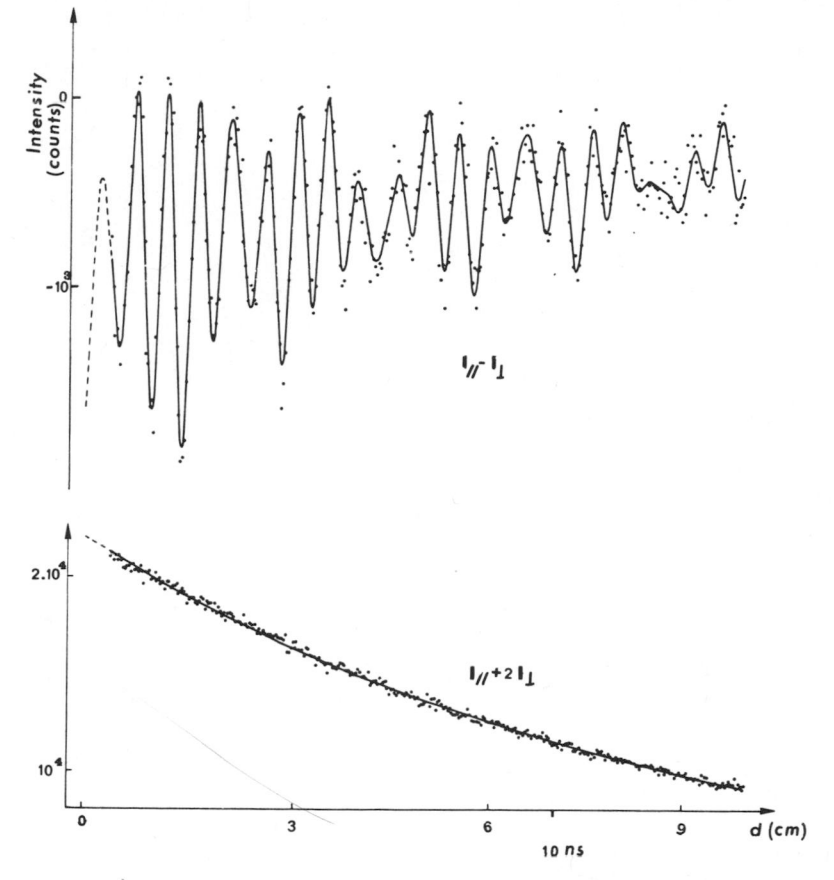

Fig. 4. Experimental and fitted curve obtained with
$I_{||} - I_{\perp}$ (top) and $I_{||} + 2I_{\perp}$ (bottom) for Hβ.

The phases of the SD beats have been determined to be 1.45 and
0.80 radians for n = 3 and n = 4 respectively. A complex expression
is thus obtained for σ_{SD}. We will comment upon this result in the
following section.

The decay constants of the 3P, 3D and 4D levels, given in Table
III, are in good agreement with the theoretical results. For the 4P
level, the decay constant is greather than the theoretical one.

For the SD beats, the decay constant is greather than the
calculated constant $(\Gamma_S + \Gamma_D)/2$. When we use the analytical expression
taking into account the hyperfine structure, to fit the experimental
data, the decay constant is closer to the theoretical value, but is
still about three times to large. This may be explained in parts by the
straggling of the atoms of the beam. This effect is more sensitive
for damped beats with high frequency and long lifetime (Table III).
Moreover, the observed anomalies may be due to the residual motional
electric field which mixes the $P_{3/2} - D_{3/2}$ and $4D_{5/2} - 4F_{5/2}$ levels.

Table II. Relative excitation cross sections.

	$n = 3$	$n = 4$
σ_{So}	0.3	0.055
σ_{Po}	0.135	0.24
σ_{P1}	0.166	0.30
$\sigma_{Do} + 2\sigma_{D1} + 2\sigma_{D2}$	0.22	0.11
$\sigma_{Do} + \sigma_{D1} - 2\sigma_{D2}$	-0.041	-0.033
$|\sigma_{SD}|$	0.0093	0.017

Table III. Decay constants of the beats in H_α and H_β.

	$n = 3$			$n = 4$		
	experiment		theory	experiment		theory
	a	b		a	b	
$\Gamma_{nP}(ns^{-1})$	0.19	0.19	0.192	0.130	0.130	0.0812
$\Gamma_{nD}(ns^{-1})$	0.068	0.068	0.065	0.0274	0.0274	0.0276
$(\Gamma_{nS}+\Gamma_{nD})/2$	0.19	0.10	0.035	0.081	0.051	0.016

a. Value with fine structure only.
b. Value with hyperfine structure.

Discussion.

A possible mechanism for coherent production of atomic states has been proposed by Eck[11]. In this theory, the atoms arrive at the foil exit surface in a incoherent superposition of $|LM_L\rangle$ states. The coherence is produced by the intense electrostatic field which is to be found in the vicinity of the foil. For example, the interaction hamiltonian for $n = 3$ may be written

	S	P	D
S	0	V_{SP}	0
P	V_{SP}	0	V_{PD}
D	0	V_{PD}	0

where the V_{ij} are real and equal to $\langle i|e\vec{E}.\vec{r}|j\rangle$. The density matrix is a solution of the evolution equation

$$i\hbar d\sigma/dt = [\, H(t),\, \sigma(t)\,]$$

which may be resolved formally by successive iteration. The solution is of the form

$$\begin{vmatrix} \sigma_S & i\sigma_{SP} & \sigma_{SD} \\ -i\sigma_{SP} & \sigma_P & i\sigma_{PD} \\ \sigma_{SD} & -i\sigma_{PD} & \sigma_D \end{vmatrix}$$

where σ_{SP}, σ_{PD} and σ_{SD} are real. This result is in disagreement with the experiment which yield a complex value for σ_{SD}. It seems then, that SD coherent states are also formed inside the foil.

References.

1. S. BASHKIN and G. BEAUCHEMIN, Canad. J. of Phys. 44, 1603 (1966).

2. S. BASHKIN, Progress in Optics XII, North Holland (1974).

3. I. MARTINSON and A. GAUPP, Physics Reports 15 C, 115 (1974).

4. H.J. ANDRÄ, Physica Scripta 9, 257 (1974).

5. P. DOBBERSTEIN, H.J. ANDRÄ, W. WITTMANN and H.H. BUKOW, Z. f. Physik 257, 272 (1972).

6. H.J. ANDRÄ, P. DOBBERSTEIN, A. GAUPP and W. WITTMANN, N.I.M. 110, 301 (1973).

7. D.J. LYNCH, C.W. DRAKE, M.J. ALGUARD and C.E. FAIRCHILD, Phys. Rev. Let. 26, 1211 (1971).

8. D.J. BURNS and W.H. HANCOCK, Phys. Rev. Let. 27, 370 (1971).

9. J. MACEK, Phys. Rev. Let. 23, 1 (1969).

10. I.C. PERCIVAL and M.J. SEATON, Phil. Trans. Roy. Soc. London A 251, 113 (1958).

11. T.G. ECK, Phys. Rev. Let. 33, 1055 (1974).

12. J. MACEK and D.H. JAECKS, Phys. Rev. 4, 2228 (1971).

13. D.G. ELLIS, J. O. S. A. 10, 1232 (1973).

14. H.H. BUKOW, H.V. BUTTLAR, D. HAAS, P.H. HECKMANN, M. HOLL, W. SCHLAGHECK, D. SCHURMANN, R. TIELERT and R. WOODRUFF, N. I. M. 110, 89 (1973).

15. M.J. ALGUARD and C.W. DRAKE, Phys. Rev. A 8, 27 (1973).

16. M.J. ALGUARD and C.W. DRAKE, N. I. M. 110, 311 (1973).

ORIENTATION AND ALIGNMENT CHANGES INDUCED BY TILTED FOILS[*]

R. M. Herman

Department of Physics, The Pennsylvania State University

University Park, Pennsylvania 16802

Recently, considerable interest has centered on the polarization properties of light emitted from atoms or ions following beam
passage through foils. Of particular interest are the experiments
which demonstrate changes in polarization resulting from tilting
the foil relative to the beam direction.[1,2] While the theory of
Eck[3] is the only one to date to have been published describing the
observed phenomena, several other researchers[4] have recently worked
on the problem. In each of these approaches, pronounced effects
induced by the "surface potential"--either through phase shifts
between magnetic substates of the levels under study, or through
inelastic transitions--have been important. The surface potential
would nominally represent a collective interaction with all atoms
and electrons of the foil such that the surface potential would
vary only with distance from the surface.

Nonetheless, there exists the problem of establishing that
such surface terms lead to typical matrix elements large enough to
influence significantly the states of the emergent atoms or ions.
In order to do this effectively, typical phase shifts differing
by as much as unity should be encountered for the various magnetic
substates, for example. Now, in order that the surface atoms
yield a collective "surface interaction potential," one must be
separated from the surface by at least a kinetic radius[5] (otherwise one is ordinarily in the field of a single atom); at these
distances, typical interaction matrix elements are of the order of
0.1 ev, and with typical beam energies (40 kev/nucleon, for example) and a range of 2 Å, say, we obtain phase shifts only of
order 10^{-2}. These are too small to lead to observed effects,
especially since they ultimately must be squared to yield

809

probabilities. At the same time, surface induced inelastic amplitudes can be no larger than the ratio of "surface potential" matrix
elements to typical electronic energy separations, which, again are
of order 10^{-2}.

    In order to obtain larger matrix elements, one indeed must
include shorter range interactions; but then one is clearly dealing
with single-atom encounters within the foil and because of this,
it is necessary to develop an atomic collision picture of the phenomena. The following represents an attempt to do this in a simple
manner. However, since the actual foil surface geometry and interactions are not as simple as we assume here, the present attempt
must be regarded only as a schematic theory.

    Consider a single atom passing through a foil, as shown in
Fig. 1. (Eventually we average over all such atom trajectories.)
The foil tilt introduces additional encounters in region ① while
eliminating encounters in region ② which would otherwise be

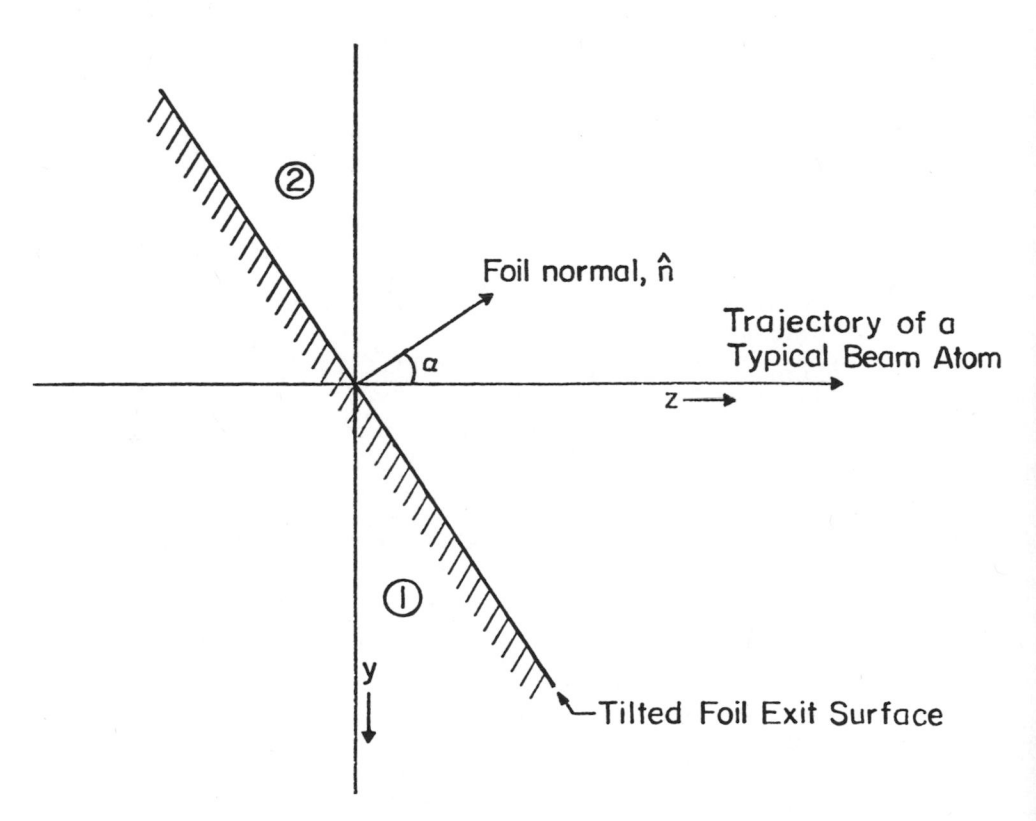

Fig 1. Side view of tilted foil geometry. The emerging atom suffers extra encounters in region ① , and lack some encounters in
region ② which would be experienced, were the foil not tilted.

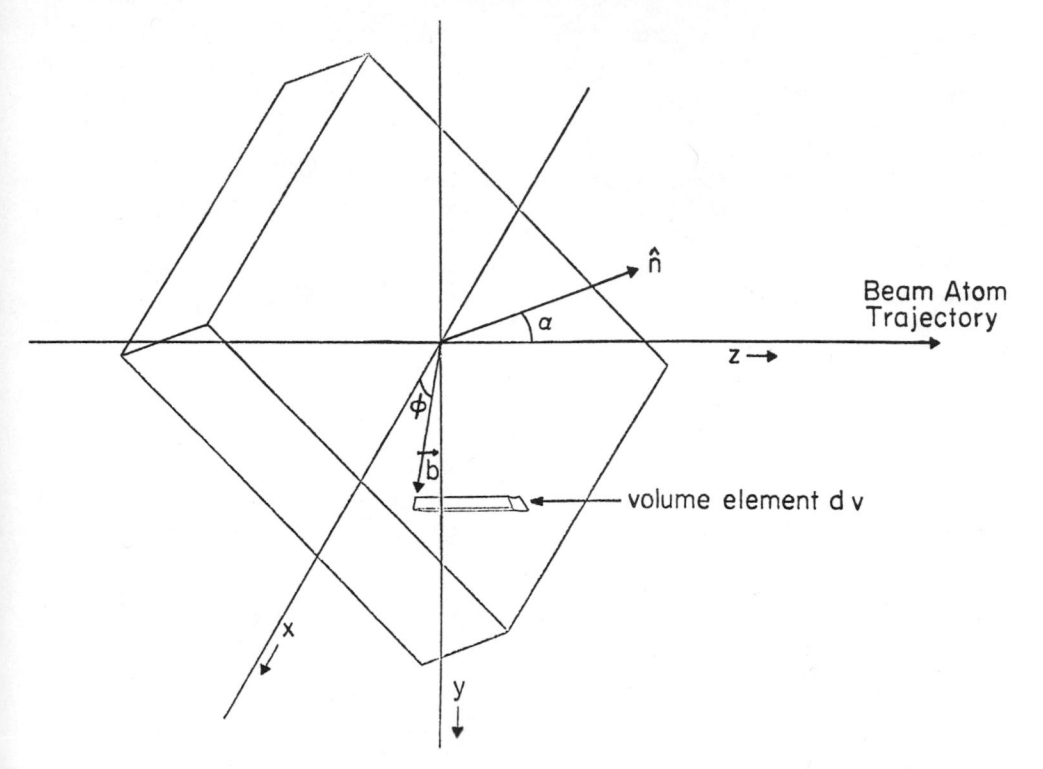

Fig. 2. View of tilted foil showing typical volume element whose
height is determined by the distance perpendicular to the x-y plane
(parallel to the beam trajectory) lying between the x-y plane and
the surface of the foil.

present in the absence of tilt. In Fig. 2, we show a typical
volume element within region ① , where the height of the element,
extending from the x-y plane to the surface is $b \tan \alpha \sin \phi$, α
being the tilt angle, ϕ the azimuthal angle measured from the
x-axis in the x-y plane and b the inpact parameter. The probability
of final encounter within the interval $b \to b + db$, $\phi \to \phi + d\phi$ is
then

$$dP(b,\phi) = n \tan \alpha \, b^2 \, db \sin \phi \, d\phi \quad , \tag{1}$$

n being the number density of foil atoms. Notice that for region
② , $\pi < \phi < 2\pi$, $dP(b,\phi)$ is negative, corresponding to the absence
of collisions mentioned above.

To lowest order in α, the density matrix may be thought of as
that existing just prior to emergence, modified by collisions which
may take place in regions ① , ② ,

$$\rho(\alpha) = \rho(\alpha = o) + \Delta\rho(\alpha) \tag{2}$$

While one is not capable of calculating $\rho(\alpha = o)$, we know from symmetry properties and experience, that it is diagonal in magnetic quantum numbers relative to the beam axis, with unequal elements representing the various substate populations, the m and -m sub-states nonetheless being equally populated.

It now remains to calculate the collisional term $\Delta\rho(\alpha)$. Because this has been discussed elsewhere,[6] we shall not go into the details at the present time. Taking into account

1) the fact that the modification of $\rho(\alpha = o)$ by a single collision is $T(b,\phi) \, \rho(\alpha = o) \, T^{-1}(b,\phi)$, where $T(b,\phi)$ is the time development matrix for completed collisions at inpact parameter b, azimuthal angle ϕ;

2) the transformation properties of magnetic substates with respect to rotations about the z-axis through angle ϕ; the facts that

3) $T(b,\phi)$ is unitary;

4) $\rho(\alpha = o)$ is diagonal;

5) $dP(b,\phi)$ is as given in Eq. (1); and finally that

6) the contributions to $\Delta\rho(\alpha)$ from all single collisions as a simple average,

$$\Delta\rho(\alpha) = \int dP(b,\phi) \; T(b,\phi) \; \rho(\alpha = o) \; T^{-1}(b,\phi)$$

we obtain the result

$$\rho(\alpha)_{mm'} = \rho(\alpha = o)_{mm} \, \delta_{mm'} + i\pi n \, \tan\alpha \int_0^\infty b^2 db$$

$$\times \left(T(b,\phi = o) \; \rho(\alpha = o) \; T^{-1}(b,\phi = o) \right)_{mm'}$$

$$\times \left(\delta_{m,m'} - 1 \; {}^{-\delta}m,m' + 1 \right) \quad, \tag{3}$$

with $\delta_{m,m'}$ being unity if $m' = m$ and zero otherwise.

We observe that the correction term $\Delta\rho(\alpha)$, in Eq. (3) vanishes if all states are initially equally populated (no alignment present for untilted foil). To the extent that multiple collisions are important, they would become so only for large values α. These effects have been ignored at this stage of development, although such terms would lead to changes in the polarization tensoral

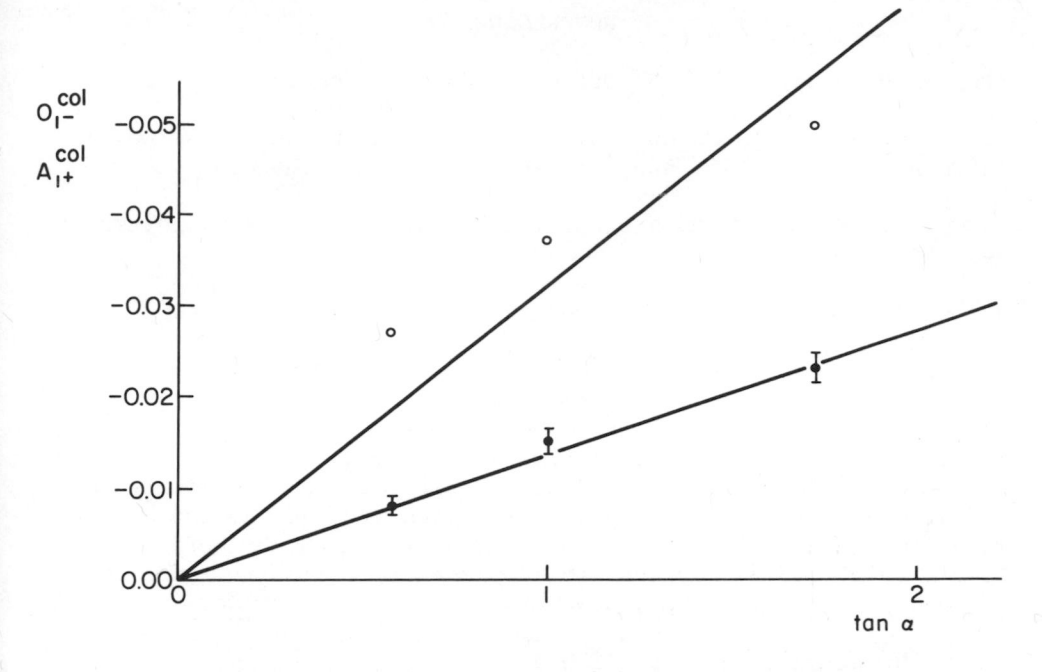

Fig. 3. Measured radiation tensor components (ref. 2) vs tan α.
• : O_{1-}^{col}; o : A_{1+}^{col}; —— : theoretically determined straight lines
having slopes determined through least squares fits to the data.

components[1,2] A_0^{col} and A_{2+}^{col} (or I and M); the lowest order changes
shown in Eq. (3) affect only O_{1-}^{col} and A_{1+}^{col} (or S and C), however,
each of which show a linear dependence on tan α.

While the α-dependence predicted by the present theory does
not yield conclusive agreement with experiment, there are some
encouraging comparisons. Possibly the most accurately measured
values of O_{1-}^{col} and A_{1+}^{col} are those determined by Church et al[2] for
the $4d^1D_2$ state in He I. In Fig. 3, we have plotted these values
vs tan α. The closeness of the measured values of O_{1-}^{col} to a
straight line passing through the origin is highly encouraging; on
the other hand, the measured values of A_{1+}^{col} (error estimates were
not given in ref. 2) do not fit the linear relationship so well.

Further comparisons between theory and experiment are given in
ref. 6. Since the results of such comparisons do not appear to be
conclusive, it is fair to say that a great deal of work remains,
within a theory based on single atom encounters, before detailed
comparisons between theory and experiment can really meaningfully
be made.

REFERENCES

*Research supported by the Office of Naval Research.

1. H. G. Berry, L. J. Curtis, and R. M. Schechtman, Phys. Rev. Letters $\underline{34}$, 509 (1975) and references contained therein.

2. D. A. Church, M. C. Michel, and W. Kolbe, Phys. Rev. Letters $\underline{34}$, 1140 (1975).

3. T. G. Eck, Phys. Rev. Letters $\underline{33}$, 1055 (1974).

4. Y. Band; E. L. Lewis and J. D. Silver; M. Lombardi; J. H. Macek; J. Jarecki, private communications.

5. Evidence from Field ion microscopy suggests that orbital over-lap between atoms in space and surfaces retain an essentially individual-atom character out to distances at least as far as 5 Å. Electric field distortions (see D. A. Nolan and R. M. Herman, Phys. Rev. B$\underline{10}$, 50 (1974) may tend to suppress the collective surface effects in those problems, however.

6. R. M. Herman, to be published.

LASER RESONANCE SPECTROSCOPY ON EXCITED STATES OF HIGH Z HYDROGENIC ATOMS

D. E. Murnick*

Bell Laboratories

Murray Hill, N. J.   07974 U.S.A.

ABSTRACT

Extension of Lamb shift measurements to high Z hydrogenic atoms is desirable and important to probe quantum electrodynamic effects in a strong Coulomb field and to search for possible new phenomena. Particle accelerators exist, or are planned, which allow production of hydrogenic species of a wide range of Z. If the metastable $2S_{1/2}$ state can be populated, high power lasers in the IR and visible can be used to study the fine structure (ΔE_{α}), $2S_{1/2}$ - $2P_{3/2}$, and Lamb shift ($\mathscr{S}$),$2S_{1/2}$ - $2P_{1/2}$, transitions in many cases. Several possible experiments are indicated with emphasis on recent work at the Rutgers-Bell Tandem Lab to study the F^{8+} ΔE-$\mathscr{S}$ splitting using a Doppler shift tuning technique with a pulsed HBr chemical laser. A double resonance technique is used where Lyman-α X-rays at 826 éV are detected synchronously with laser flashes. In order to obtain precision Lamb shift splittings from the resonance data several effects must be included. Among the most important are hyperfine structure and sizeable relativistic corrections. Experimental details of critical importance include alignment of the laser beam with respect to the particle beam and rotation axis for Doppler tuning; normalization of data to metastable state population and laser power; and accounting for slightly off resonance laser radiation. Other experiments planned or in progress are reviewed and the limits of applicability of the technique indicated.

*and Department of Physics, M. I. T. Cambridge, Mass. 02139,
  U.S.A. (1975-76)

INTRODUCTION

The level structure of the hydrogen atom is very well known and quantitatively described by Dirac's relativistically covariant wave functions. The first excited state consists of the $2P_{3/2}$ level separated by the fine structure ΔE from degenerate $2P_{1/2}$ and $2S_{1/2}$ levels. The degeneracy is removed, however, by quantum electrodynamic (QED) effects, principally vacuum fluctuations of the quantized electromagnetic field. Agreement between measurement of the $2P_{1/2}$ - $2S_{1/2}$ Lamb shift splitting, $\mathscr{L}$, and its detailed calculation has been and remains[1] one of the important tests and outstanding triumphs of QED theory.

With suitable small corrections for different finite nuclear sizes and nuclear magnetic moments, the properties of any one electron ion can be described by the wave functions and equations of the hydrogen atom by multiplying the nuclear charge by Z. Figure 1. displays the Z dependence of relevant parameters of the n=2 state. Of particular interest for resonance spectroscopy are γ_{2S}, the transition rate from the $2S_{1/2}$ to $1S_{1/2}$ level, and the Lamb shift. Electric dipole decay from $2S_{1/2}$ - $1S_{1/2}$ is forbidden leading to the well known metastability of the level, but the second order process in which two photons are emitted having a total energy equal to the Lyman α energy can take place; the transition probability being proportional to Z^6. In addition, relativistic effects allows weak single photon M1 decay having a Z^{10} dependence; M1 decay is negligible, however, for $Z \leq 20$. The "metastability" is thus only relative to the allowed electric dipole decay of the P states.

The Lamb shift splitting $\mathscr{L}$ is conventionally expressed as a series expansion in the fine structure constant α and $Z\alpha$. The convergence of the expansion has never been proven. The lowest order term is of order $(Z\alpha)^4$, and terms of order $(Z\alpha)^6$ and $\alpha^2 \times (Z\alpha)^4$ have been calculated to date. It is this strong Z dependence which challenges the experimentalist to probe systems of higher Z to test the convergence of the series and probe the limits of validity of QED. Recently Erickson[2] and Mohr[3] have obtained expressions for the Lamb shift separating the Z dependence from a series expansion in α alone. The results obtained differ from each other and from the series expansion by a significant amount, increasing from about 0.1% at Z=8.

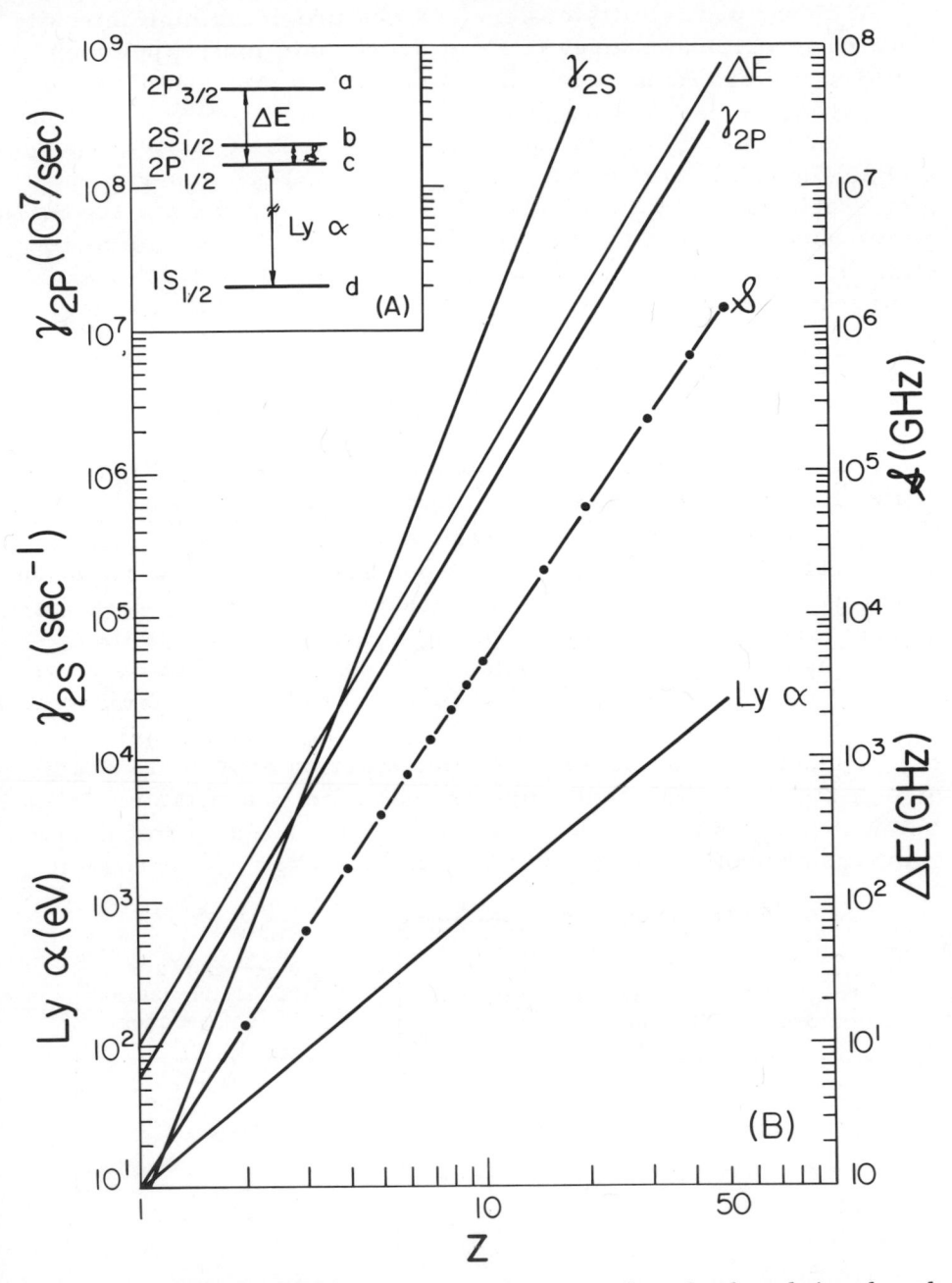

Figure 1 A) Partial Energy level diagram for the low lying levels of a hydrogenic atom with nuclear spin 0. B) Z dependence of the energy differences and widths of the levels shown in inset A.

With the availability of accelerators providing high intensity beams of sufficient energy to form hydrogenic ions, experimental activity in measuring high Z Lamb shifts has intensified in the last several years.[4] Hydrogenic carbon,[5] oxygen[6] and fluorine[7] have been studied using a D. C. quenching technique of the type pioneered by Fan, Garcia-Munoz and Sellin[8] for Z=3. In these experiments the metastable $2S_{1/2}$ state is coupled to the $2P_{1/2}$ level via the Stark effect, and the Lamb shift is obtained indirectly from the mixed state lifetime. The inherent precision limit of this type of experiment, ~0.5 to 1%, is such as to limit its usefullness for detecting small deviations from theory.

RESONANCE EXPERIMENTS

The classic experiment of Lamb and Retherford[9] (Figure 2) utilized a microwave resonance technique to drive the $S_{1/2}$ - $P_{1/2}$ transition, detecting the resonance via the absence of metastable atoms. Equivalently, spontaneous Lyman a radiation might have been observed. Rather than tune the microwave frequency directly, the energy splitting was varied about a fixed frequency using the Zeeman effect. This technique has been extended to the ionized systems He^+ and Li^{++} [10, 11]. For higher Z species, this method of tuning is not applicable as the Zeeman splitting is independent of Z and hence a smaller fraction of $\mathscr{L}$, and suitable high power tuneable oscillators are not readily available. In the Lamb experiment metastable atoms are populated by the electron bombardment of a slow atomic beam; for higher Z, hydrogenic

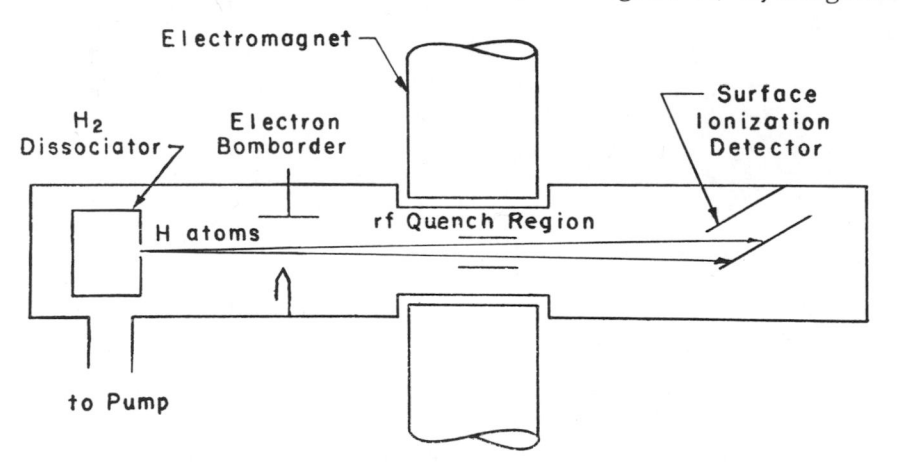

Figure 2.   Schematic arrangement of the Lamb Retherford experiment on atomic hydrogen.

ions can only be formed in dense high temperature plasmas or using high energy atomic beams. It has been shown,[5, 6, 7] that beam foil or beam gas excitation can yield significant population of the $2S_{1/2}$ metastable state in many cases.

Several criteria must be used in choosing candidates for hydrogenic ion resonance spectroscopy studies. First, of course, is being able to populate the $2S_{1/2}$ metastable state of interest. With existing heavy ion accelerators, reasonable yields of high Z hydrogenic ions is probably limited to $Z \leq 20$. In any case, as mentioned before, γ_{2S} is increasing so rapidly that the "metastable" state lifetime is less than 10^{-9} sec for $Z \geq 20$. Production of beams of hydrogenic ions requires ion velocities of order 0.1 c suggesting a convenient tuning technique provided suitable almost resonant lasers are available — the doppler shift in the frame of the moving ion:

$$\nu_f = \frac{\nu_L \left(1 - \frac{v}{c} \cos \theta\right)}{\sqrt{1 - \frac{v^2}{c^2}}} \qquad (1)$$

where ν_f is the apparent frequency to an ion moving at a velocity v with respect to the laboratory, ν_L is the laser rest frame frequency and θ is the angle between the particle beam and the laser beam.

Realizing that a precision measurement of the $2S_{1/2} - 2P_{3/2}$ splitting, $\Delta E - \mathscr{L}$, is of equal value as a direct measurement of $\mathscr{L}$, given the negligible uncertainty in the calculation of ΔE, Table I lists some candidates for laser resonance spectroscopy. Each case listed presents experimental problems with respect to laser beam or particle beam or both. The Z=2 ($\mu^- He^{++}$) case does not strictly belong on this list, but the recently reported[12] beautiful experiment on this system utilized techniques common to the high Z atomic beam case.

RUTGERS-BELL F^{8+} EXPERIMENT

An experiment in progress at our laboratory, for which preliminary results have recently been reported[13] is on $\Delta E - \mathscr{L}$ of F^{8+} studied via doppler shift tuning an almost resonant HBr chemical laser.[14] A schematic of the experimental arrangement is shown in Figure 3. An F^{7+} beam, from the Rutgers-Bell tandem vandeGraaff is post stripped to F^{9+} by a 10 $\mu gm/cm^2$ carbon foil prior to analysis by a 90° bending magnet. The F^{9+} beam passes through a ~5 $\mu gm/cm^2$ "adder foil" after which a

TABLE 1

CANDIDATES FOR HYDROGENIC ION LASER RESONANCE SPECTROSCOPY

Z	Transition (GH_z)		Laser	Tuning	Average Power
2 ($\mu^- He^{++}$)		368473.	dye	continuous	75mW
5 (B^{4+})		405.3	CH_3Br	D.S.	2mW
6 (C^{5+})		783.7	CH_3F	D.S.	6mW
7 (N^{6+})	ΔE-	25030.	Spin Flip CO_2 pumped	continuous	10mW
9 (F^{8+})	ΔE-	68832.	HBr	D.S.	10mW
10 (Ne^{9+})	ΔE-	105201.	HF	D.S.	400mW
17 (Cl^{16+})		31930.	CO_2	discrete	10W
20 (Ca^{19+})		56990	CO	D.S.	1W

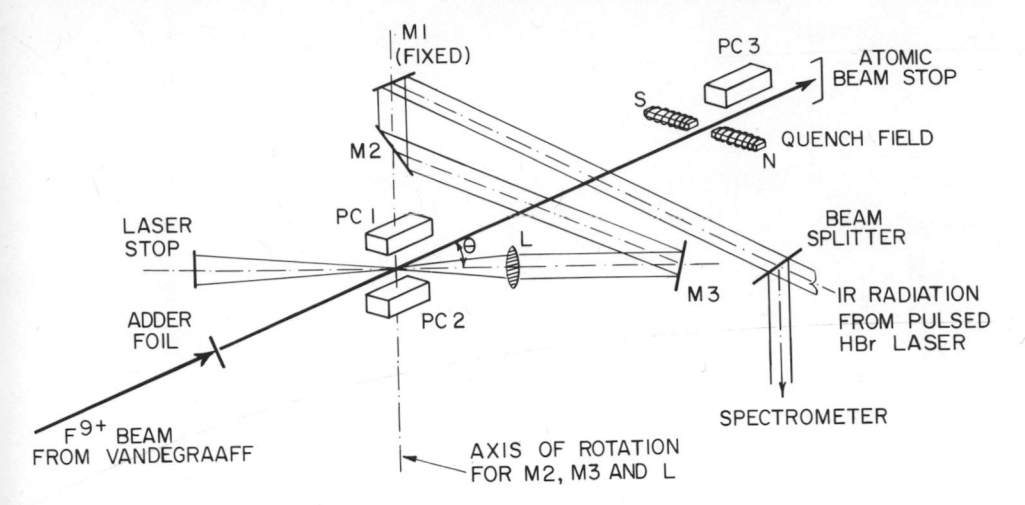

Figure 3.   Schematic arrangement of particle beam - laser beam interaction.

small fraction of the emerging ions are observed to be in the hydrogenic $2S_{1/2}$ state.

Two thin window gas flow proportional counters about 0. 5 cm from the beam, 1m beyond the adder foil detect radiation from the particle beam.   Though the energy resolution of the detectors is poor ($\Delta E/E \sim$. 30) the observed spectrum is consistent with the expected predominence of 2 photon spontaneous emission from the hydrogenic $2S_{1/2}$ state.   Figure 4 is a time to amplitude converter spectrum obtained by running the two proportional counters in coincidence with rather wide energy windows, $\sim$200 to 1000 eV.   The Lyman a energy in this case is 826 eV.   The coincidence peak verifies the presence of .the hydrogenic $2S_{1/2}$ ions.   A third proportional counter in an electromagnet downstream from the resonance apparatus detected Stark effect induced Lyman a radiation as an additional monitor of the F^{8+} metastable beam. A typical energy spectrum from that detector is shown in Figure 5.

A pulsed HBr chemical laser was used to drive the ΔE-δ transition.

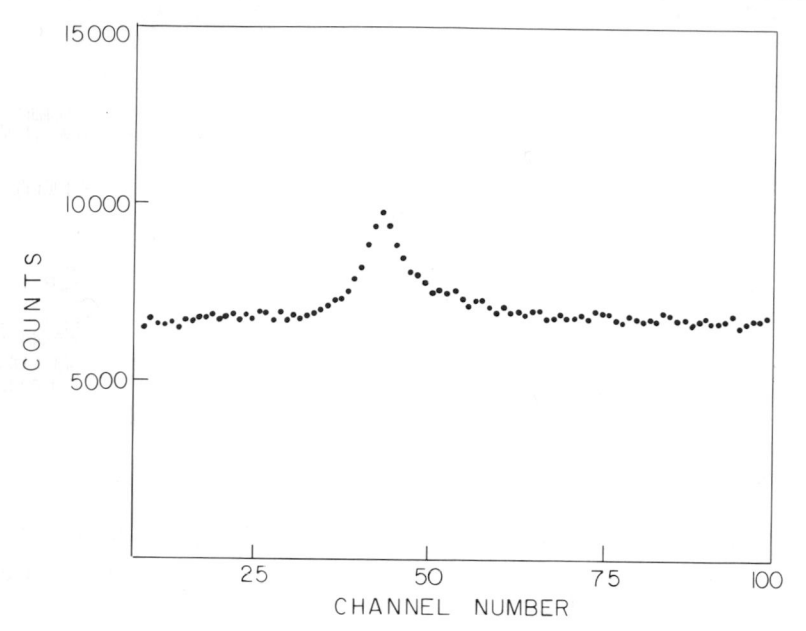

Figure 4.   Two photon coincidence TAC spectrum between proportional counters 1 and 2.   The width of the peak is about 0.7 μsec.

Figure 6 indicates the relevant laser lines and relative powers
obtained.   Each line is really a doublet arising from $H^{79}Br$ and
$H^{81}Br$ molecules.   The most intense lines, at 2383 cm^{-1}, had a
peak power of ~10kW in a 500 nsec pulse. (The laser was run at
1.5 pps).   For the expected $\Delta E-\mathcal{J}$ frequency of 68833GHz,
this line would be Doppler shifted to resonance if incident at an
angle of about 27° from normal for a 64 MeV beam.   For the present experiments, the total laser output was directed into the
interaction region through a KCl window by a rotatable mirror
system consisting of the mirrors M2, M3 and a lens L (refer to
Figure 3).   The lens was used to focus the laser beam to a 1.5 mm
spot so as to intersect the particle beam on the axis of the rotating
mirror system and between the two proportional counters.   This
optical system, together with the fixed mirror M1, allows angle
changes without realignment of the various optical components,
and is designed to assure negligible vertical motion (with respect
to the particle beam) of the laser beam throughout the angles of
interest.   Tuning about the angle where the 2383 cm^{-1} line is in
resonance with $\Delta E-\mathcal{J}$, the neighboring P_{2-1} (4) laser line at

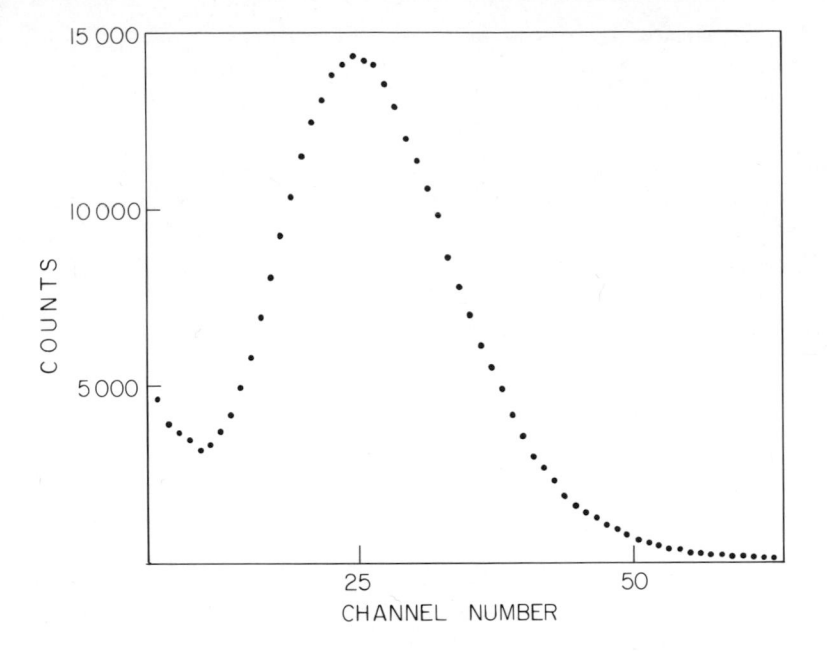

Figure 5.   Lyman a quench spectrum in proportional counter 3.
(See Fig. 3)

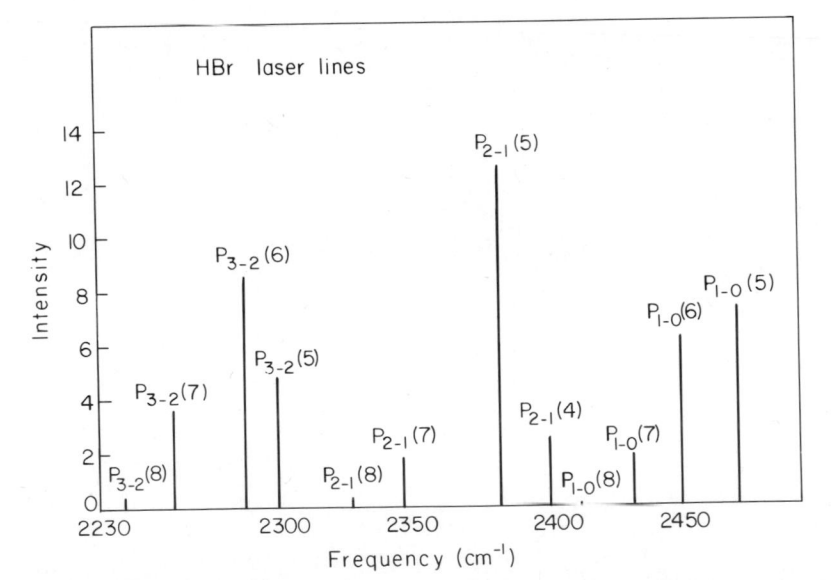

Figure 6.   Prominent HBr laser lines in the region of the resonance.

2400 cm^{-1}, though present at reduced intensity, contributes to the observed line shape. The $P_{2-1}(6)$ laser line at 2354 cm^{-1} is missing because of atmospheric CO_2 absorption.

Data were obtained, as a function of angle θ, by observing the time coincidence between laser pulses and detected X-rays in either proportional counter with a coincidence resolving time of about 1 μsec. For each laser beam-particle beam angle chosen coincidence data were accumulated for about two hours. Shown in Figure 7 are normalized time to amplitude converter (TAC)

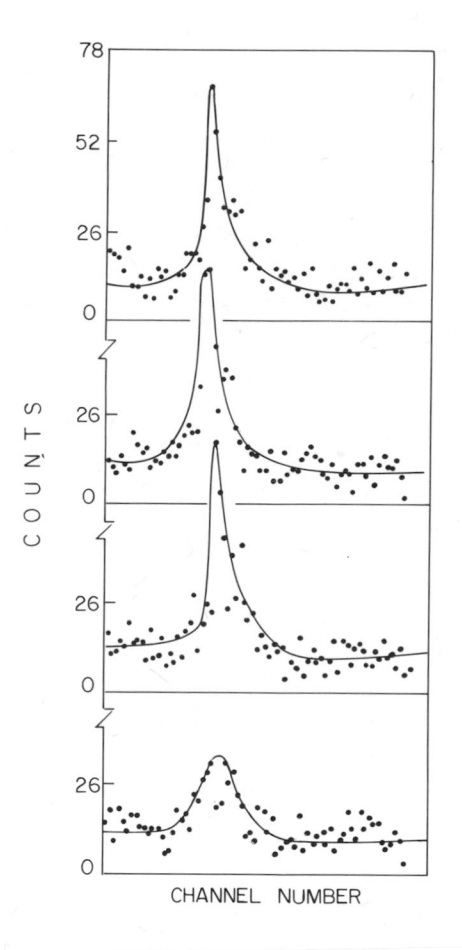

Figure 7. Typical TAC spectra at selected points on a resonance curve. Start pulses are generated by a Ge (Au) IR detector, stop pulses by a proportional counter signal. The width of the peak is about 1 μsec.

spectra obtained for one detector at a few angles around the resonance. Primary normalization is obtained by integrating the particle current at each angle. The angle was changed frequently under computer control to minimize any long term electronic shifts. In addition to the TAC spectra, coincidence spectra of the type shown in Figure 4, quench spectra from the third proportional counter and Ge(Au) detector outputs were separately stored for each angle. The Ge (Au) IR detectors monitored the power outputs through spectrometers tuned to the 2383 cm^{-1} and 2400 cm^{-1} lines.

ANALYSIS AND RESULTS

Resonance profiles obtained by integrating normalized TAC spectra such as those of Figure 7 were fit to a theoretical line shape $F(\theta)$ with $\mathcal{L}$ as a parameter:

$$F\ (\theta) = \sum_{k} \left\{ 1 - \exp\left(- \sum_{ij} \mu_{ij}^{k}\ (\theta)\ t\ (\theta)\right) \right\} \qquad (2)$$

Typical data and a best fit are shown in Figure 8.

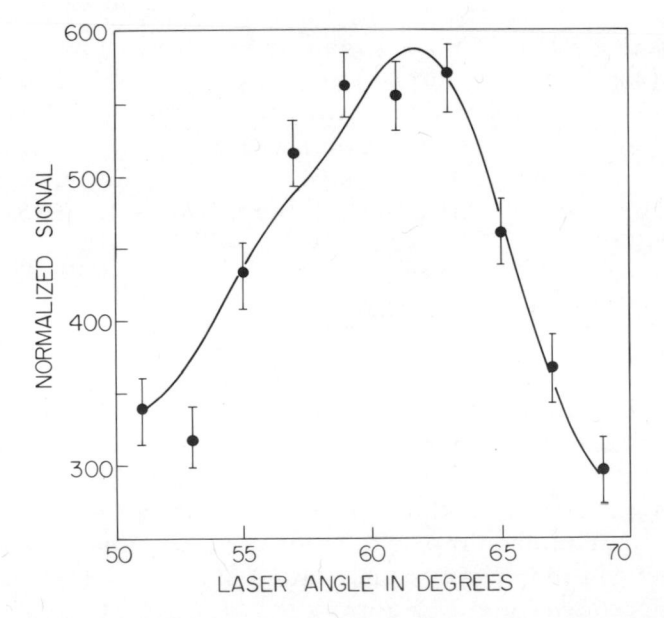

Figure 8.   One of resonance curves used to obtain $\Delta E - \mathcal{L}$ line center.

The nucleus ^{19}F has a nuclear spin (I = 1/2) so the $S_{1/2}$ and $P_{3/2}$ levels are doublets having F=0, 1 and F=1, 2 respectively. The selection rules, $\Delta F = \pm 1$, 0; energy splittings, and matrix elements are such as to lead to a broadening of the resonance line but not to a shift in its center of gravity provided that the beam foil excitation yields a statistical population of the hyperfine levels. Thus far, the data have been fit assuming such a population. The summation indicates k and i in Equation 2 are over initial S states and final P states respectively. The summation over j is over all laser frequencies present (Figure 6).

The parameter $t(\theta)$ is the relativistically contracted time that a moving atom spends in the finite spatial extent of the laser field. The transition probability $\mu(\theta)$ is given by:

$$\mu_{ki} (\theta) = \frac{e^2 \Gamma S(\theta) \left| r_{ki} \right|^2}{2\pi c \hbar^2 [(\nu_f(\theta) - \nu_o)^2 + \frac{\Gamma^2}{4\pi}]} \qquad (3)$$

where $\left| r_{ki} \right|^2$ is the dipole matrix element squared, $\nu_f(\theta)$ is the Doppler shifted frequency given by Equation (1) and $S(\theta)$ is the Lorentz transformed laser power. The natural linewidth $\Gamma/2\Pi$, determined by the P state lifetime is 654 GH$_z$ which is equivalent to about a $7.5°$ angle width.

Assuming a value for E given by the Dirac theory with corrections[15] ($\Delta E = 72,192.96GHz$) our initial data[13] yields a value for $\mathcal{J}$ of 3339 ± 35 GH$_z$. For comparison, the calculated values are 3349 using the series expansion with the fourth order term corrected[16], 3360 using Erickson's model[17] and 3342 using Mohr's calculation.[18] All calculations assume 2.8 f for the nuclear radius.

SUMMARY

The statistical uncertainty quoted above $\frac{\Delta \mathcal{J}}{\mathcal{J}} \sim 01$ can be improved by an order of magnitude as more and higher quality data become available. At the 0.1% level, given the present calculations, there will be a deviation between experiment and one or more QED calculations. Such a situation, which has not existed for several years, should stimulate more theoretical and experimental work in this area, and may lead to increased understanding of certain quantum electrodynamic effects.

Possible systematic effects in the experiment, of course, will

have to be studied in great detail. For Doppler tuning experiments, such as the F^{8+} case, subtle geometric effects may broaden or shift the resonance. Though angles are defined by carefully machined slit systems to better than $\pm 0.1°$ a continuous monitor of the perpendicular particle beam - rotation axis-laser normal direction would be desirable. For the HBr F^{8+} resonance repeating the experiment at a different beam energy and using different resonance laser lines should probe possible velocity dependent or geometric effects not considered. Thus far our experiments have been carried out with a 64 MeV fluorine beam. We have begun studies at 50 MeV, where the reduced metastable yield is somewhat compensated for by the increased interaction time of the slower beam. We also plan to obtain one or more resonance curves with the 2294 cm^{-1} line, which is resonant within $2°$ of normal to the particle beam.

Extension of resonance experiments to higher Z, as suggested by Table I, or other cases not listed, is important to check the Z dependence of $\mathscr{A}$, and possibly yield new effects. The Cl^{16+} case is especially promising due to the high laser power available and the possibility of discrete rather than Doppler shift tuning.

ACKNOWLEDGEMENT: The ideas and experiments described in this paper result from close collaboration and discussion with my colleagues H. W. Kugel, M. Leventhal, C. K. N. Patel and O. R. Wood.

REFERENCES

1. B. E. Lautrup, A. Peterman, and E. deRafael, Physics Reports 3, 193 (1972)

2. G. W. Erickson, Phys. Rev. Letters 27, 780 (1971)

3. P. J. Mohr, Phys. Rev. Letters 34, 1050 (1975)

4. M. Leventhal, Nuclear Instruments and Methods 110, 343 (1973)

5. H. W. Kugel, M. Leventhal and D. E. Murnick, Phys. Rev. A6, 1306 (1972)

6. M. Leventhal, D. E. Murnick and H. W. Kugel, Phys. Rev.
    Letters 28, 1609 (1972)
    G. P. Lawrence, C. Y. Fan and S. Bashkin, Phys. Rev.
    Letters 28, 1612 (1972)

7. D. E. Murnick, M. Leventhal and H. W. Kugel, Int'l
    Conference on Atomic Physics (1972), unpublished

8. C. Y. Fan, M. Garcia - Munoz and I. A. Sellin, Phys. Rev.
    161, 6 (1967)

9. W. E. Lamb and R. C. Retherford, Phys. Rev. 79, 549 (1950)

10. M. A. Narishimham and R. L. Strombotne, Phys. Rev. A4,
    14 (1971)

11. M. Leventhal, Phys. Rev. A11, 427 (1975)

12. A. Bertin et al,Physics Letters 55B, 411 (1975)

13. H. W. Kugel, M. Leventhal, D. E. Murnick, C.K.N. Patel
    and O. R. Wood, Phys. Rev. Letters 35, 647, (1975)

14. O. R. Wood and T. Y. Chang, Appl. Phys. Letters 20, 77 (197

15. B. N. Taylor, W. H. Parker and D. N. Langenberg, Revs.
    Mod. Phys. 141, 375 (1969)

16. T. Applequist and S. J. Brodsky, Phys. Rev. Letters 24,
    562 (1970)

17. G. W. Erickson, private communication (see reference 2)

18. P. J. Mohr, private communication (see reference 3)

PHOTOEXCITATION OF A FAST H(2s) BEAM TO HIGHLY EXCITED AND CONTINUUM STATES USING DOPPLER-TUNED CW ARGON ION LASER UV RADIATION*

P.M. Koch, L.D. Gardner, and J.E. Bayfield

J.W. Gibbs Laboratory, Yale University

New Haven, Connecticut, U.S.A.   06520

Laser excited atomic beams in the keV energy range have been employed in atomic lifetime measurements.[1,2] Resonant electric dipole laser transitions have been used to excite states of low excitation in fast beams, beginning with either ground or metastable states. We report the extension of this to highly excited states using the uv Ar III laser line at 363.789 nm (wavelength in air) to induce H(2s)→H(high n) transitions in atomic hydrogen for $40 \leq n \leq 55$. The energy of the atomic beam was varied to Doppler-tune[1] to a desired individual value of n. The laser-pumped highly excited beams were then used in higher quality studies of the strong-field microwave ionization of highly excited atoms[3] previously conducted using beams containing a band of partially-resolvable n-states.[4] Furthermore we report the first measurement, to our knowledge, of the photoionization cross section of H(2s) at any wavelength. For this experiment we used the Ar III laser line at 351.1 nm (wavelength in air).

Schematic diagrams of the photoexcitation and photoionization experiments are shown in Figure 1, parts (a) and (b) respectively. Both experiments used common H(2s) beam production and product-ion detection techniques.[3-5] The fast H(2s) beam was produced by electron transfer collisions of a proton beam passing through an H_2 gas target. The 30 mW (nominal) laser beam was directed colinear and parallel to the atomic beam, with a one meter long region of overlap prior to entering the experimental region shown in the figure. This first beam overlap region introduced into the beam laser pumped highly excited atoms which are used in experiment (a) but not in (b).

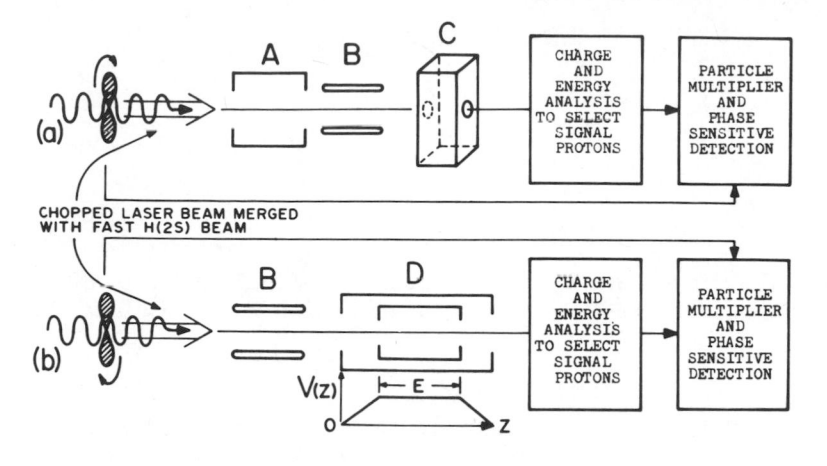

Fig. 1. Schematics of apparatus for (a) Doppler-tuned laser pumped H(high n) microwave ionization experiments and (b) H(2s) photoionization experiment. A: X band TM_{010} mode high Q cavity, B: static electric field quench plates, C: TE_{10} mode X band waveguide, D: liquid helium cryopump, E: voltage-labeled photoionization interaction region.

The common detection technique was the kinetic energy labeling of protons produced by the ionization of neutral hydrogen atoms within a region of static electric potential above apparatus ground. In experiment (a), highly excited atoms were field-ionized in field regions A or C held above ground. The resultant protons were charge analyzed by electrostatic deflectors, energy analyzed by an electrostatic filter lens, and further deflected by electrodes and a deflection magnet into a Johnston MMI particle detector. On the other hand, in the photoionization experiment (b), protons with a labeled kinetic energy were produced by laser ionization of H(2s) within the photoionization interaction region E held at static potential V. Region D here was the interior of a one meter long LHe cryopump operating at about 10^{-11} Torr. The laser beam was chopped in both experiments in order to take advantage of phase sensitive detection techniques. That the signals in both experiments originated from H(2s) atom beams was verified by quenching studies.

Portions of the Doppler tuning spectrum of the laser excitation of highly excited states are shown in Figure 2. The different peaks are highly resolved with the observed 20 eV FWHM of final ion energy corresponding to an inverse wavelength resolution of 0.12 cm^{-1}. As shown in Figure 3, the positions E(n) of the peaks agree well with expectations. The calculated energy values are computed using the relativistic formula

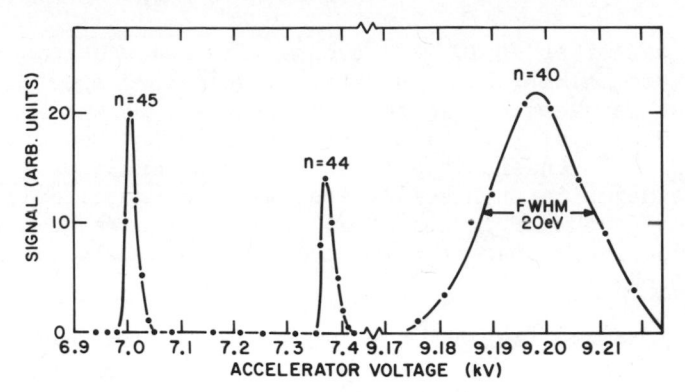

Fig. 2. Portions of the Doppler-tuned laser excitation spectrum
H(2s) + $h\nu_D \rightarrow$ H(n).

$$E(n) = \frac{mc^2}{2} \left[1 - \frac{\Delta\nu(n)}{\nu}\right]^2 \left[\frac{\nu}{\Delta\nu(n)}\right] ,$$

where the laser frequency $\nu = c/(\lambda_{vac}) = c/(N\lambda_{air})$, N is the index
of refraction of air, and λ_{air} = 363.789 nm.[6,7] Also $h\left[\Delta\nu(n)\right]$ =
$(I_{2s}-I_n)$, where I_{2s}, I_n are accurate ionization potentials
for H(2s) and H(n) atoms respectively. Our typical laser pumping

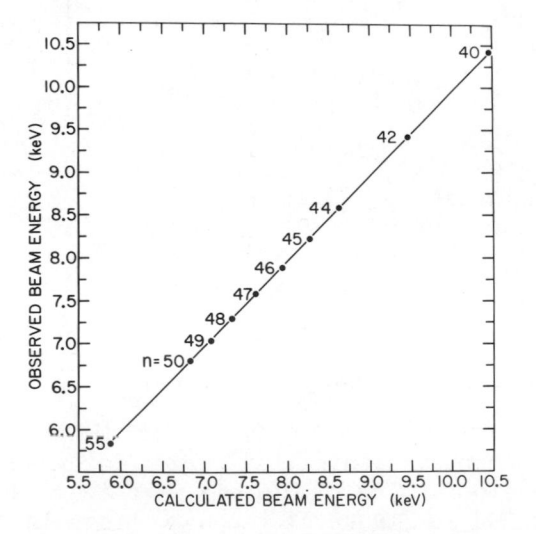

Fig. 3. A comparison of calculated and observed atomic beam
energies for the laser excitation spectrum.

efficiency was of order 10^{-5} in accord with calculations based on
formulae given in Ref. 2.   Uncertainties in the reported values
of the laser wavelength lead to an uncertainty in the horizontal
scale in Figure 3 of ±30 eV; uncertainties in the measured beam
energy lead to an uncertainty in the vertical scale of ±25 eV.
The uncertainties in both scales are small enough to rule out an
error in assignment of values of n.   A least squares fit repro-
duces the nonlinear spacing between data points with all deviations
less than 5 eV.

Some microwave ionization results for n=40 and n=50 atoms are
presented in Figure 4.   The n=50 data were obtained using the TE_{10}
waveguide structure C of Figure 1(a), whereas the higher field
n=40 data were obtained using the TM_{010} cavity A having Q~6500.
The oscillating electric field was along the beam direction in both
cases.   Although these first laser-pumped highly excited atom
microwave ionization data exhibit some scatter caused by weak
signals, the results compare well with those obtained with partially
resolved n-bands of collisionally produced highly excited atom

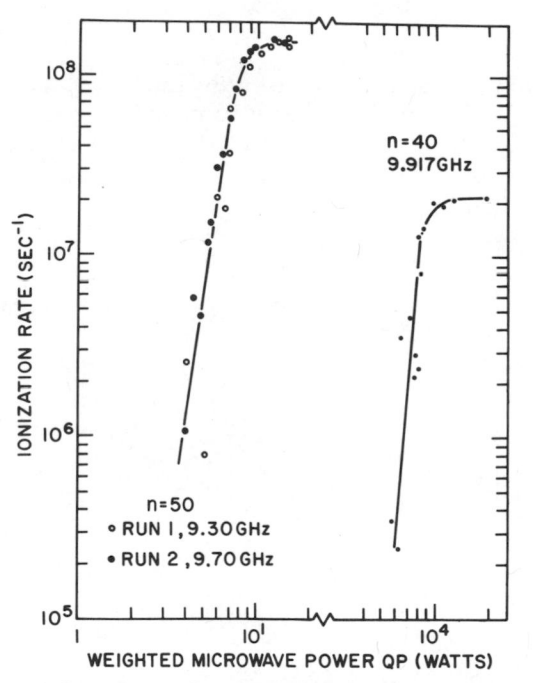

Fig. 4.   X-band microwave ionization curves for laser-pumped highly
excited hydrogen atoms.   The n=50 atoms were ionized in the TE_{10}
waveguide marked "C" in Figure 1, the n=40 atoms in the TM_{010}
cavity "A".   Smooth curves have been drawn through the data to aid
the eye.   The data experimentally saturate at high microwave powers
where the curves no longer represent ionization rate.

beams.[4] Slopes of the log-log plots of Figure 3 are in the range
7 to 12 as are the slopes found for the $40 \leq n \leq 43$ and $30 \leq n \leq 32$
bands in the work of Reference 4. A comparison of the present n=50
data with the earlier $54 \leq n \leq 56$ data indicates that apparent
slopes of n band data at very high n values may be in error because of the overlapping contributions for different n. A remaining
source of systematic experimental error is a possible distortion
of measured slope values caused by microwave field nonuniformity
across the 0.4 cm diameter of the atomic beam. Calculations applicable to the TM_{010} cavity show that the largest microwave electric
field strength experienced by all atoms in the beam occurred inside
the cavity where there was no dependence of the field strength on
axial position. Thus no anomalous ionization signal was produced
by fringe field effects near the beamline entrance and exit apertures. Inside the cavity, however, the field strength at the
radial edge of the atomic beam is calculated to be 4% lower than
that on the beamline. Since the measured ionization signal in
Figure 4 exhibits a power dependence over at least a 40% change in
microwave power level we believe this systematic effect is not a
large source of error. This will be investigated in future experiments using smaller diameter atomic beams and extremely sensitive
product ion detection techniques.

Our present result for the photoionization cross section σ_{PI}
at the Doppler-shifted wavelength of 352.4 nm is uncertain by a
factor of two because of uncertainty in the overlap integral for
the merged atom and laser beams. The results of previous metastable atom angular distribution and cross section studies[8] were
used in determining the overlap integral. The transmitted uv
laser flux was measured with an Eppley thermopile, and the fractional
flux at 351.1 nm was ascertained to be 0.5 ± 0.1 using an interference filter. The photoionization signal of about 200 Hz was
readily observable only because the experiment was conducted in a
vacuum of 10^{-11} Torr to reduce H(2s) stripping in collisions with
background gas.

The present experimental value of $\sigma_{PI} = 1.0 \times 10^{-17}$ cm^2 is in
agreement with theoretical values of 1.37×10^{-17} cm^2 and $1.43 \times$
10^{-17} cm^2 obtained using expressions in references 9 and 10 respectively.

References

*Research supported in part by the National Science Foundation and by the National Bureau of Standards.

1. H.J. Andra, A. Gaupp, and W. Wittmann, Phys. Rev. Lett. 31, 501 (1973).

2. H. Harde and G. Guthohrlein, Phys. Rev. A 10, 1488 (1974).

3. J.E. Bayfield and P.M. Koch, Phys. Rev. Lett. 33, 258 (1974).

4. P.M. Koch, L.D. Gardner, and J.E. Bayfield, Abstr. IX Int. Conf. Physics of Electronic and Atomic Collisions, Seattle, Washington, July 24-39, 1975, Univ. of Washington Press (1975), pp. 473-4.

5. P.M. Koch and J.E. Bayfield, Phys. Rev. Lett. 34, 448 (1975).

6. R.A. McFarlane, Applied Optics 3, 1196 (1964).

7. W.B. Bridges and A.N. Chester, Applied Optics 4, 573 (1965).

8. J.E. Bayfield, Phys. Rev. 182, 115 (1969).

9. H.A. Bethe and E.E. Salpeter, Quantum Mechanics of One- and Two-Electron Atoms, New York, Academic Press, 1957. See equation 71.14.

10. I.I. Sobel'man, Introduction to the Theory of Atomic Spectra, Oxford, Pergamon Press, 1972. See equation 34.65.

LASER EXCITATION IN FAST BEAM SPECTROSCOPY

H.J. Andrä[+]

Institut für Atom- und Festkörperphysik

Freie Universität Berlin, Germany

The recent development of applying lasers to fast atomic beams in order to measure cascade free lifetimes or quantum beats seems to find wide spread interest. The published material in this field[1-5] indeed looks rather promising in particular as far as lifetime measurements in the one percent accuracy domain are concerned. The initial expectation of accuracy for lifetime measurements with this technique was pointing, however, towards the 0.1% mark[2]. Therefore we felt the obligation to experimentally prove for the first time the feasibility of such high precision measurements.

As stated earlier[1,2] the use of selective laser excitation avoids the cascade problem in beam foil spectroscopy[6] but conserves all the other unprecedented properties of the fast beam source. Just how well these unprecedented properties are suited for a precision experiment will be the topic of the first half of this paper. In the second half I shall then present applications of the beam laser technique to mean live and hyperfine structure quantum beat measurements.

As our standard example we use again the BaII-$6p^2P_{3/2}$-level which is easily accessible by Doppler-tuning of the 4545Å-Ar$^+$ laser line to the BaII-4554Å resonance line as shown in Fig.1. For this transition one calculates under the assumption of broad band excitation for each atom passing through the laser beam an excitation probability of 1.6% for an interaction time of .5ns and an effective linewidth of the laser line of 4GHz at 300mW. This effective linewidth takes into account time uncertainty and Doppler broadening due to atomic or laser beam divergences or energy spread of the atomic beam. Thus with a beam of 5μA Ba$^+$ one expects $5 \cdot 10^{11}$ ions per second excited which can give rise to a peak count rate of about $2.5 \cdot 10^4$ s^{-1} at a detection effici-

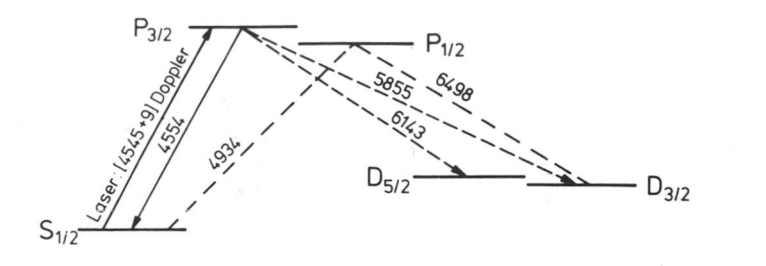

$$P_{12}\,(6^2S_{1/2} \to 6\,^2P_{3/2}) = \int_0^\infty \frac{I_0\,\lambda^3\,A_{12}}{4\pi^2\sqrt{2\pi}\;\sigma\;\hbar}\;e^{-\frac{(\omega-\omega_{12})^2}{2\sigma^2}}\;\frac{\sin^2\,[(\omega-\omega_{12})\,T/2]}{(\omega-\omega_{12})^2}\;d\omega$$

$$(I_0 = [\,\frac{\text{Laser power}}{\text{area} \times c}\,]\;;\;\sigma = [s^{-1}]\;;\;T = [s]\,\text{interaction time})$$

$$T = .5\,\text{ns}\;;\;\sigma = 4\,\text{GHz}\;;\;I_0 = \frac{300\,\text{mW}}{.07\,\text{cm}^2 \times c}\;:\;\boxed{P = 1.6\%}$$

Fig.1: Schematic leveldiagramm of BaII with excitation and detection scheme. The excitation probability is given for a broad laser bandwidth $\sigma[s^{-1}]$ and is valid only for P<5%.

ency of 10^{-6} and an observation time window of $\tau/20$. This signal guarantees good counting statistics for a precision experiment and one can start out with a standard beam-foil setup where the laser beam re places the foil in Fig.2. This setup will now be successively improved during the discussion of various sources of possible errors.

1) Time scale calibration: Fig.2 already indicated that a velocity analyzer is an absolute necessity for the exact time scale calibration in such experiments. As is shown in Fig.2 we have chosen a 90-degree electrostatic analyzer of 500 mm radius, 5 mm plate separation and focal points at 175.025 mm from the edges of the plates on both sides. The nominal calibration factor of this instrument should be f=100 (see Fig.2). However, already an error of 0.01 mm in the plate separation would cause an error of $2\cdot10^{-3}$ in the calibration! Instead of measuring this distance to such an accuracy we decided to better calibrate the analyzer in the actual experimental setup.

To this end HeI-$3p^3P_{2-1}$ beats[7] were measured by standard beam-foil technique with the energy spectrum of the He^+ particles emerging from the foil being scanned and summed during the whole measurement. This procedure is necessary since the foil thickening effect[8] leads to a steady shift of the He^+ energy (Fig.3a). As energy for

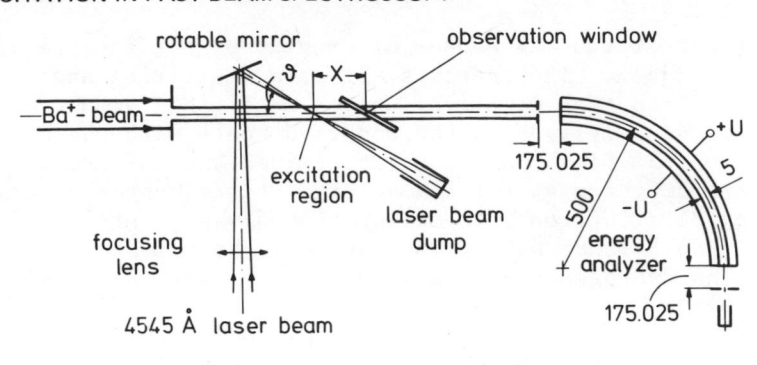

$$\text{theoretical calibration factor} \quad f = \frac{\text{true energy E [eV]}}{\text{applied voltage U [V]}} = 100$$

Fig.2: Schematic experimental setup for a Doppler tuned laser excitation experiment.

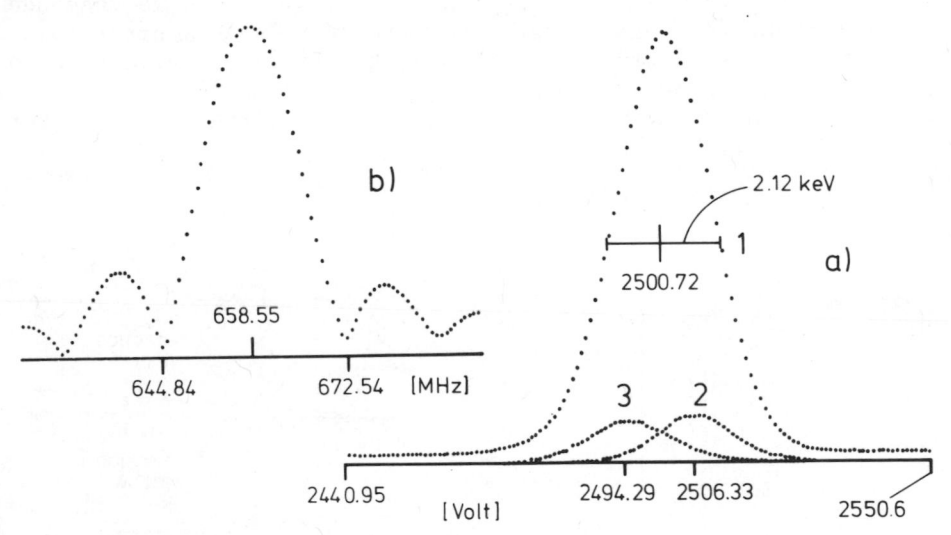

Uncertainties :

Beat frequency 658. 55 (15) :	± .023%	
Earth field :	± .025%	c)
He - He$^+$ - He^{++} volocities :	± .030 %	
Uncertainty of measurement :	± .130 %	

$$\boxed{f = 98.69 \pm .15} \qquad \Delta f = \sqrt{\sum \Delta f_i{}^2} \quad - \pm .15\%$$

Fig.3: a) He$^+$ energy distribution 1: integrated spectrum during the quantum beat run, 2: spectrum with fresh foil, 3: spectrum with old foil. b) Fourier transform of the calibrating He-3^3P_{2-1}-fs-beats. c) List of uncertainties and final value of f.

one quantum beat run the center of gravity of the integrated energy
spectrum is taken. (The energy shift at a scattering angle of 4° was
found to be less than 0.05% for only a negligible amount of beam). A
Fourier transform Fig.3b of the observed beats with the known beat
frequency of 658.55(15)MHz[9] then determined the calibration factor f,
that is the true energy for the voltage at the center of gravity of
the energy distribution.This calibration f was determined for several
energies (188,242,288 keV) giving the same result within the error
bars.Thus an average of all runs can be used for the final value of
f=98.69(15).

The whole procedure relies of course completely on the assumption that He and He^+ have the same velocity when emerging from the
foil. This assumption can be justified by the charge equilibrium[10]
obtained already after a few atomic layers at the energies used here
or can be proved experimentally by comparing the He^+ and He^{++} veloci-
ties.Since they were found to be equal to within 0.03% we conclude
that neutral He also has the same velocity within this error.In addition to this uncertainty the earth magnetic field can give rise to an
estimated error of <0.025%. These uncertainties in Fig.3c sum up to
a final uncertainty of 0.15% which has been confirmed by two later
recalibrations within these limits.

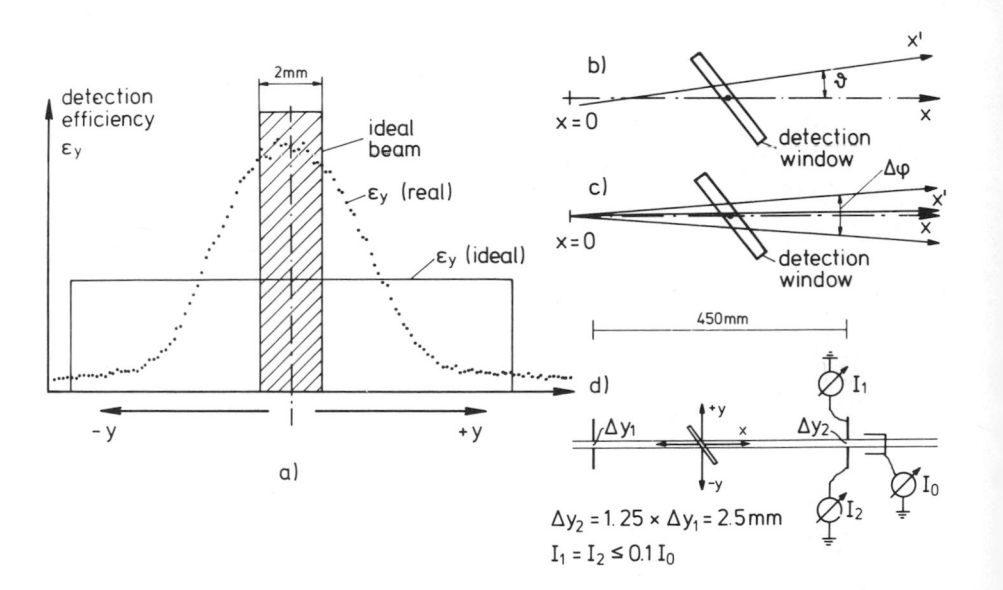

Fig.4: a) Ideal and real detection functions ε_y (ideal) and ε_y (real)
as measured across the beam (see 4c).b) Schematic beam divergence
effect in conjunction with 4a.c) Schematic non parallelism of beam
and detection. d) Beam steering and linearity control setup for the
experiment in Fig.7d.

2) Detection linearity: For a clean measurement of an intensity decay curve the detection efficiency ε_x should stay constant for the total spatial variation of the distance between excitation and detection from zero to its maximum value. This requirement is generally fulfilled if the detection efficiency ε_y (ideal) as a function of y is rectangularly shaped as shown in Fig.4a. However, imaging lenses or other optical components may change ε_y to a sharply peaked function ε_y (real) which is then sensitive to beam spreads in Fig.4c or non parallel motion of the detection window (x'-direction) in Fig.4b. In such a situation a strong variation of ε_x can be expected which introduces systematic errors in the measured decay curve.

In order to avoid such problems the beam has to be steered to within controlled limits along the x-axis. This can be achieved by using differential apertures after the measuring region in Fig. 4d which can be read separately. This also allows to determine the angular spread of the beam when Δy_1 and Δy_2 are known. In addition a mechanical motion along the y-axis should allow to actually measure ε_y at various x-positions and then ε_x with a fixed y-position by detecting the rest gas excitation of the beam with the laser off. With all these measurements a rather good control of the beam quality has actually been obtained. Furthermore decay curves can then be measured under controlled angular offset conditions (ϑ=5m rad) in order to obtain measureable estimates for such systematic errors. In Fig.4d the actual beam steering and control setup for the final measurement in Fig.7d is shown.

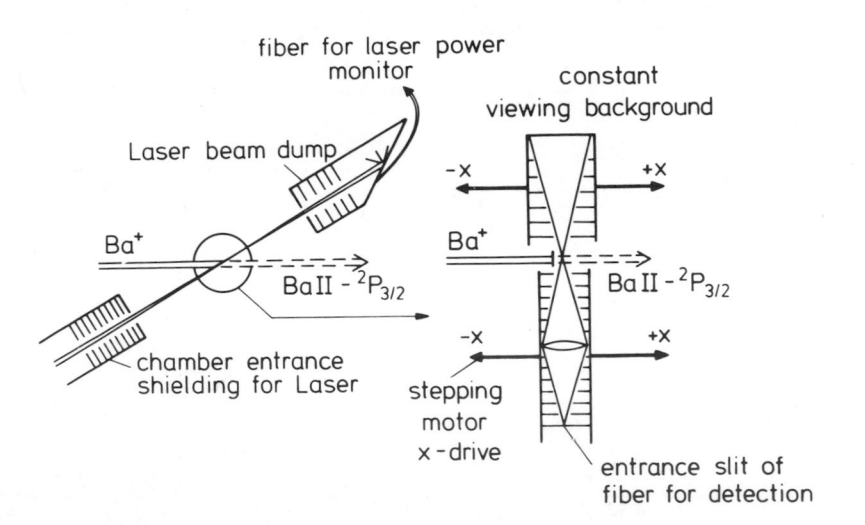

Fig.5: Schematic design of laser stray light suppressors for the laser beam and the detection optics.

3) Zeeman-beats: The interaction time of the atoms with the laser light is of the order of 0.1-2ns. Thus coherent excitation of magnetic sublevels is obtained which leads to Zeeman beats superimposed on the decay curve in the presence of external magnetic fields[!!] For a precision experiment the earth magnetic field has therefore to be compensated to a tolerable limit of <50mGauss or to be measured in order to correct for it in the data analysis.

4) Dead time correction: In photon counting mode the detection channel can in general process only a limited count rate due to deadtime limitations.If such count rates are approached the data have to be corrected in particular at the highest intensity of the decay curve where important information is contained.

5) Laser and other backgrounds: Since any least squares fit (LSF) analysis of an exponential decay curve with an underlying constant background yields a strong correlation between background and decay constant it is advisable to actually measure the background in the experiment, subtract it from the decay curve and then apply the LSF procedure.

This can easily be done for the detector noise and the rest gas beam excitation which are both constant along the beam axis. Laser stray light, however, may show pronounced structures along the beam axis and must therefore be avoided since a subtraction of such a background from the data disturbs the LSF procedure on statistical grounds. To this end sets of sharply edged apertures made of 20μm blackened aluminium along the laser beam path and at the detection optics are most effective (Fig.5). If in addition the Doppler effect is used to separate the detection from the excitation wavelength $(\Delta\lambda=9\text{Å})$ a suppression of the laser background to <1 count s^{-1} as compared to 10^4 s^{-1} maximum signal was achieved.

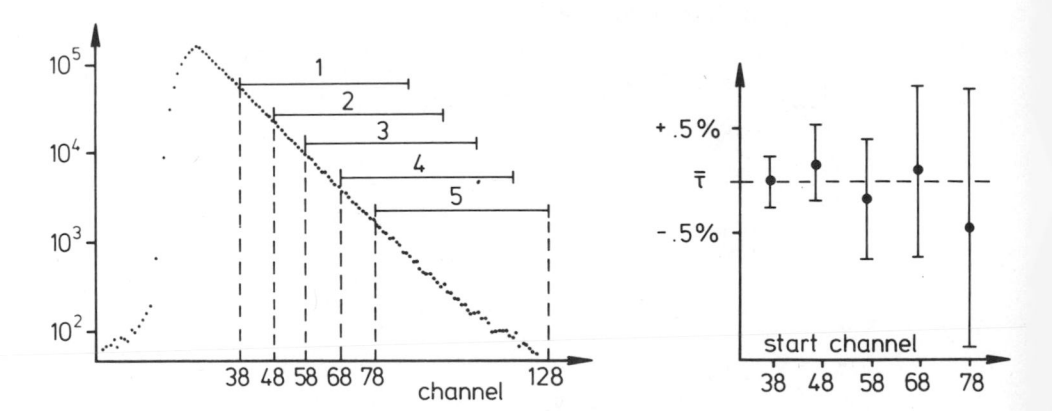

Fig.6: Sketch of data analyzing procedure.

6) Collisional deexcitation: Although the mean free path of excited atoms at 10^{-5} Torr and 10^{8} cm/s is of the order of 28m at an assumed deexcitation crossection of $\sigma=10^{-14}$ cm^2 the effect on a precision experiment cannot be neglected in particular if the pressure rises to about 10^{-5} Torr. Then the lifetime τ_0 is reduced according to the Stern-Vollmer formula to $\tau=\tau_0/(1+\tau_0\cdot\sigma\cdot\bar{c}\cdot n)$ with n being the rest gas particles cm^{-3} and $\bar{c}$ the beam velocity.

At 10^{-5} Torr, $\sigma=10^{-14}$ cm^2, and $\bar{c}=10^{8}$ cm s^{-1} a correction of 0.2% becomes necessary already. This implies that the rest gas pressure in a precision experiment is not allowed to be larger than $2\cdot10^{-6}$Torr if no information about σ is available. If possible a τ-measurement at bad pressure should be performed in order to obtain an estimate of σ.

7) Laser stability and normalization: As long as the data are taken normalized to the charge collected in a faraday cup (as was done in all our experiments so far) the stability of the laser is of crucial importance for the statistical quality of the data. Therefore a stabilized laser is to be used or the laser power has to be measured simultaneously with the data in order to correct them for any laser power fluctuations. The ideal normalization via a second optical channel monitoring the maximum fluorescence region has not yet been employed so far.

8) Data analysis: A LSF procedure has been used to analyse decay curves of the general shape in Fig.6 with a function $I=I_0e^{-t/\tau}+U$. In order to avoid the τ-lengthening influence of the convoluted excitation and detection function (four different types were used) the data set had to be cut off at about channel 38. Half the length of the rest of the data was then analysed either with the measured background subtracted or with a constant background as free parameter. In order to test the quality of the data set and in order to eliminate any residual influence from the excitation function this length of data was shifted in steps of 10 channels through the rest of the data. Each time τ_c was determined and plotted as function of the starting channel. If no systematic trend of τ_c outside the statistics was found the whole data set was accepted. The final mean live was then determined either as the weighted mean of all these τ_c, or by a single fit of the data from channel 38 to the end. These values coincide within ±0.1%.

In cases where a magnetic field had to be corrected for, the theoretical fitting function contained the Zeeman beat contribution with fixed amplitude and fixed frequency calculated from the measured field.

9) Comparison of measurements:
a) In a first set of data essentially all the discussed systematic errors were neglected thus leading to a rather poor result in Fig.7

although the statistical error was much smaller (0.3%).
b) In the second set of data the parallelism at the mechanical motion
of the detection and the beam was guaranteed by differential apertures. However, the linearity of the detection was not measured
and the earth magnetic field was not compensated. The result is
already quite accurate but the possible systematic errors sum up
to 0.5% in Fig.7b.
c) For a third set of data all major improvements had been installed
in the setup except for a control of the laser stability which was
absolutely hopeless on that day due to malfunction at the laser.
The results in Fig.7c reflect this difficulty.
d) In the fourth set of data with all improvements for avoiding
systematic errors including a simultaneous measurement of the laser
power after crossing the atomic beam. The statistical quality of
the data becomes obvious when comparing 7 different runs in Fig.7d.
Two extra measurements at large angular offset allow to limit the
uncertainty due to offset in the proper experiment to 0.1%. Also a
measurement at $1.7 \cdot 10^{-4}$Torr Ar pressure in the target chamber has
been performed leading to a reduction of τ by 2%! This suggests
(at least for Ar) indeed a deexcitation crossection of $\sigma \sim 10^{-14} \mathrm{cm}^2$
thus limiting the collisional error to less than 0.02%.

For the final result only measurement d is accepted yielding as the
mean of 7 runs $\overline{\tau_d}$=6.312±0.016ns with an uncertainty given by the

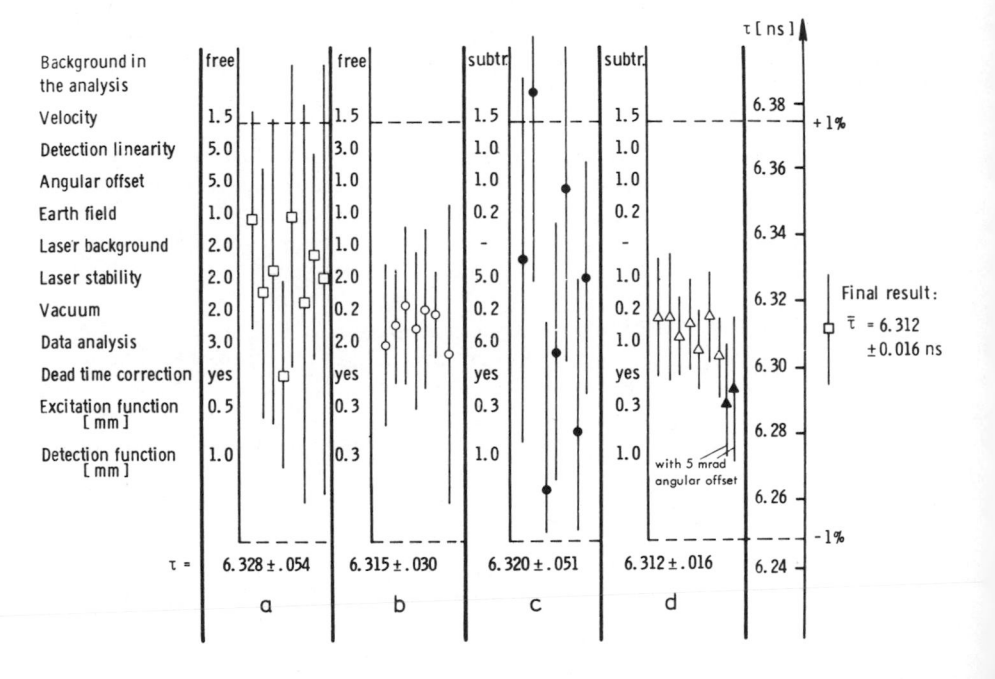

Fig.7: List of systematic errors for four different sets of measure
ments a–d (see text) and final result of the BaII – $6p^2P_{3/2}$-mean li

square root of the sum of the squares of the LSF and systematic errors
listed in Fig.7d.

Although the three other data sets a-c suffered from considerable
systematic erros it is remarkable to note that τ_a, τ_b, and τ_c lie all
within the uncertainty limits of this final result. This may support
the assumption that the possible phase correlated excitation of atoms
when passing through the coherent laser beam does not affect the lifetime measurement at this level of accuracy since the measurements a-d
were performed at widely different laser intensities (20-300 mW) and
beam particle densities ($10^{-5}-10^{7} \mathrm{cm}^{-3}$).After all no correlated emission phenomena should occur in the laser free region at these densities.

In conclusion we feel quite confident that with this systematic
error analysis finally a method has been developed which allows to
measure atomic mean lives at the sub-percent accuracy level.

In order to employ this technique we have slightly modified the
excitation scheme for lifetime measurements in MgI, SrI and RbII.
Essentially a two step excitation is used for these cases where the
first step is a collisional excitation of a metastable level, starting from which the second step is achieved by Doppler tuned laser
excitation to the level of interest. For the population of the metastable level gas excitation in a cryogenically pumped butane gas cell
in Fig.8 about 50 cm upstream from the laser excitation region was
used. The beam scattering and energy straggling introduced by this
target is well controlled by the system of apertures in Fig.4d and
by the energy analyzer respectively. However, a finite cascading was
observed in the measuring region for MgI which had to be accounted
for in the actual experiment by periodically chopping the laser and
recording directly the signal with laser on and off.

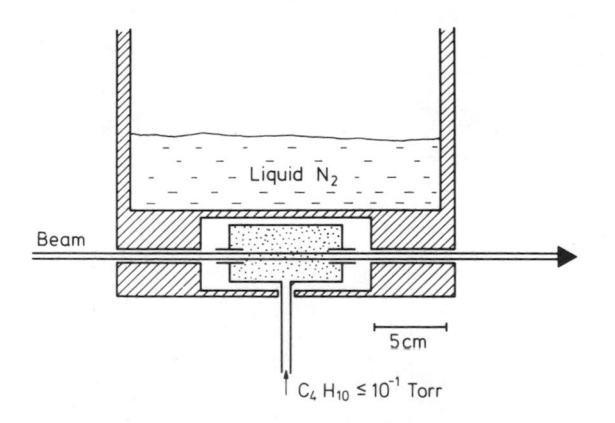

Fig.8: Sketch of cryogenically pumped butane gas target. The gas
cell is at roomtemperature whereas the surrounding walls are con-
densing butane at liquid N_2 temperature.

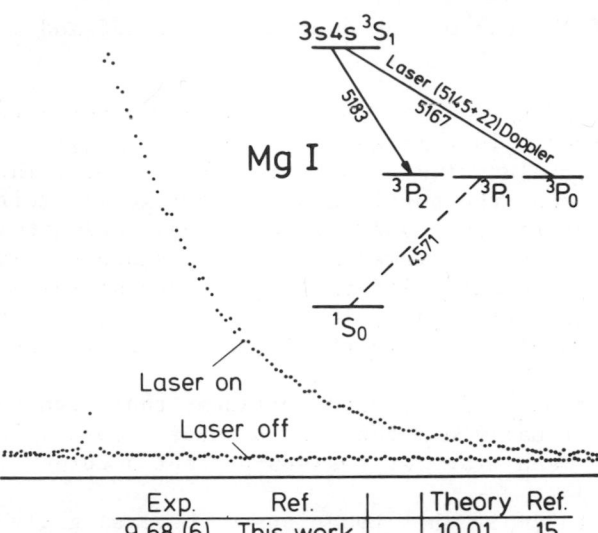

Exp.	Ref.		Theory	Ref.
9.68 (6)	This work		10.01	15
10.10 (80)	14		10.91	16
14.80 (70)	13	[ns]	10.13	17
13.00 (100)	12		9.07	18

9.64 in Atomic Transition Probabilities[19]

Fig.9: MgI excitation and detection scheme with typical decay curve for the determination of the MgI-3s4s ^{3}S$_1$ mean live and comparison of results.

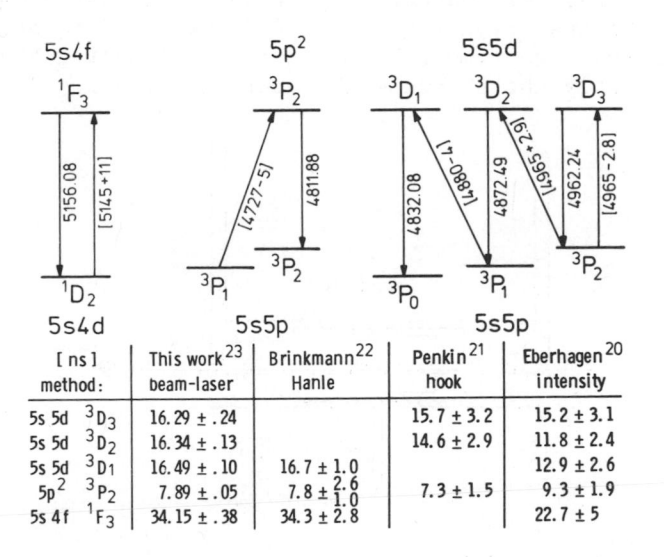

[ns] method:	This work[23] beam-laser	Brinkmann[22] Hanle	Penkin[21] hook	Eberhagen[20] intensity
5s 5d ^{3}D$_3$	16.29 ± .24		15.7 ± 3.2	15.2 ± 3.1
5s 5d ^{3}D$_2$	16.34 ± .13		14.6 ± 2.9	11.8 ± 2.4
5s 5d ^{3}D$_1$	16.49 ± .10	16.7 ± 1.0		12.9 ± 2.6
5p^2 ^{3}P$_2$	7.89 ± .05	7.8 ± $^{2.6}_{1.0}$	7.3 ± 1.5	9.3 ± 1.9
5s 4f ^{1}F$_3$	34.15 ± .38	34.3 ± 2.8		22.7 ± 5

Fig.10: SrI excitation and detection schemes for 5 levels measured and comparison of results.

In MgI the metastable $3s3p\,^3P_o$ level was populated by gas excitation at 260keV and then the Ar^+ 5145Å laser line plus 22Å Doppler shift was used to excite the $3s4s\,^3S_1$-level via the 5167Å transition. For the observation of the intensity decay curve the 5183Å transition to the $^3P_2$ state was chosen which was spectroscopically far enough away from the laser line in order to avoid any laser stray light. A typical measurement and the result as the mean of 22 such separately measured decay curves are displayed in Fig.9 together with a comparison of former experimental and theoretical values.The agreement with a beam foil result from Anderson et al.[14] is good whereas the other two experiments[12,13] are off by more than 25%. The theoretical situation is more consistent.However,the values tend to be too long except for a model potential calculation by Victor et al.[18].Under these circumstances it is somewhat surprising to find the value in the compilation of Wiese et al.[19] in excellent agreement with our result.

In SrI the $5s5p-{}^3P_{o,1,2}$ as well as the $5s4d\,^1D_2$ metastable levels were populated by gas excitation at 100 to 235keV beam energies. The laser excitation and fluorescence detection schemes are shown in Fig. 10 for 5 levels studied in SrI.Also in Fig.10 we present the results obtained in comparison to former experimental work[20-22].Although the agreement is good within the error bars a discussion is meaningless since the present accuracy[23] is by factors of 7 to 20 better than any earlier claim.Unfortunately no theoretical work on the measured levels is known to us.

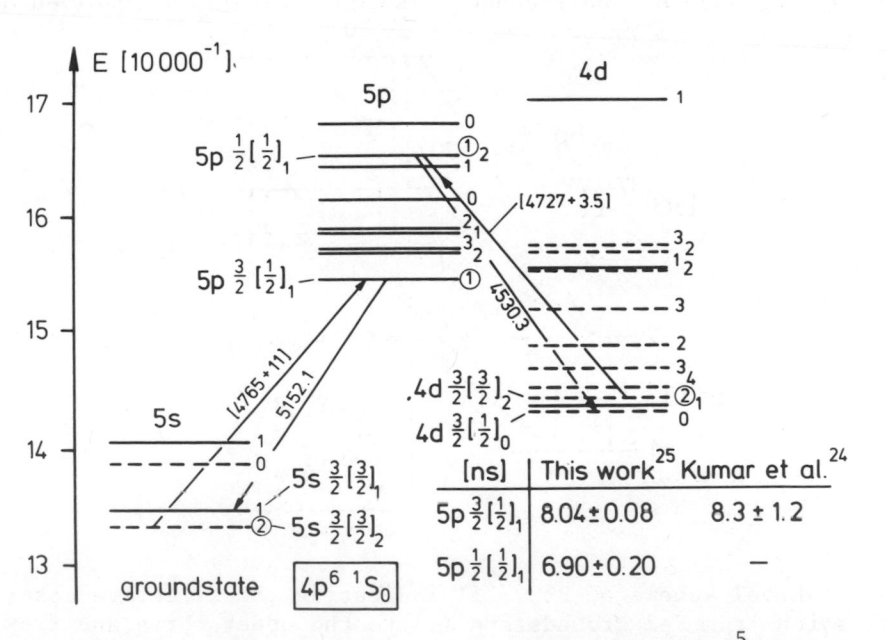

Fig.11: RbII excitation and detection schemes for 2 $4p^5 5p$ levels and comparison of results. Broken lines represent metastable levels.

In RbII the $4p^5 5s3/2 \, [3/2]_2$ and the $4p^5 4d3/2 \, [3/2]_2$ metastable
levels were gas excited at 310keV. From these levels the $4p^5 5p3/2 \, [1/2]_1$
and $4p^5 5p1/2 \, [1/2]_1$ levels could be laser excited and the fluorescence
light was easily detected according to the scheme in Fig.11. However,
the peak intensity of the 4530.3Å-line was only 250 counts s^{-1} compared to the 5152.1Å-line with 2500 counts s^{-1} leading to different
accuracies in the final results on both levels[25] in Fig.11. The comparison only with one recent beam foil experiment[24] gives good agreement but again the laser excitation method is superior in accuracy
by a factor of 15.

Besides the great improvements introduced by the laser excitatio
of fast beams to precise lifetime measurements also the measurements
of finestructures, hyperfine structures or g-values via quantum beats
can take advantage of laser excitation. Since polarized laser light
yields maximum possible alignment (or orientation) of excited states
excellent conditions for quantum beat measurements are obtained.

As a first example we chose the hfs of the ^{137}BaII-$6p^2P_{3/2}$ level
This case is of particular interest since it will give us also an
estimate of the spectral resolution which can be obtained with laser
fluorescence spectroscopy on fast beams. Considering the level scheme
in Fig.12 we find the ground state split by 8.04GHz[26], whereas the
upper states have a maximum separation of 659MHz only. When Doppler
tuning the laser resonance by changing the angle ϑ in Fig.2 and recording the 4554 Å-fluorescence emission a partially resolved double

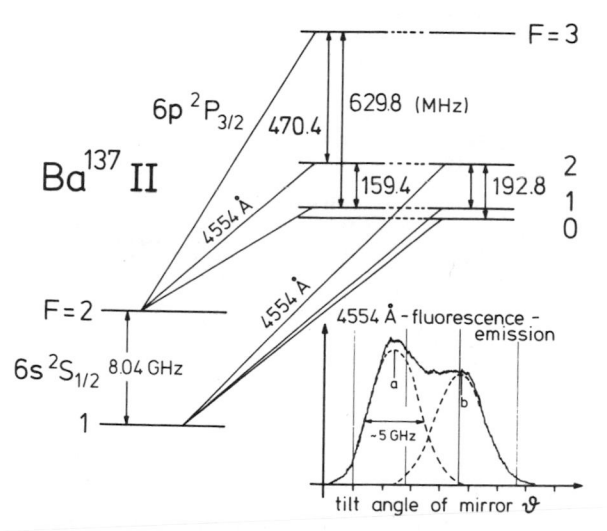

Fig.12: Level scheme of Ba ^{137}II indicating the selective laser
excitation from F=2 groundstate (a) to the upper three and from
F=1 groundstate (b) to the lower three hfs components of the excited
$6p^2P_{3/2}$ state.

peak structure as shown in Fig.12 is observed. It clearly indicates
that selective excitation from either one of the ground state hyperfine components F=1 or F=2 is achieved giving an estimate on the
resonance line width of ∿5GHz, or a resolution of about $\lambda/\Delta\lambda=150000$.
(With a single mode laser this resolution was improved to 400000 –
see Ref.4). As a consequence of the selection rules one has then
only coherent excitation of either the upper three or the lower
three hfs-components of the excited level and expects   therefore

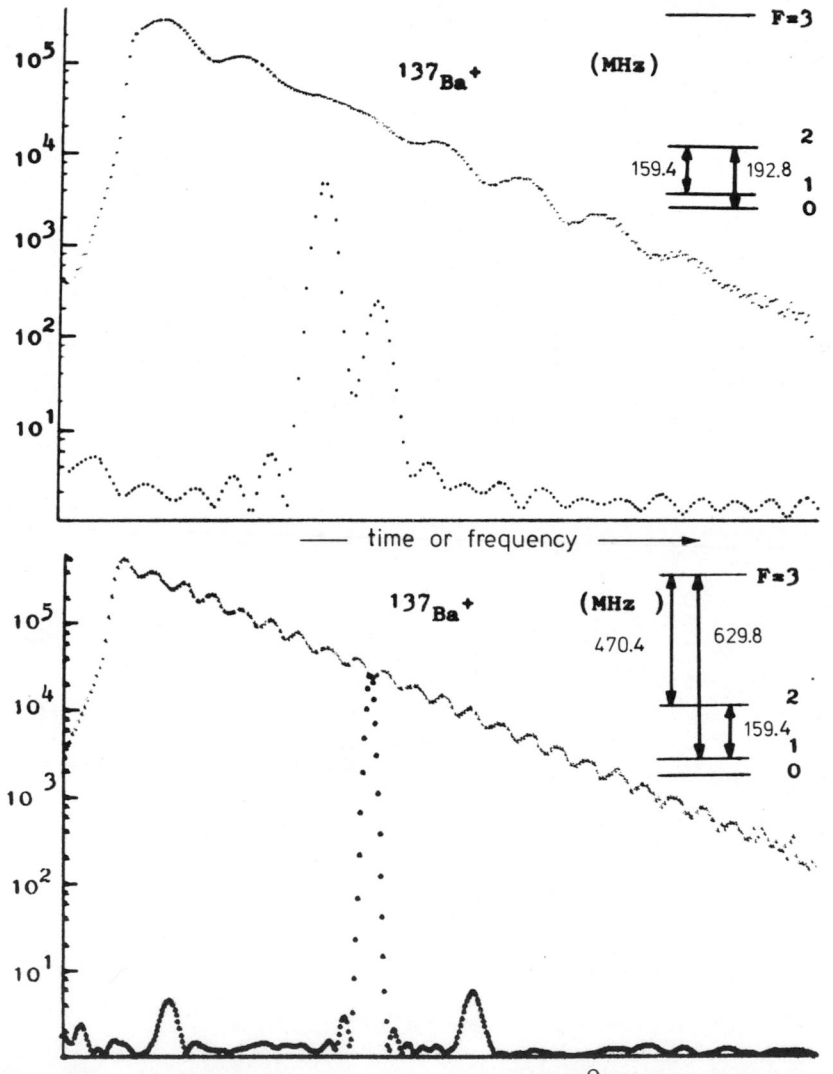

Fig.13: Observed quantum beats in the $\lambda=4554\text{Å}$ transition of
Ba^{137}II and their Fourier transforms. The inserted level schemes
indicate the measured beat frequencies. Please note that the time
and frequency scales are different in the top and bottom part.

only beats among the upper three or the lower three hfs-components depending on which angular setting (a or b in Fig.12) of ϑ has been chosen. This is exactly what has been observed in the experiment: In the upper part of Fig.13 two low frequency beats from the lower three hfs-components appear and in the lower part one low frequency plus two high frequency beats appear from the upper three hfs-components, each time with their Fourier transforms inserted.

As one would expect for broad band (non selective) excitation[11] one seems to observe at first sight only intensity minima at t=0 for the geometry of the experiment in Fig.14. However, a more detailed analysis with a direct LSF to the data clearly shows that the beat-frequency $\omega(F_1=1-F_2=2)$ starts out with a maximum at $\tau=0$. This is in full accord with the theoretical description of the hfs-quantum beats under selective excitation given by the intensity formula in Fig.14

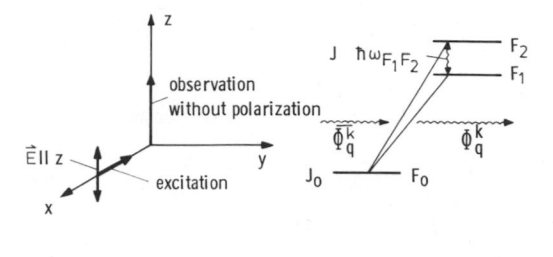

$$I(t) = \frac{1}{N} e^{-\gamma t} \sum_{kq} (-1)^{k+2J+J_0-I+F_0} (2F_0+1) \, \phi_q^k \begin{Bmatrix} 1 & 1 & k \\ J & J & J_0 \end{Bmatrix} \overline{\phi}_q^k$$

$$\times \sum_{F_1 F_2} (-1)^{F_1+F_2} (2F_1+1)(2F_2+1) \begin{Bmatrix} 1 & 1 & k \\ F_1 & F_2 & F_0 \end{Bmatrix} \begin{Bmatrix} F_1 & J & I \\ J & F_0 & 1 \end{Bmatrix} \begin{Bmatrix} F_2 & J & I \\ J & F_0 & 1 \end{Bmatrix} \begin{Bmatrix} J & F_1 & I \\ F_2 & J & k \end{Bmatrix} e^{-i\omega_{F_1 F_2} t}$$

Excitation starting from $F_0 = 1$:

$$I(t) \propto \frac{275}{7200} \, (1 - 0.1818 \cos \omega_{20} t - 0.2727 \cos \omega_{21} t)$$

Excitation starting from $F_0 = 2$:

$$I(t) \propto \frac{417}{7200} \, (1 + 0.0360 \cos \omega_{21} t - 0.1007 \cos \omega_{31} t - 0.3357 \cos \omega_{32} t)$$

A [MHz]		B [MHz]		
^{135}Ba II	^{137}Ba II	^{135}Ba II	^{137}Ba II	Authors
113.3 ± 2.1	124.76 ± 2.3	66.0 ± 9.0	97.1 ± 13.9	Becker et al.[27]
112.72 ± 0.32	125.95 ± 0.35	60.45 ± 0.35	92.53 ± 0.40	This work

Fig.14: Excitation and detection geometry with intensity formula for excitation from a selected groundstate hf-component F_0. $\overline{\phi}_q^k$ are the exciting and ϕ_q^k the detecting polarization tensors. Comparison of hfs results.

which was derived by J.Macek. Applying this formula to the specific case studied here one indeed obtains the ω_{12} beat frequency with opposite phase compared to all the others in accord with the experimental result.

From the measured frequencies one can readily deduce the hf-coupling constants which are listed in Fig.14 for Ba[137] and Ba[135] which has been measured in complete analogy to Ba[137]. The results clearly show a significant improvement over the earlier work with a Fabry Perot spectrometer[27] in particular as far as the quadrupole interaction constants are concerned. Unfortunately the theory is not yet capable of making full use of such accuracy in such a heavy atom for a better determination of the nuclear quadrupole moment. Therefore we expect this measurement to serve as a challenge to theoreticians for quite some time.

In lighter alkali-like atoms or ions the theory is much further developed and accurate experiments are of interest for tests of the theoretical results[28]. From the experimental point of view the NaI-$3p^2P_{3/2}$-hfs is particularly attractive, since the zero field hfs splittings in Fig.15c are of the same order as twice the natural width of the levels (19.9MHz) and therefore the hfs is generally

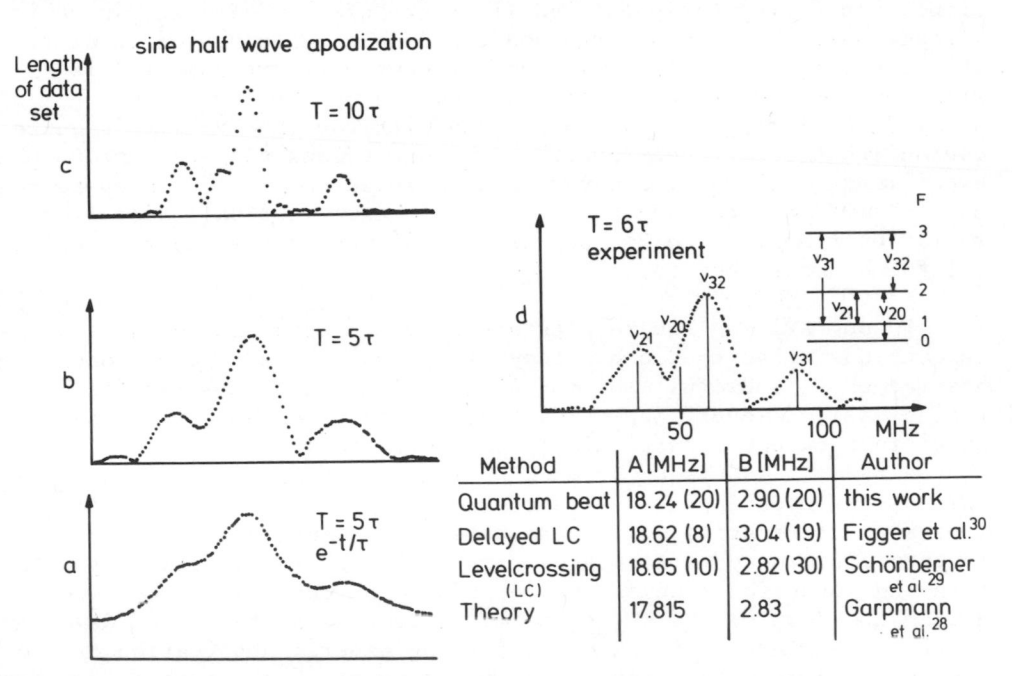

Method	A [MHz]	B [MHz]	Author
Quantum beat	18.24 (20)	2.90 (20)	this work
Delayed LC	18.62 (8)	3.04 (19)	Figger et al.[30]
Levelcrossing (LC)	18.65 (10)	2.82 (30)	Schönberner et al.[29]
Theory	17.815	2.83	Garpmann et al.[28]

Fig.15: Computer simulation of the quantum beat Fourier spectroscopy signal of the NaI-$3p^2P_{3/2}$ hfs for different lengths T of data sets a,b,c; d is an experimental result with T=6τ. A comparison with other results is given.

unresolved in zero field. However, time resolved (or delayed) spectroscopy can be used for narrowing the lines and is therefore very well suited for the NaI-3p^{3}P$_{3/2}$-hfs problem.

At first the zero field quantum beat Fourier spectroscopy seems to be unfavourable since its linewidth in the Fourier transform is $\Delta\nu=\sqrt{3}/\pi\cdot\tau$ (when a beat signal $f=\cos\omega t\cdot e^{-t/\tau}$ is transformed) as compared with optical double resonance or levelcrossing linewidths of $\Delta\nu=1/\pi\cdot\tau$. However, the advantage in quantum beat spectroscopy is the simultaneous measurement of the beat frequency and the exponential decay. This allows to divide the measured damped quantum beat signal by the measured exponential yielding an undamped beat oscillation. If this is Fourier transformed then the linewidth is basically determined by the total length of the measurement T, namely $\Delta\nu=1/T$ aside from apodization broadening. (A sine halfwave apodization causes a linewidth of $\Delta\nu\sim\pi/2T$). In Fig.15 this improvement of the linewidth is simulated with a sine halfwave apodization using the NaI-hfs example of which 3 beat frequencies become clearly resolved at T=10τ allowing an accurate determination of the hfs coupling constants.

In a preliminary measurement in Fig.15c up to only T=6τ this technique was applied and the result is compared in Fig.15a-d with the simulation giving the predicted linewidth of $\Delta\nu\sim\pi/2T=17$MHz which is already narrower than a corresponding double resonance or levelcrossing width. The extracted hfs coupling constants are compared in Fig.15 with the most accurate recent measurements by the levelcrossing[29] or delayed levelcrossing[30] technique and with the theory[28]. It is quite obvious that the present preliminary result does not yet improve the experimental accuracy. However, with a measurement under preparation up to T=10τ one can expect a very clean result not suffering from radiation trapping, collisions or ambiguities in the analysis as in all former measurements.

In summarizing the results of the examples presented here it can be concluded that the combination of laser- and fast beam spectroscopy has led us to a useful tool for high quality measurements. Its exploitation will, however, depend on the further development of lasers with sufficient cw-power particularly in the UV region, although it is clea: that already nowadays hundreds of levels can be excited and studied. Whether the use of pulsed lasers combined with frequency doubling offers an alternative will be discussed in a separate contribution to this conference[31]. Also great improvements in the optical excitation and detection schemes can be made thus reducing the cw-power requirements down to the mW region.Aside from these technical developments a breakthrough is expected for the general application of the fast beam laser spectroscopy when the "two step" excitation from short lived gas-or foil excited levels to higher levels becomes feasible. Results in this direction are foreseen for the next conference since several groups are working already on this problem.

Many of my colleagues and guests have been involved during the various stages of this work. It is therefore a great pleasure for me to accknowledge their contributions by completing the list of authors: W.Wittmann, A.Gaupp, J.Plöhn, L.Henke, M.Kraus (all Berlin,Germany), L.Curtis (Toledo,Ohio), M.L.Gaillard (Lyon,France), J.Macek (Lincoln, Nebraska), and J.O.Stoner,Jr. (Tucson,Arizona). This work is supported by the "Sonderforschungsbereich 161 der Deutschen Forschungsgemeinschaft".

REFERENCES

+ List of authors continued in the acknowledgement
1 H.J.Andrä et al., Nucl.Inst.Meth. 110,453 (1973)
2 H.J.Andrä et al., Phys.Rev.Lett. 31,501 (1973)
3 H.Harde and G.Guthörlein, Phys.Rev. A10, 1488 (1974)
4 H.J.Andrä, Atomic Physics 4, edited by G. zu Putlitz
  E.W.Weber, and A.Winnacker (Plenum Press, New York,1975) p. 635
5 A.Arnesen et al. Phys.Lett. 53A, 459 (1975)
6 Beam-Foil-Spectroscopy, Nucl.Inst.Meth. 110 (1973) and references
  quoted there.
7 W.Wittmann et al., Z.Phys. 257, 279 (1972)
8 W.S.Bickel and R.Buchta, Physica Scripta 9, 148 (1974)
  P.D.Dumont et al., to be published
9 J.Lifsitz and R.H.Sands , Bull.Am.Phys.Soc. 10, 2114 (1965)
10 H.D.Betz, Rev.Mod.Phys. 44, 465 (1972)
11 H.J.Andrä, Physica Scripta 9, 257 (1974)
12 H.G.Berry et al., Physica Scripta 1, 181 (1970)
13 A.R.Schaefer, Astrophys. J. 163, 411 (1971)
14 T.Andersen et al., Astrophys. J. 178, 577 (1972)
15 E.Trefftz, Z. Astrophys. 28, 67 (1950)
16 R.N. Zare, J.Chem.Phys. 47, 3561 (1967)
17 A.W.Weiss, J.Chem.Phys. 47, 3573 (1967)
18 G.A.Victor and C.Laughlin, Nucl.Inst. Meth. 110, 189 (1973)
19 W.L.Wiese et al., Atomic Transition Probabilities, Vol. II,
   NSDRS-NBS22 (1969)
20 A.Eberhagen, Z.Phys. 143, 392 (1955)
21 N.P.Penkin and L.N.Shabanowa, Opt. Spectr. (USSR) 12, 1 (1962)
22 U.Brinkmann, Z.Phys. 228, 440 (1969)
23 H.J.Andrä et al., J.Opt.Soc.Am., October (1975)
24 C.K.Kumar et al., Phys.Rev. A7, 112 (1973)
25 M.Gaillard et al., Phys.Rev. A, September (1975)
26 F.v.Sichart et al., Z. Phys. 236, 97 (1970)
27 W. Becker et al., Z. Phys. 216, 142 (1968)
28 S.Garpmann et al., Phys. Rev. A11, 758 (1975)
29 D.Schönberner and D.Zimmermann, Z. Phys. 216, 172 (1968)
30 H.Figger and H.Walther, Z. Phys. 267, 1 (1974)
31 M.Gaillard et al., this conference

ON THE FEASIBILITY OF PULSED LASER EXCITATION OF FAST ATOMIC BEAMS

M. Gaillard[+], H.J. Plöhn, H.J. Andrä, D. Kaiser[++],
and H.H. Schulz[++]

Institut für Atom- und Festkörperphysik
Freie Universität Berlin, W.-Germany

Laser excitation of fast atomic beams[1,2] has recently emerged as a
useful tool for precise lifetime[3-5] or quantum beat[4,6] measurements.
Excitation from neutral or ionic ground or metastable states to
higher lying levels has been successfully performed. However, limitations to this technique are imposed at present by the rather
narrow band of sufficiently powerful cw-laser action (commercially)
available covering approximately only the visible spectral region.
(Infrared lasers will be not discussed here). This represents a
severe restriction since most resonance lines of neutral and singly
ionized atoms lie already in the UV region. Although this restriction can be circumvented for higher lying levels by gas- or foil
preexcitation of some intermediate level starting from which the
laser is used as a second step to excite the level of interest, the
only solution to transitions starting from ground or metastable
states will be in general the use of frequency doubling techniques
in order to reach the UV region. Due to the high power requirements
for frequency doubling this is in general feasible only with pulsed
laser systems. The question therefore arises whether a pulsed dye
laser system can effectively be used for the excitation of fast
atomic beams.

In order to test this possibility we have made an attempt to
measure the intensity decay of the BaII-4554Å resonance transition
after excitation of the $6p^2P_{3/2}$-level with a nitrogen laser pumped
dye laser. The main difficulty when working with a pulsed laser is
the fantastically small duty cycle as can be seen in Table 1, whereas power problems generally don't occur. Essentially one can assume
that each laser pulse saturates all the atoms passing through the
laser beam while it is on. Unfortunately, however, the laser pulse
length is of the order of 10ns during which only a single pulse can

	CW-Dye-Laser	N_2-Pulsed-Dye-Laser
Power	0-2W	0-10kW
Pulse length	0.5 day	5-10ns
Repetition rate	1 day^{-1}	0-500s^{-1}
Duty cycle	1	$5\cdot10^{-6}$
Spectral range	4000-8000Å (50mW)	2300-8000Å (500W)

Table 1: Comparison of the characteristics of cw and N_2-pulsed dye
         lasers.

be processed by the detection electronics when working in photon
counting mode, no matter whether this pulse contains information
on one, two or many detected photons. Pulse heigth digitizers could
improve this situation in the future but for the time being we want
to (have to) restrict ourselves to such a simple detection electronics. The consequence is a serious dead time problem which limits the
effective detection count rate to a fraction of the repetition rate
r of the laser and which requires a correction given by $n_o = -r \cdot \ln(1-\frac{n}{r})$
where n is the measured and n_o would be the real count rate.

For an estimate of the fluorescence emission of the atomic
beam we start with the number of atoms being exposed to each laser
pulse which amounts to $6.25\cdot10^5$ Ba$^+$ions per pulse in a 10μA beam
current at 300 keV. If saturation is achieved one obtains then
$3.125\cdot10^5$ ions per pulse in the excited $6p^2P_{3/2}$ state which decays
exponentially with a time constant τ. In a time differential experiment only a time window of about $\Delta t=0.1\cdot\tau$ is observed which together
with the detection efficiency of 10^{-5} yields then a photon detection
rate of about 0.3 per laser pulse at maximum intensity. At a repetition rate of 500 Hz the total signal corresponds under these conditions to 150 photons·s^{-1} which is a reasonable signal for a cascade
free lifetime measurement. The handling of such a count rate does
not cause any problem. However, during the same short period of time
when the beam emits photons there is a large amount of laser straylight present in the target chamber due to the powerful laser pulse.
Even under favourable conditions we were not able to completely eliminate the straylight with the experimental arrangement shown in
Fig. 1. This difficulty forced us to use a beam chopping and signal
gating technique which is sketched in Fig. 2 and allows to dig out
the pure signal from the various sources of background.

The laser is running at a constant repetition rate of ~500 Hz
with pulses of 10 ns width. For obtaining the signal for one channel
the total cycle is split into two halves of about 1s each determined
by 500 laser pulses. During the first half the beam is on
and counter I registers the real beam fluorescence signal plus backgrounds in a 100ns time window correlated to the laser pulse. Also
while the beam is on counter II registers in a 100ns time window
100μs after the laser pulse only beam background and noise (almost

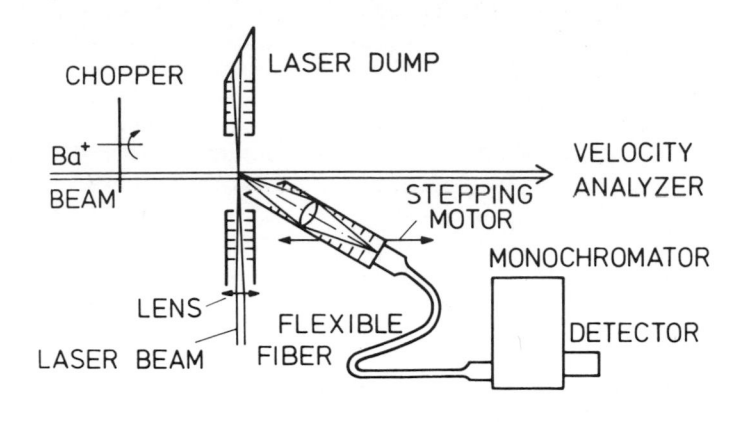

Fig.1: Schematic experimental set up for pulsed dye laser excitation of a fast Ba⁺-beam. The angle between detection optics and beam is used to spectrally Doppler shift the detection line away from the laser straylight.

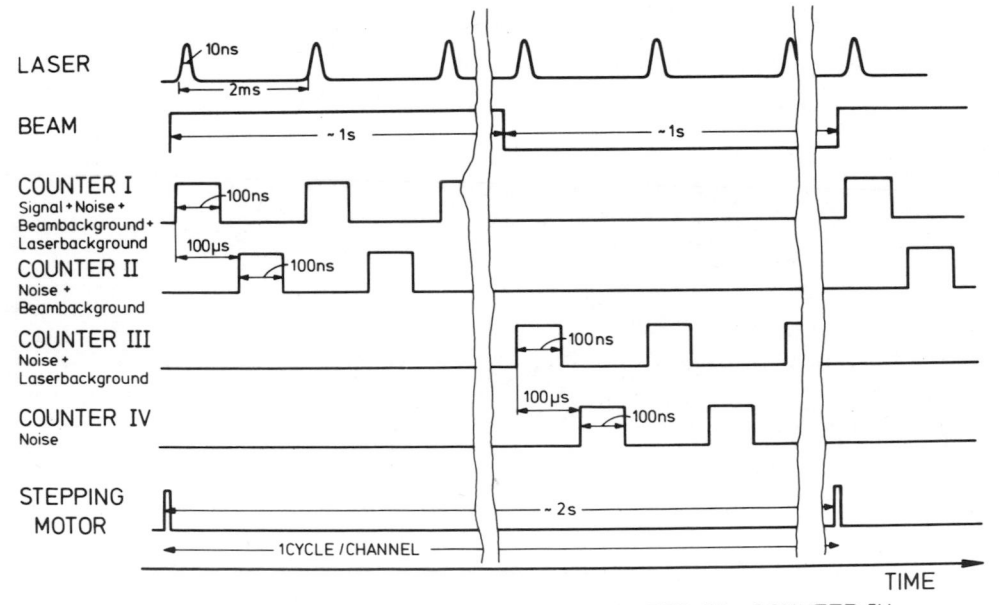

SIGNAL = COUNTER I - COUNTER II - COUNTER III + COUNTER IV

Fig.2: Laser, beam, detection, and stepping motor timing for obtaining the pure beam fluorescence signal according to the formula on the bottom of the figure.

zero!). During the second half of the cycle the beam is off and the same procedure is repeated with counter pair III and IV replacing counter pair I and II. After this second half the computer calculates the real fluorescence signal with the formula given on the bottom of Fig. 2, stores this signal in the corresponding counting channel, gives a stepping signal to the mechanical delay drive (Fig. 1) and starts a new measuring cycle for the next counting channel. Due to the time proportional normalization beam fluctuations could crucially disturb the measurement. Therefore the beam had to be kept stable to within $\pm$ 5% and the residual fluctuations were averaged out by scanning mechanically along the intensity decay at a rate of 1 scan per 2 minutes.

The result which was obtained with this arrangement at a beam current of 6μA ^{138}Ba$^+$ and 300 keV is shown in Fig. 3. It was obtained within 30 minutes at a laser repetition rate of 516 Hz and with a maximum signal rate of 130s^{-1}. One observes particularly at high intensities a significant bending over from the straight line in Fig. 3 indicating the strong influence of the dead time which

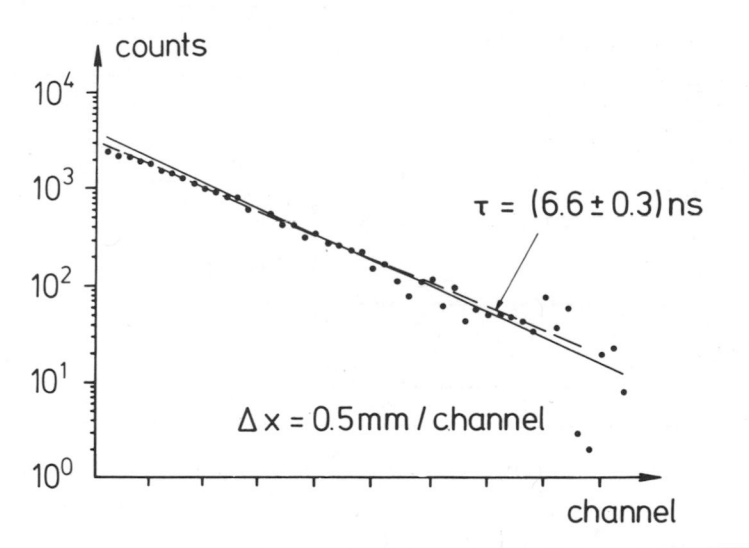

Fig. 3: Lifetime measurement of the ^{138}BaII–6p ^{2}P$_{3/2}$ resonance level without complete dead time correction using pulsed dye laser excitation. Broken line least squares fit to partially corrected data, straight line expected exponential for the known lifetime.

could only be partially corrected for. A perfect experiment as far
as dead time correction is concerned would have required a separate
storage for each counter per channel plus a registration of how
often each counter was opened per channel in order to allow for a
statistically clean dead time correction. As described before this
was not done in this first experiment which was supposed to prove
only that the excitation of fast atomic beams with pulsed lasers
is feasible and that useful data can be obtained. Therefore the
result for the mean live of the BaII-6p $^2P_{3/2}$ state in Fig. 3 was
expected to be too long compared with the best known value of τ=
6.312±0.016ns[6] since only a partial deadtime correction was made.

   Besides this deficiency the experiment has clearly shown that
there is well justified hope to successfully employ frequency
doubled pulsed dye lasers for fast beam excitation in the UV region
in the near future.

   We want to thank Prof. J. Luther for many discussions at the
beginning of this work and Prof. A. Steudel who strongly encouraged
the Berlin-Hannover collaboration on this project. This work is
supported by the Sonderforschungsbereich 161 der Deutschen Forschungsgemeinschaft.

REFERENCES

+   Université de Lyon, France
++   Technische Universität Hannover, W.-Germany
1   H.J. Andrä et al., Nucl.Instr. Meth. 110 453 (1973)
2   H.J. Andrä et al. Phys. Rev. Lett. 31 501 (1973)
3   H. Harde and G. Guthörlein, Phys. Rev. A10 1488 (1974)
4   H.J. Andrä, Atomic Physics 4, edited by G. zu Putlitz, E.W. Weber,
     and A. Winnacker (Plenum Press, New York, 1975), p. 635
5   A. Arnesen et al., Phys. Lett. 53A 459 (1975)
6   H.J. Andrä, this conference

CASCADE FREE LIFETIME MEASUREMENTS BY LASER EXCITATION OF

FOIL- OR GAS-EXCITED BEAMS

H. Harde

Hochschule der Bundeswehr Hamburg, 2 Hamburg 70

Holstenhofweg 85, Federal Republik of Germany

Introduction

Most of all beam-foil lifetime measurements are strongly influenced by the perturbing cascade effect which often will limit the accuracy of this technique and sometimes even can make impossible the evalua- tion of a lifetime of interest. Also with highly sophisticated decay curve analysis using computers, it is difficult to get satisfactory results and in general only experimental methods which directly eliminate the cascade effect, will produce reliable measurements. Therefore during the last three years different techniques have been developed and tested, one of these taking advantage of the phenomenon of quantum beats in a magnetic field[1] and another using the well-known method of cascade-coincidences[2] adopted from nuclear physics to foil excited fast-ion beams. Furthermore the principle of selective excitation has been successfully demonstrated for cascade- free lifetime measurements and was used by Andrä et al.[3] to excite a fast-ion beam by Doppler tuned beam laser resonance.

This technique of laser excitation is of particular interest in combination with a broad-band excitation as resulting from beam- foil or beam-gas interactions. Then also high lying atomic levels which cannot be populated directly by a laser from the ground state of the ion, can be investigated. Such a two-stage excitation techni- que was used by Harde et al.[4] to make high-precision time-of-flight measurements taking advantage of all the unprecedented properties of beam-foil spectroscopy (BFS) for lifetime measurements but avoiding the perturbing cascade-effect.

In this paper further lifetime measurements of highly excited atomic states of Na, Ne and Li^+ are presented and their results are compared with earlier measurements. In the same manner as for life- time experiments, the technique of stepwise excitation is also of

859

great interest for laser induced quantum beat and laser tuned beam
resonance measurements[5] opening a large field of applications for
finestructure, hyperfinestructure and Zeeman splitting investigations. Therefore first of all the main features of the excitation
process which was used in our experiments, are briefly depicted and
their essencial properties will be discussed.

Two-Stage Excitation

The principal arrangement of a two-stage excitation of fast-ion
beams is shown schematically in Fig.1. The first step consists of a
customary broad-band excitation as resulting from beam-foil or beam-
gas interactions, by which higher lying atomic states are populated.
Afterwards a selective interaction between two of these levels is
induced by a laser which is tuned to the resonance of an optically
allowed transition from a higher lying level n to a lower level k.
So with laser action the population densities of levels n and k and
just so their decay curves as a function of time after laser excitation can be changed in a manner as qualitatively indicated in Fig.1.
According to this population change, especially for level n, it is
now possible to separate the pure laser induced portion of the detected fluorescence radiation measuring the difference of the intensities with and without laser.
When this technique is used for a time-of-flight experiment in
order to measure the lifetime or an expected level splitting of the
atomic state n, the perturbing cascade-effect resulting from the
broad-band excitation can be eliminated and the observed intensity
decay-curve yields a single exponential eventually superposed by

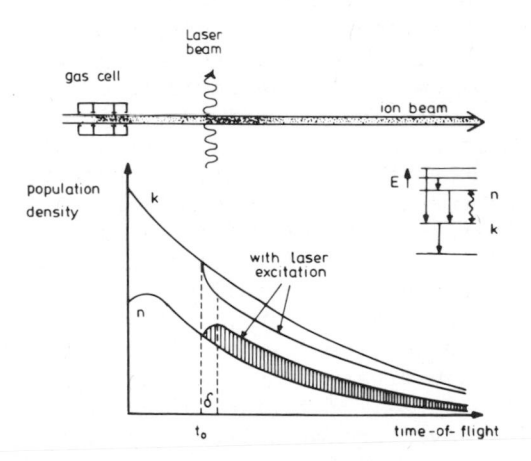

Fig. 1. Principal arrangement of beam-gas-laser excitation and resulting population change of levels n and k as a function of time
after the excitations.

quantum beats.

   The strength of the detected difference signal and by this the
applicability of the two-stage excitation is essentially determined
by the efficiency of the laser used and the population difference
$N_n(t_o)-N_k(t_o)$ of levels n and k at time t_o. These in turn are influenced by the special broad-band excitation process as well as by the
time t_o between first and second excitation according to different
lifetimes and cascades of the levels n and k. The effectiveness of
the laser depends on the available laser output power per spectral
width and on the resonance behaviour of the atoms, taking into consideration that the excitation conditions for a fast ion beam in
general are extremly different compared to a thermally flying atomic
beam. So the resonance width of a fast moving atom which may be
excited by a single mode laser is not determined by the naturel linewidth of an atomic transition, but dominantly depends on the laser
interaction time δ which for typical conditions is of the order of
10^{-9} sec and according to the energy-time uncertainty relation yields
a resonance width of about 1 GHz. Connected to this fact the usual
expressions for the transition rates of induced absorption and
emission W_{kn} and W_{nk} which have to be known to calculate the effectivity of an induced population change between levels n and k, must
be modified. In first-order perturbation theory for single-mode
laser action W_{nk} was found to be

$$W_{nk}^S = (8\pi)^{-1}A_{nk}\cdot c\cdot\lambda^2\rho\ \frac{\sin^2\left[1/2\ (\omega_{nk}-\omega)\delta\right]}{\left[1/2(\omega_{nk}-\omega)\delta\right]^2}\ \delta \tag{1}$$

which in the case of resonance reduces to

$$W_{nk}^S = (8\pi)^{-1}A_{nk}\ c\cdot\lambda^2\cdot\rho\cdot\delta \tag{2}$$

where c is the light velocity, λ the laser wavelength, ρ the photon
density, A_{nk} the spontaneous transition probability, ω_{nk} the transition frequency from level n to k, and ω the laser frequency. This
formula shows that Fermi´s "golden rule" for the induced transition
rate is no longer valid and W_{nk}^S also for the case of resonance still
is a function of the interaction time δ. That means, in general the
induced transition probabilities will be relatively small and a high
photon density produced by the laser is necessary to induce a successful population change. On the other hand the relatively extensive resonance width of an individual atom with the laser, $\Delta(\omega_{nk}-\omega)=2\pi/\delta$,
makes it possible to excite close neighboring levels such as hyper-
fine or Zeeman sublevels coherently, and also to excite such atoms
which are slightly Doppler shifted.

   Often, however, the Doppler broadening of a fast ion beam, especially after foil excitation, can be of the order of 1 Å or even more.
Then only about one percent of all incoming ions or atoms would be
in resonance with the laser and it has to be investigated, if the
effectivity of the laser excitation can be improved when the spectral
width of the laser, so far as possible, is adapted to the Doppler

broadening of the ion beam. An evident improvement will be observed
when a narrow-band laser pumping process already produces saturation
for a small class of the Doppler broadened atoms. Then multi-mode
laser operation distributes the given laser power to a larger spectral width and by this increases the number of resonant atoms without
effective loss for the individual population change of an atom.
Moreover, when dye lasers are used for selective excitation, the
laser output power in general will be increased with growing bandwidth. For such broadband laser excitation the formula for W_{nk}^s in
equation (1) is no longer applicable and has to be changed according
to the special properties of multi-mode laser operation. So for
typical conditions where the mode spacing of the laser, $\Delta\omega_L = \pi c/l$
(l-cavity length of the laser), is small compared to the resonance
width of the atoms and so different modes can contribute to the
induced transition rate,

$$W_{nk}^m = (8\pi)^{-1}A_{nk}c\cdot\lambda^2\rho_M(M+1)\delta \tag{3}$$

is obtained. ρ_M represents the mean photon density of one mode and
the resonance behaviour of an atom is approximated by a triangular
function (see Fig.2). When M - the number of modes within the resonance width $2\pi/\delta$ - is large compared to unity, the induced transition rate of a single atom, averaged over the whole spectral width
$\Delta\lambda_1$ of the laser, is found to be

$$W_{nk}^m = (8\pi\cdot\Delta\lambda_1)^{-1}A_{nk}\cdot\lambda^4\cdot\rho_t \tag{4}$$

where ρ_t is the total photon density of all laser modes within the
spectral width $\Delta\lambda_1$. This result is no longer a function of the interaction time δ, but only depends on the properties of the laser and
the particular transition of the atom. When the laser bandwidth $\Delta\lambda_1$
is limited to about 1 Å, and an induced transition rate of the
order of the spontaneous transition probability A_{nk} is sought, a
photon density of approximately 2×10^{10} photons/cm^3 which corresponds to a laser output power of about 2 W, is necessary at 6000 Å.
For typical conditions this gives a population change for the
selectively excited levels of about 5 - 10 percent and in general

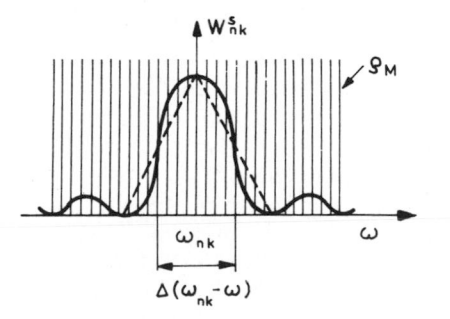

Fig. 2. Resonance curve of W_{nk}^s superposed by laser modes with a
photon density ρ_M.

this will be enough to apply the proposed technique of cascade-free lifetime measurements even to strongly Doppler broadened ion beams.

Apparatus

The experimental arrangement used for our cascade-free lifetime measurements consists of a two-stage excitation as previously discussed in detail, and is followed by a specially constructed detection system which takes into consideration the particular demands of this measuring technique.

The overall experimental set-up is shown schematically in Fig.3. A fast-ion beam which is produced by a 350-kV accelerator, passes through a differentially pumped gas cell and is highly excited by collisions with the target gas. The distance from the inner chamber to the exit of the gas cell takes 2.5 cm and corresponds to a time-of-flight for typical beam velocities of about 25 nsec. Therefore only atomic states with lifetimes of this order or longer will still be sufficiently populated just behind the cell and these levels are primarily suitable for a second selective excitation. A foil excitation will be necessary for lifetime measurements where shorter living states have to be used as intermediate levels or highly ionized atoms are to be investigated. For our measurements, however, where the laser starts from a metastable state to populate higher

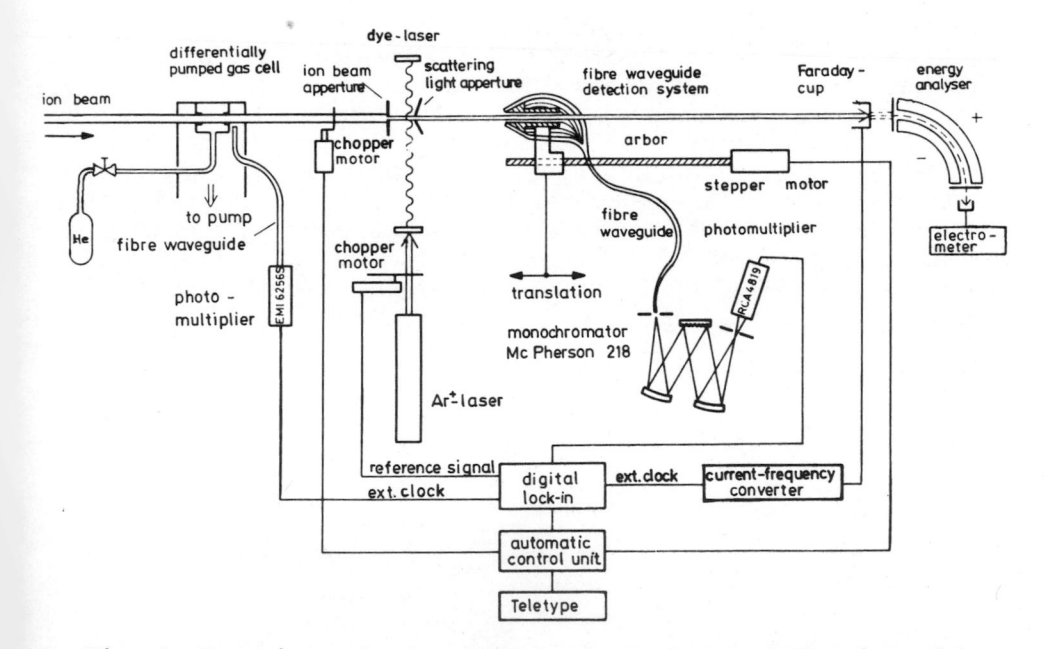

Fig. 3. Experimental arrangement of gas-laser excitation with following detection system.

lying levels selectively and the distance between broad-band and laser
excitation even can be one meter or more, it is advantageous to have
a gas cell according to a smaller Coulomb scattering and thereby a
smaller Doppler broadening of the ion beam. After passing through the
gas cell the ions or atoms created by charge exchange cross a 1-mm-
thick laser beam which is tuned to an optically allowed transition
of the ions. In order to increase the spectral power density, especially for multi-mode laser action, the atoms are excited inside the
cavity of a continuous wave dye laser. Figure 4 shows the essencial
components of the folded laser resonator[6]. The laser's active medium
is pumped by a 6-W all-lines Ar^+-laser and consists of a dye-solu-
tion jet beam. Three prisms inside the cavity allow the selection of
the wavelength and reduction of dye-laser bandwidth to about 0.5 Å,
while a further reduction can be made by a Fabry-Perot etalon. Prisms
1 and 2 are simultaneously used as windows to the vacuum system
necessary for the ion-beam. Without an etalon we obtained a maximum
photon density inside the cavity of about 1.5×10^{11} photons/cm^3 which
for an external laser beam with comparable photon density corresponds to a dye-laser output power of more than 10 W. With etalon
these values in general will be reduced by a factor of 3 to 4,
while the laser bandwidth can be reduced down to 10^{-2} Å.
    The detection system for a time-of-flight measurement is also to
be seen in Fig.3. A fibre-waveguide detection head which consists
of 16 different waveguide bundles surrounding the ion beam and
forming an envelope of an obtuse cone, observes a small element of
the radiating ion beam at an angle of 30° to the beam direction
but also over the whole azimuthal angle of 360°. This detection
head is mounted on a sliding carriage and can be moved by a stepper
motor parallel to the ion beam. The other end of the waveguide
system is formed to a narrow line equally illuminating the entrance
slit of a monochromator which separates the spectral line of
interest from scattering light. In order to assure a difference
measurement for a cascade-free signal, the laser is chopped with a
frequency of 69 Hz and the signal at the exit slit of the monochromator is detected by a digital lock-in photon-counting system.
For normalization of the signal a current-frequency converter is
used which works as an external clock for the photon counter and

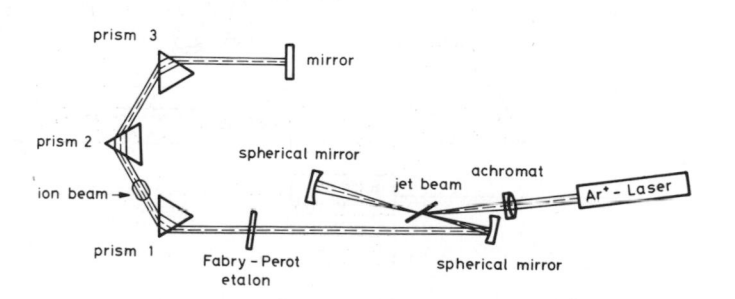

Fig. 4. Schematic drawing of the dye-laser resonator.

regulates the chopper sampling time in accordance with the ion current. Alternatively to this, another normalization can be made by a photon counter which detects the emitting radiation of the ion beam inside the gas cell. Then a further chopping of the ion beam is possible which may be necessary to eliminate scattering light from the laser.

The exact ion beam velocity, which has to be known for a lifetime measurement, is evaluated by a $90°$-electro-static energy analyser.

Lifetime Measurements

The experience of about 70 years in lifetime measurements has shown which discrepancies have to be expected, when different methods are used to measure lifetimes of the same atomic states, and up to now only very few levels are known where the lifetime is accurate within 3 percent. Therefore to get further reliable results, it is of particular importance to use methods which are far extending free from systematic errors. The experimental arrangement as described before realizes this requirement, taking advantage of all the un-precedented properties of BFS but avoiding the perturbing cascade-effect.

Sodium. A relatively simple example of cascade-free lifetime mea-surement is given for the case of Na, where the technique of two-stage excitation was applied to a 100-keV ion beam to evaluate the lifetimes of the Na 3p ^{2}P-levels (Fig.5). Here the broad-band exci-tation is used to create fast atoms by charge exchange in the gas cell. Afterwards the levels $^2P^o_{1/2}$ and $^2P^o_{3/2}$ can be populated selectively from the ground state of the atom by induced absorption, tuning the laser to a wavelength of $\lambda=5895.9$ Å resp. to $\lambda=5889.9$ Å. For this spectral region the laser is working with rhodamine - 6 G as dye and produces a photon density of about 10^{11} photons/cm^3

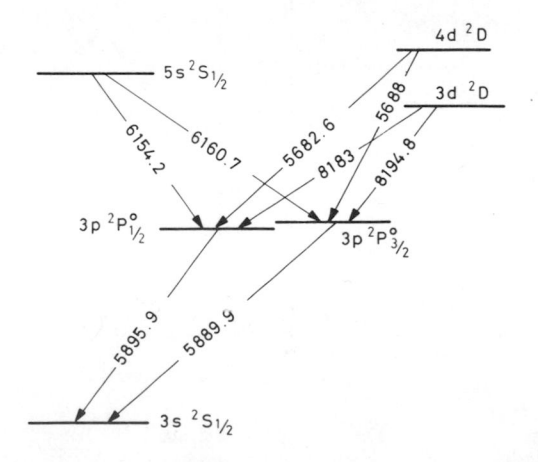

Fig. 5. Fine-structure level diagram of Na.

in multi-mode operation which approximately gives saturation for
the pumping process. So with an initial ion beam current of 3 μA
we could detect a counting rate of 6×10^5 cps in 1 mm distance
after laser excitation and such a strong signal ensures a successful lifetime measurement.

Since the fluorescence radiation is detected on the same transition which is used for laser excitation, the scattering light
of the laser which can even be of the order of 10^6 cps has to be
separated from the signal. This is far extending possible by
Doppler shifting the fluorescence radiation for about 16 Å, when
the fast beam is observed at an angle of $30°$ to the beam direction. Nevertheless for each measuring position of the decay curve
the scattering light is evaluated separately and subtracted from
the fluorescence radiation by chopping the ion beam.

The unobjectionable operation of the whole apparatus is demonstrated, when a lifetime measurement as resulting from our technique
is compared with a decay curve which is detected after beam-foil
excitation (see Fig.6). The 3p ^{2}P-levels are strongly influenced
by perturbing cascades and as already mentioned by Andersen et al.[7],
it is very difficult to get a satisfactory fit for the decay curve
in Fig.6a. On the contrary, our measurement was evaluated by a
least-squares fit with a single exponential yielding a lifetime for
the 3p ^{2}P$^o_{1/2}$ level of

$$\tau(3p\ ^2P^o_{1/2}) = 16.3 \pm 0.16 \text{ nsec.}$$

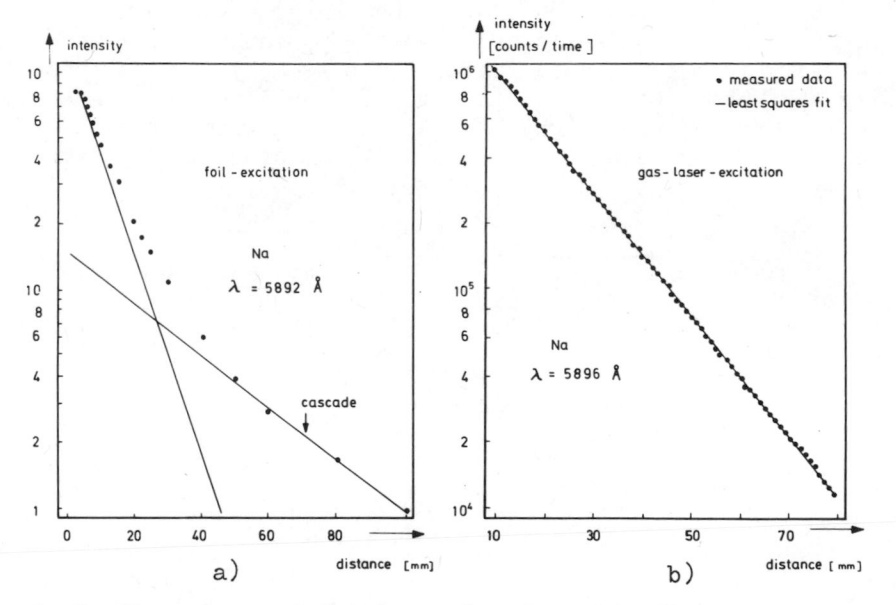

Fig. 6. Semilog plots of the intensity decay curves of Na
a) for the 3p ^{2}P^o-levels after foil-excitation[7] and b) for the
3p ^{2}P$^o_{1/2}$-level after gas-laser-excitation.

The specified error takes into consideration the counting statistics and the uncertainty of the beam velocity with 0.5 percent. Another half percent is added according to deviations which result from the indirect normalization of the fluorescence signal. However, with an overall accuracy of only one percent this technique is one of the most precise methods known for lifetime measurements.

In the same manner we investigated the 3p $^2P^0_{3/2}$ level which can be populated by the laser with a wavelength of $\lambda=5889.9$ Å. The intensity decay curve as a function of distance after excitation is shown in Fig.7. Opposite to the other level this decay is superposed by quantum beats with a beat frequency of 60 MHz which exactly corresponds to the hyperfine structure splitting between F = 3 and F = 2. These beats are observed without any polarizing medium in the detection system and result from the laser excitation process which fulfils the time-and the alignment-condition[8].

When this decay curve is fitted by a single exponential we get a lifetime of

$$\tau(3p \ ^2P^0_{3/2}) = 16.1 \pm 0.2 \text{ nsec.}$$

where the presented error already includes the uncertainties according to the beat-structure. Comparison of the measured lifetimes shows a small discrepancy within the error limits and probably can be explained by the influence of beats to the exponential decay.

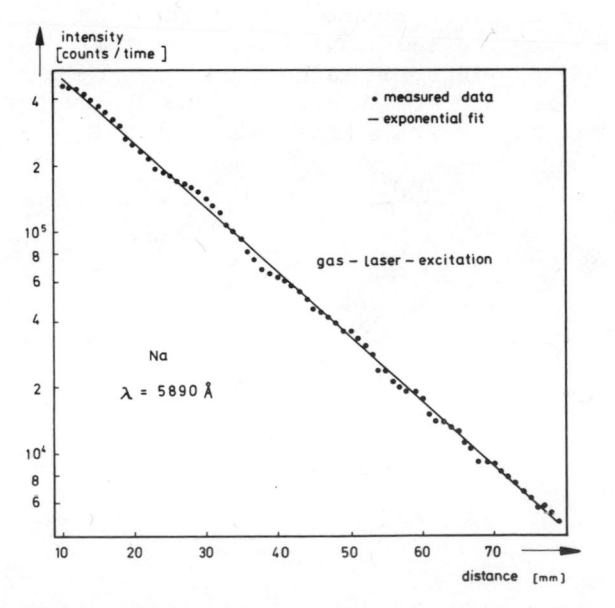

Fig. 7. Intensity decay curve of the Na 3p $^2P^0_{3/2}$ level after gas-laser excitation.

TABLE I. Comparison with earlier results for the
          lifetimes $\tau(3p \ ^2P_J)$ of Na.

J		τ(nsec)	method	Ref.(year)
	3/2	16.3 ± 0.5	Hanle-effect	9 (1964)
1/2	3/2	16.3 ± 0.4	delayed coincidence	10 (1967)
1/2	3/2	15.9 ± 0.16	phase-shift	11 (1962)
1/2	3/2	16.2 ± 0.5	beam-foil	7 (1972)
1/2		16.3 ± 0.16	beam-gas-laser	this work
	3/2	16.1 ± 0.2	beam-gas-laser	this work

Additionally, we have compared our results with earlier
measurements of other authors as to be seen in Table I and
excellent agreement was found between the different experiments.
However, it should be mentioned that these two sodium states belong
to the few exceptions where such a high consistency can be observed.

Neon. Quite different to the Na is the situation for the 3p-levels
of Ne, where the existing lifetime material of different authors
partially shows discrepancies of 40 percent and more. Since these
levels are not optically connected to the ground state and an excitation energy of about 16 eV is necessary to populate the 3p-levels,
a selective excitation is relatively difficult and only one experiment of Bennett and Kindlmann[12] is known which avoids the cascade-
effect using an electron impact excitation at threshold energy.
    In a similar way the two-stage excitation technique can be prof-
itably applied to bridge the large energy gap by gas cell excitation
and afterwards to populate selectively the 3p-levels with the laser,
starting from a higher lying intermediate state.

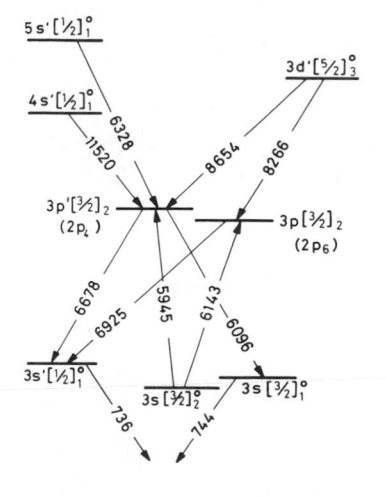

Fig. 8. Fine-structure level diagram of Ne.

TABLE II. Comparison with earlier results for the lifetimes
$2p_4$ and $2p_6$ of Ne.

τ(nsec)		method	ref.(year)
$2p_4$	$2p_6$		
19.1 ± 0.3	19.7 ± 0.2	delayed coincid.	12 (1966)
28.1 ± 1.0	22 ± 0.6	delayed coincid.	13 (1967)
22 ± 5.4	22 ± 3.2	delayed coincid.	14 (1966)
19.2 ± 1.0	18.2 ± 0.7	Hanle-effect	15 (1972)
21.5 ± 2.0	22.0 ± 2.0	beam-foil	7 (1972)
19.7 ± 0.3	19.5 ± 0.3	beam-gas-laser	this work

This has been done for the atomic levels $3p'[3/2]_2$ and $3p[3/2]_2$
($2p_4$ and $2p_6$ in Paschen notation) which can be excited from the
metastable state $3s[3/2]_2^0$ with a laser wavelength of $\lambda=5945$ Å
resp. $\lambda=6143$ Å (see Fig.8).
Since the transition probabilities as well as the production of
metastables after charge exchange are less favorable than for the
case of Na, we could only expect a counting rate for the fluorescence signal of a few thousand counts per second with a Ne^+ beam of
0.6 µA. Measuring the intensity decay curves of the transitions with
$\lambda=6096$ Å and $\lambda=6929$ Å for different runs over several hours, we
could evaluate the lifetimes of the levels $3p'[3/2]_2$ and $3p[3/2]_2$ as
presented in Table II. Although no quantum beats had to be taken into
consideration the overall error had to be specified to 1.5 percent
according to the observed deviations between different runs.
Comparing our results with previously measured ones (see Table II),
a satisfactory agreement for both levels is only found with the
measurements of Bennett and Kindlmann. Although for the other experiments least-squares-fits with partially three exponentials were used
to correct the influence of the cascades, there is still a big discrepancy to our values. These differences, however, give an impressive example for the unsatisfactory procedure of numerical analysis,
and simultaneously the superiority of measurements which experimentally eliminate the cascading effect, is demonstrated. Then also with
extremly different techniques an agreement within 1 or 2 percent can
be realized for lifetime experiments.

Lithium. A further example of greatly inconsistent lifetime experiments is found for the 1s2p ^{3}P-level of Li^+, which lies 61.2 eV above
the ground state of the ion. According to the high excitation potential and the experimental difficulties to produce sufficient ion
densities, this level has been scarcely investigated and only lifetime
measurements with the beam-gas or beam-foil technique are known. Their
results, however, differ by about 70 percent and neither of the values
agrees with the theory.
Detailed own studies of the decay curve after foil excitation have
distinctly shown the enormous influence of the cascade-effect to the

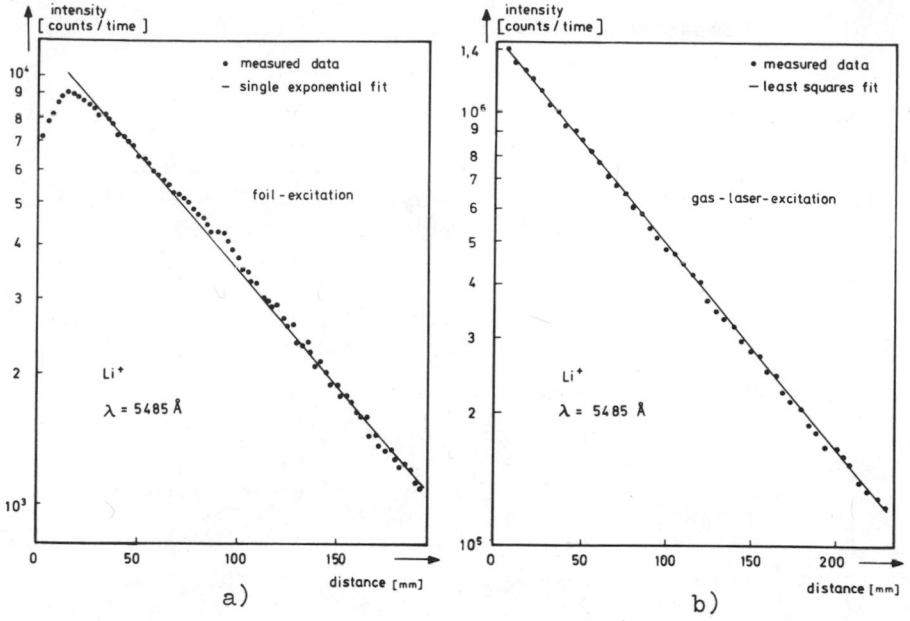

Fig. 9. Intensity decay curves of the Li$^+$ 1s2p ^{3}P-level
a) after foil-excitation and b) after gas-laser-excitation.

investigated level (see Fig.9a). So with different ion energies
from 56 to 200 keV it was possible to produce lifetime values from
33 up to 50 nsec when the decay curve was fitted with a single exponential.

According to these great uncertainties caused by the cascade perturbations after broad-band excitation, it was useful to apply our
cascade-free measuring technique to this Li-state.

The ^{3}P-level can be selectively excited with a wavelength of
λ=5485 Å starting with the laser from the metastable state 1s2s^{3}S$_1$
which is strongly populated after the gas cell (see Fig.10). In this
spectral region the laser was used with Na-fluorescein as dye and
produced a photon density corresponding to 7 W output power. With a
200-keV Li$^+$-beam of 1.5 μA beam current we observed a maximum
intensity for the fluorescence radiation of 10^{5}cps yielding a decay
curve as shown in Fig. 9b.

The evaluated lifetime which is presented in Table III together
with earlier results, shows no satisfactory agreement with one of
the other values, but according to the large influence of cascades
to the beam-foil and beam-gas experiments this could not even be
expected. Only the discrepancy to theory is surprisingly large. However, a statement of Feneuille furtheron remains true, that for a
theorist often one accurate measurement is more useful than a hundred
uncertain experiments. And this confrontation of experimental data

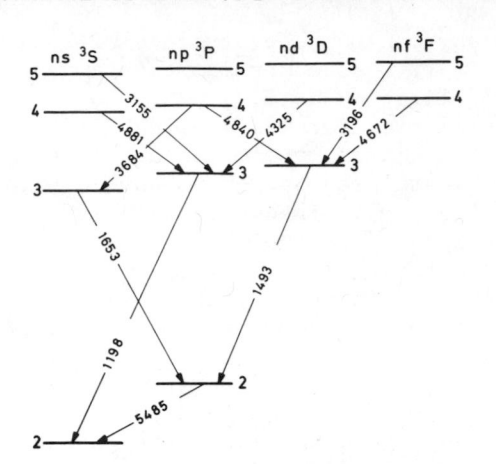

Fig. 10. Partial term scheme of Li$^+$ for the triplett levels.

again demonstrates that reliable lifetimes can only be obtained when the cascading effects are completely avoided by the measuring technique.

TABLE III. Confrontation of theoretical and experimental results for the lifetime of the 1s2p ^{3}P-level of Li$^+$.

τ(nsec)	method	ref.(year)
43.9	central-field-appr.	16 (1965)
54.4 ± 2.7	beam-gas	17 (1967)
33 ± 2	beam-foil	18 (1969)
37.1 ± 0.4	beam-gas-laser	this work

Conclusion

The different lifetime measurements presented in this paper have shown the particular advantages of the two-stage excitation and the applicability of this technique to higher lying atomic levels. However, simultaneously they illustrate the difficulties and restrictions of this technique and thereby give important hints for further improvements of the experimental set-up. So for the moment the accuracy of our measurements is primarily limited by the normalization of the fluorescence radiation and it would be more advantageous to normalize directly to the fluorescence signal after laser excitation. Then we can expect an increase of the accuracy with an overall error of only 0.5 percent. Only when quantum beats are superposed to the decay curve as in the case of

sodium, this accuracy can only be attained with difficulty and a relatively extensive fit procedure is necessary to correct the influence of beats to lifetime data.

An essential restriction for the applicability of this method is given by the available laser wavelength and laser power. For the moment it is possible to use continuous wave dye lasers from about 4000 Å up to 8000 Å and with frequency doubling and frequency mixing this can still be expanded to the ultraviolet and infrared regions. But then only substantially lower laser power can be obtained and in general this will not do for selective excitation. Also pulsed lasers which produce much higher peak powers, could be used for the excitation process, however, then the repetition rates as well as the pulse lengths still have to be substantially increased, to get sufficient counting rates for the signal.

According to these aspects we see that the two-stage excitation technique is, of course, restricted in its applicability and essentially depends on the development of new lasers. But when this technique can be successfully employed for lifetime measurements, it probably represents the most accurate method available for lifetime determinations of highly excited atomic or ionic levels.

References

[1] C.H. Liu, D.A. Church, Phys.Rev.Letters 29, 1208 (1972)
[2] K.D. Masterson, J.O. Stoner Jr., Nucl.Instrum.Methods 110, 441 (1973)
[3] H.J. Andrä, A. Gaupp, W. Wittmann, Phys. Rev. Letters 31, 501 (1973)
[4] H. Harde, G. Guthöhrlein, Phys.Rev.A 10, 1488 (1974)
[5] H.J. Andrä, this conference
[6] H.W. Kogelnik, E.P.Ippen, A. Dienes C.V.Shank,
       IEEE J. Quantum Electron. 8, 373 (1972)
[7] T. Andersen, O.H. Madsen, G. Sørensen, Physica Scripta 6, 125 (1972)
[8] H.J.Andrä, Nucl. Instrum.Methods 90, 343 (1970)
[9] G.V. Markova M.P. Chaika, Opt. Spectry (USSR) 17, 170 (1964)
[10] B.P. Kibble, G. Copley, L. Krause, Phys.Rev. 153,9 (1967)
[11] W. Demtröder, Z. Physik 166, 42 (1962)
[12] W.R. Bennett Jr., P.J. Kindlmann, Phys.Rev. 149 38 (1966)
[13] J. Bakos J. Szigeti, Opt.Specty. 23, 255 (1967)
[14] J. Klose, Phys.Rev. 141,181 (1966)
[15] C.G. Carrington, J.Phys. B 5, 1572 (1972)
[16] A.W. Weiss, "Atomic Transition Probabilities", Vol. I NS RDS-NBS 4(1966)
[17] J.P. Buchet, A. Denis, J. Desesquelles, M. Dufay,
       Compt. Rend. 265 B, 471 (1967)
[18] W.S. Bickel, I. Martinson, L. Lunden, R. Buchta, J. Bromander, I. Bergström, J. Opt. Soc. Am 59, 830 (1969)

LIFETIME OF THE 3d4p z^{1}P$^0_1$ LEVEL IN Sc II BY LASER EXCITATION OF A FAST IONIC BEAM

John O. Stoner, Jr.,[†] L. Klynning,[*] I. Martinson,
B. Engman, and L. Liljeby

Research Institute for Physics
104 05 Stockholm 50, Sweden

We have built an apparatus for the excitation of fast ionic beams by light from an argon laser, and for efficient detection of the radiation arising from the subsequent decays of the ions. This apparatus has worked satisfactorily for the measurement of the mean life of the 3d4p z^{1}P$^0_1$ level in Sc II; we expect that further applications are possible.

The apparatus is diagrammed in Fig. 1. Our principal criterion for its design was that both excitation of the beam and detection of decay radiation should be as efficient as possible, consistent with the reduction of stray light to tolerable levels.

Light from a Coherent Radiation argon laser CR-5, operated multimode on one line at a time, entered the system antiparallel to the ionic beam. The light was reflected via a mirror to cross the beam at a fixed angle near 30°; the light beam and ionic beam lay in a horizontal plane. Wavelength tuning to resonances between the laser wavelength and a Doppler-shifted ionic transition was done by varying the ions' energy about the expected value. After crossing the ionic beam the light was absorbed by a stack of razor blades inside a baffled box. The mirror and absorber were mounted on a plate that could be moved parallel to the ionic beam by a precision screw driven by a stepping motor.

Detection was accomplished by a fixed EMI 6256S photomultiplier mounted above the ionic beam. A single fused-silica lens having 75 mm focal length imaged a 3-mm segment of the beam onto the photocathode at 2:1 magnification. The solid angle subtended by the lens at the beam was masked to about 0.1 steradian. A filter could be placed in the optical path to reduce stray light. No spectrometer

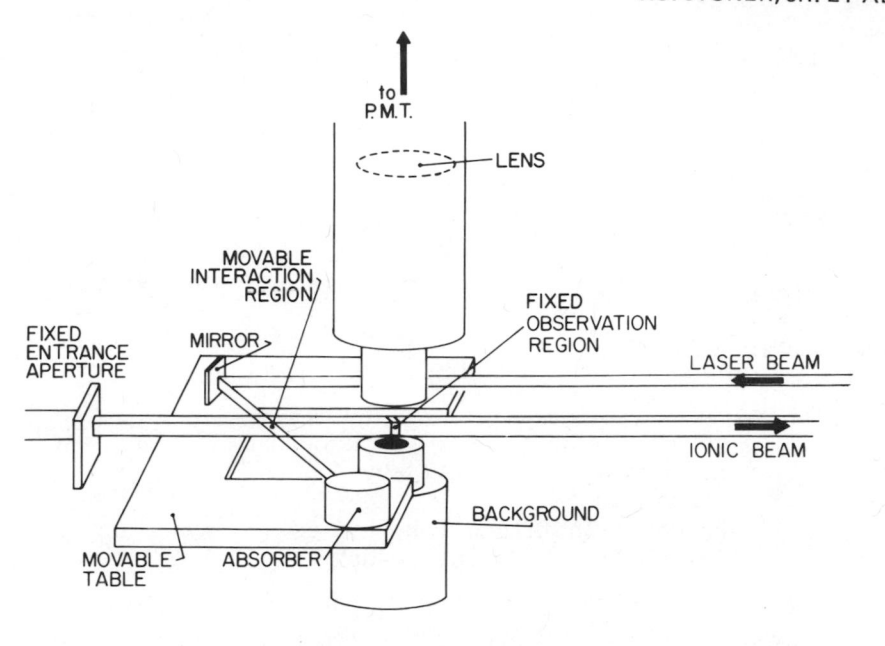

Fig. 1.   Essentials of the apparatus used to excite selectively the
          $z^1P_1^o$ level in Sc II by light from an argon-ion laser.

was used. A baffled chamber located below the detection region reduced greatly the laser light reaching the lens after being scattered from the mirror and absorber.

      Signals were handled in either of two ways. We could count in-
dividual photoelectrons, storing the number counted in successive
channels of a multiscaler as the position of the excitation region
was moved upstream. Alternately, we chopped the laser beam at
125/sec and recorded on a chart the output from the photomultiplier
at that frequency using a lock-in amplifier.

      We tested this apparatus by measuring the mean life of the
6p $^2P_{3/2}^o$ level in Ba$^+$, first studied in this way by Andrä et al.[1]
and since by several other authors.[2,3] Our result, 6.5 ± 0.5 nsec
(Table 1) agreed sufficiently well with previous measurements that
we could assume that systematic errors[4] associated with our measurements were probably less than ±10%. Our count rates indicated
that our light-collection ability was two orders of magnitude greater
than that for Andra's first system. We detected at λ5854Å, isolable
from the laser wavelength (4545Å) by a Wratten 22 gelatin filter;
as a result the background light from the laser was less than one

Table 1. Parameters Used in the Study of Transitions in Ba^+ and Sc^+

	Ba^+	Sc^+
Approx. energy	334 keV	212 keV
Typical current	10 nA	300 nA
Typical count rates	1000/sec	20000/sec
Exciting transition	6s $^2S_{1/2}$-6p $^2P^o_{3/2}$	b $^1D_2$ - z $^1P^o_1$
Wavelength	4554Å	5031Å
Laser wavelength	4545Å	5017Å
Power	50 mW	120 mW
Decay transition	5d $^2D_{3/2}$-6p $^2P^o_{3/2}$	a $^1D_2$ - z $^1P^o_1$
Wavelength	5854Å	3536Å
Upper level	6p $^2P^o_{3/2}$	z $^1P^o_1$
τ (this work)	6.5 ± 0.5 nsec	9.2 ± 0.5 nsec
τ (others)	6.31 ± 0.02 nsec[a]	5.5 ± 0.5 nsec[b]

[a] See Ref. 5
[b] See Ref. 7.

count/sec. The background due to collisional excitation of the ionic
beam and/or residual gas was less than 10 counts/sec, and pulse
counting was used for these measurements.

We then applied this apparatus to the study of the 3d4p z¹P⁰₁
level of Sc^+ (Table 1). We observed that sufficient numbers of
Sc^+ ions were produced in the ion source in the metastable level
b $^1D_2$ that no excitation after acceleration[2,4] was necessary. A
filter (5 mm of saturated $CoCl_3$ solution in H_2O)[6] was used to isolate
the strong decay transition at λ3536Å from the laser light; the background due to collisional excitation of the beam and/or residual gas
was a serious problem, typically amounting to 50 (counts/sec)/nA of
Sc^+) at a pressure of 8×10^{-6} torr in the vacuum chamber. We took
data in both detection modes to ascertain their relative merits and
concluded that in the presence of large steady backgrounds either
pulse counting or lock-in detection makes possible measurements of
lifetimes at the level of ± 5-10%; however, if either beam current
or background pressure is variable the lock-in method is much better.
It is clear that reduction of background gas pressure is of great
advantage in reducing the background light level, when, as in the
previous case, no spectrometric analysis of the detected light is
used.

Table 1 shows the results of these measurements. Our result, 9.2 ± 0.5 nsec, includes only the statistical uncertainty of the mean of seven measurements and includes no systematic errors. This lifetime for the $z^1P^0_1$ level disagrees significantly with the beam-foil measurement by Buchta et al.[7] (Table 1); the shorter mean life (5.5 ± 0.5 nsec) obtained by those authors can probably be attributed to overcompensation of the corrections applied for cascades in their work.

REFERENCES AND FOOTNOTES

[†]Present address: Department of Physics, University of Arizona, Tucson, Arizona 85721, U.S.A.

[*]Present address: Physics Department, Stockholm University, Vanadisvägen 6, Stockholm, Sweden.

1. H. J. Andrä, A. Gaupp, and W. Wittmann, Phys. Rev. Letters 31, 501 (1973); H. J. Andrä, A. Gaupp, K. Tillmann, and W. Wittmann, Nucl. Instr. and Meth. 110, 453 (1973).

2. H. Harde and G. Guthörlein, Phys. Rev. A 10, 1488 (1974).

3. A. Arnesen, A. Bengtsson, R. Hallin, S. Kandela, Tor Noreland, and R. Lidholt (to be published).

4. M. Gaillard, H. J. Andrä, A. Gaupp, W. Wittmann, H. -J. Plöhn, and J. O. Stoner, Jr., Phys. Rev. A 12, 987 (1975); H. J. Andrä, H. -J. Plöhn, W. Wittmann, A. Gaupp, J. O. Stoner, Jr. and M. Gaillard, J. Opt. Soc. Amer. 65 (October 1975).

5. H. J. Andrä, private communication.

6. S. F. Pellicori, Appl. Optics 3, 361 (1964).

7. R. Buchta, L. J. Curtis, I. Martinson, and J. Brzozowski, Physica Scripta 4, 55 (1971).

ON THE POSSIBILITY OF A PRECISE MEASUREMENT OF THE F VIII 1s2p $^3P_2$-$^3P_1$ FINESTRUCTURE SPLITTING

H.J. Andrä and J. Macek
Freie Universität Berlin, West Germany

J. Silver, N. Jelley, and L.C. McIntyre
University of Oxford, England

The interaction between experiment and theory over many decades has led to highly accurate measurements and calculations of the finestructure (fs) intervals Δ_{01} and Δ_{21} of the 1s2p ^3P-states of helium. The present situation which has culminated in a fs-constant determination to 2 ppm accuracy is summarized in a review article by M.L. Lewis[1]. It clearly shows that the theory still lacks the accuracy obtained in experiments so that further improvements in the theoretical understanding of the He-fs could improve the knowledge of the fs-constant.

In order to test higher order corrections of these calculations measurements along the isoelectronic sequence at higher Z would be very desirable. However, the experimental situation is quite unsatisfactory so far, since only standard spectroscopic techniques have been used in the higher Z ions yielding accuracies of the order of 1% only[2,3] at which the theory is easily in agreement. Therefore high resolution methods are required for an improvement of the experimental data.

This situation has led us to start a program of applying double resonance techniques to accurate measurements of the He-like 1s2p ^3P fs-intervals in higher Z ions. Unfortunately, however, our first and crude attempt on observing a resonance signal of the ^{19}F VIII Δ_{21} - fs-interval was unsuccessful. Nevertheless we believe that after further refinements on the apparatus and of the theoretical predictions we will be able to precisely determine this fs-interval. Therefore it may be worthwhile to open a general discussion on such type of measurements by communicating the present stage of our experiment.

The basic concept relies on the fact that with higher Z the
1s2p $^1P_1$ admixtures to the 1s2p $^3P_1$ state increases thus opening
at around Z=6 a direct, fast El decay channel (Fig. 1) from the
$^3P_1$ to the $^1S_0$ ground state whereas the $^3P_0$ and $^3P_2$ states can only
decay to the 1s2s $^3S_1$ state at an allowed El transition[5,6] rate of
about $10^8 s^{-1}$. (The $^3P_2 - {}^1S_0$ -M2 decay channel[7] becomes only important
above Z=14.) This behaviour of the ^3P-states can be used for the
creation of a large population difference of the two $^3P_2$, $^3P_0$ states
with respect to the $^3P_1$ state just by waiting a few $^3P_1$-decay lengths
after a statistical beam-foil or beam gas excitation. For our example
of ^{19}F VIII for instance already 11ns after excitation a population
ratio N($^3P_2$,t)/N($^3P_1$,t) of 10^4 is reached whilst the population of
the $^3P_2$ state has decreased only to N($^3P_2$,t=11ns) = 0.61 x N($^3P_2$,t=0).
These are ideal starting conditions for double resonance experiments
if sufficiently powerful radiation sources can be found for inducing
the Δ_{21} or Δ_{01} - M1 transitions in order to observe an increase in
the 1s2 $^1S_0$-1s2p $^3P_1$-intensity as a resonance signal. Under such
circumstances extremely accurate fs-measurements in higher Z He-like
ions become possible which could even outdate the present He accuracy[12,13] since the fractional line widths (FLW) in Fig. 2 have a

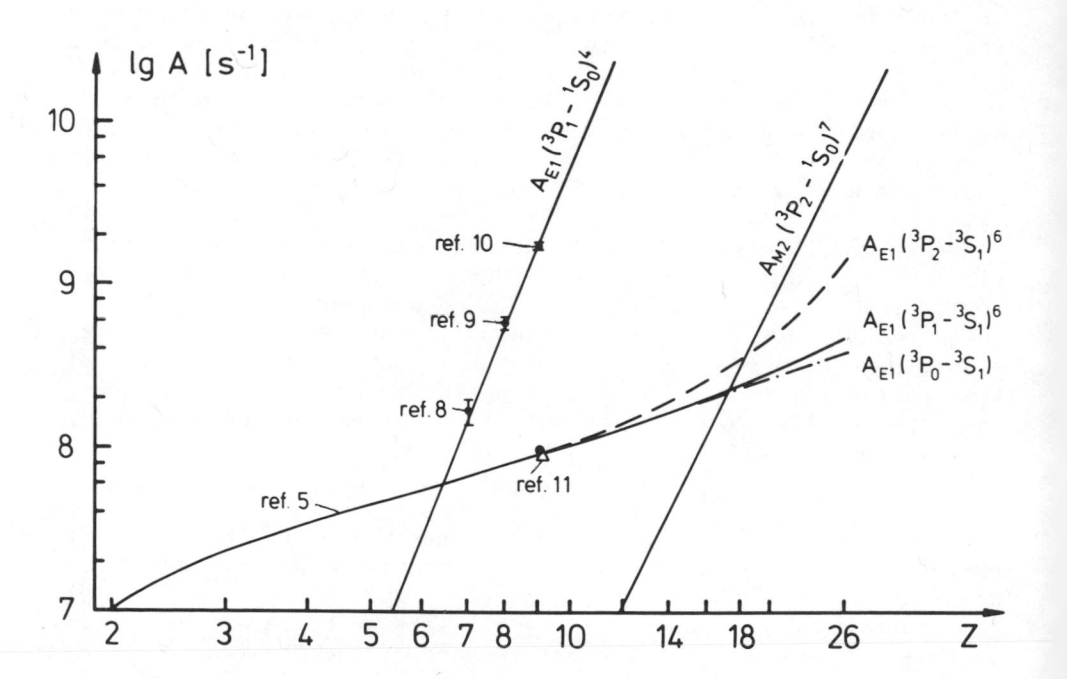

Fig.1: Transition probabilities A for the various decay channels of
the 1s2p ^3P - states of He-like ions as function of Z. Only experimental results for 5<Z<14 are included.

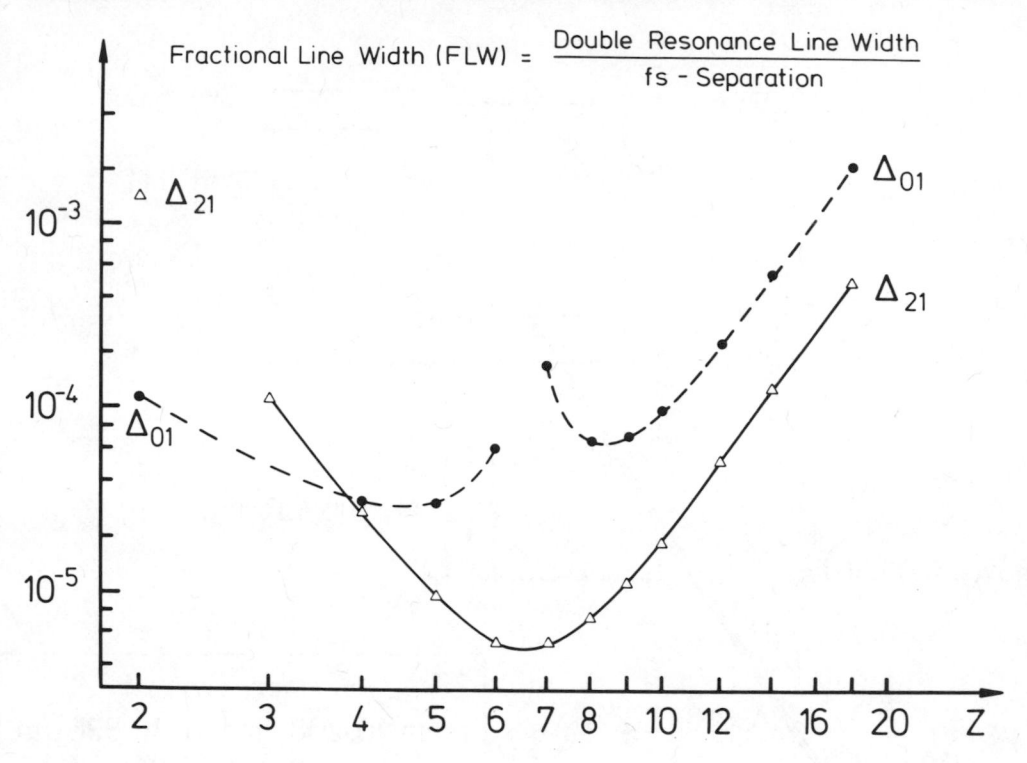

Fig.2: Fractional double resonance line widths for the Δ_{21} and Δ_{01}-fs-intervals of the He-like 1s2p ^3P states. The connecting lines are interrupted at J-J' crossovers.

pronounced minimum at Z=6,7 for Δ_{21} which is a factor of 20 better than the FLW of the He Δ_{01}-fs-interval. However, it is quite clear that experimental and theoretical problems will dominate over these prospects in the near future as shown now with our example of ^{19}F VIII.

The choice of ^{19}F VIII as the first case to be studied was solely dictated by the availability of the CO_2-laser as a high power resonance frequency source which accidentally closely coincides with the Δ_{21}-fs-separation[14] as can be seen in the level scheme in Fig. 3. The residual frequency gaps can easily be compensated by using the Doppler effect which can shift the lines up to $\Delta\lambda = \pm0.45\mu$m at a beam energy of 19 MeV where the production of the He-like ^{19}F charge state reaches it's maximum. Unfortunately, however, ^{19}F has a nuclear spin I = 1/2 giving rise to a not very well known hyperfine structure (hfs) which will cause major problems later on. Other ions (e.g. ^{16}O VII) without nuclear spin would have been much more advantageous but the proper resonance frequency sources were not available.

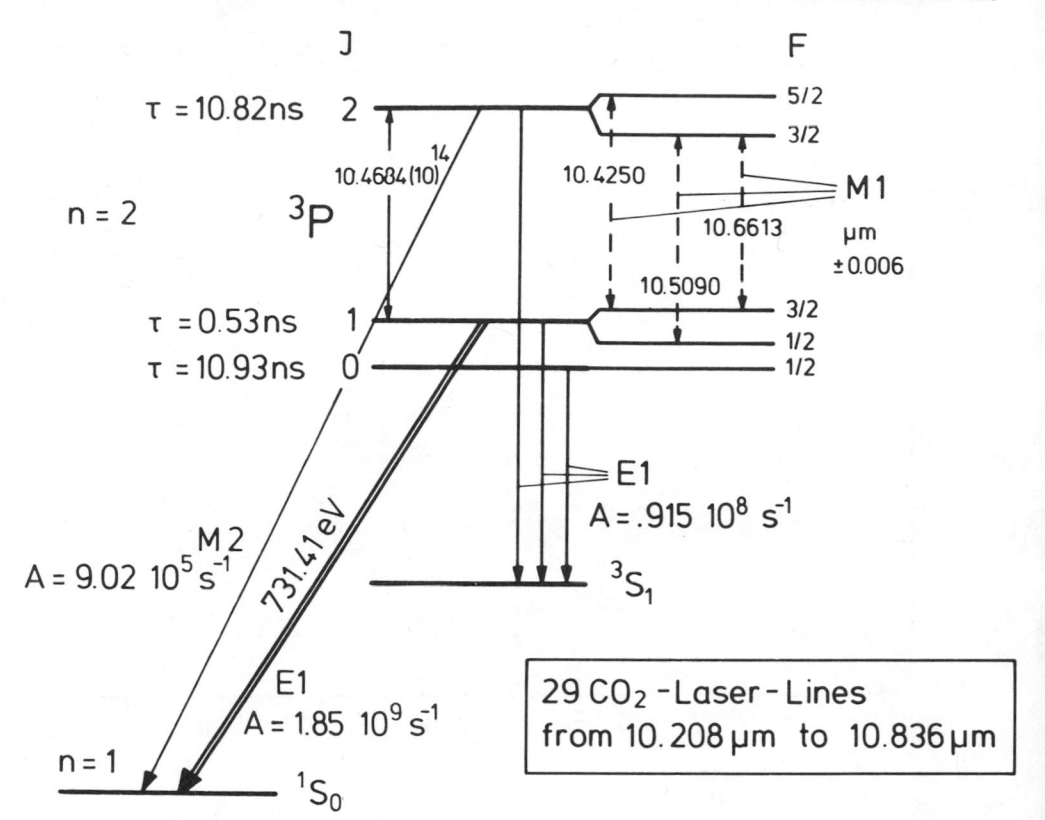

Fig.3: Schematic diagram of the ^{19}F VIII – 1s2p ^{3}P – fine – and hyperfine structure levels with their decay channels of the 1s2s ^{3}S$_1$ and 1s^2 ^{1}S$_0$ states. The broken lines indicate possible M1 double resonance transitions.

For the theoretical resonance frequency determination the best known fs-calculation[14] gives $\Delta_{21} = -955.26$ cm^{-1} which is probably accurate to 10^{-4} since the same calculation coincides with the HeI-fs measurements[12,13] only within such error bars. To this fs splitting the hfs is added based on the hamiltonian[15]

$$H_{hfs} = c \cdot \vec{S} \cdot \vec{I} + D \left\{ \vec{L} - \frac{\sqrt{10}}{2} \left[c^{(2)} {}_{x} S^{(1)} \right]^{(1)} \right\} \vec{I}$$

in which the determination of the interaction constant D for the
orbital part causes major problems without going through an elaborate calculation whereas the interaction constant C is rather well
defined (besides relativistic effects) by the contact interaction
of a hydrogenic 1s electron. Assuming a screening of the 2p electron
between the pure hydrogenic and an alkali-like situation according
to Kopfermann[16] we obtained resonance wavelengths as shown in Fig. 3.
A later estimate shifted the F=3/2-F'=3/2 resonance to 10.6639μm,
whereas a recent calculation[11] based on hydrogenic wavefunctions
would move this resonance to 10.6509μm. Thus an enormous uncertainty·
of ± 0.006μm is introduced at present and moves the main efforts to
a better theoretical prediction of the resonance.

Nevertheless we started a first experimental attempt based on
the resonance wavelengths given in Fig. 3 in order to test the signal
efficiency in such a measurement with the F = 3/2 - F'= 3/2 resonance. In order to induce the 10.6613μm resonance transition the
R 26-10.2075μm CO_2-laser transition was chosen at a fixed angle of
incidence of 5° between the atomic (4mm d) and the laser (6mm d) beam
in Fig. 4a. Due to this configuration the resonance had to be swept

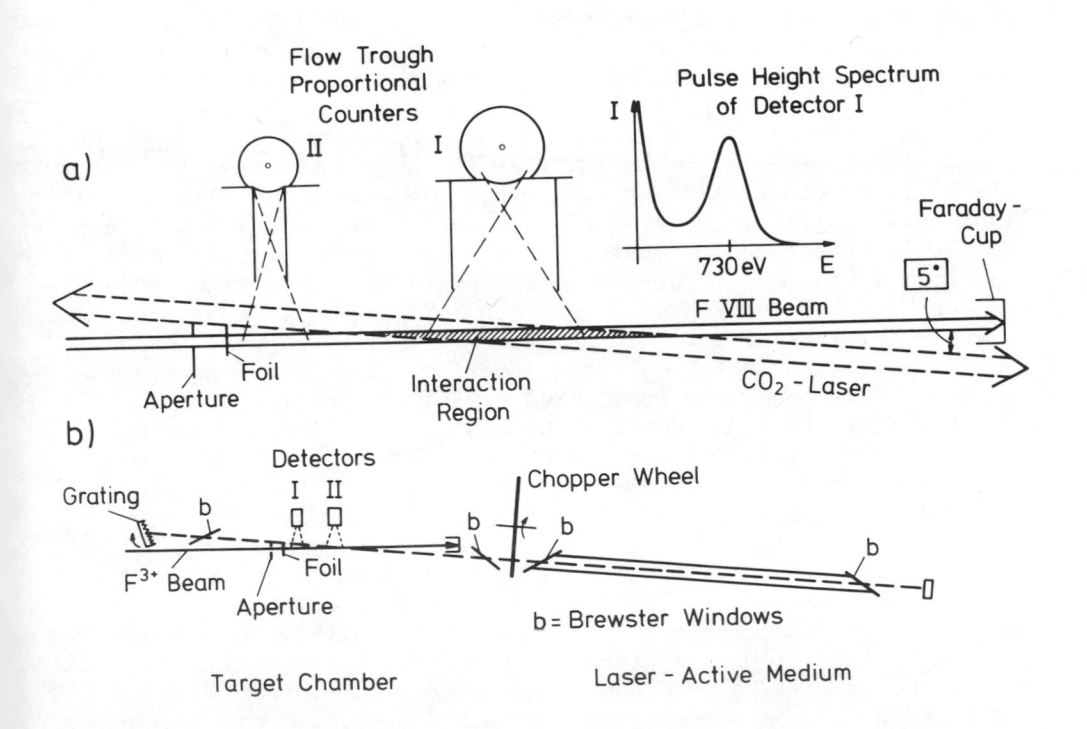

Fig.4: a) Schematic experimental set up close to the interaction
region. b) Sketch of the complete experimental arrangement.

via Doppler effect by changing the beam energy and was expected at 16.880 MeV. The center of the interaction region was located 11ns behind[17] the carbon foil (7 $\pm$ 2.5μg/cm^2) which introduced an energy loss[17] of 85 $\pm$ 30 keV to be added to the above accelerator resonance energy. With this geometry an interaction time of T = 5ns for each atom with the linearly polarized <u>intracavity</u> laser beam (see Fig.4b) was achieved at a laser power of $\overline{100}$ W/cm^2. For these conditions a transition probability from the J=2, F=3/2 to the J'=1, F'=3/2 state of P=2.85·10^{-5} was calculated. This small increase in the $^3P_1$ -population can be detected by a flow through proportional counter 1 as an intensity increase ΔI ($^3P_1 - ^1S_0$) in the interaction region due to the fast decay of the $^3P_1$ state. This counter 1 cannot separate between E1 ($^3P_1 - ^1S_0$) and M2 ($^3P_2 - ^1S_0$) photons which occur with a small transition probability[7] (see Fig. 3) but still large enough intensity to give rise to a strong background I_b in the detector 1. Although 11ns behind the excitation a 731 eV background I_b (see inserted spectrum of detector 1 in Fig. 4a) of 40.000 counts s^{-1} was obtained at a 3μA -17 MeV-F^{3+} beam which we assumed was stemming at least to 50% from M2-photons. From this background the population of the J=2, F=3/2 state can be estimated which in turn allows then to calculate the expected signal of ΔI($^3P_1 - ^1S_0$)/I_b=1.6·10^{-3} assuming a natural resonance width of 0.315 GHz. Doppler broadening due to the atomic and laser beam divergences and due to the beam velocity spread causes, however, an effective linewidth of the order of 1 GHz corresponding to 24 keV (!), which reduces the expected signal to ΔI($^3P_1 - ^1S_0$)/I_b=5·10^{-4}.

In order to detect such a small signal a fast chopping (1ms on, 1ms off) of the laser by an intracavity chopper wheel in Fig. 4b and the corresponding gating of two counters was used for obtaining I (on) and I (off) of detector 1 at a fixed beam energy. After reaching a certain preset count   number from a monitoring detector 2 close to the foil I (on) and I (off) were stored in two channels of a multi-scaler, the beam energy stepped by 10 keV, and a new counting cycle started into the next two channels. 25 energy steps covering an energy range of 240 keV corresponded   to one scan and many scans were added to one run which lasted typically 3 hours. After each run the foil was changed (foil thickening[18]!) in order to obtain reproducible conditions for 7 such runs. Finally the 7 runs were added and the ratio [I(on) - I(off)] /I(off) of the sums was computed as the result in Fig. 5·

The energy range chosen was centered around the expected (16.880+0.085)MeV resonance with $\pm$(.030+0.022) MeV experimental uncertainties stemming from the foil thickness ($\pm$2.5μg/cm^2) the angular setting ($\pm$0.17 degrees) respectively. Considering the quality of the statistics obtained a resonance signal of 5·10^{-4} with a half width of 2.4 channels could probably have been observed already in this first attempt, so that further improvements on the apparatus will guarantee the detection of the resonance as soon as better theoretical predictions reduce the energy region to search for.

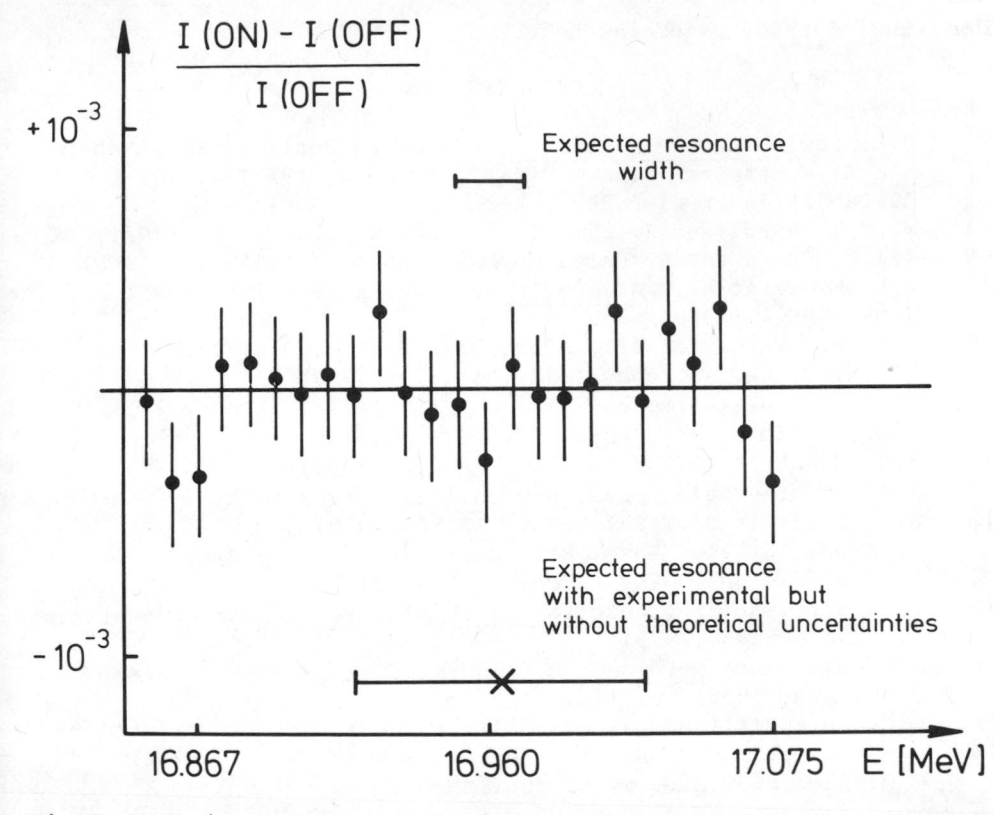

Fig.5: Experimental result obtained as the mean of 7 runs of 3 hours each. Each run was started with a new foil.

At present we can summarize the situation as follows. The fs-prediction is accurate to $\pm\ 10^{-4}$ giving rise to a tolerable $\pm$ 80 keV uncertainty for the resonance. The estimates for the hfs can set, however, only lower and upper limits of 10.650μm and 10.665μm respectively which require a search over an untolerable energy range of 1.1 MeV. Compared to these huge theoretical uncertainties the expected line width is only 24 keV and the experimental uncertainty can be limited to $\pm$ 25 keV thus giving right away an accuracy of $\pm 3.5\cdot 10^{-5}$ in the resonance frequency as soon as a resonance is observed. These very promising prospects, we hope, will encourage the theoreticians to calculate the resonance frequencies to a precision of about $\pm\ 10^{-4}$, since a successful experiment would represent the most accurate fs determination in a higher Z-ion to date.

One of us (H.J.A) wishes to express his appreciation to the others for their hospitality during the course of the experimental work at the University of Oxford Nuclear Physics Laboratory. This work is partially supported by the Sonderforschungsbereich 161 der

H.J. ANDRÄ ET AL.

Deutschen Forschungsgemeinschaft.

REFERENCES

1   M.L. Lewis, Atomic Physics 4, ed. G. zu Putlitz, E.W. Weber, and A. Winnacker (Plenum Press, New York, 1975) p. 105
2   K. Bockasten et al., Phys. Lett. 8 181 (1964)
3   W. Engelhardt and J. Sommer, Astrophys. J. 167 201 (1971)
4   G.W.F. Drake and A. Dalgarno, Astrophys. J. 157 459 (1969)
5   W.L. Wiese et al., Atomic Transition Probabilities, Vol. I, NSDRS-NBS 4 (1966)
6   G.R. Blumenthal et al., Astrophys. J.172 205 (1972)
7   G.W.F. Drake, Astrophys. J. 158 1199 (1969)
8   I.A. Sellin et al., Phys. Rev. Lett. 21 717 (1968)
9   I.A. Sellin et al., Phys. Rev. A2 1189 (1970)
10 J.R. Mowat et al., Phys. Rev. A8 145 (1973)
11 J.R. Mowat et al., Phys. Rev. A11 2198 (1975)
12 S.A. Lewis et al., Phys. Rev. A2 86 (1970)
13 A. Kponou et al., Phys. Rev. Lett. 26 1613 (1971)
14 B. Schiff et al., Phys. Rev. A8 2272 (1973)
15 I.I. Sobelman, Introduction to the Theory of Atomic Spectra, (Pergamon Press,1972)
16 H. Kopfermann, Kernmomente (Akadem. Verlagsgesellschaft, Frankfurt, 1956), p. 111.
17 L.C. Northcliff and R.F. Schilling, Nucl. Data 7A 233 (1970)
18 W.S. Bickel and R. Buchta, Physica Scripta 9 148 (1974)
    P.D. Dumont et al., to be published

EUV SOLAR SPECTROSCOPY FROM SKYLAB AND SOME IMPLICATIONS FOR

ATOMIC PHYSICS

E.M. Reeves and A.K. Dupree

Center for Astrophysics, 60 Garden Street, Cambridge,

Massachusetts   02138

INTRODUCTION

This fourth Beam Foil Spectroscopy meeting presents a particularly appropriate forum to review some of the more recent advances
in solar observations as represented by the ultraviolet experiments
of the Apollo Telescope Mount (ATM) on Skylab. The emission lines
and continua in the solar spectrum require for their interpretation
a quantitative understanding of the fundamental collisional and
radiative processes occuring among atoms, ions, electrons, and
photons. The physics of these interactions and the required inter-
pretative methods are similar to those necessary to analyze stellar
spectra. The interaction between stellar and solar physics contin-
ues to increase, spurred on in part by the substantial observations of stars in the ultraviolet energy range as will be discussed by D. Morton in the following paper.

The Sun however holds a rather special interest for many
astronomers because it is the only star on which fine scale
structure is observable. This permits detailed study of the complex processes that can take place at the radiating surface, yet
simultaneously complicates the objective of achieving a simple
model for the atmosphere.

The results which will be discussed here have been selected
to indicate the spatial detail associated with certain types of
solar phenomena, and to stress both the importance of the time
scales for solar features, and the prominent role of the solar
magnetic field. Because of our focus on the ultraviolet, we
exclude the energetic radiations observed by the two X-ray experiments (American Science and Engineering and Marshall Space Flight

Center) associated with solar flares and active regions, and the
observations of the spectacular transient events in the outer
solar corona observed by the white light coronagraph on ATM
(High Altitude Observatory). Ultraviolet instruments were provided by the U.S. Naval Research Laboratory (NRL) and by the Harvard
College Observatory (HCO).

While ATM represents a program currently in the stages of
intense data analysis, we will look briefly at some future
directions solar research will take in the next few years. Of
particular interest here are the implications of the ATM results
and the objectives of future solar missions on the requirements
for experimental and theoretical atomic physics investigations.

1. SOLAR STRUCTURE FROM EUV DATA

Emissions in the EUV reveal the structures of the solar
chromosphere, transition region and corona, the high temperature
($T \sim 10^4$ to $\sim 10^6$ K) regions that extend above the solar photosphere or visible disk of the Sun. Figure 1 shows an EUV image
of the Sun from a rocket payload flown by the Naval Research
Laboratory. The image is taken with an objective grating instrument through a thin aluminum-tellurium filter, which includes
contributions from all wavelengths in the range 150-630 Å (with
the helium lines partially suppressed) at a spatial resolution of
several seconds of arc. The Sun was photographed early in 1974,
but with an uncharacteristic amount of activity for the period
near the minimum of the 22 year solar cycle. This single image
illustrates most of the general features of the Sun as seen in
the extreme ultraviolet: active regions with the associated loop
structures; the highly structured quiet chromosphere; large
regions of reduced coronal intensity on the disk and at the poles
referred to as coronal holes; a scattering of small X-Ray/EUV
bright points; and the general limb brightening of the emission
around the edge, evidence of the concentration of the higher
ionized species in a relatively thin layer above the solar
photosphere, where the temperature is increasing.

Much of the interpretation of solar EUV data in terms of
models for the course of density and temperature in the solar
chromosphere, transition region, and corona with height has
used data from the series of Orbiting Solar Observatories and
rocket experiments over the past decade, where the intensity measurements have not been characterized by particularly good spatial resolution. The empirical models are normally derived under
the assumptions of a plane parallel atmosphere in hydrostatic
equilibrium that is in a state of time independent ionization
equilibrium.

Figure 1.   Extreme ultraviolet image of the Sun in the waveband
            150-630 Å. (Photo Courtesy of the U.S. Naval Research
            Laboratory.)

     In Figure 2 we show empirical models derived for some general
types of solar features, where fine scale spatial detail has been
averaged together. The general progression of temperature above
the solar photosphere is well-known. On the basis of a plane
parallel model, the temperature at first decreases outward from
the 6500 K of the visible photosphere, to the temperature minimum
about 1000 km above the photosphere where the temperature is
characteristically about 4500 K. Concurrent with a continuing
decrease in the average density with height above the photosphere,
the temperature rises slowly through the range $1-3\times10^4$ K in the
chromosphere, and then abruptly increases to 6×10^5 K in a height
of several tens or several hundreds of kilometers through the
transition region. The temperature subsequently approaches a
plateau of several million degrees in the corona. If the curve
for the quiet Sun is considered as representative of the average
solar atmosphere, then in active regions the temperature gradient
in the transition region is approximately five times that in quiet
regions, the electron density is increased by about the same
factor of five, and the average coronal temperature increases from
1.5×10^6 to 2.5×10^6 K (Noyes *et al.*, 1970). The large coronal hole
regions are characterized by a lower temperature gradient, reduced

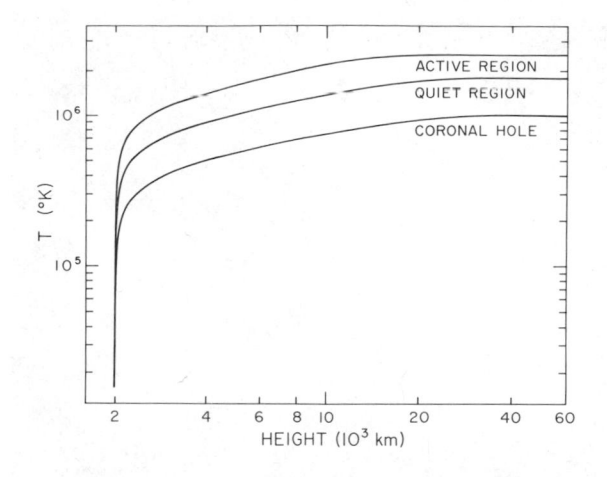

Figure 2.   Average course of temperature with height in broad
            categories of solar structure.

electron density in the transition region, and a lower coronal
temperature (Munro and Withbroe,1972).

    In spite of the progress that has been made in establishing
models of the solar atmosphere which can be used to calculate the
emission measure, these models fall considerably short of being
applicable to the known three-dimensional structure.   Much of the

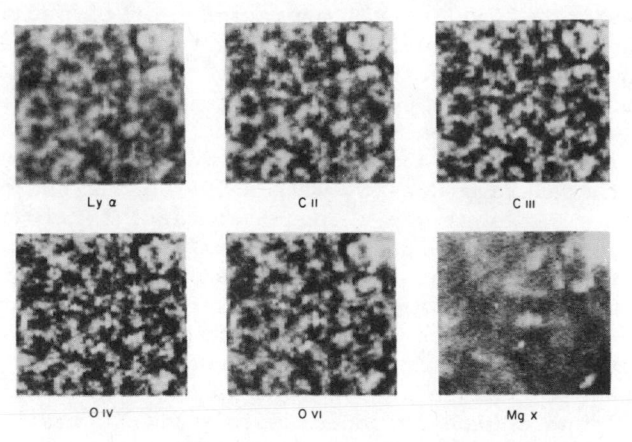

QUIET CHROMOSPHERIC NETWORK
AUG 13,1973 16:00 UT

Figure 3.   The quiet solar chromosphere on August 13, 1973,
            1600 UT.

emphasis of the current data analysis programs at HCO and NRL is
directed toward the gradual development of two dimensional
models, which will more adequately reflect the known physical
structures in the Sun's atmosphere. As these more complex
models emerge to interpret the physical structures, it will also
be of great interest to determine the extent to which the simpler
models provide a representative approximation to the real
physical state, and can thus be used to understand the general
energy balance of the Sun for application to other stars.

2.   SOME FEATURES OF THE SOLAR ATMOSPHERE

The following sections focus on a few selected features
that are currently under study in solar physics, beginning with
the simplest and most general solar feature - the quiet Sun.

a.   The Chromospheric Network

The "network" appearance of the chromosphere and transition
region dominates the EUV emission of the quiet Sun. Figure 3
shows a set of observations of a quiet solar region approximately
halfway to the limb on August 13, 1973. The data were taken with
the HCO photoelectric polychromter on Skylab which here made
simultaneous observations in the set of resonance lines H Lyman α
(1216 $\overset{\circ}{A}$), C II (1335 $\overset{\circ}{A}$), C III (977 $\overset{\circ}{A}$), O IV (554 $\overset{\circ}{A}$), O VI
(1032 $\overset{\circ}{A}$), Mg X (625 $\overset{\circ}{A}$) with a spatial resolution of 5 seconds
of arc. The instrumentation and its operation are described
completely elsewhere (Reeves *et al.*, 1974). The photographic
presentations are conversions to grey scale from the digital
data, with each spectroheliogram encompassing a 5 x 5 arc minute
field (approximately 1/30 of the Sun). The wavelengths illustrated cover the temperature range from 2×10^4 K for Ly α and
C II, through C III at 7×10^4 K, O IV at 1.5×10^5 K, O VI at
3×10^5 K to the coronal line of Mg X at 1.4×10^6 K. The chromospheric network whose boundaries delineate the regions of
increased magnetic field is observed in these EUV lines to
correspond directly with the chromospheric network photographed
from the ground in Ca II K; the width of the network, as observed normal to the surface with 5 seconds of arc resolution,
remains substantially unchanged through the temperature range
to 3×10^5 K. There is some indication of broadening when the
network is observed in Ne VIII ($T_c = 7 \times 10^5$ K); while in coronal
lines such as Mg X (1.4×10^6 K) there is no consistent indication
of the underlying network. The mean fluctuation in the intensity
of Mg X over a quiet 5 arc minute region near disk center is about
25-40% of the average intensity. This may be due to a combination
of spatial and temporal variations. Since the intensity of

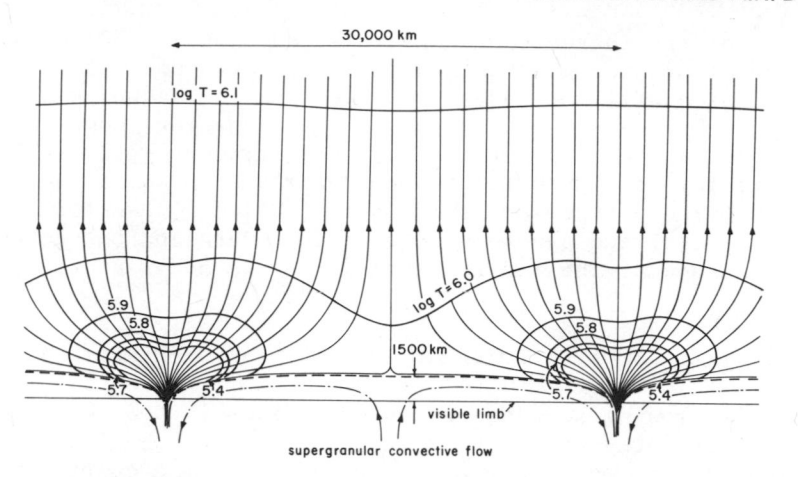

Figure 4.   Model of the supergranulation network; Gabriel (1975).

coronal lines such as Mg X are proportional to n_e^2, then we can
infer that the corona overlying the quiet chromospheric network
is uniform to approximately 10-20% (Reeves, Vernazza, and Withbroe,
1975; Mariska and Withbroe, 1975).

Figure 4 illustrates a recent model (Gabriel, 1975) that
allows for inhomogeneity on the solar surface. Gabriel assumes
a pattern of supergranulation cells in the photosphere and low
chromosphere in which the magnetic field lines are concentrated
in the network or boundaries of the supergranulation cells. The
cells have an average size of about 30 arc seconds corresponding
to about 24000 km on the solar surface. For lines formed in the
chromosphere and transition region, the network would have the
observed constant width and would begin to show evidence of
broadening for Ne VIII near 7×10^5 K. We must remember, of course,
that while EUV emission lines are frequently referred to as being
formed at a specific temperature, the emitting ions are really
formed over a temperature range, easily a factor of 2 or more.
Thus when comparing the observed width of the network to the
models, care must be taken to avoid a too literal interpretation
of *the* temperature at which the structure would be observed to
change - especially for $T > 3 \times 10^5$ K. However, this simple model
is still in one dimension and does not take into account the
observed structure along the network elements (an intensity fluctuation on the order of a factor of 2-4); it excludes as well the
known contribution from spicules which are embedded in clumps
along the network. The model does predict that the corona begins
at h ∿ 10000 km over the network, and at 5000 km over the interior

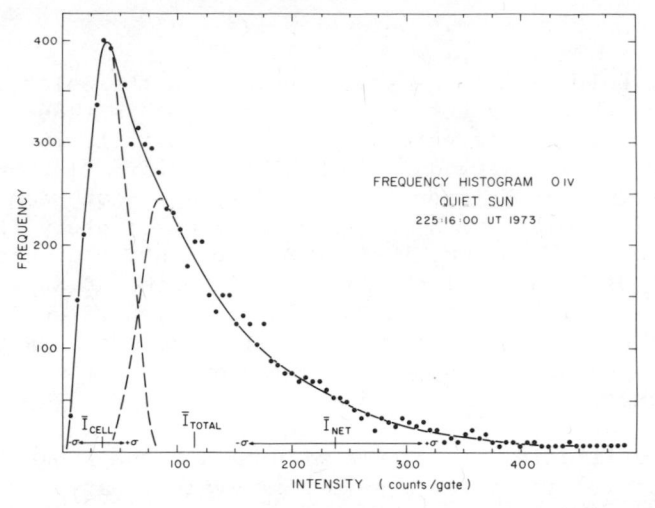

Figure 5.   Distribution of intensities of O IV in the quiet Sun.

of cells which is consistent with the structure inferred by
Withbroe and Mariska (1975) from center-to-limb observations. A
thinner transition region over the center of cells as compared to
the network boundary is suggested also by analysis of the EUV data
(Dupree *et al.*, 1976).

What are the ranges of local intensity change which must be
accounted for in the chromospheric network structure? Figure 5
shows the results of a frequency analysis of the intensities of
the quiet Sun in the line O IV $\lambda554$ Å corresponding to a temperature of 1.5×10^5 K. The dashed lines show the approximate contributions from the network and the centers of the supergranulation
cells. The observed intensity contrast between the network and
the cell interiors is a maximum near 10^5 K. From the distribution, we can see that the intensity for O IV can range from a few
counts in the centers of supergranulation cells to over 500 counts
in the brighter elements of the network. The contribution of the
network to the average quiet Sun can be evaluated as a function of
temperature of line formation, and indicates that the network
contributes about 75% of the intensity near $T=2\times10^5$ K and only
about 60% at temperatures near Lyman-α (Reeves, 1976). Since the
network structure does not extend to coronal temperatures, then
for lines such as Mg X 625 Å there is an approximately equal
contribution from image elements overlying network and cell interiors. Thus a simple division of the quiet Sun into two
separate components, cells and network, is a better approximation
of the physical state of the quiet Sun, but the models must be
able to permit a wide range of conditions.

b.   Coronal Holes

Coronal holes are large scale regions on the solar disk and at the poles, where the EUV and X-ray emission is reduced over the quiet Sun or average values by a factor of 5 to 10.   These regions were first observed in rocket experiments (Burton, 1968; Tousey *et al*., 1968; Krieger *et al*., 1971) and have been observed and studied in the EUV with Orbiting Solar Observatory instrumentation (Munro and Withbroe, 1972).   Figure 6 shows a photograph of the solar disk taken by the American Science and Engineering soft X-ray experiment on ATM.   This large region extending from the North Pole down to the southern hemisphere was a particularly long lived structure during ATM, lasting over a period in excess of five months.

The long lifetime of the ATM mission and the range of instrumentation carried permitted a number of evolutionary investigations on coronal holes and other types of solar structures.   The coronal holes are but one of the many types of solar structures where we can infer the influence of the solar magnetic field. Coronal holes are considered to be regions in which the magnetic

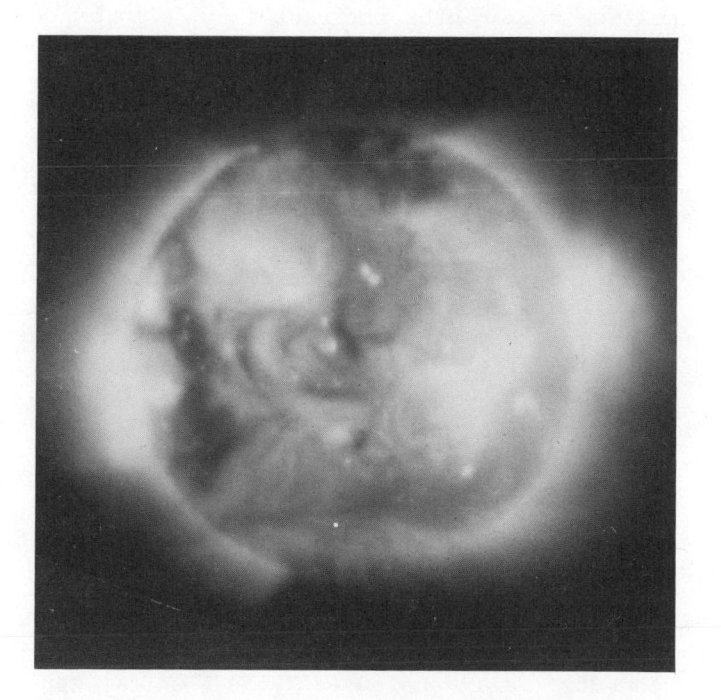

Figure 6.   Soft X-Ray image of the Sun in the wavebands 2-32 Å and 44-54 Å for 0336 UT on 28 May, 1973.   (Photo Courtesy of American Science and Engineering).

field is an open configuration rather than the large closed field
structures characterized by the active regions. The nearly rigid
rotation of coronal holes, independent of latitude (Timothy *et al.*,
1975) is at variance with the differential rotation apparent in
sunspots (Newton and Nunn, 1951) and EUV data for the quiet Sun
or active regions (Dupree and Henze, 1972; Simon and Noyes, 1972),
and is representative of the range of newly emerging data from
both space and ground-based instruments which demonstrate the
inherent complexity within a concept as simple as solar rotation.
Several investigations also indicate that the coronal holes are
the likely source of recurrent high velocity wind streams
(Krieger *et al.*, 1975).

The different structure of the atmosphere above coronal holes
is immediately apparent from a series of EUV spectroheliograms
over the South Pole of the sun obtained by the HCO instrument
during the Skylab mission (Huber *et al.*, 1974). Mg X is formed in
the corona at a temperature of 1.4×10^6 K and shows most of the
same features as observed in soft X-ray images. The spectroheliograms at the solar limb show the quiet corona formed above the
visible limb of the Sun, and the rather sharply defined edges of
the coronal holes. Above the photographic representation, the
height of the half intensity point of Ne VII 465 Å ($T=6\times10^5$ K) is
plotted relative to the height of formation of the Lyman continuum
($T=10^4$ K). The figure illustrates schematically the structure
of coronal holes shown in Figure 2, namely that in coronal holes
the temperature rise in the transition region is less steep than
in quiet regions and hence lines such as Ne VII are formed
higher in the solar atmosphere. Similar results have also been
reported from the NRL spectroheliograms in lines of the transition
region (Bohlin *et al.*, 1975).

c.   Bright Points

The Mg X spectroheliogram of Figure 7 also shows the presence of small bright points. The more intense bright points in
the EUV are also observed in X-ray images such as the one shown
in Figure 6. The bright points are most easily discerned in the
relatively less intense and uniform areas of coronal holes. These
structures are envisaged as relatively small scale, approximately
10,000 km loop structures with a closed magnetic field configuration. The bright points also show up as enhancements in
chromospheric and transition region lines, although they are not
clearly discernible in quiet Sun images of the chromospheric
network, because point-to-point intensity fluctuations in the
network mask their appearance. The bright points are distributed
over the entire solar disc with an increased concentration near
the equator (Golub *et al.*, 1975) and can be recognized in ground-
based magnetograms as small magnetic bi-polar regions in the

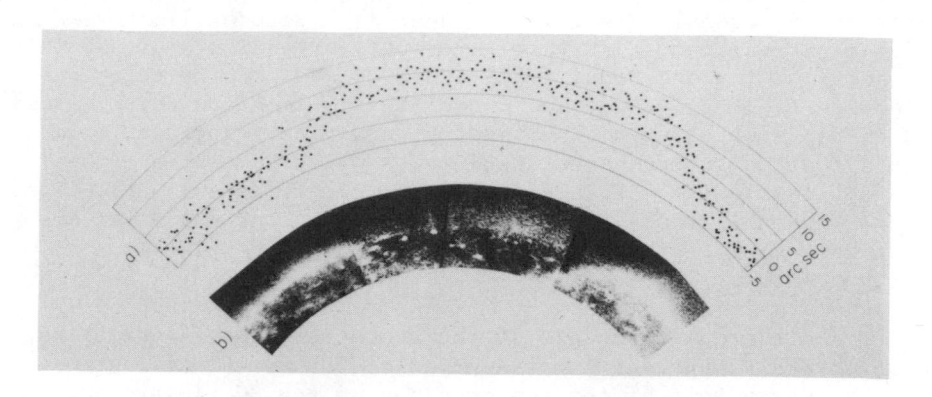

BEHAVIOR OF TRANSITION ZONE OVER SOUTH POLAR HOLE
SL 4, DAY 349, 16:56 TO 22:01 GMT

Figure 7. The corona at the South Pole. (a) indicates the
height to which Ne VII 465 Å is formed relative to the
Ly continuum limb and (b) shows a composite spectroheliogram in Mg X 625 Å in late 1973.

photosphere.

The flare-like behavior of bright points is seen in a series
of the HCO spectroheliograms in Mg X 625 Å for August 27, 1973
(Figure 8). In the sequence of three spectroheliograms at
approximately 5 minute intervals, one of the bright points
emerges, reaches maximum intensity, and disappears, with a lifetime shorter than the 10 minute interval between the frames.
Golub *et al*., (1974) suggest that 5-10% of all X-ray bright points
exhibit flaring phenomena. It is possible that the mechanisms
responsible for the sudden disappearance of bright points may
be related closely to the formation processes of flares in active
regions, and may thus provide an example of how potential energy--
possibly magnetic--is converted into radiant energy in a short
time scale. Since the structure of bright points is much simpler
than that in complex active regions, these structures are the
source of much interest for their possible relevance to the understanding of solar flares and active regions.

d. Active Region Loops

The role played by the magnetic field in structuring the
observed features in the solar atmosphere is displayed most

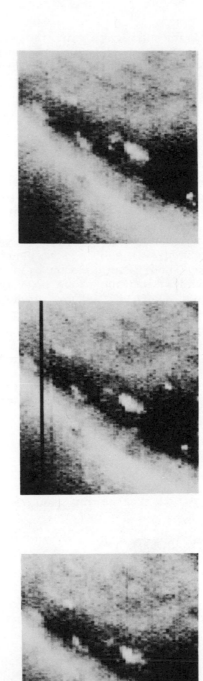

01:38 UT 01:43 UT 01:54 UT

Figure 8. The emergence of a flaring bright point, observed in Mg X·625 Å on August 27, 1973.

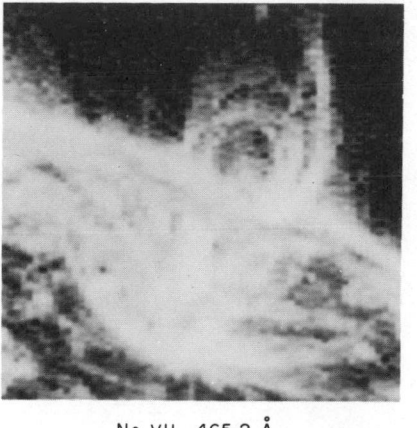
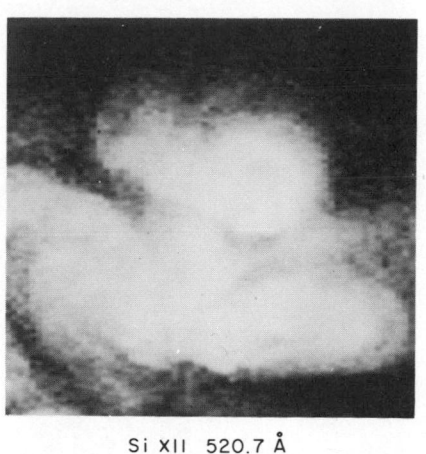

Ne VII 465.2 Å   Si XII 520.7 Å

Figure 9. Loop structures near active regions. Observations are
shown near the limb in Ne VII and Si XII at 0445 UT
on November 23, 1973.

picturesquely in the observations of loop-like structures associated with active regions. Coronal loops begin to appear at
temperatures in excess of 10^5 K and are very pronounced in many
lines of the transition region and corona. Loops are frequently
very fine in structure, down to the 5 arc second (3600 km) resolution of the instrument for lines of the upper transition region and
lower corona such as Ne VII and O VI. For coronal lines such as
Mg X and Si XII, where the temperature exceeds 10^6 K, the loops
appear more diffuse. Current models for these loops envisage a
cylindrical structure or tube where hot gas is present to indicate
the geometry of the magnetic tube. These loops frequently consist
of cooler material embedded in the corona although high temperature
loops are observed in Si XII and Fe XV. Along the length of the
tube, conduction should maintain the loop material at nearly
constant temperature, but gravity would produce a density (and
hence pressure) gradient extending upwards along the arches from
the footpoints and in the absence of some mechanism to replenish
the material, the loops would fade beginning near the top. Some
loops show significant brightening near the foot points, and for
some loops the arches appear to be rather uniformly filled with
material (e.g. Ne VII of Figure 9).

Figure 9 shows an active region in the simultaneous observations of Ne VII and Si XII. Ne VII at 6×10^5 K is particularly
well suited for illustrating the active region loops, and in this
case a large plume of gas can be observed directly over the
sunspot in the active region; this plume is nearly absent in

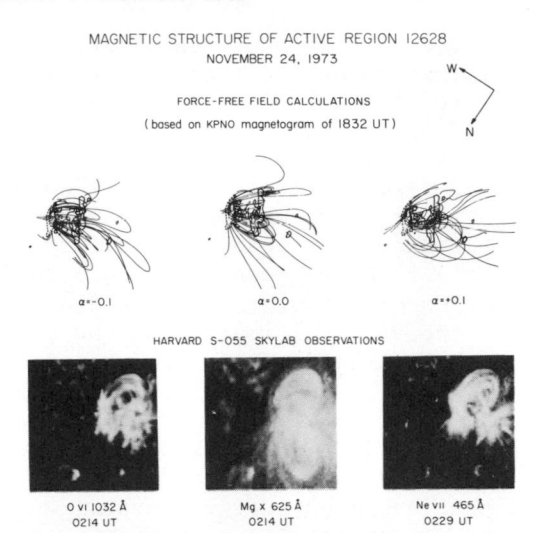

Figure 10. The relationship between loop structures observed at
            different temperatures and the magnetic field config-
            uration projected from the photosphere under different
            values of α where α is the ratio of the current
            to the magnetic field.

other EUV lines. The plume of gas appears to be almost isothermal
since it is observed in ions from a narrow range of temperatures;
the plume remains stationary for a period of several hours.

    Active region loops also are observed to be nearly isothermal,
and can remain stable for many hours when they represent material at
high temperatures ($T > 10^6$ K), and when associated with older active
regions. The material in the loops is frequently observed to
fall out of the suspended arch in a period of a few hours. Loops
also reform over active regions in a few days--although whether
in the same magnetic flux tube or not would be conjecture, since
the magnetic field lines cannot be observed unless the region is
filled with plasma. More recent observations and modelling with
magnetic field map projections into the corona give an indication
(Levine, 1975) that loops in younger active regions are flatter
to the surface, and are observed at higher inclinations in later
stages of development. Attempts by Levine to fit the projected
photospheric fields with force-free calculations suggest that
the loops spread to opposite sides of the dividing neutral line,
lie at approximately the same inclination (see Figure 10), and
may result from oppositely directed current flow.

    Good plasma diagnostics of temperature and density over a
wide range of temperatures will be necessary to arrive at a
complete understanding of this type of loop structure, for it will

be the state of the plasma combined with theoretical models that
provide us with information on the scale of the magnetic fields
associated with active regions. Certainly the earlier concept of
a general solar field of one or so gauss is now being replaced with
the reality of magnetic fields over very small regions of ~1 arc
second (~720 km) with field strength of 1200-2000 gauss measured
at the photosphere. Energy storage and release mechanisms postulated for these two cases can lead to significantly different
conclusions, and until methods become available for measuring
coronal magnetic fields directly, the plasma diagnostics will
continue to provide the only information on the strength and
structure of the magnetic fields above the photosphere.

3.   FUTURE SOLAR PROGRAMS

     The material that has been used in preceding sections to
illustrate the dynamics and complex nature of even the more
quiescent types of solar plasma represents the current level of
observational detail which can be achieved from satellites in the
EUV. Spatial resolution becomes comparable with average scale
heights and the known physical scale of the plasma elements in the
solar atmosphere. Although improvements in observational parameters such as spatial, temporal, or spectral resolution will
still be integral advantages of future solar missions, the primary
thrust of the solar program will be in specific problem-oriented
areas. Following this brief review of the ATM data, we can
discuss one illustration of the direction of solar space research
for the next few years.

     The most probable next solar satellite mission will be the
Solar Maximum Mission (SMM), a Shuttle-launched unmanned satellite
devoted to the study of solar flares and related phenomena during
the period near maximum in the solar cycle in 1979-1980. Currently the proposed instrumentation is in a second stage of competition
for selection following initial selection and a 4 month design
definition study. The final instrumentation will be selected
with the objective of determining the energy and mass balance in
the chromospheric and coronal domains of active regions, and
hence the potential energy storage and release processes responsible for the solar flares. For instance, the proposed HCO
instrumentation in the XUV region primarily addresses the question
of the electron temperature determined from the measured intensity ratio of pairs of lines in beryllium and lithium-like ions
which are separated in energy, and the comparison of the electron temperature with that calculated on the basis of ionization
equilibrium. The instrumentation, a 12 channel polychrometer-
spectrometer with 4 arc seconds spatial resolution, will also
make measurements of spectral line ratios for determination of the

electron density, that can lead to an understanding of the magnetic field configuration in the corona.

Solar flares and other dynamic phenomena in active regions have very time-dependent characteristics. Lines in the EUV region are observed to increase by more than an order of magnitude in a few seconds. The questions associated with the state of ioniza- tion in transient solar phenomena are not unrelated to similar problems associated with ionization equilibrium in pulsed plasma devices. McWhirter (1960) has considered the time dependence for the latter case and has devised an approximation

$$\tau \simeq \frac{10^{12}}{n_e} \ \text{sec}$$

as a <u>lower limit</u> to the ionization time. Actual ionization times for elements of intermediate Z may be greater than this value by a factor of 5 or more. For the expected electron densi- ties in solar flare plasmas ($n_e \simeq 10^{13}$ cm), the time for ioniza- tion equilibrium is several tenths of a second, comparable to known time scales in solar flares. It is clear, then, that there is reason to consider that the assumption of ionization equilibrium might not be completely valid. The investigation of any signifi- cant departure from equilibrium between electron and ion tempera- ture becomes a significant parameter in the understanding of the flare process and a prime objective for SMM.

Other instrumentation proposed for SMM includes a white light coronagraph for observing the effects of flares on the outer extension of the solar atmosphere, instrumentation for the EUV, soft X-ray and hard X-ray instruments for diagnostics of plasma states over a broad range of temperatures. These various experi- ments will be closely coordinated in an attempt to attain a quan- titative understanding of the energy storage and release mech- anisms associated with solar flares. Initial plans are also being developed for larger Shuttle-launched solar facilities with much higher spatial resolution, again with the specific objective of quantitative interpretation of the observed features.

4.   RELEVANT ATOMIC PHYSICS

The success of the Solar Maximum Mission and other future problem-oriented missions will depend in large measure on the reliability of diagnostic techniques employed to interpret the measured spectral intensities in terms of the relevant physical characteristics of the solar plasma. Radiative transition probabilities and collision cross-sections are important not only for the particular transitions in the EUV and XUV that are observed, but also for all the important related transitions

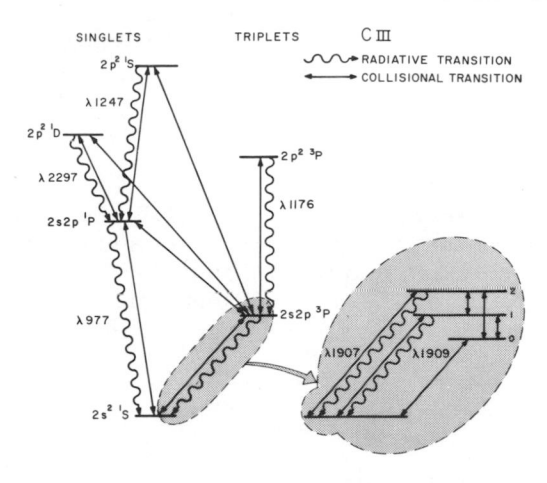

Figure 11.   Partial term diagram for the lines of C III used for
              plasma density diagnostics.

involving the upper and lower excited states for the observed
transitions. This problem of interpretation of intensities can
be illustrated by examination of several transitions in the
beryllium isoelectronic sequences; a notable example is the use
of the ratio of the C III 1176 Å line ($2s2p$ ^3P $-$ $2p^2$ ^3P) to
the ground state C III 977 Å transition ($2s^2$ ^1S $-$ $2s2p$ ^1P) shown
in Figure 11. Both lines are collisionally excited from their
respective lower levels and are observed in emission in the solar
atmosphere. Because the 1176 Å line arises from a metastable
level, whose population is directly proportional to the electron
density, the ratio of the lines is sensitive to the electron
density. Thus the observed line ratio λ1176 Å/λ977 Å when compared to theoretical calculations is a powerful diagnostic technique for the ambient electron density in the line forming region
near $T_e \approx 8 \times 10^4$ K (Munro, Dupree, Withbroe,1971; Jordan,1971).
However, it is apparent that existing values of the atomic parameters used to infer the level populations and hence the electron density, are unsatisfactory. And indeed, if these values
are adopted, lead to theoretical line ratios that are not in
harmony with observations (Dupree *et al.*, 1976).

Figure 12 shows the line ratios found from pairs of sequential spectra taken near Sun-center from the Harvard College
Observatory scanning spectrometer on ATM. The improved spatial
resolution of the Harvard instrument (5x5 arc seconds) shows a
larger range in intensity of the λ977 Å transition than previously

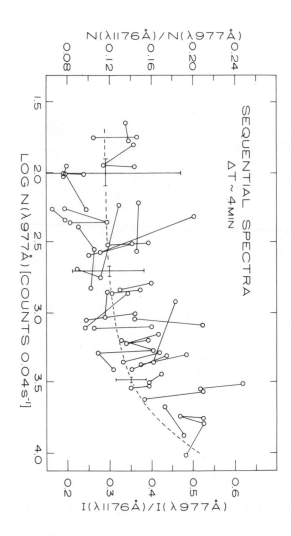

Figure 12. C III line ratios (1176 Å/977 Å) for successive wavelength scans plotted against the intensity of the 977 Å $^1P-^1S$ transition.

observed.   The increase in the ratio with $\lambda977$ Å intensity points
to a generally increasing value of electron density with activity in
the Sun.   There is a statistically significant range of the ratio,
and hence the electron density for a given value of the $\lambda977$ Å
intensity.   Since the observed line intensity (I) is proportional
to n_e^2L where L is an effective path length, an increase in the
ratio at constant intensity of 977 Å suggests an increasing
density and thinner transition region.   If many such ratios are
averaged from raster spectroheliograms, active regions show, on
average, a higher value of the ratio, and hence density than
quiet regions (Dupree *et al.*, 1976).   In addition the transition
region is thinner in active regions than in the quiet Sun.   The
connected circles represent values of the ratio and $\lambda977$ Å intensity from spectra taken at the same point on the solar disk, but
separated in time by 3^m46^s.   Intensity variations of $\sim$50 per cent
occur, as well as changes in the ratio that imply a variation
in the electron density.

Now, to match these ratios with theoretical calculations
requires a detailed knowledge of atomic parameters such as
collisional excitation rates for inelastic collisions by electrons and protons, both for allowed and forbidden transitions.
Also, we require the spontaneous emission rates for these same
transitions.   In order to fit the observed values of the ratio
to the theoretical predictions, it is necessary to change the
existing collision cross sections by amounts just consistent
with, but at the limit of, their predicted uncertainties (Jordan,
1974; Dupree *et al.*, 1976).   In particular, the $2s^2\ ^1S$ – $2s2p\ ^1P$
electron collision rate needs to be increased by 25%, and the
$2s2p\ ^3P$ – $2p^2\ ^3P$ and $2s^2\ ^1S$ – $2s2p\ ^3P$ rates must be simultaneously decreased by 25%.   Then, it is possible to interpret
measured line ratios in C III as arising from regions where
$10^9 \lesssim N_e \lesssim 2\text{x}10^{10}$ cm^{-3}.   A 25% uncertainty in cross section can
lead to an order of magnitude uncertainty in the implied density.
Atomic parameters such as these are now needed with accuracies
more nearly like 10%.

Another result of consequence from Figure 12 is the presence
of line ratios with values in excess of 0.5.   Large values have
been observed previously (Munro, Dupree, and Withbroe,1970) and
attributed to flaring regions or rapidly developing active regions.
It is difficult to predict such large ratios from existing theory,
and these values may suggest that a steady-state formulation is
particularly inapplicable here where transient phenomena are
known to be present.   As a matter of fact, the large ratio would
be consistent with a cooling plasma, where the recombining electrons are preferentially in the metastable level of higher statistical weight.

Although the C III line has been selected to illustrate the application of measured excitation cross sections and transition probabilities to the problem of determining the density in solar plasma at $T_e \sim 8 \times 10^4$ K, the problem is in many ways typical of other diagnostic techniques.

Electron temperatures in the solar plasma between 10^5 and 5×10^6 K can be determined from the intensity ratios of lines corresponding to the 2s–2p and 2s–3p transitions in lithium-like (N V, O VI, Ne VIII, Mg X, Si XII) and beryllium-like (O V, Ne VII, Mg IX, Si XI) transitions. These lines fall in the XUV part of the solar spectrum 40–700 Å. Since the fractional concentration of beryllium-like ions is a sharply peaked function of temperature (in calculations of ionization equilibrium), these lines provide a good method for determining the electron temperature. However, these methods too rely directly on the measured excitation cross sections and transition probabilities for many of the astrophysically important elements in intermediate stages of ionization. Hence we again return to the need for more complete and better atomic parameters for those systems used for solar plasma diagnostics.

5.   SUMMARY

In these remarks we illustrate some of the current directions of analysis for the EUV data from the recent ATM solar experiments from Skylab. In particular we emphasize the interrelationships of the spatial structure, temporal behavior, and the magnetic field that are apparent even in the less dynamic phenomena in the solar atmosphere. The interpretation of measured intensities in terms of the physical plasma parameters relies completely on our knowledge of the atomic parameters such as cross sections for electron excitation and radiative transition probabilities. Much work remains to be done both experimentally and theoretically to provide these parameters with much improved accuracy not only for the direct transitions observed, but also for many other states connecting with the upper and lower states of the measured transitions. This larger array of parameters becomes especially important for future solar missions such as the Solar Maximum Mission, where the dynamic character of the solar flare will likely require that certain of the diagnostics be carried out in a time-dependent manner. In this latter respect, at least, the solar flare diagnostic problem has many attributes in common with laser-produced and pinch plasmas.

ACKNOWLEDGEMENTS

The authors wish to acknowledge American Science and Engineering and the Naval Research Laboratory for the use of the ATM data for some of the illustrations. The authors also acknowledge the helpful comments on the manuscript provided by Dr. G.L. Withbroe. This work has been supported in part by the National Aeronautics and Space Administration under contract NAS 5-3949.

REFERENCES

Bohlin, J.D., Sheeley, N.R., and Tousey, R., (1976) Space Research XV (in press).
Brueckner, G.E. and Bartoe, J.-D.F. (1974) Solar Physics 38, 143.
Burton, W.M., (1968) in Structure and Development of Active Regions, ed. K.O. Kippenheuer (Dordrecht: D. Reidel Publishing Co.) p. 393.
Dupree, A.K., Foukal, P.V., and Jordan, C., (1976), Astrophys. J. (in preparation).
Dupree, A.K., and Henze, W. Jr. (1972) Solar Physics 27, 271.
Gabriel, A.H., (1975) Phil. Trans. Roy. Soc. (in press).
Golub, L., Krieger, A.S., Silk, J.K., Timothy, A.F., and Vaiana, G.S. (1974) Astrophys. J. 189, L93.
Golub, L., Krieger, A.S., and Vaiana, G.S. (1971) Solar Physics 42, 131.
Huber, M.C.E., Foukal, P.V., Noyes, R.W., Reeves, E.M., Schmahl, E.J., Timothy, J.G., Vaiana, J.E., and Withbroe, G.L. (1974) Astrophys. J. 194, L115.
Jordan, C. (1971), in Highlights of Astronomy, ed. C. de Jager, (Dordrecht: D. Reidel Publ. Co.), p. 519.
Jordan, C. (1974) Astr. and Astrophys. 34, 69.
Krieger, A.S., Timothy, A.F., and Roeloff, E.L., (1973) Solar Physics 29, 505.
Krieger, A.S., Vaiana, G.S., and Von Speybroeck, L.P., (1971), Solar Magnetic Fields, I.A.U. Symposium 43, Ed. R. Howard, p. 397.
Levine, R. (1975) private communication.
Mariska, J.T., and Withbroe, G.L. (1975) Solar Physics (in press).
McWhirter, R.W.P., (1960) Proc. Phys. Soc. 75, 520.
Munro, R.H., Dupree, A.K., and Withbroe, G.L. (1971) Solar Physics 19, 347.
Munro, R. and Withbroe, G.L. (1972) Ap. J. 176, 511.
Newton, H.W. and Nunn, M.L. (1951) Mon. Not. Roy. Astron. Soc. 111, 413.
Noyes, R.W., Withbroe, G.L. and Kirshner, R.P. (1970) Solar Physics 11, 388.
Reeves, E.M. (1976) Solar Physics (submitted).
Reeves, E.M., Timothy, J.G., and Huber, M.C.E. (1974) in

Instrumentation in Astronomy – II, SPIE Publications,
Vol. 44 (Redondo Beach, Calif., Soc. Photo.-Opt. Instrumentation Engineers), 159.

Reeves, E.M., Vernazza, J. and Withbroe, G.L. (1975) Phil. Trans.
Roy. Soc. (in press)

Simon, G.W. and Noyes, R.W. (1972), Solar Physics 22, 450.

Timothy, A.F., Krieger, A.S., and Vaiana, G.S. (1975) Solar
Physics 42, 135.

Tousey, R., Sandlin, G.D. and Purcell, J.D. (1968) in Structure
and Development of Active Regions, ed. K.O. Kippenheuer
(Dordrecht: D. Reidel Publishing Co.) p. 393.

Withbroe, G.L. and Mariska, J. (1975) Solar Phys. (in press).

RECENT ADVANCES IN ULTRAVIOLET ASTRONOMY

Donald C. Morton

Princeton University Observatory

Princeton, New Jersey

I.   INTRODUCTION

Laboratory measurements of transition probabilities are vitally important in modern astrophysics. Of course accurate wavelengths are essential for the identification of spectral lines and the determination of radial velocities, but as soon as these results are available we want to obtain the populations of the energy levels of each ion producing a line in the spectrum. For this purpose we need oscillator strengths of as many transitions as possible for each ion in order to sort out the saturation effects in the curve of growth as well as to average over the observational errors. In addition, for the strongest lines, which may have damping wings, we need the radiative and collisional damping constants of the levels. Beam-foil spectroscopy and other lifetime experiments provide the radiative damping constants directly, and the oscillator strengths follow if we know the fraction decaying to other lower levels. We are interested in the number of absorbers producing each line because the relative populations of two levels in an ion and the number ratio of two ion states permit the determination of the temperature and particle densities in the gas. Furthermore, addition of all ion states of an element and comparison with hydrogen or another standard gives an abundance which provides information about the nuclear or chemical history of the object.

In the following sections I shall discuss some examples of how transition probabilities have been used to interpret the spectral lines formed in three different types of objects - interstellar clouds, expanding stellar atmospheres, and the absorbing regions in the directions of some quasars. In each case it is primarily resonance lines that are observed because usually neither the density of

particles nor the flux of photons is large enough to sustain a significant population fraction in excited levels. However, within
the ground term absorptions from the excited fine-structure levels
up to 0.1 eV are frequently seen, and often the density in an expanding shell around a star is high enough to populate higher levels,
particularly metastable ones. Atomic lifetimes are specially
useful in these applications because many resonance lines require
little or no correction for branches to other levels than the ground
state. It was the importance of resonance lines that led Morton and
Smith (1973) to compile a list of their f-values and radiation
damping constants from the best available experiments and theoretical calculations.

     A few of these resonance lines such as the D lines of Na I and
the H and K lines of Ca II, occur at visual wavelengths, but most
appear in the ultraviolet shortward of the earth's atmospheric cut-
off at 3000 Å. Consequently to observe these lines, we need either
the extreme redshifts of the quasars, or instruments on spacecraft.
In this review I shall refer particularly to the Copernicus satellite which Princeton University Observatory operates for the National Aeronautics and Space Administration. This satellite contains
an 80 cm mirror and a guidance system that can keep a 7.5 mag star
stable on a 0.3 wide slit of an ultraviolet scanning spectrometer.
Reflective optics overcoated with lithium fluoride and windowless
photomultipliers were used to permit the detection of wavelengths
in the important region between 912 and 1150 Å. The highest resolutions available are 0.05 Å at 1200 Å and 0.1 Å at 2400 Å, corresponding to about 12 km s^{-1}. Normally our investigations are confined to wavelengths longward of the Lyman limit because of the
strong continuous absorption of the interstellar atomic hydrogen.
However, I shall mention some cases where oscillator strengths shortward of 912 Å also are of interest.

II. INTERSTELLAR CLOUDS

     Interstellar clouds usually reveal themselves by sharp absorp-
tion lines superposed on the continuous spectra of O and B-type stars
since these objects are distant enough that the line of sight inter-
sects one or more clouds. In most cases we can separate the interstellar spectrum from the star spectrum by differences in radial
velocity and line width. The star lines often are wider due to rotation and pressure broadening. We know that the interstellar lines
generally are not formed in low density shells associated with the
O and B stars because the lines are stationary in spectroscopic
binaries and the radial velocities of the clouds on the average mimic
the effect of galactic rotation expected for objects at half the distance of the stars. When the interstellar lines are observed with
very high resolutions such as 0.5 km s^{-1} FWHM which Hobbs (1969)

used for the Na D lines, the profiles show a complex structure which implies that there are several clouds along most lines of sight.

With the Copernicus resolution, only the components with large relative velocities are separated, so that most analyses of the atomic lines have been based on curves of growth using the total equivalent widths W_λ (usually quoted in wavelength units). Since many of the lines are saturated, it is necessary to sort out the combined effects of column density N, and the velocity distribution of the absorbers. Frequently it is possible to assume that certain ion stages of different elements have a similar distribution over the clouds so that the curve of growth determined for one ion can be used to interpret another. For example, in interstellar clouds outside the ionization zone of any hot star, the hydrogen is mainly H I or H_2 and the dominant ion stage of the heavier elements is likely to be the highest one which can be produced by the radiation less energetic than 13.6 eV ($\lambda > 912$ Å), which can penetrate the opacity of the hydrogen Lyman continuum. Therefore, if we also assume that the relative abundances of the elements are the same in each cloud, the lines of C II, N I, O I, Mg II, Si II, P II, S II, Ar I, Mn II, Fe II, and several more first ions all should lie on the same curve of growth. The radial velocities of the blended lines usually are consistent with this picture.

Such a composite curve of growth is shown in figure 1 for some of these ions observed in the direction of the O star ♄ Ophiuchi by Morton (1975). The f-values are those adopted by Morton and Smith (1973) except for two points of P II which have been based on lifetimes recently reported by Livingston, Kernahan, Irwin, and Pinnington (1975); the excellent agreement with the other ions

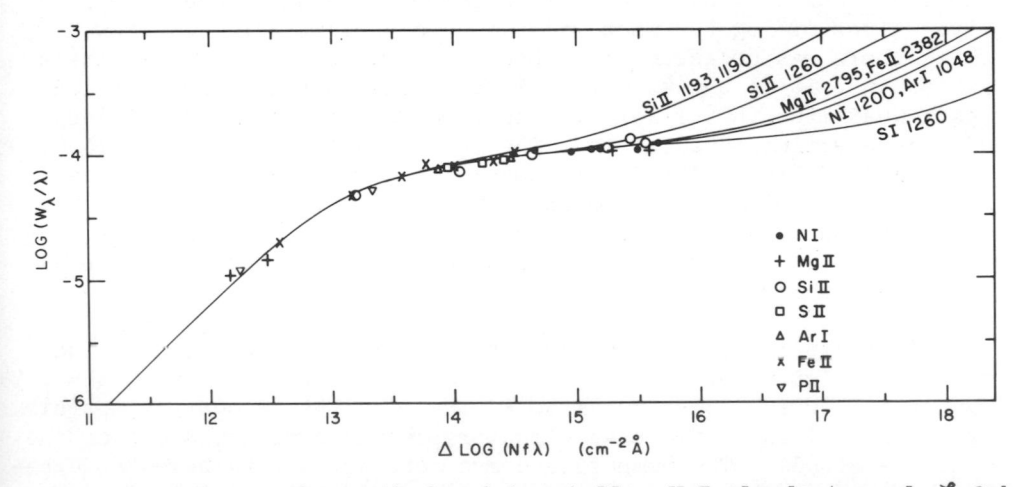

Fig. 1. - Curve of growth for interstellar H I clouds toward ♄ Oph, reproduced from figure 5 of Morton (1975) with points added for P II.

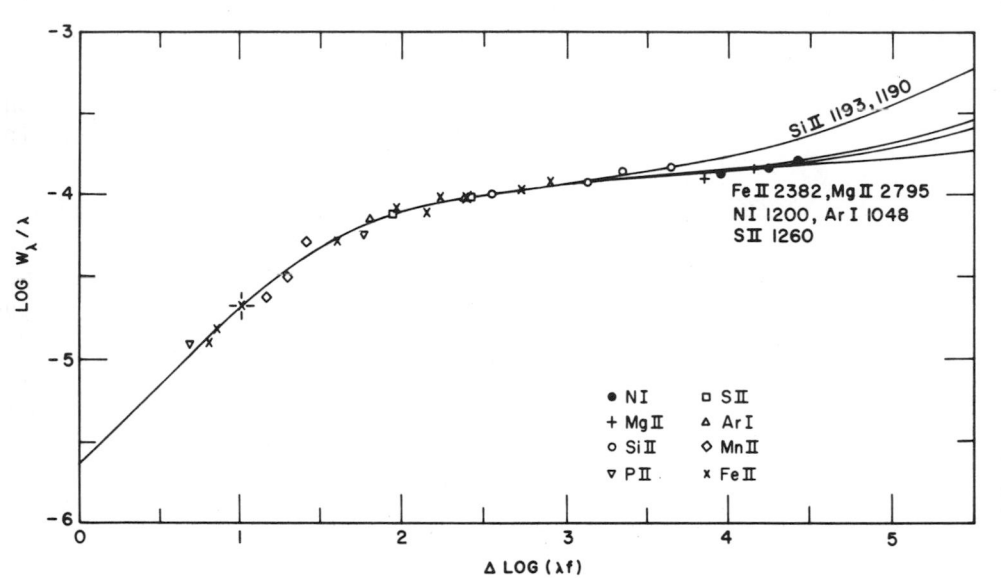

Fig. 2. - Curve of growth for interstellar H I clouds towards γ Ara, reproduced from the <u>Astrophysical Journal</u>, © American Astronomical Society.

supports the reduction by a factor 2.4 that they obtained compared with earlier values of $\underline{f}(\lambda 1301.87)$. For each ion in figure 1, log W_λ/λ was plotted against log $\lambda\underline{f}$, where λ is the wavelength and $\underline{f}$ is the absorption oscillator strength. Then the curves were slid along the log $\lambda\underline{f}$ axis to give the best superposition and hence the relative values of log $\underline{N}$. Thus we have an empirical curve of growth in which there was no assumption about the distribution of velocities except that it is the same for all the ions. If at least one ion has a weak line that likely is unsaturated, all column densities can be determined absolutely. The Mg II doublet $\lambda\lambda 1239.925$, 1240.395 and the P II line $\lambda 1301.87$ are specially useful for this purpose. A check on the P II lifetime is very desirable, and an experimental confirmation of the theoretical calculation by Black, Weisheit, and Laviana (1972) for the weak Mg II lines would be extremely helpful, though probably very difficult.

This empirical curve of growth actually fits rather well the curve calculated for a Maxwellian velocity distribution with $b = 6.5\pm 0.5$ km s^{-1} which corresponds to a line-of-sight velocity dispersion of $\sigma = b/2^{\frac{1}{2}}$. This effective $\underline{b}$ must represent some average over the multiple clouds. The composite curve need not always have the classical shape; de Boer and Morton (1974) found that the C I lines

towards ζ Oph lie on a curve with no unique b-value.

Once the shape of the curve of growth has been established, it also can be used to estimate f-values of lines whose equivalent widths are known, provided one f-value is already available. Thus de Boer, Morton, Pottasch, and York (1974) obtained oscillator strengths for 13 lines of Fe II and Mn II between 1050 and 1300 Å. In figure 2, the curve of growth towards the B star Gamma Ara obtained by Morton and Hu (1975) shows that these new points are reasonably consistent with other ions, indicating that the f-values probably are reliable. However we would welcome a confirmation from laboratory measurements. Comparison of figures 1 and 2 can help in the choice between two lifetime measurements for the Si II resonance multiplet at 1194 Å. Figure 1 was based on the lifetime of 0.11 ns measured by Curtis and Smith (1974), while figure 2 used 0.28 ns obtained by Livingston, Kernahan, Irwin, and Pinnington (1976). The curves in the upper right corner show the effect of radiation damping on the equivalent width of each transition. The two open circles farthest to the right in figure 2 represent the Si II absorptions $\lambda\lambda 1190.416$, 1193.289 from the ground state. They fit the damping curve rather well, whereas these lines, which are the first and third open circles from the right in figure 1, deviate considerably from their expected position on the damping curve thus favoring the longer lifetime. In contrast with the astrophysical evidence, Smith has confirmed his phase-shift lifetime with another measurement at Princeton, but he suspects the value for $\lambda 1304.372$ may be in error. Clearly an independent beam-foil experiment is desirable, provided it has adequate time resolution. Another determination for the weak line at $\lambda 1808.012$ also would be useful since this line can be unsaturated in many stars.

An empirical curve of growth is useful too for the estimation of a column density when only one line is available, as was the case for C II, O I, and Al II towards ζ Oph, though some extrapolation often is needed for the strong lines $\lambda 1302.169$ of O I and $\lambda 1334.532$ of C II. The column density of O I would be more reliable if good f-values, were available for $\lambda 1039.230$ and the intersystem line $\lambda 1355.598$, both of which have been recorded with the Copernicus spectrometer.

There are a number of other resonance transitions still requiring measurements of oscillator strengths. Equivalent widths are available for Cr II $\lambda 2055.59$, Ni II $\lambda\lambda 1370.136$, $1393,330$, and Cu II $\lambda 1358.773$ in the directions of various stars, but the abundances of these elements are based on rough calculations of the f-values by Kurucz and Peytremann (1975). For the analysis of higher ion states, presumably located in H II regions, two important lines lacking laboratory f-values are S III $\lambda 1012.504$ and Fe III $\lambda 1122.526$. None of the ions V III, Cr III, Co II, or Co III has been detected

yet in the interstellar gas, but laboratory checks on the calculated
f-values would be helpful in the estimation of upper limits for the
column densities.

Neutrals usually are more difficult to measure with beam-foil
spectroscopy so that other lifetime techniques often are preferred.
However, for the record, there are several transitions urgently in
need of experimental determinations. Some 82 lines of C I representing absorptions from both the ground level and the two excited
fine-structure levels were detected in ζ Oph. Many f-values were
derived from these data by de Boer and Morton (1974), but laboratory
confirmation of the multiplets that are not seriously blended, as
well as additional measurements of the calibration line λ1260.736,
would be very worthwhile. Since the excited levels usually are
populated by collisions with hydrogen atoms or molecules, the relative numbers in the three C I levels can be used to estimate both
the particle density and the kinetic temperature. In addition to
the O I transitions already mentioned, there are several N I and O I
lines between 912 and 1000 Å which have no f-values. The UV lines
of S I shortward of 1420 Å have been observed with Copernicus, but
none of their f-values has an accuracy better than ±50%, and some
Cl I lines lack any experimental values. The two Ar I lines
λλ1048.218, 1066.660 ought to be well established, but experience
with less saturated lines than those towards ζ Oph or γ Ara implies
that f(1048)/f(1066) is somewhat smaller than the ratio 3.9 given
by the lifetimes of Lawrence (1968). The weak K I lines λλ4044.136,
4047.206 have been observed in ζ Oph (Shulman, Bortolot, and
Thaddeus 1974) but Morton (1975) has shown that their strengths are
seriously inconsistent with λ7698.959 unless f(7698)/f(4044) considerably exceeds the factor of 5.6 obtained from the available
measurements. Smith and his colleagues at Princeton have obtained
some new data on lines of C I, N I, O I, Cl I, and Ar I, as well as
Si II, Fe II, and Fe III already mentioned, but parallel work by
other investigators would provide useful confirmations.

Figure 3 shows the results for element abundances relative the
hydrogen in the H I clouds towards ζ Oph compared with the solar
ratios, which mainly were adopted from Withbroe (1971). Since the
sun burns lithium, its cosmic abundance was obtained from the car-
bonaceous - chondrite meteorites, which usually agree well with the
sun. However, in the case of boron, the high meteorite abundance
N_B/N_H = 4 x 10^{-10} by number was rejected in favor of 1 x 10^{-10} found
in the photosphere of α Lyrae from the Copernicus measurement of
B II λ1362.461 by Boesgaard et al. (1974). This abundance is consistent with the solar upper limit of 1.2 x 10^{-10} obtained by Hall
and Engvöld (1975) and resolves a difficulty in the interstellar
limit discussed by Morton, Smith, and Stecher (1974).

A point below the horizontal line in figure 3 indicates that

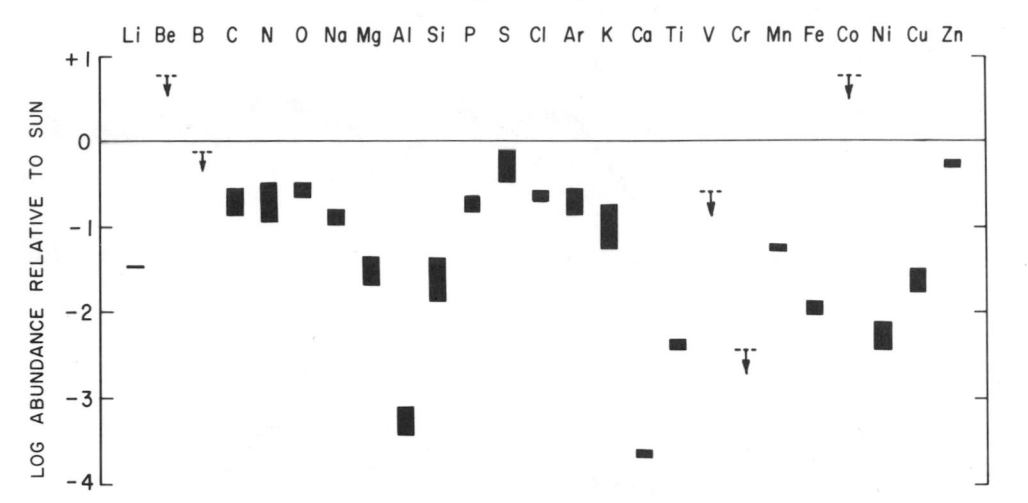

Fig. 3. - Interstellar element abundances by number towards ζ Oph, relative to the sun except as noted in the text. This diagram is reproduced from figure 7 of Morton (1975) with corrections for boron and phosphorus.

an element is depleted in the interstellar gas with respect to the sun, and the length of a bar shows the possible range of values due to poor observations, inaccurate f-values, and uncertainty in the curve of growth. In addition, since the dominant ion stages Li II, Na II, and K II cannot be observed, large corrections for the ionization equilibrium were applied to the column densities of the neutrals. It is clear that 50% or more of most elements are missing from the gas. Only S, Zn, and possibly B are near normal. If anything the gas should be richer than the sun in heavy elements due to nucleosynthesis since the time the sun was formed. Very likely the rest of the material is located in the solid interstellar grains whose presence has been inferred from the continuous absorption and polarization of the light from distant stars. The interpretation of these depletions in terms of condensation in gas ejected from cool stars has been discussed by Field (1974) and Morton (1974).

III.   STELLAR WINDS

One of the early results of stellar rocket spectroscopy was the discovery of large Doppler shifts in the resonance lines of some ions in luminous hot stars such as δ , ϵ , and ζ Orionis

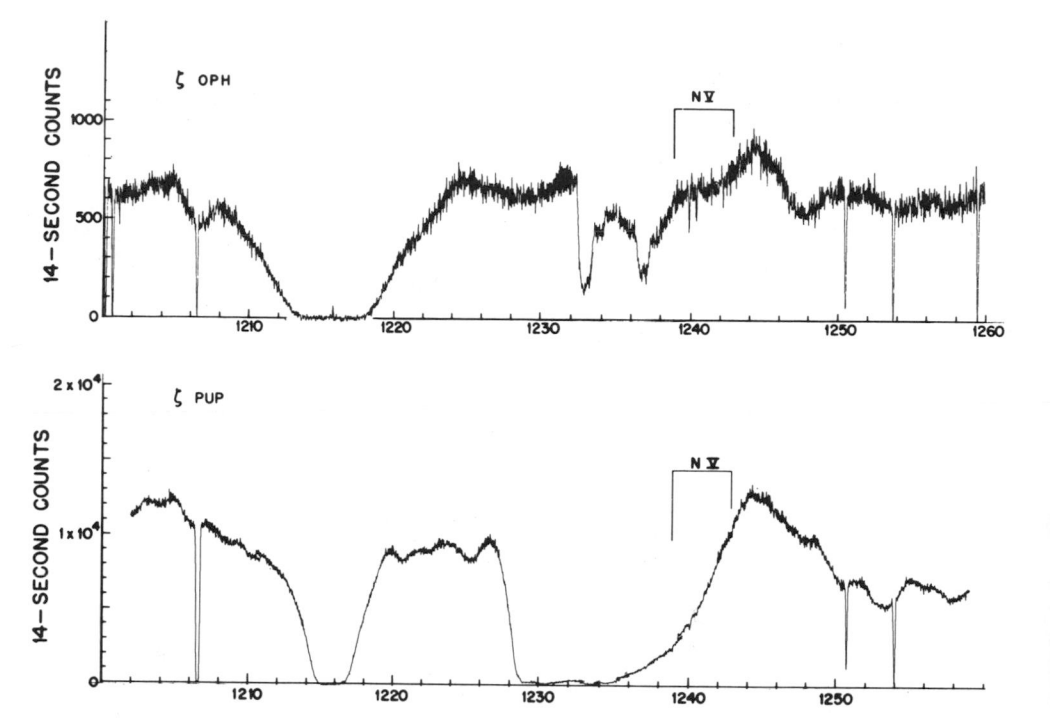

Fig. 4. - Displaced profiles of the N V resonance doublet in the O stars ζ Oph and ζ Pup. Narrow interstellar absorption lines of N I, Mn II, Si III, Mg II, and S II are present as well as the strong Lα line. The narrow night-sky Lα emission is present in the ζ Oph spectrum.

and ζ Puppis. The absorption lines were shifted to shorter wave-lengths by 1000 to 3000 km s^{-1}, while an associated emission component was near the laboratory wavelength. Copernicus spectra of the N V profile in the O9.5 main-sequence star ζ Oph and the O4 supergiant ζ Pup are shown in figure 4, where the rest wavelengths are indicated by the vertical lines. In the cooler, lower luminosity star ζ Oph, the centers of the components are shifted by 1440 km s^{-1} and they are narrow enough to be resolved, in contrast with the hotter, more luminous ζ Pup, where the lines are saturated and the short-wavelength edge has a shift of 2880 km s^{-1}. Since the escape velocities from ζ Oph and ζ Pup are about 1000 and 1400 km s^{-1} respectively, these stars must be ejecting material into the interstellar medium (Morton 1976). This pattern of emission and displaced absorption is called a P-Cygni profile after the star which has similar profiles with shifts of about 100 km s^{-1} in many of the subordinate lines in the visible spectrum. Presumably there

is an expanding shell around such a star with the displaced absorption originating in the portion projected against the photosphere, which is emitting a continuous spectrum, while the emission comes from all the shell not occulted by this opaque part of the star. As with the interstellar clouds, the phenomenon is seen best in the resonance lines, though sometimes the density and temperature in the expanding atmosphere are high enough to excite the lower levels of some subordinate lines such as Hα, He I λ5876, He II λ4685.682, C III λ1175.7, and N IV $\lambda\lambda$955.335, 1718.551.

In ζ Pup Lamers and Morton (1976) have compared the observed UV profiles of 12 ions from C III to O VI and S VI with calculations based on a model for a spherically symmetric expanding atmosphere. The shapes of the lines and the relative ion populations can be reproduced best with a terminal velocity of 2660 km s^{-1}, a kinetic temperature near 2×10^5 K, and a rate of mass ejection of about 10^{-5} solar masses per year. This mechanical energy flux is 0.25% of the total luminosity of the star. Castor, McCray, and Weaver (1975) have shown how this wind will produce expanding shock waves in the surrounding interstellar gas and inject kinetic energy at a rate comparable with that expected from supernovae shells. During a typical lifetime of 10^6 yr for such a hot star, the total mass loss could represent a significant fraction of an initial mass of some 50 to 100 suns. Of course, the analysis of the line profiles depends on reliable f-values. Most of these transition have been measured several times, with good agreement, but only a theoretical value was available for the S VI doublet $\lambda\lambda$933.382, 944.517.

Lucy and Solomon (1970) proposed that the mass flow is driven by the force imparted to the ions in the shell by the resonance - line absorptions of the photons from the continuous spectrum of the stellar photosphere. If the line is not totally opaque before the motion is started, the acceleration gradually Doppler shifts the material, permitting it to absorb again at another wavelength, thus continuing the acceleration. In the case of ζ Pup, Lamers and Morton (1976) have shown that summation over all the observed lines with velocity shifts produces only 5% of the required force. However, a star as hot as ζ Pup must have considerable emission between the Lyman limit of H I at 912 Å and the similar limit of He II at 227 Å where there are at least 500 resonance lines of the well populated ion states of the abundant elements. A preliminary estimate of the acceleration produced by these lines indicates that it is sufficient to drive the stellar wind. Oscillator strengths are known for about 85% of these extreme UV lines, but a number of them were obtained from theoretical calculations. Therefore additional lifetime measurements of ions producing these resonance lines would be useful.

Another interesting result from this investigation of ζ Pup

is the kinetic temperature of 2×10^5 K in the wind compared with
a maximum of 55 000 K in the photosphere. Evidence for a similar
high temperature corona around the BO main-sequence star Tau Scorpii
was found by Rogerson and Lamers (1975). Although its visible spectrum is noted for its narrow lines, all with the same radial velocity,
the Copernicus scans revealed broad absorptions due to the resonance
lines of C IV, N V, O VI, and Si IV with displacements up to 1000
km s^{-1} away from the star. In this case the 30 000 K photosphere
definitely cannot produce the O VI. The source of mechanical energy
to heat the corona is not known yet, but such a corona may be necessary to produce ion states whose lines are not saturated like the
photospheric ones so that wind can be started.

IV. QUASAR ABSORPTION LINES

Many quasi-stellar objects have narrow absorption lines in
their spectra. To date only resonance lines, including some absorptions from excited fine-structure levels, have been found. The
large redshifts have transformed the UV lines common in the interstellar gas and stellar winds into visible wavelengths, though the
identifications often are complicated by multiple systems with considerably different shifts. Some of the quasar absorption lines
must represent material flowing away from the object since there
are many cases where an absorption redshift is just slightly less
than the shift of the emission lines. However, the line widths of
50 km s^{-1} FWHM or less at relative velocities of the order of 10^4
km s^{-1} are difficult to understand. Other systems of absorption
lines are shifted considerably less than the emission lines; it has
been proposed that some of these absorptions originate in the gas
in intervening galaxies. If the presence of galaxies at intermediate
redshifts can be demonstrated conclusively, it would show that the
quasars are located at the large distances implied by their emission-
line redshifts. Of course we would still have the problem of explaining how some quasars produce 100 times the luminosity of a
bright galaxy, but that may be less difficult than accounting for
how quasars can be ejected from galaxies having much smaller redshifts.

Some information on element abundances in one of these absorbing
regions is available from the high resolution spectra of the quasar
PHL 957 obtained with an SEC integrating television sensor and the
Hale telescope on Mt. Palomar. As described by Lowrance et al.
(1972) and Wingert (1975), the emission lines are redshifted by z =
$\Delta \lambda / \lambda$ = 2.69 and the absorption systems have shifts between 2.664
and 0.278. One of these, with a shift of 2.309, has strong Lα and
Lβ lines, corresponding to a H I column density of of 1×10^{21} atoms
cm^{-2}, as well as numerous narrow lines identified with C II, N I, O I,
Si II, S II, and Fe II, which are just the ions prevelant in an interstellar cloud. Since the hydrogen density is reasonable for a

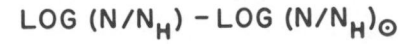

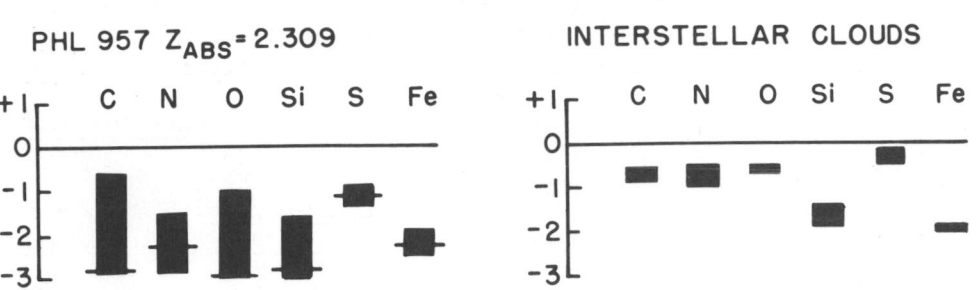

Fig. 5. - Element abundances by number relative to the sun for the
indicated absorption-line system in the quasar PHL 957 compared with
the interstellar clouds in the direction of ζ Oph.

line of sight through a galaxy, this absorption system very possibly
could be due to an intervening galaxy. From a curve-of-growth analysis of the line strengths, Wingert derived abundances which are
plotted relative to the sun in figure 5, where again the length of
a bar indicates the uncertainty. For comparison, the right-hand
side shows the abundances in the interstellar gas towards ζ Oph,
taken from figure 3. Although the errors are large, it is clear
that the heavy elements are considerably more depleted in the quasar
absorption system than in the interstellar gas, but some of the relative values are rather similar. In particular the ratio of S to
Fe in this quasar cloud is of the same order as along a variety of
directions in the gas in our galaxy (York 1975, Morton and Hu 1975).
Such a result by itself does not settle the location of the absorbing
material, but it does provide an important constraint on any model.

Several quasars have a sufficient redshift that the Lyman limit
of the emission system is displaced to a wavelength well longward of
the atmospheric cutoff at 3000 Å. Usually there is a detectable
continuum signal shortward of the limit and sometimes absorption
lines have been reported there, as in the case of 4C 05.34 with
z(em) = 2.877 (Lynds 1971) and OQ 172 with z(em) = 3.53 (Baldwin
et al. 1974). Since both of these quasars have absorption systems
with the Lyman limit longward of some of these absorption lines,
transitions whose laboratory wavelengths are less than 912 Å must
be considered for the identifications. Consequently we soon may be
looking for more $\underline{f}$ values in this wavelength region.

REFERENCES

Baldwin, J. A., Burbidge, E. M., Burbidge, G. R., Hazard, C.,
    Robinson, L. B., and Wampler, E. J. 1974, Astrophys. J., 193,
    513.
Black, J. H., Weisheit, J. C., and Laviana, E. 1972, Astrophys. J.,
    177, 567.
Boesgaard, A. M., Praderie, R., Leckrone, D. S., Faraggiana, R.,
    and Hack, M. 1974, Astrophys. J. (Letters), 194, L143.
Castor, J., McCray, R., and Weaver R. 1975, Astrophys. J. (Letters),
    200 in press.
Curtis, L. J., and Smith, W. H. 1974, Phys. Rev. A, 9, 1537.
deBoer, K. S., and Morton, D. C. 1974, Astron. and Astrophys., 37,
    305.
deBoer, K. S., Morton, D. C., Pottasch, S. R., and York, D. G. 1974,
    Astron. and Astrophys., 31, 405.
Field, G. B. 1974, Astrophys. J., 187, 453.
Hall, D. N. B., and Engvöld, O. 1975, Astrophys. J. 197, 513.
Hobbs, L. M. 1969, Astrophys. J., 157, 135.
Kurucz, R. L., and Peytremann, E. 1975, A Table of Semi-Empirical
    gf Values, Smithsonian Astrophys. Obs. Special Rep. 362.
Lamers, H.J.G.L.M., and Morton, D. C. 1976, submitted to Astrophys. J.
Lawrence, G. M. 1968, Phys. Rev., 175, 40.
Livingston, A. E., Kernahan, J. A., Irwin, D.J.G., and Pinnington,
    E. H. 1975, Physica Scripta, in press.
——————————. 1976, J. Phys. B, submitted.
Lowrance, J. L., Morton, D. C., Zucchino, P., Oke, J. B., and Schmidt,
    M. 1972, Astrophys. J., 171, 233.
Lucy, L. B., and Solomon, P. M. 1970, Astrophys. J., 159, 879.
Lynds, R. 1971, Astrophys. J. (Letters), 164, L73.
Morton, D. C. 1975, Astrophys. J., 197, 85.
——————————. 1976, ibid, 203, in press.
Morton, D. C. and Hu, E. M. 1975, Astrophys. J., 212, in press.
Morton, D. C., Smith, A. M., and Stecher, T. P. 1974, Astrophys. J.
    (Letters), 189, L109.
Morton, D. C., and Smith, W. H. 1973, Astrophys. J. Suppl., 26, 333.
Rogerson, J. B., and Lamers, H.J.G.L.M. 1975, Nature, 256, 19.
Shulman, S., Bortolot, V. J., and Thaddeus, P. 1974, Astrophys. J.,
    193, 97.
Wingert, D. W. 1975, Astrophys. J., 198, 267.
Withbroe, G. L. 1971, The Menzel Symposium (NBS Special Pub. 353),
    ed. K. B. Gebbie (Washington: Government Printing Office).
York, D. G. 1975, Astrophys. J. (Letters), 196, L103.

HIGH MAGNETIC FIELD SPECTROSCOPY

R. H. Garstang

Joint Institute for Laboratory Astrophysics
University of Colorado and National Bureau of Standards
Boulder, Colorado 80302

INTRODUCTION

Astrophysical interest in magnetic fields dates from the discovery by Hale in 1908 of magnetic fields in sunspots from the Zeeman effect of their spectral lines. In 1947 Babcock demonstrated the presence of magnetic fields in certain A-type stars. The possibility of much larger magnetic fields being present in neutron stars has stimulated new interest in the behavior of atoms under high fields. Finally, the discovery by Kemp, Swedlund, Landstreet and Angel[1] of circularly polarized continuum radiation from a white dwarf star and its interpretation as being due to a magnetic field of 10^7 gauss has led to renewed interest in the behavior of free atoms in high magnetic fields. Work in solid-state physics has also stimulated studies of hydrogenic spectra under conditions which mimic the effects of much higher fields on free hydrogen atoms.

SPECTROSCOPIC REGIMES

When an atom is present in a magnetic field a number of regimes can be distinguished.

(a) The (linear) Zeeman effect. This is the usual situation for small magnetic fields, with the first order magnetic splitting of the energy levels much smaller than the spin-orbit splitting.

(b) The first intermediate regime. This occurs when the magnetic interactions are comparable to the spin-orbit interactions. Elaborate calculations are needed to predict the spectrum

919

in this situation, which occurs in several spectra of laboratory
interest, the classic example being Li I.

(c) The Paschen-Back effect. This occurs when the magnetic
interactions are large compared with the spin-orbit interactions,
but small compared with the Coulomb interactions. In the ideal
limit all transitions are split symmetrically into exactly three
components. These are Paschen-Back triplets. They are identical
to those predicted by classical Lorentz theory. If we define the
cyclotron frequency $\omega = eB/mc$, the outer components of the triplets are displaced by $\frac{1}{2}\hbar\omega$.

(d) The quadratic Zeeman effect. This arises from terms in
the Hamiltonian which depend on the square of the magnetic field
(B). The most likely situation in light atoms is when B is
large enough to give a Paschen-Back effect, and the quadratic
effects are superposed on the triplet splitting. If the spin-orbit
interactions are large (in heavy atoms) the quadratic effects may
become appreciable before the complete transition to the Paschen-
Back effect has taken place. For a hydrogen-like level with
$\ell = 1$ and states described by $n\ell m_\ell m_s$ the quadratic Zeeman energy
shifts are given by

$$\Delta E_Q = 4.97 \times 10^{-15} \, B^2 n^4 (1+m_\ell^2) \tag{1}$$

where E_Q is in cm^{-1} and B in gauss, the formula in this form
being due to Jenkins and Segrè,[2] The important points to notice
are that ΔE_Q increases as n^4 with increasing n, so that the
quadratic shifts are important for the high members of a series,
and that for the σ components of an s - np transition ($m_p = \pm 1$)
the shifts are twice those for π components ($m_p = 0$), and all
are positive energy shifts so that the triplets become asymmetrical.
The shifts of the lines are towards the blue.

(e) The second intermediate régime. This occurs when the
magnetic fields are so large that the magnetic interactions are
comparable to the Coulomb interactions. This region is difficult
to study and variational calculations have been used in work on
hydrogen. If we define the parameter $\gamma = \hbar^3 B/m^2 ce^3$, where γ
is also equal to $\frac{1}{2}\hbar\omega$ expressed by Rydbergs, then the second intermediate régime is characterized by $\gamma \sim 1$. The value $\gamma = 1$
corresponds to a field $B = B_0 = 2.35 \times 10^9$ gauss.

(f) The Landau régime. This is the extreme high field régime
in which the magnetic interactions are much larger than the Coulomb
interactions. The electrons move in quantized orbits around the
magnetic field, and are confined in the field direction by the
Coulomb forces. The energy levels were first derived by Landau in
1930, and of the many later investigations we mention Johnson and
Lippmann[3] as perhaps the most valuable. The energy levels are

$$E = \left\{n + \tfrac{1}{2} \ell + \tfrac{1}{2}|\ell| + \sigma + \tfrac{1}{2}\right\}\hbar\omega + p_z^2/2m \qquad (2)$$

where ℓ is the angular momentum of the electron, σ is its spin component ($\sigma = \pm 1/2$), n is any integer (or zero), and the last term takes the motion along the field into account (unquantized for a free electron). The root-mean-square radius of the lowest electron orbit ($n = 0$, $\ell = 0$, $\sigma = -1/2$, $p_z = 0$) is $\sqrt{2}\,R$, where $R = (c\hbar/eB)^{1/2}$ is the cyclotron radius (some authors define this to be $\sqrt{2}\,R$). In terms of R the parameter $\gamma = a_0^2/R^2$, where a_0 is the first Bohr radius.

SPECTRA OF ALKALI ELEMENTS

The earliest studies of high field effects were concerned with the Paschen-Back effect, work on Li I dating back to 1914. Subsequently interest attached to studies of the quadratic Zeeman effect in the high members of the principal series of the alkalis. Equation (1) shows the strong dependence on the principal quantum number. The general correctness of the theory was verified in observations by Jenkins and Segrè[2] on Na I and K I and by Harting and Klinkenberg[4] on K I, Rb I and Cs I, using fields up to 27000 gauss and n ranging upwards from 10 to beyond 30. At the higher values of n several complications occur, firstly inter-ℓ mixing, and for still higher n inter-n mixing. Thus accurate calculations would be more complex than the simple theory.

SPECTRUM OF Ba I

A more recent experiment by Garton and Tomkins[5] studied the quadratic Zeeman effect in the principal series ($6s^2 - 6snp\ {}^1P_1$ of Ba I up to $n \sim 75$ with a field of 24,000 gauss. From $n = 26$ to $n \sim 38$ the quadratic shifts were clearly seen. From $n = 31$ satellite lines appear, due to inter-ℓ mixing, their intensities increasing rapidly with increasing n. Above $n = 40$ inter-n mixing becomes appreciable, and the pattern changes to one of a series of almost equally spaced resonances which cross the series limit and continue beyond. These resonances are spaced by 1.5 times the cyclotron frequency. This curious effect was explained on a simple Bohr theory by O'Connell,[6] and on a semi-classical basis by Starace.[7]

THEORETICAL CALCULATIONS

There have been a considerable number of attempts at calculating energy levels of atoms in high fields. Much work on hydrogen has been stimulated by solid-state physics, and most recently by the astronomical application to white dwarf stars. Generally speaking there are two approaches to the theory. One is to make use of perturbation theory. This will be valid only for some restricted range of the magnetic field, the precise range depending

on the circumstances. This approach has been used in work on
hydrogen by Garstang and Kemic[8] and by Hamada and Nakamura.[9] They
considered fields up to roughly 2×10^7 gauss. The Balmer series
lines become split into their magnetic components, with quadratic
shifts, and the splittings are so large that at fields of 10^7
gauss or more the spectra would become unrecognizable. Garstang
and Kemic[8,10] studied He I and Kemic[11] studied Ca II. In Ca II
the spin interactions are larger than in hydrogen and helium, and
the Paschen–Back splitting does not become dominant until the magnetic field is about 10^7 gauss. For smaller fields we are in the
first intermediate régime mentioned above.

     The second approach to the theory is to use variational
methods. This is unavoidable for fields of 10^8 gauss or more.
Examples of work of this kind include Smith et al.[12] on hydrogen,
Edmonds[13] on hydrogen, and Surmelian and O'Connell[14] on He II, and
there are many other papers, particularly in the solid–state literature and in connection with pulsar fields (up to 10^{12} gauss).
The general conclusion is that one can now predict the spectra of
at least the simpler atoms, in any magnetic field strength which
might be of interest, with at least moderate accuracy.

ASTRONOMICAL APPLICATIONS

     The discovery of magnetic white dwarf stars stimulated much
of the recent interest in spectra at fields of 10^7 gauss. An important difficulty is that in a real star we observe an integration of the spectrum over the visible hemisphere of the star. The
magnetic field will vary from point to point in the stellar atmosphere, both in field strength and in direction. Even if we take
a simple dipole model for the field it may be inclined to the line
of sight at any angle, and it may also be inclined to the axis of
rotation of the star. At each point the emission from an atom has
an angular distribution relative to the field direction which depends on the ΔM value of the particular line component. Every
Zeeman component has its own wavelength, which will vary over the
surface of the star. Thus we may expect to observe a very complex
spectrum which results from the convolution of all the above factors. A beginning in the study of these problems has been made by
Borra[15,16] and Kemic.[17] A total of 8 magnetic white dwarf stars
are now known. All show spectral peculiarities or unidentified
features as well as polarized continua. So far a complete resolution of the problem of the unidentified weak spectral features in
the magnetic white dwarfs has not been attained, but the discovery
by Angel et al.[18] of a DA magnetic white dwarf, GD 90, showing
Paschen–Back splitting of the hydrogen Balmer series puts beyond
doubt the essential importance of the presence of a magnetic field
of 5×10^6 gauss in that star.

ACKNOWLEDGEMENTS

A much longer version of this review, with a more comprehensive bibliography, is in preparation and will soon be submitted to Reports on Progress in Physics. The preparation of these reviews has been supported in part by the National Science Foundation (Grant MPS73-04867). I am also indebted to my former student S. B. Kemic for many hours of discussion on the problems surveyed in these reviews.

REFERENCES

1. Kemp, J. C., Swedlund, J. B., Landstreet, J. D. and Angel, J. R. P., Astrophys. J. (Lett)., 161, L77, 1970.

2. Jenkins, F. A. and Segrè, E., Phys. Rev., 55, 52, 1939.

3. Johnson, M. H. and Lippmann, B. A., Phys. Rev., 76, 828, 1949.

4. Harting, D. and Klinkenberg, P. F. A., Physica, 14, 669, 1949.

5. Garton, W. R. S. and Tomkins, F. S., Astrophys. J., 158, 839, 1969.

6. O'Connell, R. F., Astrophys. J., 187, 275, 1974.

7. Starace, A. F., J. Phys. B, 6, 585, 1973.

8. Garstang, R. H. and Kemic, S. B., Astrophys. Space Sci., 31, 103, 1974.

9. Hamada, T. and Nakamura, Y., Publ. Astron. Soc. Japan, 25, 527, 1973.

10. Garstang, R. H. and Kemic, S. B., in The Structure of Matter, ed. B. G. Wybourne, Univ. Canterbury Press, New Zealand, 1972, p. 396.

11. Kemic, S. B., Astrophys. Space Sci. (in press).

12. Smith, E. R., Henry, R. J. W., Surmelian, G. L., O'Connell, R. F. and Rajagopal, A. K. Phys. Rev. D, 6, 3700, 1972.

13. Edmonds, A. R., J. Phys. B., 6, 1603, 1973.

14. Surmelian, G. L. and O'Connell, R. F., Astrophys. Space Sci. 20, 85, 1973.

15. Borra, E. F., Astrophys. J., 183, 587, 1973.

16. Borra, E. F., Astrophys. J., 193, 699, 1974.

17. Kemic, S. B., Astrophys. J., 193, 213, 1974.

18. Angel, J. R. P., Carswell, R. F., Strittmatter, P. A., Beaver, E. A. and Harms, R., Astrophys. J. (Lett), 194, L47, 1974.

SPECTROSCOPY OF HIGHLY-STRIPPED IONS IN LASER-INDUCED PLASMAS

N. J. PEACOCK

CULHAM LABORATORY, (EURA.-UKAEA Assn. Fusion Research)

ABINGDON, OXON, OX14 3DB, U.K.

Abstract

Consideration is given to the degree of electron stripping in ions produced at the surface of solid targets irradiated by high power lasers. Results are presented for elements irradiated by the neodymium laser focussed to an intensity in the range 10^{11} to 10^{15} watts cm^{-2} at the target surface. Attention is directed to some topical features of the x-ray emission spectrum such as satellite lines, line broadening and optical opacity, recombination and possible use of the target plasma as an ion source, and effects arising from ultra-high densities.

1. INTRODUCTION

The XUV and X-Ray spectrum from Laser-Irradiated targets has been studied now for about a decade. Amongst the earliest experiments were those of Ehler and Weissler (1966) and Fawcett, Gabriel et al., (1966), both groups using pulsed ruby lasers. The former authors employed a modest laser power in a relatively long, 40 nanosecond, pulse while the latter group having the advantage of a higher laser intensity, were able to record appreciable degrees of electron stripping with ion stages like Fe $\underline{XV} \rightarrow \underline{XVII}$ and Ni $\underline{XVII} \rightarrow \underline{XVIII}$. Rather similar ion species were identified by Basov and his co-workers at the Lebedev Institute, Moscow, but using a neodymium laser with an output photon energy of 10 joules in a pulse duration of 15 nanoseconds; Basov, Boika et Al., (1967). Laser powers of the order of 10 to 500 Megawatts with focussed intensities $\simeq 10^{11}$ watts cm^{-2} at the target surface were typical of these early experiments.

More recently, Peacock, Fawcett and their colleagues at the Culham Laboratory, England, using focussed light intensities $< 10^{13}$ watt cm^{-2}, have studied the atomic term structure of He-like ions and their satellites in elements of atomic number up to Al, ie., Al XII, and of ions with charge states up to + 20, ie. V XXI, in the first long period; Peacock et al., (1973), Fawcett et al., (1974 a, b,).

With the development of high power neodymium glass laser systems for laser-interaction experiments and for laser-initiated thermonuclear fusion research , Nuckolls, Wood et Al., (1972), considerably higher light flux intensities have become available for the spectroscopic study of highly-stripped ions. At the Lebedev Institute for example ions such as Ti XXI, Cr XXII and Fe XXIV have been produced in the first long period, while charge states of the order of +30 eg., Y XXX, Zr XXXI, Nb XXXII and Mo XXXIII are reported for ions in the second long period, see for example Aglitskii, Boika et Al., (1974 a b c). At the Naval Research Laboratory, Washington D.C., Burkhalter, Nagel and Whitlock (1974) have reported similar ion stages going as high as +38 for the element Gd.

This review paper describes the physical processes by which these highly-stripped ions are produced at the surface of laser-irradiated solid targets. The relationship between the radiant flux intensity and the charge state is derived for a wide range of irradiation conditions. Finally, some of the more topical features of the X-Ray emission spectrum such as the existence of satellite lines and the effects of the extreme plasma pressure are discussed.

2.   THE PHYSICAL MODEL FOR ELECTRON STRIPPING

The ion populations in a collisional plasma, such as that produced by laser-irradiation of a solid surface, are related to the plasma parameters through the density and temperature dependent ionisation and recombination rates. The first requirement therefore in studying the ion species is to deduce the density and temperature of the free electrons, N_e and T_e respectively, as a function of laser intensity.

The relative ion populations also depends on the transit time, τ, of the ions through the plasma layer and are related to the steady-state, ionisation-recombination balance by the relation,

$$\frac{N_z}{N_{z-1}} - \left(\frac{N_z}{N_{z-1}}\right)_s = \left(\frac{N_z}{N_{z-1}}+1\right) \exp\ (-\ \tau/\tau_s) \quad \ldots \ldots \ (1)$$

where the subscript 's' refers to the steady state. In the absence
of three—body effects which cannot always be neglected, τ_s is given
by

$$n_e \tau_s = \frac{1}{\alpha(N_z) + S(N_{z-1})} , \qquad \dots (2)$$

where α and S are the relevant recombination and ionisation rate
coefficients. When N_{z-1} is H-like, $N_e \tau_s$ lies between 10^{11} and
10^{12} cm^{-3} sec.

Since the confinement of the ions is inertial, we can
relate τ to their escape time from the ionising region, viz

$$\tau = \frac{\ell_i}{v_{is}} , \qquad \dots (3)$$

where again, ℓ_i is the thickness of this region and the ion acoustic
velocity is

$$v_{is} = \left[(Z k T_e + k T_i) / M(Z) \right]^{\frac{1}{2}} \qquad \dots (4)$$

The problem therefore reduces to identifying relations
between the incident laser intensity Φ_o, watts cm^{-2}, and the
parameters T_e and $n_e \tau$ in the ionising plasma region. This region,
we will assume initially, has a density close to the critical
density n_{ec} where, classically, the light is reflected. We have
$n_{ec} = \pi(r_e \lambda^2)^{-1}$, ie., $n_e \lambda_o^2 = 1 \cdot 1 \times 10^{13}$ and for neodymium laser
light with a wavelength $\lambda_o = 1.06$ µm, $n_{ec} = 10^{21}$cm^{-3}.

The electron temperature dependence on Φ_o can be deduced
approximately by balancing the light energy absorbed within a
few wavelengths of the critical density layer, Φ_A, with the mean
energy, $\sim kT_e$, carried away by the ions at their thermal speed,
$\simeq v_{is}$.

$$n_{ec} \cdot v_{is} \cdot kT_e = \Phi_A \qquad \dots (5)$$

this leads to the relationship,

$$T_e (eV) = 2.6 \times 10^{-6} \cdot A^{1/3} \cdot \Phi_A^{2/3} \qquad \dots (6)$$

A is the atomic number of the target element. Generally
$\Phi_A \simeq 3.0 \cdot \Phi_o$, Galanti & Peacock (1975). The dependence in
equation (6) is close to that derived for a 1—dimensional, time—
independent model in which the region of the tenuous plasma which
spills away from the solid surface towards the incident light
beam, i.e. the 'corona', takes no part in the light absorption;

Fauquignon and Floux (1970), Bobin (1971). When radiation losses
are taken into account, Colombant et Al., (1972) note that
dependence should be modified, viz.,

$$T_e(eV) = 5.2 \times 10^{-6} \ A^{1/5} . \ (\lambda_o^2 \Phi_o)^{3/5} \qquad \qquad \ldots \ldots (7)$$

Alternative, time-dependent models where the competing processes
are plasma opacity to the incident light and gas-dynamic expansion,
have been proposed by other authors eg., Dawson (1964), Krokhin
(1971). In these cases, the solution involves the conservation
equations for energy, momentum and mass flow coupled with the
equation for absorption given by,

$$\nabla \Phi = K_a \Phi \qquad \qquad \ldots \ldots (8)$$

where K_a, the collisional (Inverse Bremsstrahlung) absorption
coefficient is

$$K_a = \frac{c}{v_o^2} \ Z. \ n_e^2 \ \frac{\ln \Lambda}{(kT_e)} \ 3/2 \ \frac{1}{[1 - (n_e/n_{ec})]^{\frac{1}{2}}} \qquad \ldots \ldots (9)$$

In the case of a plasma with constant parameters, the absorption
length for the laser light is

$$\ell_a = \frac{5 \times 10^{27} . \ T_e^{3/2}(ev). \ [1 - (\lambda/\lambda_c)^2]^2}{n_e^2 \ Z \ \lambda^2} \ cm, \qquad \ldots \ldots (10)$$

while in the case of the time- dependent model we have

$$\int_{x=o}^{\infty} \overline{K_a}(x).dx = 1.$$

Krokhin (1971) derives the following relationships for the time-
dependent absorption , viz,

$$v_{iz} = \Phi_o^{\frac{1}{4}} \ t^{1/8} \qquad \qquad \ldots \ldots (11)$$

$$T_e = 0.74 \ \Phi^{\frac{1}{2}}. \ t^{\frac{1}{4}} K_a^{\frac{1}{4}} \qquad \qquad \ldots \ldots (12)$$

By inspection of the above relations T_e depends on the
duration and intensity of the irradiation pulse. In practice
however, the relation is considerably more complicated. Even
the angle of the irradiating beam to the target normal and its
state of polarisation can affect the light absorption, for example,
by resonance effects between the light and plasma waves,
Vinogradov A.V. and Pustovalov (1971), Freidburg et al., (1972).
At flux intensities exceeding 10^{12} watts cm^{-2}, for a 1 μm

wavelength light beam, the high frequency resisivity of the plasma and therefore the heating may be enhanced by parametric instabilities of the type described by Kaw and Dawson (1969).

For short pulses < 1 nanosecond, the spread of energy from the critical density region may well be determined by electron heat conduction into the core rather than by particle ablation into the 'corona', as discussed eg., by Bobin (1971), Floux (1971).

Experimental values of the power-law dependence, $T_e \propto \Phi^P$, with p in the range from 0.3 to 0.66 have been reported. Figure 1 shows the dependence at relatively low flux intensities in the Culham experiments where p $\simeq$ 0.5. For this case, Galanti and Peacock (1975), a 1-dimensional expansion is a reasonable description for the motion of the ablated material, $(CH_2)_n$, and the density and temperature profiles in the light-absorption region shown in Figure 2, are good agreement with a theoretical model in which the laser energy not absorbed by inverse Bremsstrahlung is deposited at the critical density. Typically, most of the incident radiation is absorbed in a region whose dimension is of the order of ten wavelengths of the incident light. The region of the plasma whose density is less than the critical density, ie, $n_e \ll 10^{21} cm^{-3}$, is termed the 'corona'.

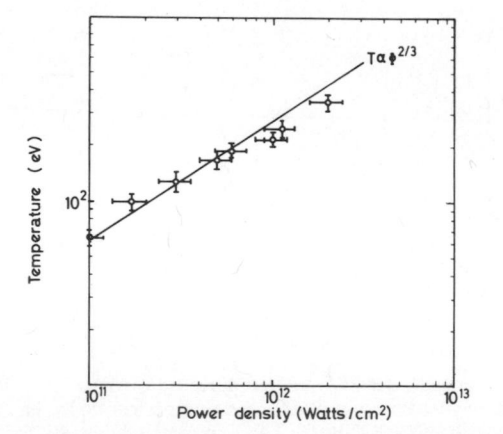

Figure 1. Plasma temperature (deduced from the Lyman continuum) against laser power density at the surface of a solid polyethylene target.

3.   ION CHARGE AS A FUNCTION OF LASER LIGHT INTENSITY

Having commented on the relationship between the incident light flux Φ_0 and the electron temperature, T_e, we can immediately evaluate the mean ionisation stage on the provisional assumption that the relative ion populations are determined by a steady-state balance between ionisation and recombination.

In the case where the He-like ions of charge Z-2 have their maximum emission, Gabriel (1972)

$$Z-2(\text{He-like}) \;=\; 1.13 \; T_e (\text{eV})^{0.37} \;+\; 0.6 \;;\; Te \text{ in eV} \;\; \ldots\ldots \; (13)$$

Thus at a plasma temperature of 1KeV and with a silicon target most of the resonance lines will come from Si <u>XIII</u>. For ion stages iso-electronic with the K, L and M-shells elements a scaling due to Colombant and Tonon (1973) may be more appropriate viz,

$$Z \;=\; \frac{2}{3} \cdot \left[A.T_e \right]^{1/3} \qquad ;\quad Te \text{ in eV} \qquad\qquad \ldots\ldots \; (14)$$

A is the atomic weight of the target element. Both of the expressions (13) and (14) are, of course, only approximate. Taking into account the relation between Φ_0 and T_e, we have

$$Z \;=\; \text{Const. } T_e^{1/3} \;;\; Z \;=\; \text{Const.}(\Phi^{2/9} \to \Phi^{1/9}) \; \ldots\ldots \; (15)$$

ie. the charge state only weakly increases with the flux intensity.

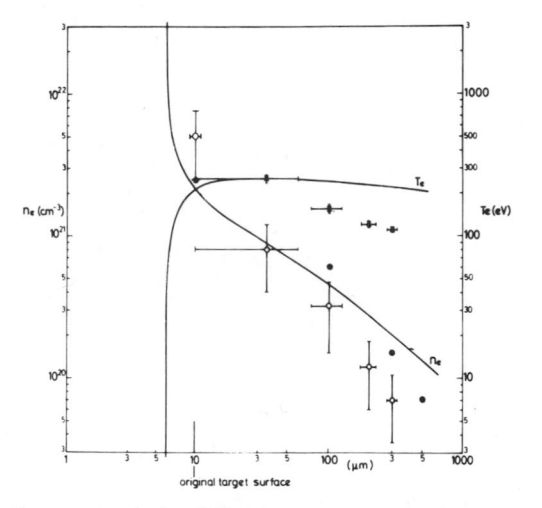

Figure 2. Predicted, time-averaged, density and temperature profiles along the normal to a poly-ethylene surface using the one-dimensional (MEDUSA) numerical code. The peak N_d laser power density is 10^{12} W cm^{-2} in a 4.5 ns long pulse. The experimental results for the same irradiation conditions are deduced from the spectroscopic data and are shown for comparison with the theoretical model profiles. X temperature; ● density by Inglis-Teller method; ○ density by absolute intensity of recombination continuum into C^{6+} ions (5—25 Å).

X ray photo of laser produced
plasma

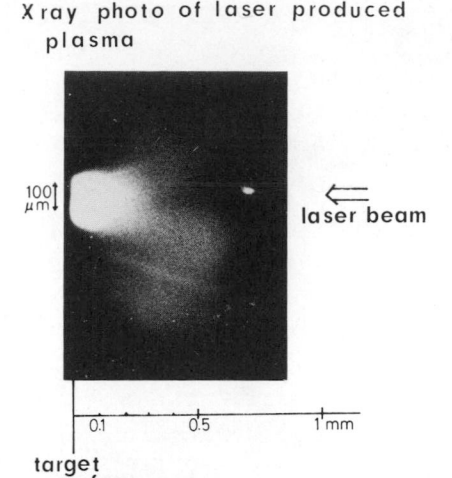

100
μm

laser beam

0.1   0.5   1 mm

target
surface

Figure 3.   Time-integrated X-ray image of C^{6+} Lyman continuum from laser-irradiated polyethylene target. Mean energy of X-rays is 1 keV. Spatial resolution is ~ 25 microns.

4.   THE IONISATION TIME: THEORETICAL MODELS

An upper limit for the time, τ, to reach a given stage of ionisation is determined by the life-time of the ions in the ionising region, ℓ, equation (3). τ may be less than τ_s, the steady-state ionisation time, equation (2) depending on the parameters ℓ_i and v_{is}.

Several different interpretations for the scale length ℓ are conceivable. The absorption length of the incoming light $\ell = (K_a)^{-1}$, equation (10) is one possibility. Yet again, ℓ has been identified with the focal spot diameter, Donaldson et Al. (1973). Both of these models would predict several tens of microns for ℓ, and certainly the dimensions of the X-ray emitting region can be of this order, see Figure 3; also, NRL report 7838, (1974). If, however, the spatial extent of the ionising region were to remain constant as the flux intensity increases, then $n_e\tau$ and thus the ionising efficiency must <u>decrease</u> with Φ.

Evidence to the contrary comes from the work on $(CH_2)_n$ targets by Galanti and Peacock (1975) and is shown in Figure 4. The degree of ionisation, represented here by the ratio $N(C^{6+})/N(C^{5+})$ actually increases with the electron temperature in the critical density region, over a range of Φ from 10^{11} watts cm^{-2} to 2×10^{12} watts^{-2}. Over this range $T_e(eV) = 1.4 \times 10^{-4} \Phi^{0.52}$ (watts cm^{-2}); Galanti and Peacock (1975).

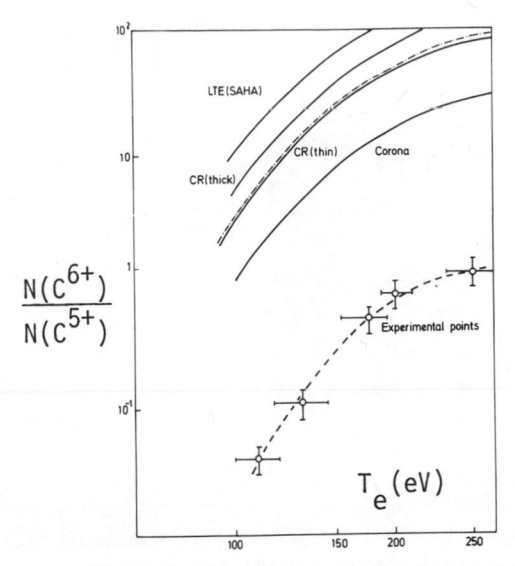

Figure 4. Theoretical prediction of the relative ion population of C^{5+} and C^{6+}, in the critical density region of the plasma. The calculations are based on a steady-state model. —·—·—, corona with ionization potential reduced to the collisional limit. The ground-state experimental ion population ratios are also shown.

An alternative scale length for the ionising region is the scale length for thermal conduction ℓ_H, at densities <u>above</u> the critical density. A scale length of ℓ_H may be more appropriate since ionisation is biased heavily towards high pressure regions, Section 5.

The heat conduction length, ℓ_H is given by,

$$\frac{\partial k T_e}{\partial \tau} = \frac{K_H}{n_e} \frac{\partial^2 T_e}{\partial x^2} \qquad \qquad \ldots \ldots (16)$$

where K_H is the heat conduction coefficient, Spitzer (1956).

The heat propagation length is therefore

$$\ell_H = \left(\frac{K_H \cdot \tau}{k \, n_e} \right)^{\frac{1}{2}} \qquad \qquad \ldots \ldots (17)$$

From Dawson (1964), we have, with T_e in $^{\circ}K$,

$$\ell_H^2 = \frac{6.4 \times 10^9 \cdot T_e^{5/2} \cdot \tau}{Z(1 + \frac{1}{Z}) \, n_e} \qquad \qquad \ldots \ldots (18)$$

Hence,

$$n_e \, \tau_H \; = \; \frac{n_e \, \ell_H}{v_{is}} \; \simeq \; 3.8 \times 10^1 \cdot \frac{A}{Z^2} \cdot T_e^{3/2} \qquad \ldots \ldots (19)$$

T_e in $^\circ K$, A the atomic number and Z the ion charge. Thus identifying ℓ_H with ℓ_i, then $n_e \, \tau_H \propto \Phi$, and the degree of ionisation should <u>increase</u> linearly as the radiant laser intensity is increased.

Inserting values appropriate to the Culham experiments, Galanti, Peacock et Al. (1974), then $\Phi = 2 \times 10^{12}$ watts cm^{-2}, $T_e = 4 \times 10^{6} \,^\circ K$, $n_e = 3 \times 10^{21} \, cm^{-3}$ we find $n_e \tau_H \simeq 10^{11} \, cm^{-3}$ sec, that is, the ion population should reflect a balance between ionisation and recombination. Figure 5 shows the various ion species of the element iron as a function of $n_e \tau$ and T_e calculated by Colombant and Tonon (1973), assuming a steady-state collisional-radiative model for the ionisation balance for each ion. Plotted on Figure 5 are also $n_e \tau_H$ values derived in a similar manner to equation (19). It is to be noted that the intersection points correspond to $n_e \tau_s \simeq n_e \tau_H \simeq 10^{11 \to 12} \, cm^{-3}$ sec. Therefore, theoretically, steady-state, ionisation-recombination balance should prevail.

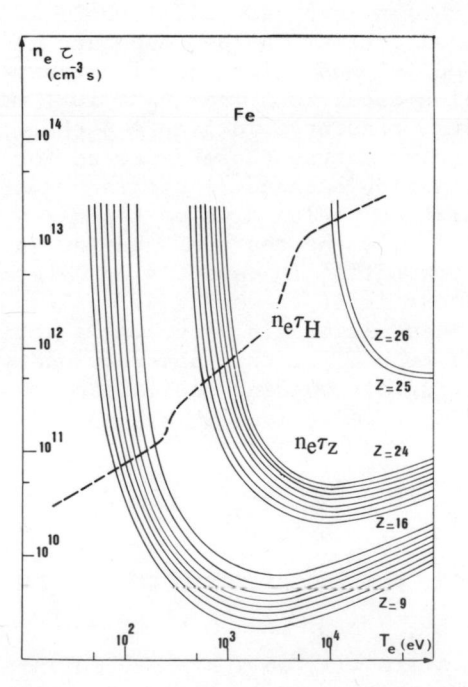

Figure 5. Comparison of $n_e \tau_z$ with $n_e \tau_H$ for laser-produced iron plasma; Colombant and Tonon (1973).

With a longer wavelength laser, such as the CO_2 laser, operating at 10.6 μm, the critical density is only 10^{19} cm^{-3}; and so the time to reach stationary ionisation conditions in the critical density region is two orders of magnitude longer than with the neodymium laser. Moreover, with far infra-red lasers a large fraction of the input energy may appear as energetic ions in the 'corona'. Heat conduction into the core may also be frustrated by decoupling of the pressure profile, near the critical density as discussed by Kidder and Zink (1972). For these reasons it is likely that a stationary ionisation balance characteristic of a single temperature cannot be attained with infra-red lasers. On the other hand there are definite advantages, associated with high N_{ec}, with laser wavelengths shorter than 1μm – provided such lasers could be produced with peak powers equivalent to the neodymium systems.

5.   EXPERIMENTAL EVIDENCE FOR NON-STATIONARY IONISATION

In the case of experiments on the irradiation of $(CH_2)_n$ targets with a neodymium laser focussed to an intensity of between 10^{11} and 10^{13} watts cm^{-2} at the target surface, Galanti and Peacock (1975) find that the most appropriate model to describe the state of ionisation involves collisional-radiative recombination and includes optical opacity in the resonance lines. This model is likely to hold good for most elements of higher Z than carbon and for most irradiation conditions when $\Phi_0 > 10^{11}$ watts cm^{-2}. It is clear from the data, presented in figure 4 that the ion populations are less highly stripped than those expected for a steady-state ionisation-recombination balance. Applying equation (1) to the relative ion population yields a value for $n_e\tau < 10^{10}$ cm^{-3} sec while from equation (3), the depth of the ionisation layer $\ell_i \sim 1$ to 0.1 μm. Thus ionisation is essentially determined by a plasma layer whose thickness is of the order of or less than the wavelength of the incident light (in vacuo). Galanti and Peacock (1975) point out that the reason for this non-stationary ionisation could well be the weighting of ionising collisions towards the high density, high temperature regions, viz,

$$\frac{\partial N_z(x)}{\partial x} \sim n_e^3(x)\exp(\text{const.}\psi.x^{\frac{1}{2}}) \qquad \qquad \ldots\ldots (20)$$

where x is measured from the highest pressure point in the plasma, Figure 2.

A stationary state will only be reached in a scale length determined by the slowest process ie., recombination. This scale length is of the order

$$\ell_s = \frac{V_{is}}{n_e \cdot \alpha_Z^{CR}} \quad (n_e, T_e, \psi) \qquad \cdots (21)$$

but ℓ_s is of the order of the scale length for the density inhomogeneity, a, where

$$a = n_e (\partial n_e / \partial x)^{-1}{}_{x = o} \qquad \cdots (22)$$

Thus stationary ionising conditions might not ever be achieved.
An accurate description of the ion populations can only be calculated
with a full numerical treatment of the hydrodynamics of the expansion,
the energy balance in the light absorption layer, the heat and
radiation transport to the core and the relevant ionisation and
recombination processes, including opacity, McGill, (1975).
The physical picture must change somewhat with laser wavelength,
target material and with laser flux intensity and pulse duration.

6.   ION CHARGE STATES PRODUCED IN LASER--TARGET EXPERIMENTS

The most commonly used laser system for spectroscopic studies
of highly stripped ions has been the neodymium laser, operated in
a pulsed mode of duration 1 nanosecond or less. Typical peak powers
have varied between 10 and 100 gigawatts. In Table 1 we list the
predominant ion species observed spectroscopically at the surface
of irradiated targets as a function of the focussed laser light
intensity and pulse duration. Using targets of the lighter elements,
say with atomic number up to Ti, only two or three ion species,
generally H-like, He-like or Li-like ions, tend to dominate the
spectrum. For elements of higher atomic number the ion stages
become more numerous and are iso-electronic with the L-shell elements,
or even with the M-shell in the case of high atomic number like Sn.

Those ion stages whose lowest resonance lines have been observed
spectroscopically with a 10 gigawatt neodymium laser are summarized
in Figure 6. The three straight lines represent ion species isoelectronic with the K, L and M shell elements. The points on these
graphs are taken from laser-irradiation data while the crosses +
and x, show ion stages produced in other laboratory sources eg.,
the vacuum spark; Lee, (1974). The laser source can be seen to have
contributed much of the recent K-shell data for ions with atomic
number up to Ti; L-shell up to Zr and M-shell up to Dy. The fall-
off in ionisation potential with atomic number represented by the
dotted line is probably due to the increasing energy and time which
must be expended in stripping the heavier elements.

In the K-shell sequence, the important astrophysical ions
Fe XXV, Ni XXVII have not yet been observed spectroscopically at

Table I

NEODYMIUM LASER, FLUX INTENSITY Φ, watts/cm² / PULSE DURATION (NANOSEC)		ION SPECIES PRODUCED AT TARGET SURFACE
+	7×10^{11} / 4.5	C^{6+}
+	8×10^{12} / 4.5	$F8+$ Mg^{10+} Al^{11+} V^{20+} Ti^{19+}; Fe^{19+} mainly H-, He-, Li-like ions;
*	$\sim 3 \times 10^{13}$ / ~0.9	Na^{10+} Mg^{11+} Al^{11+} Si^{12+} S^{12+} mainly H- and He-like ions
o	$\sim 5 \times 10^{14}$ / 1.5	Mg^{11+} Al^{12+} Si^{12+} Ti^{20+} mainly H- and He-like ions
o	$\sim 5 \times 10^{14}$ / 1.5 ?	Ti^{20+} V^{+21} Cr^{22+} Fe^{23+} mainly Li-like ions
o	$\sim 5 \times 10^{14}$ / 1.5	Y^{29+} Zr^{30+} Mo^{32+} Ta^{50+} W^{55+} mainly Ne-like ions
*	$\sim 10^{14}$ / < 1.0	Ge^{22+} Se^{24+} L-shell ions, typically 3 ion stages present
*	$\sim 10^{14}$ / <1.0	$Sn^{23+} \rightarrow 31+$ M-shell ions, typically 10 ion stages present

+ Culham Laboratory, see M. Galanti and N. J. Peacock CLM-P401(1975). N. J. Peacock, M. G. Hobby and M. Galanti (1973) J. Phys B. 6 L.298 - L304. B. C. Fawcett, M. Galanti and N. J. Peacock (1974) J. Phys B. No. 7, 1149-1153, also No. 4 L.106-L107

* NRL. Washington D.C. see NRL Report (1974) No. 7838

o Lebedev Institute, Moscow, see Aglitski E.V., Boika V.A., et al (1974) Kvontovaya Elektron (Moscow) 1, 908-935, 1 1731- , 1 2067-2069.

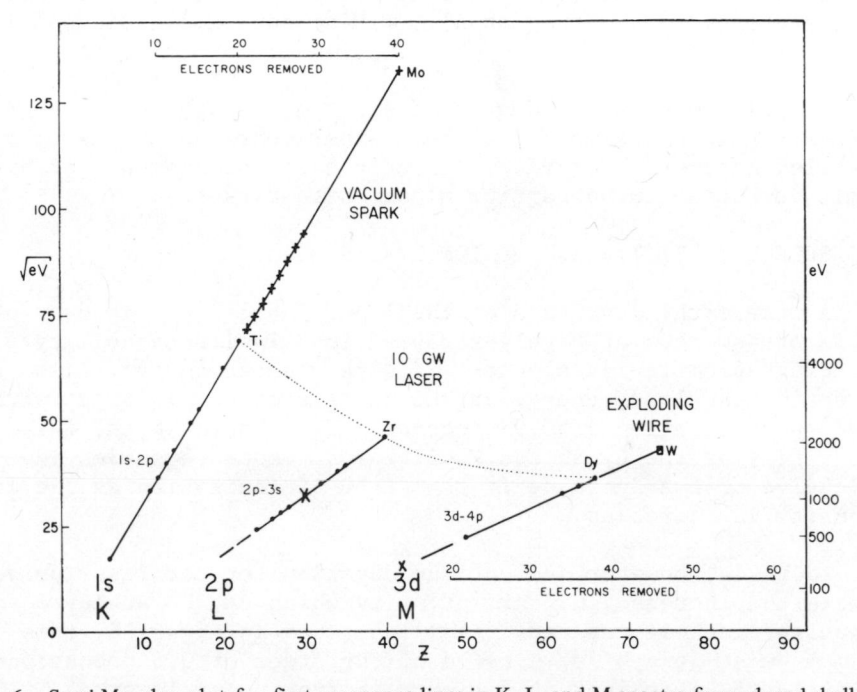

Figure 6. Semi-Moseley plot for first resonance lines in K, L, and M spectra from closed shell ions in high-temperature plasmas. Ionization stages are indicated for K and M series. The dotted line connects the heaviest element for which lines can be excited and measured with a crystal-film spectrograph using a 10 GW, 1 nsec laser pulse. Nagel et al. (1974).

irradiated targets. From our previous considerations, section (3), temperatures $\sim$ 5keV with consequent flux intensities > 10^{15} watts/ cm^2 would be required. While an extension of the dotted line, Figure 6, to higher charged states can be anticipated using the terrawatt (10^{12} watts) neodymium glass systems currently being developed (see e.g., Lawrence Livermore Laboratory annual reports (1973), (1974)),there is likely to be some fall-off in the degree of ionisation when $\Phi \gg 10^{15}$ watts/cm^2 due to unavoidable instabil- ities in the laser beam and thus in the density profile. Again, ion-acoustic turbulence at the critical density surface at high irradiation intensities can inhibit electron heat conduction to the core as discussed by Kruer et Al., (1975) and Ehler, Giovanielli et Al., (1974).

The ultimate in the degree of electron stripping would occur in laser-ignited thermonuclear burn experiments as suggested by Nuckolls, Wood et Al. (1972). In this case the elements of high atomic number would be part of a composite target.

7.   SELECTIVE EXCITATION OF ION STAGES

A noteworthy advantage of the laser-induced plasma over other laboratory sources of highly-stripped ions is its capability for producing specific ion species. Figure 7 shows the effect of increasing the laser energy at the surface of an Al target with a fixed pulse duration of 0.9 nanosecs. The change in the relative intensities of the Lyman and He-like ion resonance series is indicative of the increase in the state of ionisation as the laser intensity is increased.

There is, however, an optimum duration for the laser pulse despite the increase in peak intensity which can be achieved as the pulse width is reduced. As illustrated in Figure 8, this optimum pulse length seems to be of the order of 0.5 nanoseconds. The fall-off in ionisation for much shorter pulse lengths, $\sim 10^{11}$ to 10^{-12} seconds, may be due to the finite equivalent time required for ionisation and to set up the plasma profile by heat conduction, section 4.

Spatial resolution of the light absorption region and the corona is another useful way of isolating different ion species. Figure 9, for example, shows a separation of the resonance lines of H- and He-like carbon from their He- and Li-like satellites, the latter being confined spatially to the continuum emitting region, only a few microns wide, close to the solid surface. The width of this continuum is exaggerated for clarity of presenta- tion in Figure 9 by the limited resolution of the pinhole optics.

Since the decrease in density due to volume expansion in the 'corona' is considerably faster than that due to recombination,

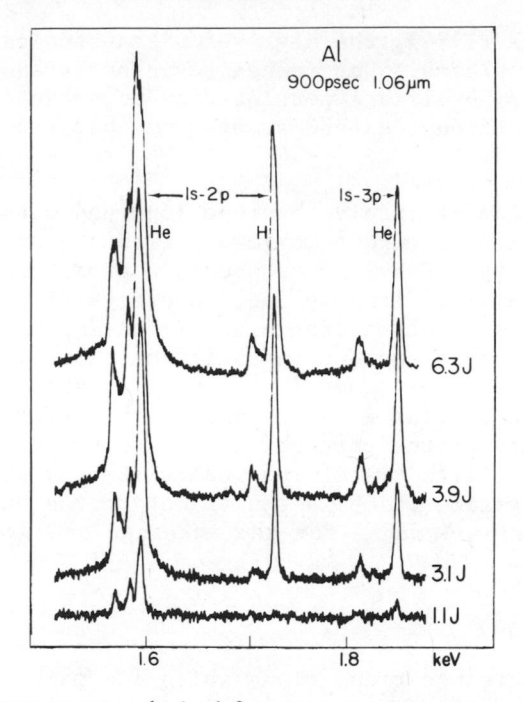

Figure 7. Al K resonance spectra obtained for constant pulse width and variable laser energy. Nagel et al. (1974).

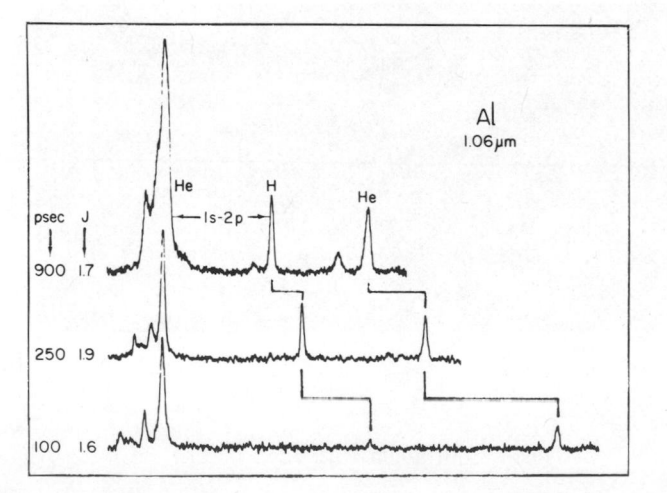

Figure 8. Al K resonance spectra obtained for nearly constant laser energy and variable pulse width. Nagel et al. (1974).

Irons and Peacock (1974), the highly-ionised species remain abundant
in the 'coronal' plasma. Unscreened resonance transitions in these
ions, whose upper levels are populated by recombination from the
free-electron continuum, extend several mms out from the solid surface, see Figure 9.

    Identification of rarely-observed ions and classification of
their atomic term structure sometimes presents a problem in laser-
target spectroscopy. For ions of moderate atomic number, say up
to the end of the first long period, isoelectronic extrapolation
from existing data is often feasible. In more complex ions, atomic
structure calculations, Herman and Skillman (1963), of the Hartree-
Fock-Slater type are most useful. Figure 10, shows an example of
the spectrum from Se ions isoelectronic with Ne-, Na- and Mg-like
ions. The heights of the vertical lines are proportional to the gf
values of the satellite transitions calculated by R.D.Cowan using
his multiple programme which is equivalent to the Hartree theory
plus a statistical allowance for the exchange energy term, Cowan
(1967).

8.   SATELLITE LINES

    Partially-screened transitions which lie maily to the long
wavelength side of the resonance lines, are prominent features of

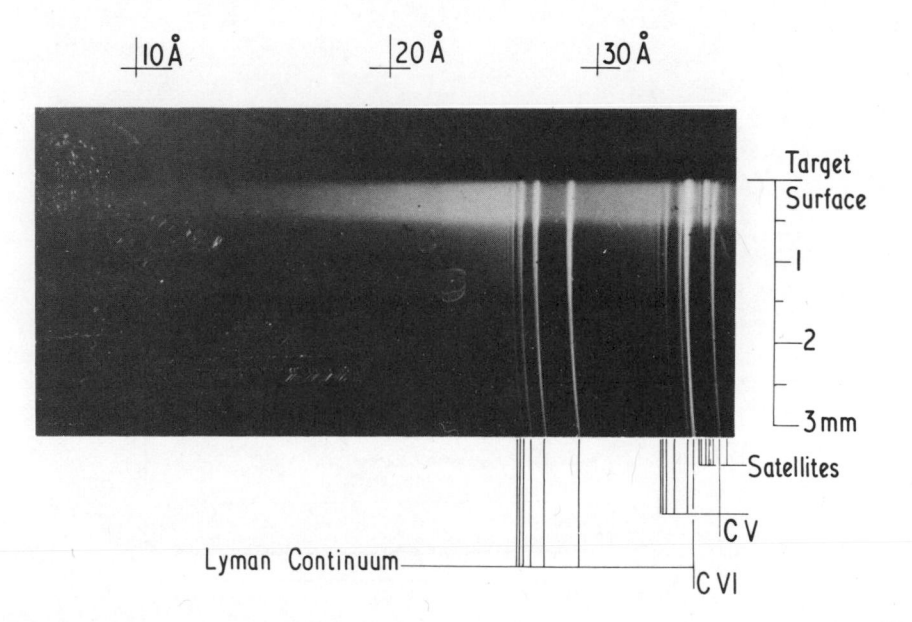

Figure 9. Spatially-resolved spectrum from laser-irradiated polyethylene target. Note satellite
lines are confined to region of continuum emission.

the X-ray spectrum from the sun and from laboratory plasmas, as we
have already observed in Figure 10.

Satellites to the He-like ions are particularly important
because of the persistence of these ions in non-stationary conditions of ionisation and recombination. A relatively complete study
of the term structure of the He-like ion satellites has been made
by Aglitskii, Boika et Al. (1974) using a focussed neodymium laser
with intensities $\Phi_0 \sim 5 \times 10^{14}$ watts cm^{-2} on the elements Na through V.
Gabriel, in a series of papers, e.g. Gabriel (1972), has pointed
out that the relative intensities of the He-like ion emission
features, such as those illustrated for Mg $\underline{XI}$ in Figure 11, can be
used to determine the parameters T_e and n_e and the extent of
departure from a steady-state ionisation balance. The basis for
such an analysis is a comparison of the intensity of the resonance
line, $I_0 = n_e N_{z-2}.X(1,2)$ (where X_{z-2} $(1,2)$ is the collisional
excitation coefficient) with the satellite intensities $I_s = n_e N_{z-2}$
$.\alpha_d(2n\acute{}) + n_e N_{z-3}.X_{z-3}(1,2n\acute{})$; $\alpha_d(2n\acute{})$ being the dielectronic
recombination rate. The ratio, $I_s (2n\acute{})/I_0$, is independent of n_e
and depends only on matrix elements and T_e, varying approximately
as T_e^{-1}. The ratio of I_0 to the intercombination line intensity
on the other hand depends linearly on density where the inter-system
collision rate $n_e.X_{z-2}$ $(2,3)$ and the spontaneous decay of the 3p

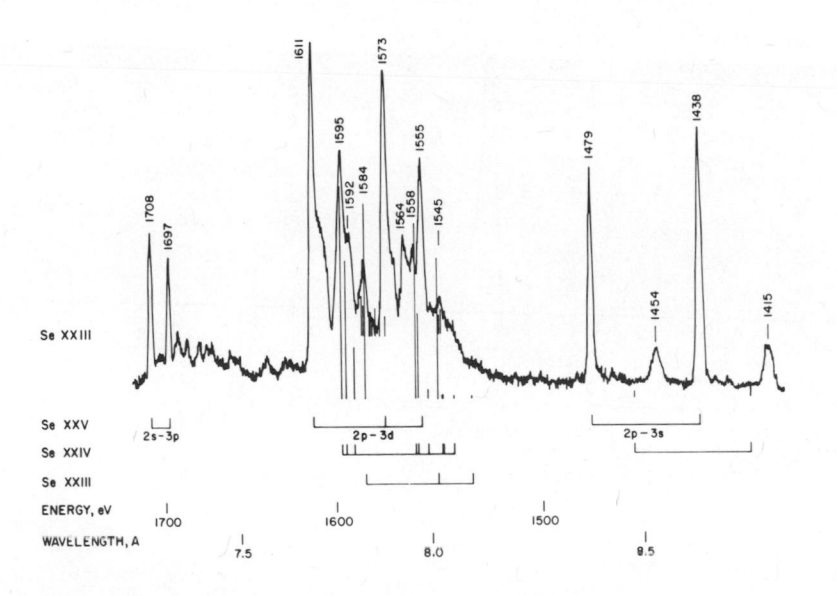

Figure 10. Laser-produced spectrum for Se. Transition energies and relative oscillator strengths
for the Ne-like Se XXV and Na- and Mg-like satellites obtained by atomic structure calculations
are indicated as vertical lines. Burkhalter et al. (1975).

levels, A_{31}, are of the same order, Gabriel et Al., (1975). Applications of these concepts to the satellite lines of various spectral
sources is given by Bhalla et Al., (1975). These authors derive the
following parameters from the emission spectrum in Figure 11, viz.,
$T_e=1.8 \times 10^6$ °K, $T_z=0.8 \times 10^6$ °K, and $n_e=2 \times 10^{19}$ cm^{-3}; T_z is the temperature
at which the derived ion population would be in a stationary state.
Since $T_z < T_e$, the laser plasma is concluded to be in a transient
ionising situation, a conclusion noted earlier by Galanti and
Peacock (1975) from an analysis of the relative intensities of the
free-bound continua, section 5.

It should be noted that the parameters n_e, T_e derived from the
satellite lines are somewhat less than those to be expected from
the irradiation conditions. With $\Phi_o \sim 5 \times 10^{12}$ watts cm^{-2} we might
expect $\overline{T_e} \sim 3 \times 10^6$ °K and $\overline{n_e}$ 1×10^{21} cm^{-3}. However the spectrum
shown in Figure 11 is spatially integrated over part of the 'corona'.
In contrast to the ionising region at the critical density and
above, the 'corona' is a recombining plasma, thus smearing over these
two spatial regions could affect the analysis of the satellite
intensities. Also Gabriel's theory is not completely applicable to
very high density conditions $n_e > 10^{21}$ cm^{-3}, where collisional
depopulation competes with radiative decay as the stabilising
mechanism in dielectronic recombination, Weisheit (1975).

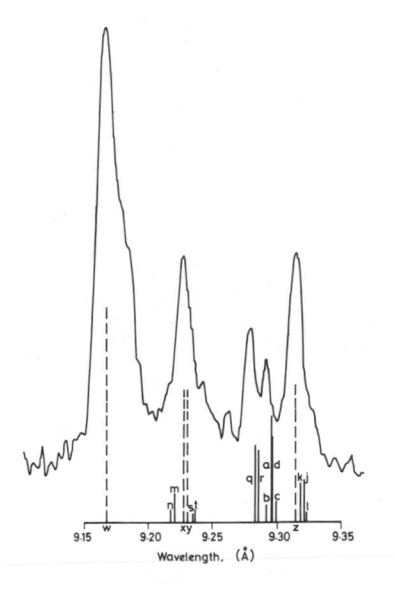

Figure 11. Curved mica crystal spectrum from a laser-produced plasma comparing the MgXI He-
like ion emission and associated long wavelength satellites with the computed spectrum. Laser flux
$\sim 2 \times 10^{12}$ watts/cm^2 on a pure magnesium target.

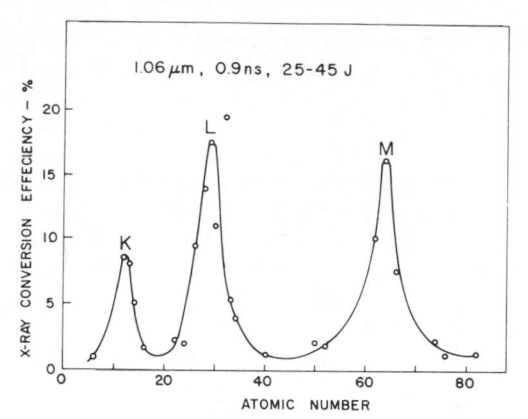

Figure 12. X-ray intensity from various elemental targets excited by pulses with the indicated characteristics as measured through a 25 μm window (800 eV cutoff). The measured intensities were integrated over 4π solid angle and divided by the incident pulse energy in order to be expressed as conversion efficiencies. Nagel et al. (1974).

Satellite lines have also been observed on the short wavelength wings of the resonance lines in laser target plasmas. Such lines have been interpreted as due to double quantum transitions involving a forbidden-line photon plus a plasmon of energy $\hbar\omega_{pe}$. If this interpretation, Boika, Krokhin et Al., (1974), is correct the plasmon energy corresponds to a plasma density about equal to that of the solid state.

9.   OPACITY AND LINE BROADENING

X-ray radiation losses from laser-irradiated targets can amount to a substantial fraction of the input laser energy. The efficiency in converting near-visible laser light into K-, L-, and M-shell X-ray emission is shown, by the NRL data in Figure 12, to be of the order of 20%. Peaks in the energy conversion occur when the electron temperature, equation (6), is matched to the binding energies in the appropriate quantum shell. Typically, this occurs when the temperature is a factor of between two and five times lower than the electron binding energy.

Galanti, Peacock et Al., (1974) find that with a polyethylene target most of the X-ray energy is radiated in the lower resonance lines which are broadened due to opacity. The source profiles are given by

$$I(\nu) = S_P(1 - e^{-\tau(\nu)}) \qquad \qquad \dots\dots (23)$$

$$\tau(\nu) = N_1 \cdot L(\nu) \cdot \frac{h\nu_0}{c} \cdot B_{12} \, D \qquad \qquad \dots\dots (24)$$

where $\tau(\nu)$, the optical depth, is itself a function of N_1 cm^{-3} the ion density, D cm is the plasma dimension, and most important, $L(\nu)$ is the line-shape factor within the plasma, determined by thermal and Stark broadening.

The resonance lines at the target surface are observed to be broadened and in some cases self-reversed due to a combination of opacity, mass motion and Stark broadening.   Such is the case for Lyman α C $\overline{\text{VI}}$, shown in Figure 13, where the experimental profile is compared with $I(\nu)$, equation (23), with S$_p$ taken as the Planck function at the temperature of the critical density layer.   Self-reversal precludes a fit at the line core but there is reasonable agreement on the outer wings when $\tau(\nu) \simeq 55$.   Spectral broadening due to opacity is clearly an important factor in assessing the X-ray conversion factor in laser-irradiated targets.

Opacity in the 'coronal' region of the target plasma is much less severe, although here again central line intensities can be altered by about an order of magnitude due to self-absorption, Irons and Peacock (1974).

While the lower resonance lines are most prone to "opacity broadening", equation (24), the upper members tend to be broadened by the electric microfields in the plasma, and for high enough

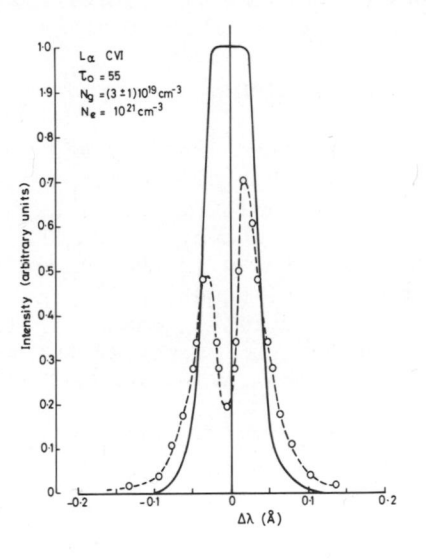

Figure 13.   Experimental profile (dotted line) of CVI Lyman a within 100 microns of polyethylene target surface irradiated by neodymium laser with flux intensity of 3×10^{11} W cm^{-2}. Full line is the calculated intensity from an optically thick line with optical depth $\tau = 55$ and with peak intensity corresponding to a Black-body temperature of 120 eV. For comparison the measured free-bound temperature derived from the Lyman continuum is 126 eV.

quantum number the broadening is so severe as to merge the lines
apparently with the continuum (Inglis–Teller effect). In a comparison of the line shapes with Stark theory at high densities there
exists in practice only a relatively few suitable transitions namely
those from intermediate quantum levels such as Lγ of CVI shown in
Figure 14. The theoretical profile, also inserted in Figure 14,
is computed using the average electric microfield due to the carbon
ion perturbers corrected for the screening effect of the electrons
while electron collisions are taken into account in the impact
approximation. The theory, Richards (1975), has not yet been tested
at the much higher densities expected in laser compression experiments.

10. HIGH DENSITY EFFECTS

It can be anticipated in future laser-irradiation experiments
associated with the compression of matter, Nuckolls et Al., (1972),
that plasma densities more typical of the solid state or above will
be reached. At these densities the emission spectrum will reflect
the influence of polarisation of the particles by the plasma medium,
usually an insignificant effect at low densities. Screening of the
ions by the electrons for example will result in a lowering of the
ionisation potential in the plasma given by

$$\Delta E_z \simeq \frac{Ze^2}{\lambda_D} \qquad\qquad\qquad \ldots\ldots (25)$$

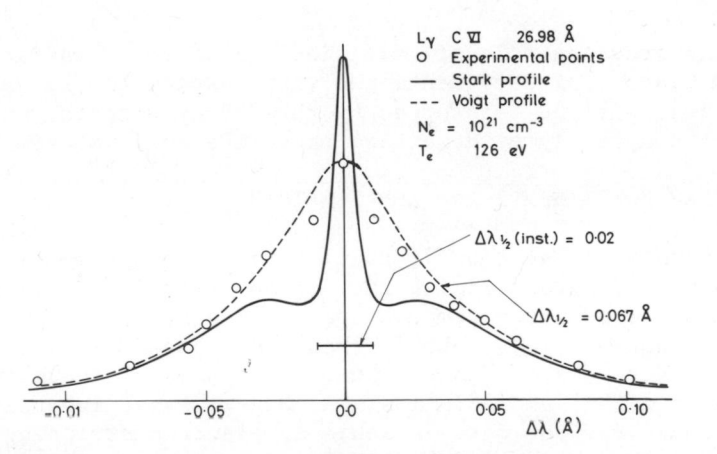

Figure 14. The full line is the Stark profile of CVI L_γ (26.98 Å) calculated according to Richards
(1974) for a carbon plasma at 126 eV and an electron density of 10^{21} cm^{-3}. The Stark profile is
folded with the instrument function to give the dotted line which is compared to the experimental
profile represented by the circles.

where λ_D is the Debye screening distance. Equating ΔE_z to the potential energy of a bound electron in state n in a H-like ion, we have

$$\Delta E_z = Ry \frac{z^2}{n^2}, \qquad \text{(Ry in Rydbergs.)}$$

Numerically,

$$n_e = 4.9 \times 10^{21}.kT_e.(1 + z^2).n^{-4} \qquad \qquad \dots \dots (26)$$

where kT_e is in eV. At electron densities $\sim 10^{24}$ cm^{-3} and $T_e \sim$ 100 eV in the core of the irradiated target, a hydrogen atom will suffer 'pressure' ionisation. In other words, the interparticle distance is of the order $a_o = e^2/2Ry$. At these densities also, ΔE_z is of the order of the thermal energy so that a description of the particles based on binary Coulomb collisions is invalid. At frequencies of the order of the plasma frequency, ω_{pe}, the continuum emission will be reduced by correlations between the electrons due to ion motion, Griem, (1964). A further complication is that the electrons themselves become degenerate when ΔE_z is less than their Fermi energy ε_F, the latter being given by

$$\varepsilon_F(ev) = 3.6 \times 10^{-15} n_e^{2/3}. \qquad \qquad \dots \dots (27)$$

This occurs approximately at the same or higher densities i.e. when $n_e \, a_o^3 > 1$.

The increasing role of collisions relative to radiative decay of bound states, already mentioned with respect to the satellite intensities, will be reinforced by three-body effects, which may take over completely at densities above the solid state.

11.   FUTURE EXPERIMENTS AND APPLICATIONS

We have seen, section 7, that there is an optimum laser pulse duration in order to maximise the degree of ionisation. For a fixed pulse length, the temperature and the degree of ionisation increase monotonically, both in theory and experiment, with the focussed laser power. Table 1 summarizes some of the more recent experiments with 1 μm radiation. At higher laser intensity than has been used for the data in Table 1, electron stripping may well saturate however due to dissipation of the input radiation by those collisionless processes which frustrate core heating. The scaling of <Z> with laser wavelength and with Φ_o in the range >> 10^{15} watts cm^{-2}, where the limit for energy transport to the core is set by free-streaming of the electrons, is one of the more interesting

questions still to be answered.

The intense focussed power and efficient conversion of the electrons to X-ray energies mean that it is possible to distort specific atomic level populations transiently from their thermal densities. Laser action between such 'inverted' levels is a serious possibility and is discussed by R.C. Elton (see proceedings of this conference). Optical pumping experiments in the X-ray region using resonance lines which are broadened by opacity has been suggested by Norton and Peacock (1975). In such experiments it is necessary to extend in frequency the core of the line, which will saturate at the black body intensity, Figure 13, so as to overlap the level spacing in a neighboring ion species. For the specific case where the pumping line has a dispersion shape factor whose width due to impact broadening in the plasma is $\Delta\lambda_{st}$, then the half-intensity, half width is given by

$$\Delta\lambda_{\frac{1}{2}\tau} = \Delta\lambda_{st} \cdot \tau^{\frac{1}{2}} \qquad\qquad \ldots\ldots (28)$$

For optical pumping experiments we might need τ, the optical depth, in the range 10^3 to 10^4, or greater. These values can be achieved in principle along a line-focus whose dimension in the critical density region exceeds 1 cm or so. Pumping of the n=4 level of He-like ions of C, N etc. with the appropriate, broadened Lyman α transistion has been suggested, Norton and Peacock (1975). The resulting population inversion between the n=4 and lower quantum levels can best be verified by the observation of super-radiant emission.

An important application of the rapid ionisation in the critical density region followed by relatively slow recombination and the 'freezing-in' of ion stages in the corona, section 7, is the use of laser-target plasmas as ion sources for accelerator experiments. Peacock and Pease (1969), calculate that pulses, $\simeq 10^{14}$ to 10^{15} ions of charged-state +20, are available for extraction from the 'corona'. These calculations may be somewhat of an underestimate since they do not take into account enhanced non-linear absorption of the laser light nor preferential acceleration of those ions with the highest charge.

While there are some practical difficulties associated with the transient nature of the plasma, the ease and flexibility with which highly-charged ion species can be selected makes laser-irradiated targets potentially an attractive but so far unexploited source of highly-stripped ions for accelerators.

ACKNOWLEDGEMENTS

The author has drawn widely from the work at the Naval Research

Laboratory, Washington, D. C. for some of his illustrations.
Permission to reproduce figures from NRL Report No. 7838 (1974)
is greatly appreciated.

REFERENCES

Aglitskii E. V., Boika et Al., (1974). Kvantovaya Elektron.
(Moscow) (a) 1, 908-936; (b) 1, 1731-; (c) 1, 2067-2069.

Basov N. G., Boika V. A., Voinov Yu. P., Kononov E.Ya.,
Mandelshtam S. L., (1967). J.E.T.P. Letters 5, 141.

Bhalla C.P., Gabriel A.H., and Presnyakov L.P., (1975). Submitted
Mon. Not Roy. Astr. Soc.

Bobin J.L. (1971). Physics Fluids 14, No. 11, 2341-2354.

Boika V.A., Krokhin O.N., Pikuz S.A., Faenov A.Ya. (1974).
J.E.T.P. Letters 20, No. 2, 50-51.

Burkhalter P.G., Nagel D.J. and Whitlock R.R. (1974). Phys. Rev.
A.9. 2331-6.

Burkhalter P.G., Nagel D.J. and Cowan R.D. (1975). Phys. Rev. A.
11, 782-8.

Colombant D. and Tonon G.F., (1973) J. Appl. Phys. 44, No. 8,
3524-3537.

Colombant D., Perez A. and Tonon G., (1972). Proc. Conference on
Ion Sources, Vienna.

Cowan R.D., (1967). Phys. Rev. 163, 54-61.

Dawson J.M. (1964). Physics Fluids 7, 981-7.

Donaldson T.P., Hutcheon R.J. and Key M. (1973). J. Phys. B. 6,
1525-1534.

Ehler A. W., Giovanielli D.V., Godwin R.P., et Al., (1974).
LASL Report LA-5611-MS.

Ehler A.W. and Weissler G.L. (1966). Appl. Phys. Letters 8, 89.

Fauquignon C. and Floux F., (1970). Physics Fluids 13, 386.

Fawcett B.C., Galanti M. and Peacock N.J., (1974). (a) J. Phys. B.
7, No. 10, 1149-1153. (b) J. Phys. B. 7, No. 4, L106-107.

Fawcett B.C., Gabriel A.H., Irons F.E., Peacock N.J., Sanders P.,
(1966). Proc. Phys. Soc. 88, 1051.

Floux F. (1971). Nuclear Fusion 11., 635-47.

Friedberg J.P. et Al., (1972). Phys. Rev. Letters 28, 795-9.

Gabriel A.H. (1972). Mon. Not Roy. Astron. Soc. 160, 99–119.

Gabriel A.H., Paget T.M., Kunze H.J., (1975). Submitted J. Phys. B.

Galanti M., Peacock N.J., Norton B.A., Puric J. (1975). Proc. 4th
Int. Conf. on Plasma Physics and Controlled Nuclear Fusion Research.
Tokyo (1974). Paper IAEA-CN-33/F-3/4. To be published (1975) IAEA
Vienna.

Galanti M. and Peacock N.J., (1975). To be published J. Phys. B.
Vol. 8.

Griem H.R. (1964)."Plasma Spectroscopy". McGraw Hill, New York.

Herman F. and Skillman S., (1963). "Atomic Structure Calculations",
Prentice Hall, Englewood Cliffs.

Irons F.E. and Peacock N.J. (1974). J. Phys. B. (At. and Mol. Phys.).
7, No. 15, 2084–2099.

Kaw P.K. and Dawson J.M. (1969). Physics Fluids 12, No. 12, 2586–2591.

Kidder R.E. and Zink J.W. (1972). Nuclear Fusion 12, No. 3, 325–328.

Krokhin O.N. (1971). 'Enrico Fermi' School of Physics. Corso 48.
(New York Academic Press), 278–305.

Kruer W. and Valeo E. et Al., (1975). Proc. 4th Int. Conf. on
Plasma Physics and Nuclear Fusion Research. Tokyo (1974).
Paper CN/33/F5-3. To be published IAEA Vienna.

Lawrence Livermore Laboratory Laser Program – Annual Reports
(1973) UCRL-50021-73; (1974) UCRL-50021-74

Lee T.N. (1974). Astrophys. Jnl. 190, 467–479.

McGill J. (1975). Ph.D. Thesis, (to be submitted), University of
Glasgow.

Nagel D. J. et al., (1974). Naval Research Laboratory
(Washington, D.C.) Report 7838.

Norton B.A. and Peacock N.J. (1975). J. Phys. B., Atom. Molec. Phys.
Vol. 8, No. 6, 989–996.

Nuckolls J., Wood L., Theissen A., Zimmerman G., (1972).
Nature 239, 139–142.

Peacock N.J. and Pease R.S. (1969). J. Phys. D. 2, 1705–

Peacock N.J., Hobby M.G. and Galanti M. (1973) J. Phys. B. 6, 298-304.

Spitzer L. Jnr., (1956). Physics of Fully Ionised Gases. Inter-science Inc., New York.

Vinogradov A.V. and Pustovalov V.V. (1971). Zh. E.T.F. Pis. Red. 13, 317-320.

Weisheit J., (1975) UCRL Report No. 76660, to be published J. Phys. B. (At. and Mol. Phys.).

ATOMIC OSCILLATOR STRENGTHS IN FUSION PLASMA RESEARCH[†]

W. L. Wiese and S. M. Younger

National Bureau of Standards

Washington, D.C.   20234

INTRODUCTION

Recent successes in fusion research with magnetically confined plasmas, especially the increase in confinement time for Tokamak-type devices, have made possible detailed spectral observations of the plasmas. These investigations have, among other things, revealed the presence of highly ionized, heavy element impurities, which are apparently released from the walls of the vessels due to intense bombardment by plasma particles. While light element impurities--particularly oxygen--are present in much larger numbers, it has now been realized [1-4] that heavy element impurities, even in very small concentrations, have a much more detrimental effect on the plasma behavior. This is primarily due to the enormous amounts of radiative energy that are released from the very highly ionized heavy atoms.

EFFECTS OF HEAVY ION IMPURITIES IN TOKAMAK PLASMAS

To provide some perspective on specific atomic data needs, a few relevant facts concerning a typical Tokamak are summarized below. One of the Tokamak devices which has been extensively investigated spectroscopically has been the ST Tokamak of the Princeton Plasma Physics Laboratory [1-3]. The experimentally determined composition for a typical ST Tokamak "hydrogen" plasma discharge is given in Table I. Presented are two quantities,

---

[†]Supported by Energy Research and Development Administration.

the number densities n_i and the densities multiplied by an
(effective) ionic charge Z_i (the sum $\sum_i n_i Z_i$ represents the
electron density). It may be inferred from these numbers that
oxygen is mostly in the form of its very high ions, i.e., with
(effectively) 7 electrons stripped off, while the metallic
impurities are as high as 20 times ionized.

Also presented in Table I are the radiative energy losses.
The contributions of oxygen and hydrogen are lumped together in
order to emphasize the very appreciable contribution of the heavy
metal impurities. Even though heavy metals constitute only a
minute fraction (0.2%) of the total plasma composition, they
contribute about 40% of the radiative losses, mainly in the form
of discrete lines [1].

Table I. Composition and radiation losses for a typical
ST Tokamak plasma [1].

Species	Number density $n_i (cm^{-3})$	% of total	Contribution to electron density[a] $(Z'_i n_i (cm^{-3}))$	Total power radiated[b]
H^+	22×10^{12}	96 %	22×10^{12}	30 kW
O	1×10^{12}	4 %	7×10^{12}	
Metals (Fe,Mo)	0.05×10^{12}	0.2%	1×10^{12}	20 kW

[a]Electron density = $\sum_i n_i Z'_i = 30 \times 10^{12} cm^{-3}$; ($Z'_i$ = ionic charge).

[b]Power input (steady phase) = 200 kW.

In addition to the radiative energy losses, there are other detrimental effects of heavy ion impurities which, according to Hinnov [3], become significant at impurity concentrations above 0.1%, comparable to the number quoted above. Furthermore, one has to consider the temporal and spatial distributions of the impurities in the plasma since heavy ion impurities tend to be drawn into the hot center [2] and will accumulate there, accentuating the above mentioned effects, unless a mechanism to continuously remove them is devised.

In view of these heavy ion impurity effects in Tokamak plasmas, it has become essential for the progress in this field to (a) thoroughly analyze the role of the impurities, i.e., identify them, measure their concentrations and their buildup in time and space, and quantitatively understand their effects, and then (b) find ways to control them to the required degree.

The most important prerequisite for heavy ion impurity analyses is a reliable atomic data base. Two of the principal types of data needed are: (a) atomic oscillator strengths (f-values) and (b) excitation rate coefficients. Both are needed to analyze the impurity problem quantitatively, i.e., to determine (or to model) the concentrations of the various species and their effects on the plasma. These two quantities are related through the widely used Seaton-Van Regemorter formula [5], i.e., excitation rate coefficients depend on the atomic oscillator strength.

With respect to the elements involved, the data required are for those materials which are used as first walls, including the limiters and divertors. Presently most favored are stainless steel (Cr, Fe, Ni), Ti, Zr, Nb, Mo, Ta, W, Re, Pt, and Au. The impurity atoms will become highly ionized as they diffuse into the center of the plasma where they have their greatest impact. Line identifications on the ST Tokamak have shown that metallic species become 20-30 times ionized, e.g., lithium-like iron has been observed, which is 23 times ionized. The spectra for the various ions are usually highly complex, but in certain ionization stages--for example, when the ions are members of the alkali or alkaline earth isoelectronic sequences--the spectra become rather simple. These spectra are also important for another reason: the excitation rates for the principal resonance lines are very large due to the relatively low excitation energies of the first excited states and the fairly large oscillator strengths. Thus, in the regime of coronal equilibrium, these transitions give very large contributions to the radiative energy losses. The atomic data work therefore needs to be concentrated first on such species and their principal lines. A

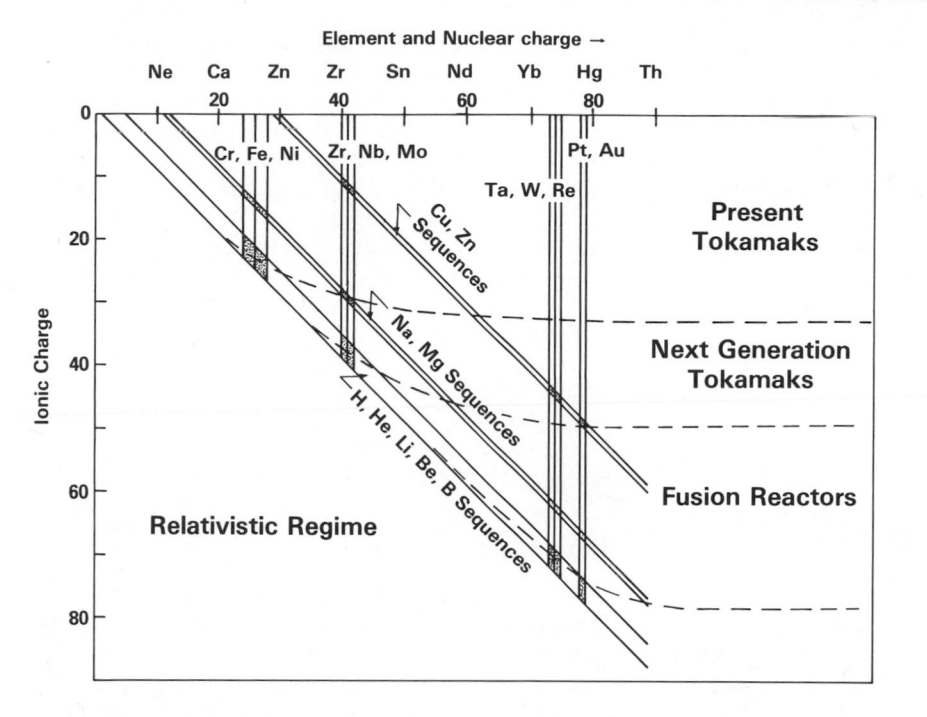

FIG. 1.   Graphical overview of principal atomic data needs for
fusion research.   Elements which are used or are under consideration as first wall materials are given as vertical lines.
The isoelectronic sequences shown (diagonal lines) are for simple
atomic systems which, due to low-lying resonance transitions,
give large contributions to radiative energy losses.

good orientation on what specifically is needed may be obtained
from Fig. 1.   The elements are plotted in order of their nuclear
charge versus the stage of ionization.   The elements identified
earlier as wall materials are shown as the vertical lines.
Isoelectronic sequences are represented by diagonal lines.   The
diagonals shown represent the sequences for the simple atomic
systems, mostly alkalis and alkaline earths.   The lowest diagonal
line is for the hydrogen sequence, beyond which atoms are fully
stripped.   The intersections between the vertical and diagonal
lines identify the species for which atomic data are most urgently
needed.

To set priorities it is useful to distinguish between
presently operating Tokamaks, next generation machines (which
will become operational very soon), and ultimately the projected
fusion reactors.   The dashed lines on the right side of Fig. 1

indicate the ranges of ions which are estimated to be present in
the different devices. The immediate data needs concern the
present and next generation Tokamaks, where Cr, Fe, and Ni are
expected to be fully ionized. The heavier element wall materials
will be in various intermediate stages of ionization, with Zr,
Nb, and Mo expected to be close to the fully ionized state in
next generation devices. The heaviest materials under consideration (W, Ta) are estimated to reach the lithium-like state only
in a full size fusion reactor.

METHODS FOR THE DETERMINATION OF OSCILLATOR STRENGTHS OF HIGHLY
IONIZED SPECTRA

     For most of the important species singled out in Fig. 1,
little or no reliable atomic data exist as yet. This is particularly true for the atomic oscillator strengths with which we
shall be exclusively concerned. (The situation for the wavelengths and atomic energy levels is only slightly better.)

     In considering the question of which approaches are most
promising for obtaining the relevant oscillator strengths,
theoretical methods appear to have the greatest potential. Some
of these are indeed quite powerful for determining efficiently
many oscillator strengths of the higher ions, as discussed by
A. Weiss at this conference [6]. Calculations may generally be
performed for selected transitions along an entire isoelectronic
sequence, since this is readily accomplished by proper scaling
of energies and distances with the nuclear charge Z. Configuration interaction effects, perturbations by interacting terms,
and cancellation in the transition integral, which are often
very pronounced for prominent transitions of neutral or weakly
ionized atoms, gradually fade away along the sequence as the
ionic charge increases, and the spectrum settles down to a more
orderly fashion, arranged primarily according to the principal
quantum numbers. The process of rearrangement is rather slow
for more complex atomic structures, however, and accidental
level crossings may occur far into the isoelectronic sequence.
Also, effects such as core polarization appear in certain cases
to remain strong along an isoelectronic sequence. The principal
problem encountered in the theoretical approaches, however, is
the appearance of increasing relativistic effects for higher Z
ions. As seen from A. Weiss' discussion, the subject of relativistic corrections is just now starting to receive some attention and is still largely unexplored for more complex atomic
systems.

     Thus, the problems in determining f-values for highly
ionized atoms by theoretical methods are sufficiently serious to
make it necessary to find other independent approaches. Unfor-

tunately, there is not much hope at present on the experimental
side to determine such f-values directly. Beam-foil spectroscopy
is the most promising method, but rough estimates according to
a formula by Nikolaev and Dmitriev [7] show that the beam
energies needed for the excitation of very highly ionized species
are extremely high and are for many of the needed species (such
as indicated in Fig. 1) essentially beyond the range of the
method. Furthermore, an additional and very difficult problem
arises insofar as the lifetimes associated with most transitions
in very highly ionized species are so short that extremely good
spatial resolution is required, which must be far better than
presently achieved values. The important "in-shell" transitions
(i.e., no change in the principal quantum number) from the
ground state are, however, an exception. Since the transition
energies remain low, the lifetimes of these important lines
should remain relatively long.

     Beam-foil spectroscopy is nevertheless very useful in an
indirect way when it is utilized in the analysis of systematic
trends of f-values along isoelectronic sequences [8,9]. Such
analyses, which should include any additional (especially
theoretically determined) data, appear to be a very promising
approach, which can provide the needed data for very high ions
by extrapolation or interpolation and also assure a high degree
of reliability. Furthermore, discrepancies and inconsistencies
in the data along the isoelectronic sequence become immediately
apparent, and reasons for this may often be analyzed. We shall
now apply this combination of approaches to the case of two
important transitions in the Na and Cu sequences and show that
it leads to fairly reliable determinations of much needed f-
value data for very high ions.

     (a) The principal resonance line of the sodium sequence.
Figure 2 shows the present f-value data situation for the 3s-3p
resonance transition of the sodium isoelectronic sequence. The
oscillator strength f is plotted against the inverse nuclear
charge, Z^{-1}. The data available for this transition are so
numerous, especially for the neutral sodium atom, that only the
most recent and advanced material has been presented. The
experimental data are exclusively from lifetime measurements,
which must generally be considered as being very accurate. For
the neutral sodium atom, the data are from the Hanle effect-,
phase shift-, and delayed coincidence-techniques, while data for
the ions up to Ar VIII are from the beam-foil spectroscopy
technique. On the theoretical side, the data are restricted to
very recent calculations, some of which are explicitly concerned
with the relativistic regime, i.e., the high Z end of the sequence. Because of space limitations, only references to papers
essential to the following discussion are given. All other

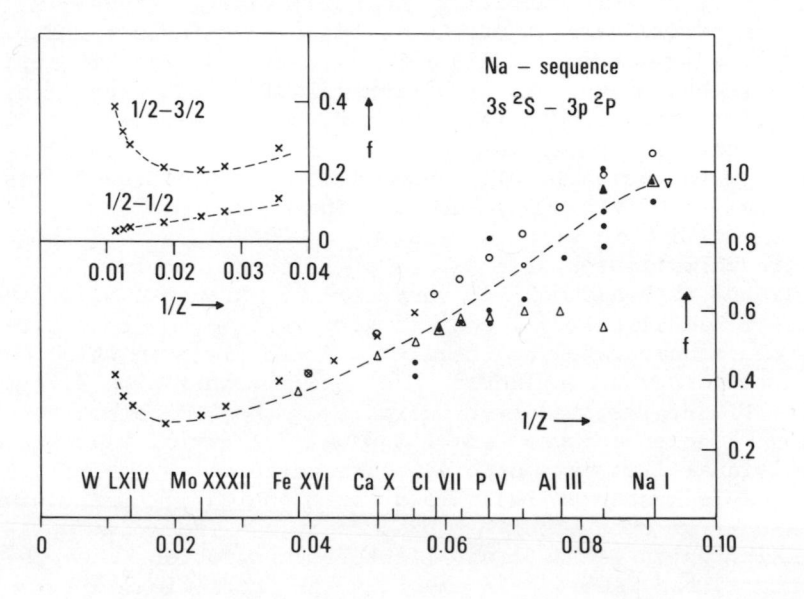

FIG. 2. Systematic trend for the resonance transition $3s\ ^2S$ -
$3p\ ^2P$ of the Na sequence. ▲ weighted average of lifetime
data; ▲ phase shift experiment; ● beam-foil data; O SCF
calculation [11]; △ nuclear charge expansion method with core
polarization [12]; ▽ SCF with core polarization [13]; ✕
relativistic SCF [14]. In the insert, the f-value trend for the
two component lines $3s_{1/2}$-$3p_{1/2}$ and $3s_{1/2}$-$3p_{3/2}$ is shown for the
relativistic regime of very high Z ions, as calculated by
Desclaux and Kim [14].

references are found in the NBS Bibliography on Atomic Transition
Probabilities [10].

The principal problem of interest on this graph is the
almost constant discrepancy, along the sequence, of Biemont's
self-consistent field (SCF) calculations [11] (and other similar
theoretical results which are not listed) with recent experimental lifetime data, especially from beam-foil experiments.
For this very strong transition, large cascading effects in the
lifetime experiments seem unlikely, but such effects may still
contribute somewhat to the discrepancy. In addition, it appears
that the recent advanced Z-expansion calculation by Laughlin et
al. [12], in which core polarization effects are included,
provides part of the explanation. Their results, which agree
very well with the beam-foil measurements for the higher ions--
as well as Weisheit's SCF value [13] for neutral sodium where
core polarization is also taken into account--appear to indicate
that core polarization effects are significant far into the
sequence and decrease only at very high Z. Around Z = 30, the
results of Laughlin et al. connect very well to the relativistic
SCF f-values calculated by Desclaux and Kim [14], in which core
polarization is not considered. In the extremely high Z,
relativistic regime, the f-value ratio of the two components of
the 3s-3p doublet changes from its usual 2:1 ratio, and, therefore,
the individual line component data are given as an insert to
Fig. 2. This graph readily allows the graphical determination
of f-values for such highly ionized species as Fe XVI, Mo XXXII,
and W LXIV, which are of central interest in fusion research.
Furthermore, the recent refinement in the theoretical data
situation due to the inclusion of core polarization effects
permits a better understanding of the less well-developed situation
for the copper sequence resonance transition, which shall now be
discussed.

(b) The principal resonance line of the copper sequence.
Figure 3 shows a plot of f versus 1/Z for the 4s-4p multiplet of
the Cu sequence. Immediately apparent from this graph is a
systematic discrepancy of about 30-40% between theoretically and
experimentally determined data. An analysis of the possible
causes for this discrepancy was undertaken and centered on two
principal possibilities. For the experimental data, the problem
of cascading was investigated since all data are obtained from
lifetime techniques, and, for the higher ions, exclusively by
beam-foil spectroscopy. The principal cascading to the 4p state
occurs from the 4d state, with a lifetime roughly half that of
the 4p state. A "growing-in" of the decay curves due to this
fast cascade was indeed observed by Sørensen [15], and his
spatial resolution of 0.2 mm should be fully adequate to account
for this cascade. Livingston's recent beam-foil measurement for
Kr VIII [16], which also includes cascade corrections, is an

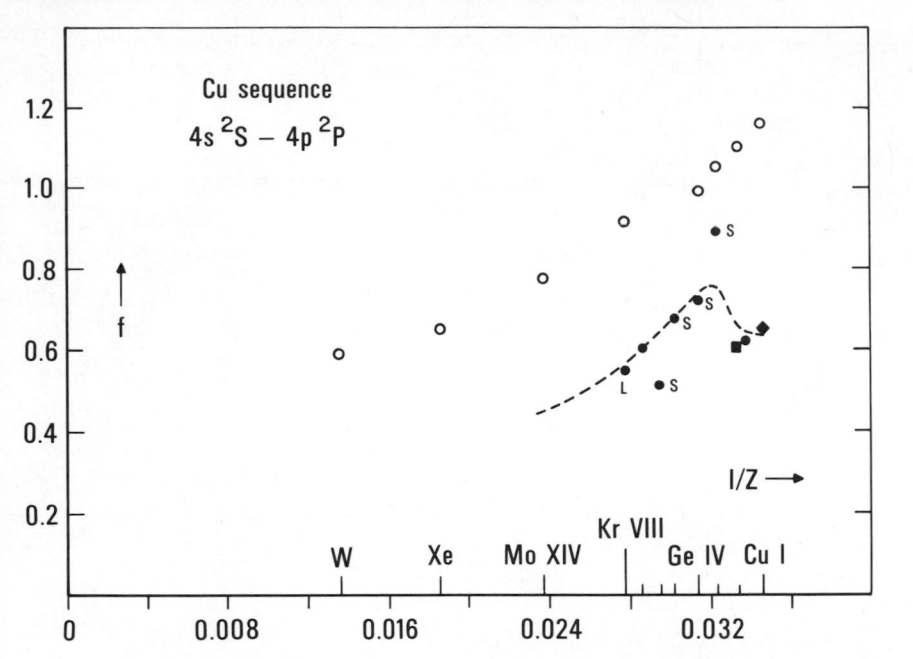

FIG. 3. Systematic trend for the resonance transition $4s\ ^2S$ – $4p\ ^2P$ of the Cu isoelectronic sequence. ◆ critical compilation [18]; ○ SCF calculation with relativistic corrections [17]; ● beam-foil data: L = Ref. [16]; S = Ref. [15].

independent check of the trend established by Sørensen. We also considered energy losses in the foil but concluded that, at the incident energy of 350 kV in Sørensen's experiment, even significant uncertainties in the energy loss result only in small changes in the lifetimes.

It appears unlikely that the beam-foil data are responsible for such a large experimental-theoretical discrepancy. Turning to the theoretical data, Fig. 3 shows that the self-consistent field calculations of Weiss [17], with one-electron relativistic corrections and spin-orbit coupling included, produce a curve generally similar to the beam-foil data curve but offset by about a factor of 1.4. At the lowest Z values where core polarization is estimated to be strong, and at high Z, where relativistic effects begin to dominate, the two curves (broken lines) are different in shape. The discrepancy between theory and beam-foil results for intermediate values of Z (Ga III – Kr VIII) could be in large part due to the effect of core polarization, which was not considered by Weiss. A similar, less pronounced phenomenon was noted above for the Na sequence. The increase in magnitude of the effect for the Cu sequence appears

very likely in view of the fact that the copper-like core contains ten 3d electrons, whereas the sodium-like core contains
only six 2p electrons. Also, the 4s electron for the Cu sequence
is somewhat closer to the $3d^{10}$ core than the 3s electron for the
Na sequence is to the $2p^6$ core. It is thus conceivable that
copper-like ions display considerable polarization effects even
at moderately high Z.

Based on the above considerations, an admittedly very
tentative systematic trend for the Cu sequence resonance multiplet
may be established, which is shown as the broken line in Fig. 3.
We essentially follow the experimental data very closely and
indicate that for high Z, beyond about Mo XIV, relativistic
effects should start to become noticeable. Refined theoretical
calculations which include core polarization as well as configuration interaction effects would be very helpful to better
establish this important systematic trend.

REFERENCES

[1] E. Hinnov, Princeton Plasma Phys. Lab. Rpt. MATT-1024
     (1974).
[2] E. Hinnov, Princeton Plasma Phys. Lab. Rpt. MATT-1022
     (1974).
[3] D. M. Meade, Nucl. Fusion 14, 289 (1974).
[4] N. Bretz, D. L. Dimock, E. Hinnov, and E. B. Meservey,
     Nucl. Fusion 15, 313 (1975).
[5] R. C. Elton, in Methods of Experimental Physics (Academic
     Press, New York, 1970), H. R. Griem and R. H. Lovberg,
     eds., Vol. 9, p. 135.
[6] A. W. Weiss, Proceedings of this conference.
[7] V. S. Nikolaev and I. S. Dmitriev, Phys. Lett. A 28, 277
     (1968).
[8] W. L. Wiese and A. W. Weiss, Phys. Rev. 175, 50 (1968).
[9] M. W. Smith and W. L. Wiese, Astrophys. J. Suppl. 23,
     No. 196, 103 (1971).
[10] Bibliography on Atomic Transition Probabilities, NBS
     Spec. Publ. 320 (1970); Suppl. 1 (1971); Suppl. 2 (1973)
     (U.S. Government Printing Office, Washington, D.C.).
[11] E. Biemont, J. Quant. Spectrosc. Radiat. Transfer 15,
     531 (1975).
[12] C. Laughlin, M. N. Lewis, and Z. J. Horak, Astrophys.
     J. 197, 799 (1975).
[13] J. C. Weisheit, Phys. Rev. A 5, 1621 (1972).
[14] J. P. Desclaux and Y.-K. Kim, private communication.
[15] G. Sørensen, Phys. Rev. A 7, 85 (1973).
[16] A. E. Livingston, Thesis, University of Alberta, Canada (1974).
[17] A. W. Weiss, private communication.
[18] T. M. Bieniewski and T. K. Krueger, Aerospace Res. Lab.
     Rpt. ARL 71-0135, Project No. 7114 (1974).

PLASMA AND PROJECTILE STRIPPING: A COMPARISON

D. J. Nagel

Naval Research Laboratory

Washington, D. C. 20375

ABSTRACT

High ionization stages result when fast electrons strip slow ions in plasmas and when slow electrons strip fast projectiles inside targets. This reciprocity was examined quantitatively by comparing ionization states computed from the plasma coronal equilibrium model with experimental projectile charge states. At equal relative electron-ion velocities, substantially higher charge states are predicted by the plasma model compared to the ion beam data. However, use of a lower plasma temperature, which brings the plasma and ion beam ionization curves into agreement, is justifiable. As a rule of thumb, plasma and beam-gas experiments produce comparable stripping when the plasma electron temperature (eV) is 90 times the projectile energy (MeV/nucleon). The plasma-projectile reciprocity indicates conditions for maximization of resonance radiation in suggested beam-gas experiments. A further comparison was made between ionization stages observed to date in plasmas and in ion beams. Presently, plasma excitation has produced consistently higher stripping for heavy elements. However, it should be possible to produce similar ionization states with planned accelerators. Whether or not plasma excitation will remain ahead in this "ionization race" is contingent on success in attainment of higher temperatures and longer plasma lifetimes.

I. INTRODUCTION

Because the relative (center-of-mass) velocity between ambient electrons and ions can be comparable in plasma and ion beam experiments (1), there are many points of similarity in the

stripping and excitation of atomic electrons in the two cases (2).
Several of the mechanisms, the excitation states, and the spectra
which result are similar. For example, collisional ionization
(charge loss) and recombination (charge pickup and radiative
electron capture) are processes common both to ions in plasmas
and to ions moving through targets.

Similarities and points of contrast between plasma and projectile excitation were discussed qualitatively in an earlier
article (2). The present paper has two primary parts. The first
(section II) is a more quantitative comparison of atomic stripping
in plasma and ion beam experiments. A plasma equilibrium
model is used to compare ionization stages resulting from plasma
and beam-gas interactions. The result is a simple, approximate
relationship between plasma and ion-beam experiments. The
plasma-projectile reciprocity which is exhibited suggests that
resonance radiation may be maximum when the projectile energy
is suitably related to atomic binding energies. In the second
major part (section III), ionization stages attainable in present
and planned plasma generators and ion accelerators are discussed.

II. ESTIMATION OF CHARGE STATES

Study of ionization states for plasma and projectile experiments
requires a combination of theory and experiment, with due attention
to time dependence. We conveniently begin this section with a discussion of equilibrium charge states (averages and distributions),
first for plasmas and then for ion beams (II.A). Next the reciprocity between plasma and projectile stripping is examined (II.B).
Finally, density-dependent temporal effects are considered for
each method of excitation (II.C).

A. Equilibrium Conditions

The charge state distribution for a plasma in or near equilibrium depends most strongly on the plasma electron temperature
(T_e). For high-temperature plasmas ($T_e > 10^5 K$), it is uncommon
to have independent determinations of temperature and charge state
in the same experiment. Furthermore, the short lifetimes
(frequently submicrosecond) of very hot plasmas usually restrict
temperature determinations to time-integrated, "average" T_e
values. In most cases, plasma theory is used to compute a relationship between T_e and the average charge state. This is used to estimate T_e from observed charge states which are gotten from
analysis of time-integrated spectra.

A variety of plasma equilibrium models has been developed
for use in different density regimes (3). We will mention two
models that bracket most of those in use. The first is the coronal
model which, although it is not density dependent, applies for

relatively low densities. In the coronal model, the ionization
state distribution is determined by a balance between collisional
excitation (ionization) and radiative deexcitation (recombination).
The second model is the LTE (local thermodynamic equilibrium)
picture. It is applicable for high densities which cause collisional
(rather than radiative) deexcitation. Radiation is still emitted in
the LTE model but it does not significantly influence the charge
state distribution.

Although a proper treatment of ionization in plasma requires
the solution of complex rate equations, the coronal equilibrium
model has been found useful for approximate temperature determinations. The LTE model is somewhat more complex and has
not been as widely used for very high T_e plasmas. The relations
between average charge state Q and T_e computed from the coronal
model (4) are given in figure 1 for iodine. Also plotted in this figure is projectile data which will be discussed later. In general, Q
from the coronal model increases monotonically with T_e, with un-
dulations in the curve arising from shell effects. A Q-T_e curve for
the LTE model lies above that for the coronal model in figure 1 (5).

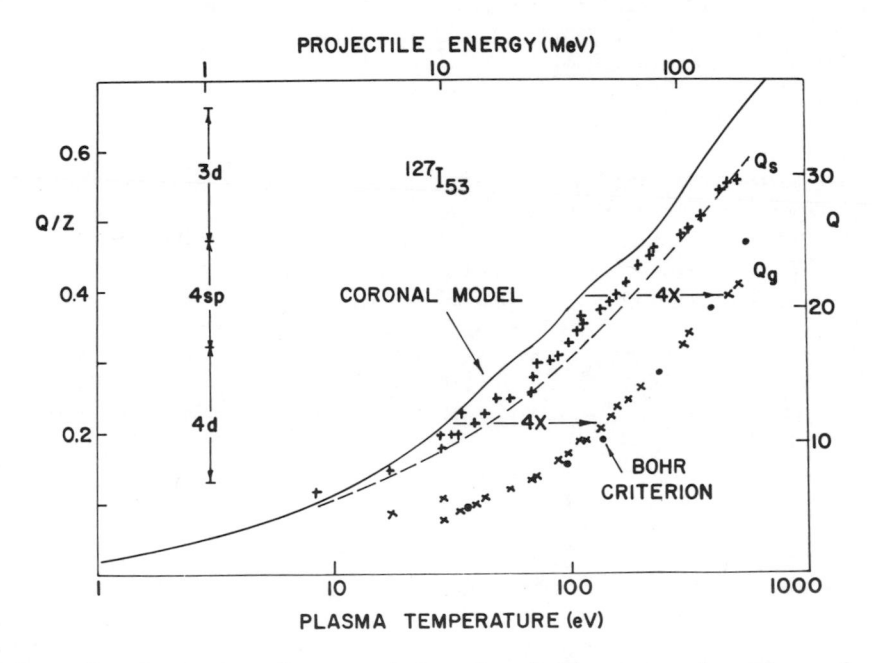

Figure 1. Average charge states for iodine as a function of
plasma electron temperature computed from the coronal equilibrium model (——) and as a function of projectile energy. The projectile data includes observed values for foil (+) and gas (x)
targets and values computed from the Nikolaev-Dimitriev equation
(---) for solid targets and from the Bohr criterion (•). The arrows
indicate a factor of 4 difference between the coronal curve and Q_g.

The situation for projectile charge state information is the re-
verse of that for plasmas. It is rather complex to compute the
charge distribution for a projectile traveling through a target.
More charge-changing mechanisms are active because of projectile-atom interactions and the needed cross sections are often
inaccurate or unavailable. Hence, projectile charge state information largely consists of experimental values and semiemperical
equations (6, 7).

Projectile charge states mainly depend on the beam energy (E)
and the target state (solid or gas). Observed values for iodine are
plotted in figure 1 for both foil (Q_s) and gas (Q_g) targets, along
with Q values gotten from the Nikolaev-Dimitriev (8) relation for
solid targets and from the Bohr criterion* (9). The projectile
charge data exhibits the well-known characteristics, namely (a)
monotonic increase with ion energy, (b) Q_s is greater than Q_g at the
same ion energy, (c) semiempirical equations fit the observations
quite well and (d) the Bohr criterion gives Q values close to Q_g.

A comparison of the plasma and projectile equilibrium charge
distributions is made in figure 2 for iodine for two T_e-E combinations. The T_e-E relationship used in both figures 1 and 2 is
derived and discussed below.

B.   Plasma-Projectile Reciprocity

The similarity in terms of relative electron-ion velocity between slow ions in a plasma of fast electrons and a fast projectile
in a target containing relatively slow electrons prompts a more
quantitative examination of charge states attained in the two
methods of excitation. A simple way to obtain a relationship between the two is to equate the (average) plasma electron velocity
to the projectile velocity (neglecting both plasma ion motion and
target electron motion). For the plasma, $M_e V_e^2/2 = 3 kT_e/2$ and
in the target $M_p V_p^2/2 = E$. Equating V_e and V_p gives

$$T_e \text{ (eV)} = \quad 363 \; \frac{E}{A} \left(\frac{MeV}{nuc.} \right) \tag{1}$$

This equation was used to relate the abscissa scales in figure 1
and to compute the T_e corresponding to the projectile energy
values in figure 2.

---

*The Bohr criterion states that all electrons with orbital velocities
less than the projectile velocity (or equivalently, with binding
energies less than one-half the electron mass times the square of
the projectile velocity) are removed from the projectile.

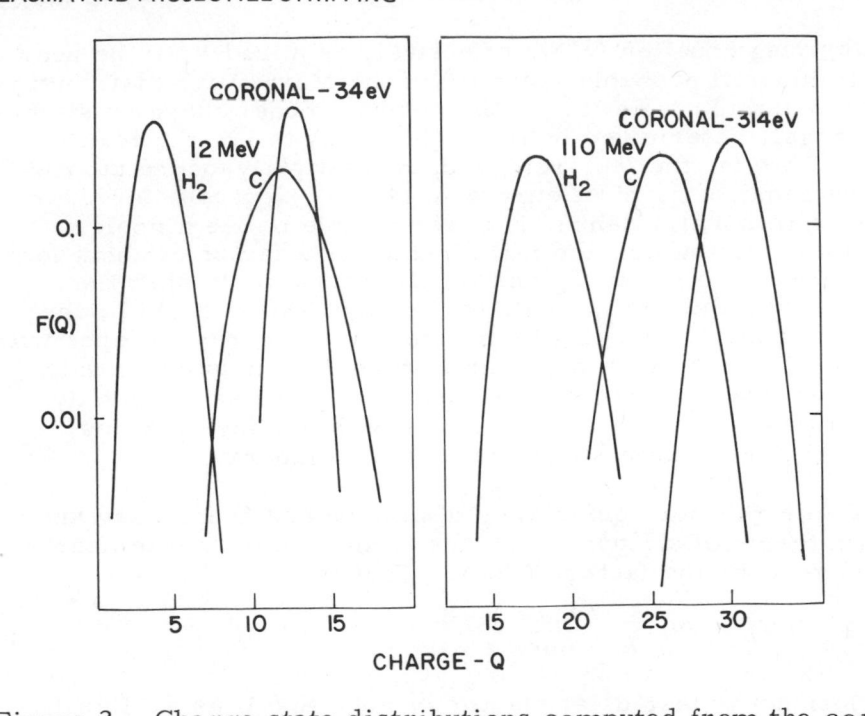

Figure 2. Charge state distributions computed from the coronal
model at two temperatures related to projectile energies used in
ion-target experiments. The shapes (widths) of the coronal
model distributions can be compared with those observed for the
gas and solid targets.

We can now reexamine figures 1 and 2 to judge the utility of
equating V_e and V_p to compare Q values for plasmas and ion
beams. The coronal model Q (continuous line in figure 1) is substantially greater than both Q_s (+) and Q_g (x). Agreement between
the plasma charge and Q_s improves with atomic number, being
poor for S, fair for I (figure 1) and good for U. However, the
Betz-Grodzins picture (6) attributes a substantial fraction of Q_s
(e.g., most of the excess over Q_g) to autoionization after exit of
the beam foil. Hence, we should compare the coronal model Q
values with Q_g. Figure 1 shows that the coronal curve is shifted
to lower T_e and E values than Q_g by a consistent factor of four.
That is, we would have gotten a good match between the coronal
and Q_g curves (which are reasonably similar in shape) if we had
taken a lower temperature in computing T_e. Also, use of a T_e
four times lower than those in figure 2 would bring the coronal
curve nearly into correspondence with the H_2 data.

In deriving equation (1) we effectively required V_p to be near the peak (most of probable velocity) of the Maxwellian distribution associated with T_e. However, the coronal model charge distribution is largely determined by the high energy tail of the Maxwellian. That is, the last-removed, most-tightly-bound atomic electrons require the most energetic plasma electrons for their collisional removal. Hence, it is reasonable to use a cooler Maxwellian distribution, the tail of which is a factor of about four above the peak, to relate T_e and E. Doing so would shift the coronal curves and the T_e scale to the right as indicated by the arrows in figure 1, bringing them into closer correspondence with the Q_g curves. The factor of four is roughly consistent with the observation that resonant x-ray lines from plasmas commonly have energies (i.e., arise from levels with binding energies) around a factor of three above the plasma temperature.

A simple rule for equivalence of plasma and beam-gas experiments (in terms of stripping) follows from dividing the temperature in equation (1) by the factor of four. That is,

$$T_e \text{ (eV)} \simeq 90 \ \frac{E}{A} \left(\frac{MeV}{nuc.} \right) \tag{2}$$

This relation was tested satisfactorily for S and U as well as I.

We note, in passing, that the Bohr criterion points follow the coronal curve in shape and agree well with Q_g (for the heavier elements). This is nice since the Bohr criterion is easy to apply: an outer electron is removed when $M_e V_p^2/2$ is equal to its binding energy as tabulated by Carlson et al. (10).

This discussion of reciprocity has ignored ion motion in plasmas. If the ions and electrons are equilibrated, $V_{ion} = V_{el}/\sqrt{1836 \ A}$ and is negligible. Target electron motion was ignored also. Velocities of conduction electrons ($< 2 \times 10^8$ cm/sec) are low compared to V_p for most cases of interest in charge state experiments. Core electrons have substantial velocities, but they are tightly bound and do not remove projectile electrons in the same manner as the outer (bonding) electrons of the target. However, projectile-atom interactions, such as molecular orbital (MO) formation (11), result in core level ionization which leads to Auger cascades, raising the charge state of projectiles. Hence, the reciprocity between plasma and projectile stripping is flawed in principle even though we have a relationship between the two methods of electron removal.

An additional point of comparison between plasma and projectile ionization and spectroscopy has arisen lately: oscillations as a function of atomic number Z are observed in both cases. For plasmas, strong resonance radiation was obtained from elements

which have electrons bound in some shell with energies related to the plasma temperature (12) (i.e., about a factor of about three greater than T_e, as already noted). Hence peaks occur at those Z for which the K, L or M binding energies (10) are about three times T_e. In the ion beam case, peaks in ionization cross section and x-ray intensity occur when projectile and target atom levels match, leading to MO formation and efficient electron promotion (11). The point is that, although peaks in x-ray intensity as a function of Z are found in both plasma and ion beam work, the mechanisms causing the observed variation is different. Nonetheless, the plasma-projectile reciprocity discussed here leads to the suggestion that resonance radiation peaks should also occur as a function of projectile Z_1 in beam experiments. For a fixed velocity V_p (equivalent to fixed T_e), stripping to closed shell configurations (He-like for 1s, Ne-like for 2p, Ni-like for 3d, ...) should occur for particular Z_1, leading to peaks in the intensity due to outer-shell, resonant transitions as a function of Z_1. Maxima should appear for projectiles with binding energies $\mathcal{E}$ related to E by

$$\frac{E}{A} \left(\frac{MeV}{nuc.} \right) = 1.836 \times 10^{-3} \, \mathcal{E} \, (eV) \tag{3}$$

Peaking is also expected for fixed Z_1 as E is varied. A gas target would have to be used to reduce collisional depopulation of the excited states and to permit observation of the soft (~ 1 keV) radiation (i.e., to avoid self absorption). As noted in reference 2, very similar Rydberg series spectra can occur in both plasma and beam foil experiments. The Z_1 and velocity-dependent peaking in intensity due to outer-shell transitions which is anticipated here is similar to the Z and temperature-dependent peaking in plasma radiation. It is not analogous to the level-matching oscillations in intensity due to inner-shell transitions.

C. Density and Temporal (Non-Equilibrium) Effects

While equilibrium models are useful in many instances, the short lifetimes of very hot plasmas require attention to transient effects. For example, less than a nanosecond is available for ionization in plasmas produced by high-power pulsed lasers. It is necessary to consider whether or not there is time to achieve the equilibrium ionization state. An estimate of the ionization time τ (in sec) can be computed for a plasma of electron density n (in cm^{-3}) containing atomic number Z (13):

$$n\tau = \frac{4 \times 10^6 \, Z^2 \, \mathcal{E} \, \exp (\mathcal{E}/T_e)}{J \sqrt{T_e}} \tag{4}$$

Here, the ionization energy $\mathcal{E}$ of the last-removed electron and the plasma electron temperature T_e are in eV. J is the number of equivalent electrons in the outer shell prior to the last ionization step. Equation (4) includes only collisional ionization and not recombination. Hence, it gives a lower bound on the ionization time. For example, the coronal model Q for Au at a T_e of 1000 eV is 49. Equation (4) yields $n\tau \sim 10^{13}$ cm^{-3} sec in this case (14). For a Nd laser, for which $n \lesssim 10^{21}$ cm^{-3}, a time of 10^{-8} sec is needed to achieve the equilibrium ionization state, but a laser pulse length $\lesssim 10^{-9}$ sec must be used to obtain $T_e \sim 1000$ eV with present lasers. Hence, equilibrium ionization cannot be achieved in this plasma situation.

Density and non-equilibrium effects in ion beam experiments have received extensive study also (5). As measurements are extended to higher beam energies, higher Q values will result if there is time available to achieve them. Again using Au as an example, equation (4) requires about 10^{-10} sec for equilibrium at solid density. Using 10 MeV/nucleon, corresponding to $T_e =$ 1000 eV, figure 3 shows that only about 10^{-12} sec is available inside a carbon-foil as thick as 10 mg/cm^2. Dropping the use of equation (4), a less pessimistic estimate of attainable stripping is gotten if charge loss cross sections are taken equal to orbital areas. However, the question remains: will limits on available ionization time restrict achievable charge states in beam as well as in plasma experiments?

III. ATTAINABLE CHARGE STATES - PRESENT AND FUTURE

There are several reasons for studying high ionization states. Atomic spectroscopy and potential ion source development are among the more compelling. Hence, we inquire about which method of excitation, plasma or ion beam, is yielding higher charge states now and how this status will change in the forseeable future. To do so, we set aside the question concerning the time available for equilibration which was just discussed. It remains important, but it makes the present discussion more complex.

Present and projected charge states are given in figure 4. Plasmas with (actual or apparent) temperatures of 1 keV can be routinely produced now so a coronal model curve is shown for this temperature. The plasma data in figure 4 is from laser-produced plasmas (13-19) and exploded wire plasmas (20). If controlled thermonuclear fusion is achieved on the planned scale (i.e., with significant burn), T_e values of around 10 keV will result. Hence, a coronal curve for this temperature is also shown in figure 4.

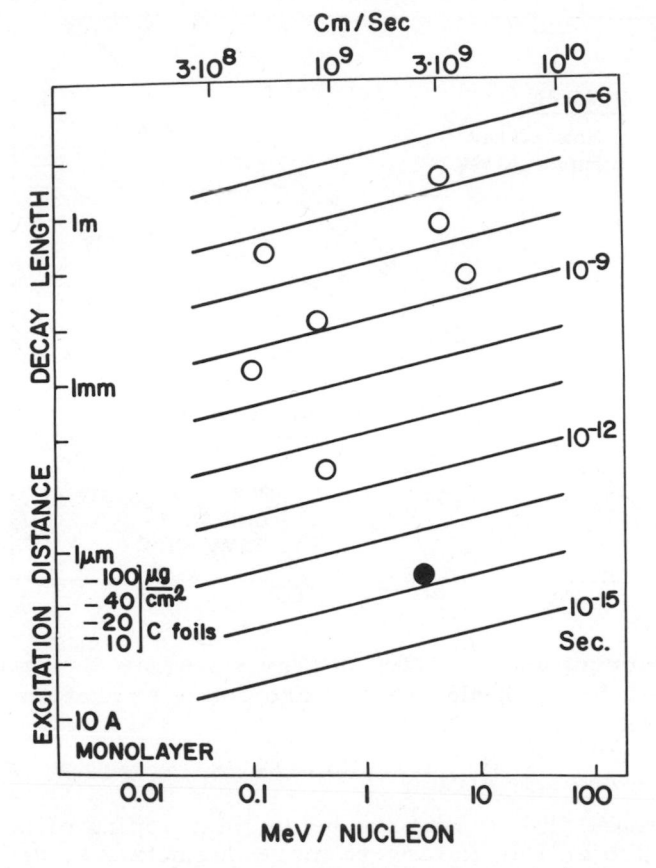

Figure 3. Beam foil diagram giving excitation and decay distances
as a function of projectile energy (velocity) and the excitation and
decay time. The circles indicate a few of the many lifetime experiments which have been reported. In the new method (15) of
determining lifetimes by varying the foil thickness (solid circle),
the excitation and decay time scales are comparable.

The highest projectile ionization states given in a recent review
(7), plus recently-reported work with Kr^{+34} (21), are plotted in
figure 4. Expected ion charge states are also plotted for two new
accelerators (UNILAC and the 25 MV tandem-ORIC combination)
using the Nikolaev-Dimitriev (8) formula.

Examination of figure 4 shows that plasma excitation is presently producing much higher Q values than accelerator work for Z
above about 40. However, the new beam machines should cause ion-
beam results to surpass the present plasma data. If so, only success
in planned plasma heating and confinement experiments will keep the
plasma spectroscopists "in the lead". But even in this case, the
plasma-projectile gap may not be so wide as it is now.

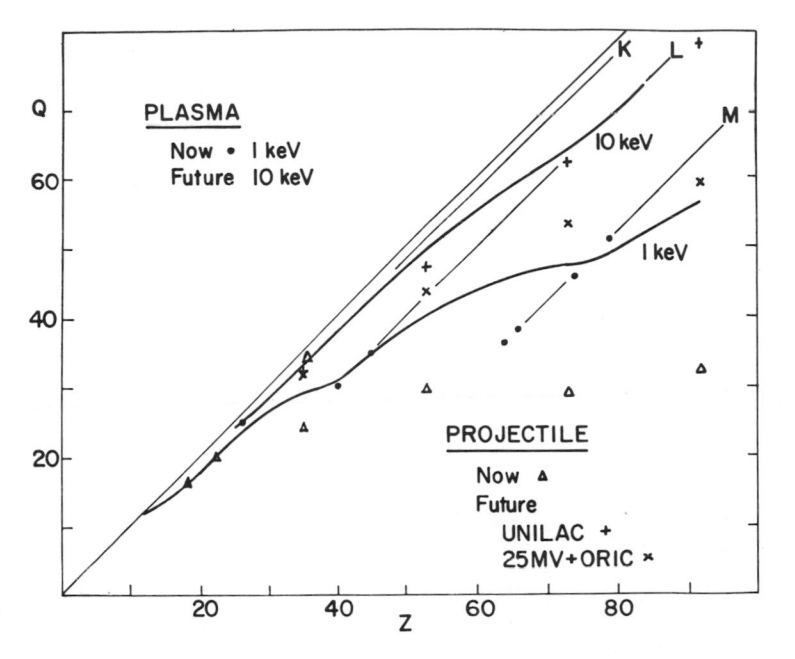

Figure 4. Current and expected charge states as a function of atomic number for both plasma and projectile excitation.

IV. DISCUSSION

We have made the reciprocity between stripping of ions in plasmas and projectiles in targets more quantitative, and have reviewed what is and will be possible with the two means of excitation. A few points of comparison remain.

It seems worthwhile to note that the rate equation (non-equilibrium) approach to projectile stripping will be used increasingly, as in plasma physics where it is used routinely. Doing so requires abandoning the fluorescence yield viewpoint in favor of rate coefficients and solving many simultaneous equations. However, it is both necessary and possible.

Finally, we pause to consider the interaction between energetic ions and plasmas. Of course, the bonding electrons in a solid target are a plasma, and their many-body response to passage of a projectile has been studied (22). But, ion interactions with hot plasmas are also of interest. Consider the interaction between the solar wind ions and the earth's magnetosphere. In the laboratory, ion beams are being used to probe (diagnose) and to heat plasmas. Recently, consideration has been given to using ion beams to compress and heat dense fusionable material to the ignition temperature (23).

The cases considered in this paper are really end points in a spectrum of possibilities, namely hot electrons with a Maxwellian distribution interacting with stationary ions and monoenergetic, monodirectional projectiles interacting with cold targets. Future work will involve mixed plasma and ion velocity distributions for which the absolute plasma electron velocities and absolute projectile velocity are comparable.

ACKNOWLEDGEMENTS

D. Mosher is thanked for the use of his coronal model program and a helpful conversation. This work was supported by the Office of Naval Research and the Energy Research and Development Administration.

REFERENCES

1. J. Bromander, Nucl. Inst. & Methods, 110, 11 (1973).
2. D. J. Nagel in S. Datz et al. (Eds), Atomic Collisions in Solids, Plenum (1975), p. 433.
3. R. P. McWherter in R. H. Huddlestone and S. L. Leonard (Eds.), Plasma Diagnostic Techniques, Academic Press (1965), p. 201.
4. D. M. Mosher, Phys. Rev. A10, 2330 (1974).
5. R. D. Cowan, private communication.
6. H. D. Betz, Rev. Mod. Phys. 44, 465 (1972).
7. A. B. Wittkower and H. D. Betz, Atomic Data 5, 113 (1973).
8. V. S. Nikolaev and I. S. Dimitriev, Phys. Lett. 28A, 277 (1968).
9. N. Bohr and J. Lindhard, Kgl. Dan. Videnskab. Selskab, Mat.-Fys. Medd. 28, No. 7 (1954).
10. T. A. Carlson et al., Atomic Data 2, 63 (1970).
11. J. D. Garcia, R. W. Fortner and T. M. Kavanagh, Rev. Mod. Phys. 45, 111 (1973).
12. D. J. Nagel et al., Bull. Am. Phys. Soc. 19, 557 (1974).
13. H. Griem, Plasma Spectroscopy, McGraw-Hill (1964), p. 292.
14. D. J. Nagel and R. E. Latham, Bull. Am. Phys. Soc. 20, 1318 (1975).
15. H. D. Betz et al., Phys. Rev. Lett. 33, 807 (1974).
16. G. A. Doschek, U. Feldman and P. G. Burkhalter, this conference.
17. P. G. Burkhalter et al., Phys. Rev. A9, 2331 (1974).
18. P. G. Burkhalter et al., Phys. Rev. A11, 782 (1975).
19. E. V. Aglitsky et al., Kvantovaya Elektron, 1, 2067   (1974).
20. C. M. Dozier et al., Bull. Am. Phys. Soc. 20, 1303 (1975).
21. R. Marrus and H. Gould, Bull. Am. Phys. Soc. 20, 818 (1975).
22. V. V. Neelavathe et al., Phys. Rev. Lett. 33, 302 (1974).
23. F. Winterberg, Pl. Phys. 17, 69 (1975).

SPECTROSCOPY OF PLASMAS FOR SHORT WAVELENGTH LASERS[*]

R. C. Elton and R. H. Dixon

Naval Research Laboratory

Washington, D. C.   20375

ABSTRACT

The achievement of significant amplified spontaneous emission at short wavelengths requires either very high densities or very long lengths. The former approach appears more promising at present for concentrating the large pumping power required, and plasma media are anticipated under such conditions. Most often lasing times are short and extension with metastable states does not appear possible at the high density required and with competing dipole transitions. Focused high power laser beams offer the most promising source of concentrated pump energy at present. An inversion scheme of high pump probability for short wavelength lasing involving resonance charge transfer of an electron from a neutral atom to an ion has been identified, and an experiment designed to test this scheme is described. Initial space-resolved spectra are presented, as is the distribution of the measured photographic spectral line density with distance from the surface of a laser heated carbon target as obtained with a space-resolved grazing-incidence spectrograph. Early results indicate optically thick resonance lines extending to approximately 0.8mm from the target. Charge transfer pumping is expected at distances $\geqslant$ 2 mm. No definitive data are so far available with a neutral gas background.

INTRODUCTION

It is the primary intention in this paper to present some recent spectroscopic results from helium-like and hydrogenic

[*]Supported in part by the Defense Advance Research Project Agency, DARPA Order 2694.

carbon ions obtained in a laser-produced plasma. The experiment is
being conducted to investigate a potentially promising approach to
achieving laser action at very short wavelengths, namely resonance
charge transfer of electrons from a neutral atom into excited
states of a highly-stripped ion. Although a number of articles
have recently been written on the subject of short wavelength
lasers and a comprehensive review article is in preparation,[1] it is
appropriate to include some brief introductory remarks in this
paper to accent those areas of particular interest to beam foil
spectroscopists and other specialists at this conference.

At present the existence of undisputed vacuum ultraviolet
lasing has been achieved for wavelengths as short as 1098 Å with
power levels varying from 10 kW to 500 MW, mostly with molecular
transitions. Frequency multiplication of a coherent infrared
laser beam has been successfully extended to wavelengths as short
as 887 Å. At shorter wavelengths small degrees of population inversion have been reported for hydrogenic ion lines following
recombination, but so far no undisputed claims of measured net
amplification are available.

The difficulties encountered in extending lasing to wavelengths
shorter than 1000 Å and into the x-ray region can easily be seen
from simple scaling laws.[2] The problems begin with a lack of
efficient resonanting cavities so that a measurable gain given by
exp (αL) for an amplifying length L is only achieved for a positive
gain coefficient α exceeding unity. For Doppler broadened spectral
lines the linear gain coefficient α_D scales as

$$\alpha_D \propto \frac{\lambda^3}{\bar{v}} \left[A_{u\ell} \, N_o \left(\frac{N_u}{N_\ell} \right) \right] \left\{ 1 - \frac{N_\ell}{N_u} \frac{g_u}{g_\ell} \right\} , \qquad (1)$$

where $\bar{v}$ represents the mean velocity of the lasant ion; N_o, N_u, and
N_ℓ, represent the population densities of the initial state and
the upper and lower laser states, respectively; g_u and g_ℓ are the
appropriate statistical weights; $A_{u\ell}$ is the transition probability
for the laser transition; and λ is the wavelength of the lasing
transition. Positive values of α_D are required for net gain.
Written in this form, it is obvious that the achievement of
population inversion is possible as indicated in the last factor
(shown in braces) irregardless of the absolute value of the gain
coefficient; thus the early reports of evidence of population inversion without measurable gain. A high gain coefficient at short
wavelengths therefore depends primarily on the first factor (indicate in brackets) in Eq. (1). This factor may be written as

$$\left[\left(\frac{P}{aL} \right) \frac{\lambda_{ou}}{hc} \right] , \qquad (2)$$

where P/aL is the pump power density for a cross-sectional area a and length L of laser volume, and hc/λ_{ou} represents the pumping energy. When λ_{ou} is assumed equal to 10 λ and α_D is set equal to 5, absolute values for P/aL become approximately 10^{11} W/cm^2 for a wavelength λ = 100 Å, and it is clear that this power density scales as λ^{-4}. The bracketed factor in Eq. (2) can also be written as

$$\left[N_o \, N_p \, r_{ou} \right] , \qquad\qquad (3)$$

where $r_{ou}(\lambda)$ is the rate coefficient for pumping from some initial state o to upper state laser u, and N_p is the density of pumping particles or photons. The rate coefficient r_{ou} depends upon the particular pumping atomic process and varies with wavelength according to the particular cross section as well as the conditions of the pumping source. Independent of the wavelength dependence of r_{ou}, the density product $N_o \, N_p$ varies as λ^{-3} for a fixed value of α_D. It is desirable to keep these densities low, both to avoid additional collisional effects and also to permit higher gain coefficients achievable at increased densities. Therefore, a pumping process with a rate coefficient r_{ou} bearing a strong inverse dependence on wavelength and a large absolute cross section is most desirable. From this point of view, the most promising process identified so far for pumping at short wavelengths is resonance charge transfer of an electron from a neutral atom to an ion in a collision, where the cross section[3] is given approximately by 10^{-16} πa_o^2 cm^2 and the rate coefficient r_{ou} is given approximately by $10^6 \lambda^{-5/4}$, which is several orders of magnitude larger than other known pumping processes.[2] With this process a gain coefficient α_D = 5 is predicted at a density of 10^{19} cm^{-3} and a wavelength as short as 8 Å. Further details of this scaling for other processes is published elsewhere.[2,4]

In addition to considerations of the pulse power and the lasant density requirements, the time available for lasing is also a serious consideration at short wavelengths; since for many processes it is expected to be as short as the lifetime of the upper laser state which scales as λ^{-2} and varies from 10^{-9} sec. at 1000 Å to 10^{-15} sec. in the x-ray region. Thus, there is also a search for cw or quasi-cw laser schemes with extended lasing times. It is natural, particularly for participants at this conference, to think first of metastable states for this purpose. The above relations show that both the gain coefficient and the pump power density are proportional to the product of a density and the upper to lower state transition probability. Thus, proportionally higher densities are required for the low probability of metastable states, for a given gain coefficient, and collisional effects begin to dominate over radiative effects. (However, it is to be noted that the pump power density is independent of the lasant density or the transition probability for a fixed gain coefficient.) Also it

has so far not been possible to identify   a transition from a
metastable state that is not in direct competition with a dipole
transition to the same lower state, which will rapidly destroy
population inversion and thereby negate any advantage of the long-
lived metastable state. Therefore, the only proposal
published[5] for the use of metastable states in short wavelength
lasing is for the storage of electrons in the 2^1S metastable
state of helium-like ions for subsequent rapid transfer (pumping)
to the 2^1P state, followed by lasing on the 2^1P $\rightarrow$ 1^1S resonance
transition.

RESONANCE CHARGE TRANSFER EXPERIMENT

The high pumping flux required in small volumes and (most
often) the short pump pulse risetime required demand the use of
plasmas as a lasant and focused laser beams as a pumping source for
the shorter wavelength regions. A typical plasma approach is the
resonance charge transfer pump scheme, as originally proposed[3]
by Vinogradov and Sobel'man and illustrated schematically in
Fig. 1 for a particular ion atom combination, namely C^{5+} and
helium, which has been shown by a simple Landau-Zener theory[4,6]
for exothermic reactions to have a large cross section for the
particular plasma temperatures expected in the presence of these
ions. The n = 3 states of the helium-like ions formed are expected

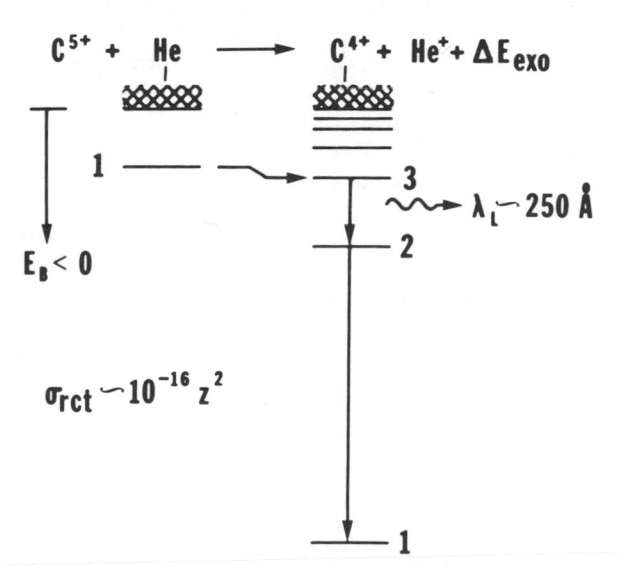

Fig. 1--
Schematic diagram of the exothermic resonance charge transfer reaction leading to population inversion between n=3 and n=2 levels in
C^{4+} helium-like ions following collisions with helium atoms. E_B
represents the binding energy.

to be preferentially populated at a high rate from the $1s^2\ {}^1S$
ground state of neutral helium atoms in collisions, with subsequent lasing between n=3 and n=2 states and rapid lower n=2
laser state depletion, with inversion obtained as long as the
initial ions are maintained at a sufficient density. An experiment designed to test this scheme is shown schematically in
Fig. 2, where a laser beam is focused onto a carbon slab target
before the entrance slit of a grazing incidence spectrograph capable of recording the resonance series of hydrogenic and helium-
like carbon lines in the soft x-ray spectral region. The slot
shown between the entrance slit and the grating in Fig. 2 provides
spatial resolution of the emission from the plasma plume produced,
as indicated along the length of the spectral lines recorded.
The neutral-atom background, namely helium in the present case,
is not indicated in Fig. 2. The experiment is also shown in the
photograph in Fig. 3 where the lucite chamber holding the rotatable
disk carbon target is shown attached to the entrance of the
grazing incidence spectrograph. A ruby laser capable of delivering
6 J, 20 ns FWHM pulses to a 500 μm focal spot is shown mounted
above the spectrograph.

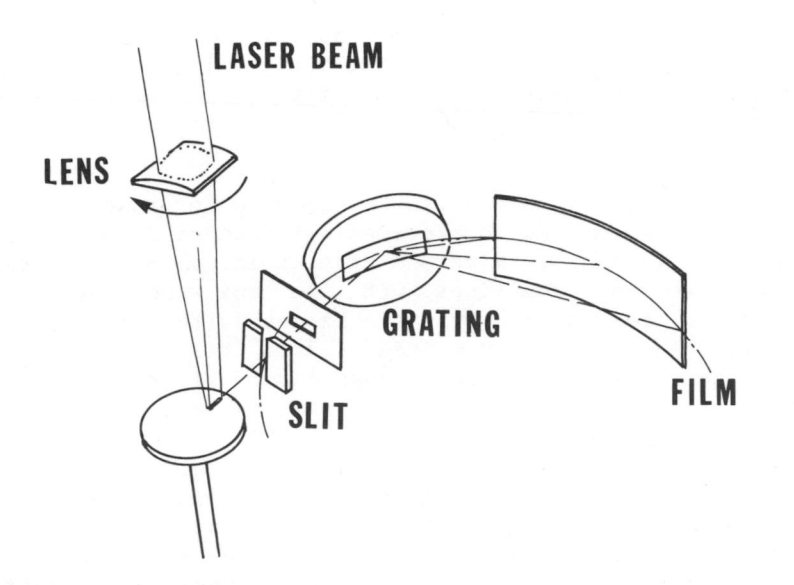

Fig. 2--
Schematic diagram of the resonance charge transfer experiment, including the grazing incidence vacuum spectrograph. The orthogonal
slot shown provides spatial resolution along the direction of
plasma expansion from the target surface, as traced in Fig. 5.

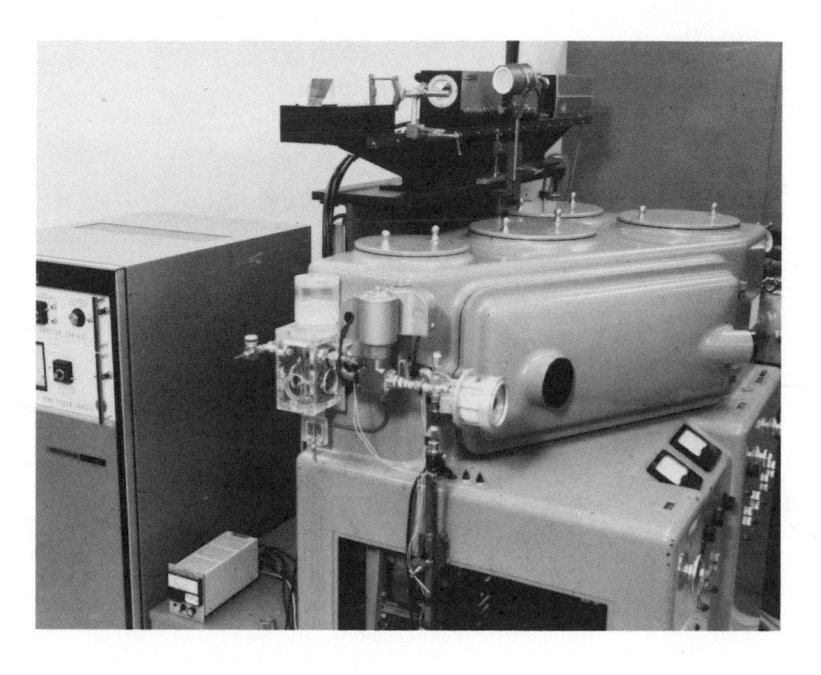

Fig. 3
Photograph of the resonance charge transfer experiment showing
the lucite target chamber attached to the grazing incidence
spectrograph and the laser in current use.

        The initial results that can be reported at present are for
expansion into vacuum, which provides a reference spectrum, as
shown in Fig. 4. This is a second order spectrum of hydrogenic
and helium-like resonance lines, with the true wavelength indicated
in Å units and the distance from the target surface indicated in
millimeters. The weaker lines indicated in this spectrum have been
identified as other second order lines, known satellite lines,
or in a few cases impurity lines. Significant Stark broadening of
the spectral lines can be observed in the high electron density
region near the target surface, and is an indicator of the electron
density. At present, microdensitometer scans along the wavelength
axis at various distances from the target surface are being
compared with similar spectra obtained in the presence of helium
background gas, in a search for evidence of enhanced population of
specific levels as indicated by the resonance line emission from
these levels. While the data for such comparisons is at present
sparce and inconclusive, it is possible to report on the variation
of the photographic densities with distance from the target surface for various spectral lines important to this experiment. The
results are plotted in Fig. 5 for a vacuum expansion, where an

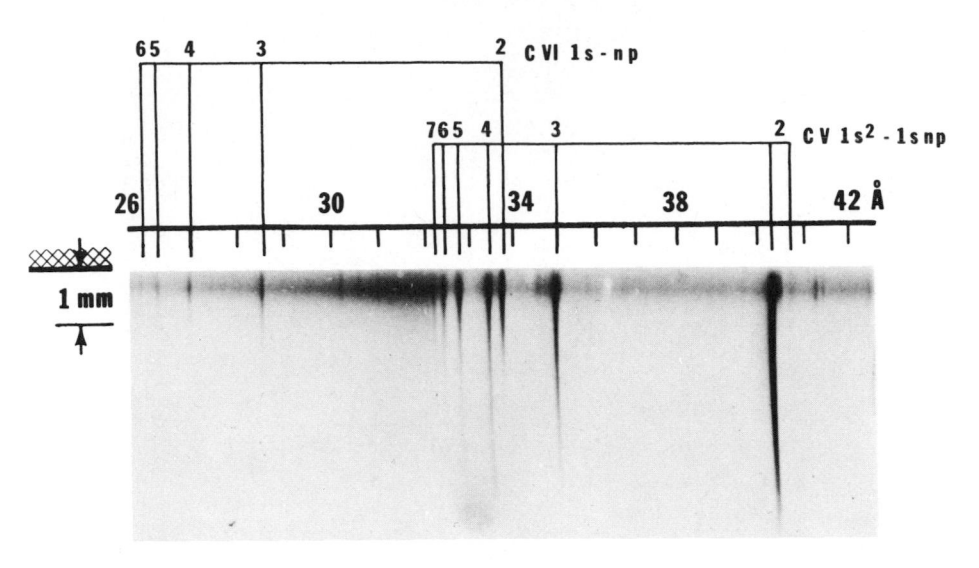

Fig. 4--
Spatially resolved CV and CVI grazing incidence spectrum using a
400 μm slot as shown in Fig. 2. No background gas was present.

instrumental resolution in distance of 1 mm is expected at the plasma.

INITIAL RESULTS

The preliminary results obtained comparing spectra with and
without the neutral helium atmosphere are to date inconclusive
and will not be discussed here. Under vacuum conditions, both
the line (Fig. 5) and the continuum emission are observed to
initially increase with distance from the target surface. This
is consistent with a rising temperature and the formation during
the rising laser pulse of an expanding high density critical
absorption layer for the laser radiation. Association of the
increasing continuum emission with density through bremsstrahlung
and recombination processes is consistent with the increasing
Stark broadening[7] in the same region, as observed in the spectrum
of Fig. 4.

Compared in Fig. 5 are the photographic densities (with the
continuum background subtracted) of the resonance series lines[8]

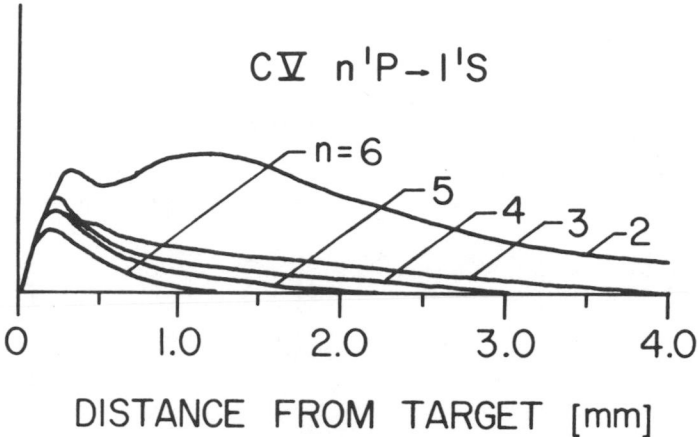

Fig. 5--
Photographic density versus distance from target from space-resolved grazing incidence spectrum for the C^{4+} (CV) resonance series.

from the helium-like C^{4+} ions in vacuum (CV reference spectrum). While these densities have so far not been converted to exposures, it is clear that at distances greater than 0.8 mm the total line intensity ratios scale approximately as expected, particularly for those lines associated with transitions originating on levels with principal quantum number n ≥ 3. It is this "outer" region (⪞ 2 mm), where the density decreases[7] to ⩽ 10^{18} cm^{-3}, that is of interest to the present carbon-ion/helium-atom resonance charge transfer experiment. The $2^1P \to 1^1S$ first resonance line appears to have an enhanced emission which may be associated with a considerably higher oscillator strength compared to the other lines in the series, as evidenced from data available for neutral helium[9].

Close to the target surface, the relative line emissions are more nearly equal, for some of the series members. This can most likely be attributed to opacity effects[10] in the higher density region, with perhaps the emission of lines of varying strength from separate layers. Near 0.5 mm the first series member shows a distinct dip in emission which is also evident on

the second member of the series. This also is most likely
associated with increased opacity for these stronger members of the
series. Indeed, the first member may be suppressed in the closest
regions by reabsorption in cooler outer layers of the plasma plume.
Again, it is fortunate that the resonance charge transfer effects
sought will become most evident at greater distances where it
appears that reabsorption is less severe a problem.

REFERENCES

1.   R. W. Waynant and R. C. Elton, Proceedings IEEE (in pre-
      paration).

2.   R. C. Elton and R. H. Dixon "X-Ray Laser Research; Guide-
      lines and Progress at NRL", Proceeding Third Conference on
      Lasers, New York Academy of Sciences, April 1975 (to be
      published).

3.   A. V. Vinogradov and I. I. Sobel'man, Soviet Physics JETP,
      36, 1115 (1973).

4.   R. C. Elton, in Progress in Lasers and Laser Fusion,
      A. Pearlmutter and S. M. Widmayer, eds., p. 117 (Plenum Press,
      New York, N.Y., 1975).

5.   H. Mahr and V. Roeder, Optics Comm. 10, 227 (1974).

6.   H. J. Zwally and D. W. Koopman, Physical Review A 2, 1851
      (1970).

7.   A. M. Malvezzi, E. Jannitti and G. Tondello (to be published).

8.   R. L. Kelly and L. J. Palumbo, "Atomic and Ionic Emission
      Lines Below 2000 Ångstroms", Naval Research Laboratory
      Report No. 7599 (1973).

9.   W. L. Wiese, M. W. Smith, and B. M. Glennon, "Atomic Transition
      Probabilities", National Bureau of Standards Report No.
      NSRDS-NBS 4 (1966).

10.   R. C. Elton, "Atomic Processes", in Methods of Experimental
      Physics-Plasma Physics, Vol. 9A, p. 142, H. R. Griem and
      R. H. Lovberg, eds. (Academic Press, New York, N. Y., 1970).

SUBJECT INDEX

(Pp. 1-460 are found in Volume 1, pp. 461-982 in Volume 2)

Active region loops, solar, 894
Adiabaticity parameter, 727-729
Adler formula, 788
Alignment, 731
    ellipsoid, 734
    orbital, 274
    spin, 784
    tensor, 782-788
Astrophysical resonance lines, 908
Atomic alignment, 756
Atomic beam, 829-833
Atoms in high magnetic fields, 914-923
Auger electrons, 788-789
Autoionization electrons, 786, 788, 789
Autoionizing states, possible non-exponential decay, 77

Back scattering, 665-670
Beam velocity determination Doppler shift method, 253
Binary encounter approximation, 519-520, 525, 556, 583, 596
Branching ratio method, 706-712
Bremsstrahlung, 465, 500, 520-521
    nuclear, 463
Bright points, solar, 893

Cascade population, 788

Cascade radiations, 136, 203-245, 712-719
Charge exchange, 408, 424
Charge state distributions, 402-403, 657-663, 665-670, 671
Charge states, attainable, 968-970
Charge states in laser targets, 935-938
Charge transfer, 284, 650
Charge transfer laser pumping, 973, 975-979
Chromosperic network, solar, 889
Coherent excitation, 785, 786
Collisions, beam-foil, 781-789
Collisional quenching, 545-552
Cometary spectra, 207
Copernicus scanning spectrometer, 908
Core-excited states of three-electron ions, 121
Coronal holes, solar, 892
Correlation effects, 54
Coulomb deflection correction factor, 530, 531
Coulomb ionization, 519-520, 525, 534
    multiple, 568
Cross section, photoionization, 833
Curves of growth, 909-911

Delta electrons, 581-582
Density matrix, 776, 780, 800-803, 807, 811

Detection efficiency, UV, 705
Dielectric recombination, 380
Dissociation
    magnetic, 740
    lifetimes, 741
Doppler broadening, 260
Doppler shift measurements, 175, 236
Doppler tuned spectrometer, 284, 299
Doppler tuning, 829, 873
Dye-laser, 863, 864
Dynamic fluorescence yield, 546

Eck-beats, 760
Electron density vs. temperature in plasmas, 926-942
Electron gas, 509
Electron loss cross sections, 643-648
Emission and displaced absorption profiles, 914-916
Electron temperatures, solar, 903
Energy loss, 683-687
Energy straggling, 175
Equipartition rule, 506, 509
Excitation, two stages, 860, 863, 865, 868-871
Extreme ultraviolet solar spectroscopy, 885-904

Fifth period element lifetimes, 191
Fine and hyperfine structure in alkali sequences, anomalies, 105
Fine structure
    intervals, 188
    F VIII, 877-883
First period element lifetimes, 183
Fluorescence yield, 629, 631
    core-excited states of three-electron ions, 121
Foil excitation matrix, 765
Foil thickening, 264-345

Forbidden transitions, beryllium sequence intercombination line, 38
Fourth period element lifetimes, 129, 155, 169, 217
Franck-Condon factors, 711
Fusion plasma oscillator strengths, 951-960

Grazing incidence spectrometer, 322, 347, 355, 368

Heavy ion impurities in fusion plasmas, 951-959
Heteronuclear cluster, 513
High density effects in laser irradiations, 945-947
High, laser-induced ionization states, 925-948
High intensity method, 719-725
High power neodymium laser irradiation, 935-948
High velocity polarization effect, 530
Homonuclear cluster, 508
Hydrogenic and helium-like carbon ions in laser plasmas, 973-981
Hydrogenic transitions, 129
Hyperfine quenching of 2^3P_0 state in heliumlike ions, 97
Hyperfine structure, carbon-13, 791-798

Interstellar clouds, 902
Ionization, microwave, 832
Ionization, multiple, 629-630
Ionization-recombination balance in plasmas, 926-929
Ionization time in plasmas, 931
Isoelectronic, isoionic, and Isonuclear systematics, 637-640

Isoelectronic sequences
    in oscillator strengths,
        955-960
    regularities, 52
    term analysis, 1
Isotope ratios, 208
Isotope shifts, 208

Jackson & Schiff scaling rule,
    263, 268

Lamb shift, 89, 271, 305, 308,
    319, 332, 815-827
Lamb shift in hydrogenlike ions,
    89
Larmor frequency, 780
Laser-irradiated targets, 925-
    942
Laser excitation
    barium, 835-839, 846, 853-
        857, 874, 875
    fluorine, 879-881
    lithium, 869-871
    magnesium, 844
    neon, 868,869
    rubidium, 845
    scandium, 873-876
    sodium, 865-868
    strontium, 844
Laser
    photoexcitation, 829-833
    photoionization, 829-833
Laser resonance spectroscopy,
    819-827
Level population studies, 231
Lifetime measurement techniques,
    129, 165, 191
    energy loss corrections, 174
    electron excitation, 203
    foil effection. 176
    high frequency deflection,
        200
    laser excitation, 199
    molecular levels, 206
Ligand atoms, 573

Line broadening, kinematic, 424,
    445, 448, 453, 462

Metastable states, 649-655
    quenching, 272-274, 317
Molecular cluster, 505
Molecular orbital radiation
    angular distributions, 477
    in single and double
        collisions, 503
    widths, 472
Multiple photon transitions, 278
Muonic molecules, 507

Non-stationary ionization in
    laser-induced plasmas, 934

Opacity and line broadening
    effects in laser-induced
    plasmas, 943-945
Orientation parameters, 776
Orientation vector, 786
Oscillator strengths
    beryllium isoelectronic
        sequence, 29
    correlation effects in
        sodium sequence, 69
    in atomic and ionic oxygen
        and nitrogen, 115
    in CaI, ScII, TiIII, 43
    isoelectronic trends, 235
    model potential method, 43

Pair production, 491
Particle clusters, 505
Partition constant, 509
Pauli excitation, 520
Penning ion source, 409
Percival-Seaton hypothesis, 784,
    785
Photoionization 414
    cross sections, 640

Plane wave born approximation, 263, 519-520, 525, 526, 530, 583, 596
Plasmon frequency, 505
    excitation, 598
Polarization, 731, 734-736
    ellipse, 761
Position sensitive detectors, 289
Potential wake, 505
Predissociation, 209
Projectile charge states in solids and gases, 962-964
Pumping, off-resonant, 744

Quantum beats, 250, 749-759, 755-769
    barium, 847
    carbon-13, 793-796
    HeI, 773-780
    HeII, 759
    H_α and H_β, 799-808
    lamb shift, 756
    magnetic, 773-780
    magnetic field, 762
    N IV, 763, 764
Quasar absorption lines, 916-917
Quenching, 415, 545-552

Radiation tensor, 813
Radiative auger process, 609-614
Radiative electron capture, 461, 466, 501, 545-552
Rare earth element lifetimes, 172
Reduced binding energy, 530
Relativistic effects, continuous to transition energies in Ni and Cu sequences, 111
Relativistic effects on transition probabilities, 59
Resonance lines, sodium and copper sequences, 956-960
Rutherford scattering, 522
Rydberg levels, 129

Second period element lifetimes, 134
Secondary electron emission, 577-592
Secondary electron yield, 595-602
Selective ionization in laser-induced plasmas, 938-940
Shake-up, 443
Short wavelength laser phenomenology, 973-981
Single hole parameter, 526
Solar chromosphere, 886
Solar corona, 886
Solar photosphere, 886
Solar transition region, 886
Spatially correlated clusters, 505-506, 511
Standard lamps, 706
Stark beats, 749-754
Stark quenching, 728
Stark shifted energy separations, 759
Stellar atmospheres, 907
Stellar particle densities and abundances, 907-913
Stellar winds, 913
Stokes parameters, 760-769
Stopping power, 505, 507-509, 514, 679-686, 687
Stripping in foils vs. plasmas, 961-971
Supergranulation cells, solar, 890
Surface interaction, 755-769
Surface potential, 809

Third period element lifetimes, 147, 168, 217, 223
Tilted foil, 760-769, 783, 809-813
Transition probabilities
    charge expansion method, 52
    Coulomb methods, 83
    configuration interaction, 29, 84

intermediate coupling, 62
trivalent atoms, 85

Wake riding states, 505, 510,
    512, 514-515

Valence stopping number, 505
Vicinage function, 508-509

X-ray sattelite lines, 567-575
    in laser induced plasmas,
    941-943

Yrast states, 553-557

Zeeman effect, 919-923
    first and second
        intermediate, 919-920
    linear and quadratic, 919-
        920
    London regime, 920
    of pulsars, 922
    spectra of alkalis, 921
    splitting, 778